MICROPHONE

GENERATOR

ELEPHONE
ECEIVER

Thomas Alva Edison (1847-1931) brought the
age of technology to the twentieth century
through his development of numerous designs
and inventions. His work has had enormous
impact on the lifestyle of the world. Edison
was awarded 1033 patents.

TICKER
TAPE

ROELECTRIC
LANT

EARLE
84

GRAPHICS
FOR
ENGINEERS

GRAPHICS
FOR
ENGINEERS

JAMES H. EARLE
Texas A&M University

Addison-Wesley Publishing Company

Reading, Massachusetts • Menlo Park, California
Don Mills, Ontario • Wokingham, England • Amsterdam
Sydney • Singapore • Tokyo • Mexico City • Bogotá
Santiago • San Juan

Thomas Robbins   Sponsoring Editor

Ann E. Delacey   Manufacturing Supervisor
James H. Earle   Illustrator and
   Cover Designer
Karen Guardino   Production Manager
Lorraine Hodsdon   Layout Artist
Maureen Langer   Text Designer
Laura Skinger and Natasha Wei   Production Editors
Marcia Strykowski   Art Editor

Library of Congress Cataloging in Publication Data

Earle, James H.
   Graphics for Engineers.

   Includes index.
   1. Engineering graphics.   2. Mechanical drawing.
I. Title.
T353.E3 1985   604.2′4   84-9202
ISBN 0-201-11430-5

CDEFGHIJ-HA-89876

Dedicated to Donald L. Earle
An outstanding welterweight

The following supplements are compatible with the following textbooks from Addison-Wesley Publishing Co.:

Drafting Technology,

Engineering Design Graphics,

Design Drafting,

Graphics for Engineers,

Geometry for Engineers

DRAFTING TECHNOLOGY PROBLEMS is a new problem book designed to cover basic graphics, descriptive geometry, and specialty drafting areas. It is designed to accompany a new text, *Drafting Technology,* that is available from Addison-Wesley Publ. Co., Reading, Mass. 01867. 131 pages.

BASIC DRAFTING is a problem book that covers the basics for a one-semester high school drafting course. 67 pages.

CREATIVE DRAFTING is a problem book that covers mechanical drawing and architectural drafting for the high school. 106 pages.

DRAFTING & DESIGN is a problem book for mechanical drawing for a high school or college course. 106 pages.

DRAFTING FUNDAMENTALS 1 is a problem book for a one-year high school course in mechanical drawing. 94 pages.

DRAFTING FUNDAMENTALS 2 is a second version of **DRAFTING FUNDAMENTALS 1** for the same level and content. 94 pages.

TECHNICAL ILLUSTRATION is a problem book for a course in pictorial drawing for the college or high school. 70 pages.

ARCHITECTURAL DRAFTING is a problem book for a first course in architectural drafting. 71 pages.

GRAPHICS FOR ENGINEERS 1 is a problem book for a first course in engineering graphics for the college student. 100 pages.

GRAPHICS FOR ENGINEERS 2 is a problem book for a first course in engineering graphics for the college student. 100 pages.

GRAPHICS FOR ENGINEERS 3 is a problem book for a first course in engineering graphics for the college student. 108 pages.

GEOMETRY FOR ENGINEERS 1 is a problem book for a college-level descriptive geometry course. 100 pages.

GEOMETRY FOR ENGINEERS 2 is a problem book for a college-level descriptive geometry course. 100 pages.

GEOMETRY FOR ENGINEERS 3 is a problem book for a college-level descriptive geometry course. 100 pages.

GRAPHICS & GEOMETRY 1 is a problem book for a college course in graphics and descriptive geometry. 119 pages.

GRAPHICS & GEOMETRY 2 is a problem book for a college course in graphics and descriptive geometry. 121 pages.

GRAPHICS & GEOMETRY 3 is a problem book for a college course in graphics and descriptive geometry. 138 pages.

COMPUTER GRAPHICS is a problem book that is supplemented with extensive notes to serve as the basis for a first college-level course in computer graphics. 92 pages.

Creative Publishing Co.
BOX 9292   COLLEGE STATION, TEXAS   77840
PHONE 409-775-6047

Preface

Graphics for Engineers has been written to cover the basic and more advanced principles of engineering graphics for a course at the college level. The topics covered range from the metric system to computer graphics.

Rationale

The major areas presented in the text are communications, working drawings, design, and computer graphics. Each of these important areas is essential to the background of the engineer, technologist, or technician.

Communications is covered as it relates to the presentation of ideas, graphical analysis of data, and the preparation of pictorials. The methods of making working drawings are given in accordance with the latest ANSI standards and include materials, tolerances, welding specifications, pipe drafting, and electrical/electronic drafting.

An introduction to design is given in Chapter 2 to show the value of engineering graphics as a vehicle for developing creative solutions to technical problems. Computer graphics by microcomputer is covered in Chapter 22 to give an introduction to this expanding medium of engineering graphics.

Objectives

The objective of this book is to support a course in which the student learns the various standards and techniques of preparing engineering drawings and specifications, learns how to read and interpret drawings, and learns how to supervise the preparation of drawings by others. Above all, this textbook is designed to help students learn to expand their creative talents and to communicate their ideas in a meaningful manner.

Format

Graphics for Engineers can be used as a self-instructional format to enable the student to work independently. Many problems are presented in a step-by-step sequence that shows the progressive steps in solving a problem, a second color is used to emphasize important points of construction, and many practical applications from industry are included as examples. Problems at the ends of chapters offer a wide range of assignments that can be solved by students to aid them in grasping the principles covered in each chapter.

Supplementary problems

Nine different problem books are available that can be used with this textbook. A listing of these manuals and their descriptions can be obtained from Creative Publishing Company, Box 9292, College Station, Texas 77840.

Acknowledgments

We are grateful for the assistance of many who have influenced the development of this volume. We have been significantly aided by the use of many illustrations developed by the late William E. Street and Carl L. Svensen for their earlier publications. Many industries have furnished photographs and drawings that have been acknowledged in the legends where they are used. The staff of the Engineering Design Graphics Department of Texas A&M University has been helpful in making suggestions for the preparation of this book.

Professor Tom Pollock provided information on various metals and their designations. Bill Zaggle organized the material used in the chapter on computer graphics.

Above all, we are appreciative of the many institutions who have thought enough of our publications to adopt them for classroom use. It is indeed an honor for one's publications to be accepted by his colleagues. We are hopeful that this textbook will fill the needs of engineering and technology programs. As always, comments and suggestions for improvement and revision of this book will be appreciated.

College Station, Texas
January 1985   J. H. E.

Contents

Introduction to Engineering and Technology

1

1.1
Introduction

Essentially all our daily activities are assisted by products, systems, and services made possible by the engineer. Our utilities, heating and cooling equipment, automobiles, machinery, and consumer products have been provided at an economical rate to our population by the engineering profession.

The engineer must function as a member of a team composed of other related, and sometimes unrelated, disciplines. Many engineers have been responsible for innovations of life-saving mechanisms used in

FIG. 1.1 Problems in this text require a total engineering approach with the engineering problem as the central theme.

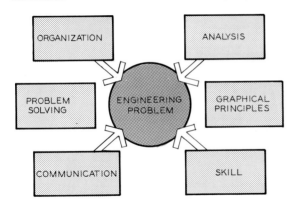

medicine, which were designed in cooperation with members of the medical profession. Other engineers are technical representatives or salespeople who explain and demonstrate applications of technical products to a specialized segment of the market. The engineer is basically a *designer*. This is the activity that most distinguishes him or her from other members of the technological team.

This book is devoted to the introduction of elementary design concepts related to the field of engineering and to the application of engineering graphics to the design process. Examples are given that have an engineering problem at the core and that require organization, analysis, problem solving, graphical principles, communication, and skill (Fig. 1.1).

Creativity and imagination are encouraged as essential ingredients in the engineer's professional activities. As Albert Einstein, the famous physicist, said, "Imagination is more important than knowledge, for knowledge is limited, whereas imagination embraces the entire world . . . stimulating progress, or, giving birth to evolution" (Fig. 1.2).

1.2
Engineering graphics

Engineering graphics is the total field of graphical problem solving and includes two major areas of specialization: descriptive geometry and working drawings. Other areas that can be utilized for a wide variety

FIG. 1.2 Albert Einstein, the famous physicist, said, "Imagination is more important than knowledge."

of scientific and engineering applications are nomography, graphical mathematics, empirical equations, technical illustration, vector analysis, and data analysis.

Engineering graphics is not limited to drafting, since it is more extensive than the communication of an idea in the form of a working drawing. Graphics is the designer's method of thinking, solving, and communicating ideas throughout the design process.

Humankind's progress can be attributed to a great extent to the area of engineering graphics. Even the simplest of structures could not have been designed or built without drawings, diagrams, and details that explained their construction (Fig. 1.3). Gradually, graphical methods were developed to show three related views of an object to simulate its three-dimensional representation. A most significant development in the engineering graphics area was descriptive geometry.

FIG. 1.3 Leonardo da Vinci developed many creative designs through the use of graphical methods.

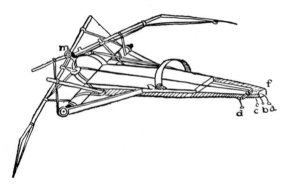

Descriptive geometry

Gaspard Monge (1746–1818) is considered the "father of descriptive geometry" (Fig. 1.4). Young Monge used this graphical method to solve design problems related to fortifications and battlements while he was a military student in France. He was scolded by his headmaster for not solving a design problem by the usual, long, tedious mathematical process traditionally used for problems of this type. It was only after long explanations and comparisons of the solutions of both methods that he was able to convince the faculty that his graphical methods could be used to solve the problem in considerably less time. This was such an improvement over the mathematical solution that it was kept a military secret for 15 years before it was allowed to be taught as part of the technical curriculum. Monge became a scientific and mathematical aide to Napoleon when he was general and emperor of France.

FIG. 1.4 Gaspard Monge, the "father of descriptive geometry."

Descriptive geometry can be defined as the projection of three-dimensional figures onto a two-dimensional plane of paper in such a manner as to allow geometric manipulations to determine lengths, angles, shapes, and other descriptive information concerning the figures.

1.3
The technological team

Technology has broadened at a rapid rate during recent years with the introduction of many new processes and fields of specialty unheard of a decade ago.

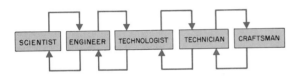

FIG. 1.5 The technological team.

It has become necessary for professional responsibilities to be performed by highly qualified people with specialized training. Thus technology has become a team effort involving the scientist, engineer, technologist, technician, and craftsman (Fig. 1.5). A project may include mechanical design, an advanced electronics system, a structure, and chemical processes; therefore it may require many engineers, technologists, technicians, and craftsmen to complete the design.

Scientist

The scientist is primarily a researcher who is seeking to establish new theories and principles through experimentation and testing (Figs. 1.6 and 1.7). Often he or she has little concern for the application of specific principles that are being developed, being primarily interested in isolating significant relationships. Scientific discoveries are used as the basis for related re-

FIG. 1.6 Scientists such as this chemist conduct research to establish fundamental relationships. This chemist is trying to determine the best combination of ingredients to yield a better tire. (Courtesy of Uniroyal, Inc.)

FIG. 1.7 A researcher solders fine wires to each strain gauge in an experimental model. (Courtesy of the Bureau of Reclamation, U.S. Department of the Interior.)

search and for the development of practical applications for future discoveries that may not come into existence until years after the initial discovery.

Engineer

Engineers are trained in areas of science, mathematics, and industrial processes. This prepares them to apply the basic principles discovered by the scientist to practical problems (Fig. 1.8). They are concerned with the conversion of raw materials and power sources into needed products and services. The emphasis on the practical application of principles distinguishes the engineer from the scientist.

The application of scientific principles to new products or systems in a creative manner is the process of designing, which is the engineer's most unique function and requires the highest degree of creativity. Engineers must always be concerned with efficiency and economy in their designs to provide the greatest service to society. In general terms, the engineer practices the art of using available principles and resources to achieve a practical end at a reasonable cost.

Technologist

The technologist is a technically trained person who assists the engineer at a semiprofessional level slightly below the level of the engineer and above that of the technician. While the engineer is concerned with the application of theoretical concepts, the technologist is

FIG. 1.8 One of the modern engineering marvels is the Astrodome of Houston, Texas, shown here in the early stages of construction.

FIG. 1.9 Engineering technologists combine their knowledge of production techniques with the design talents of engineers to produce a product. (Courtesy of Omark Industries, Inc.)

Technician

Technicians assist the engineer and technologist at a technical level below that of the technologist. Their work may vary from conducting routine laboratory experiments to supervising craftsmen involved in manufacturing or construction. In general, technicians work as liaisons between the technologist and the craftsman. Technicians must exercise a degree of judg-

FIG. 1.10 Technicians are skilled in performing routine jobs requiring technical knowledge and skill. This technician is responsible for inspecting chain assemblies to ensure that quality standards are met. (Courtesy of Omark Industries, Inc.)

more concerned with routine, practical aspects of engineering whether at the planning or the production stage (Fig. 1.9). Several technologists working with an engineer will greatly increase the engineer's capability of performance, since he or she will be relieved of many duties by the technologist. Likewise, the technologist can coordinate the activities of several technicians, thereby sharing supervisory responsibility with the engineer. The technologist usually has a four-year engineering technology background at the college level.

ment and imagination in their work and assume supervisory responsibilities beyond those required of the craftsman (Fig. 1.10). They are usually required to have a two-year technical-training background beyond the high-school level to be qualified for their assignment.

Craftsman

Craftsmen are vital members of the engineering team since they must see that the engineering design is implemented by producing it according to the specifications of the engineer. They may be machinists who fabricate the various components of the product or craftsmen who assemble electrical components. Craftsmen supply technical skills that cannot be provided by the engineer or the technician. The ability to produce a given part in accordance with design specifications is as necessary as the act of designing the part. Craftsmen include electricians, welders, machinists, fabricators, drafters, and members of many other professions (Fig. 1.11).

FIG. 1.11 This craftsman is a machinist who is skilled in the making of metal parts in accordance with the specifications that he is given. (Courtesy of the Clausing Corporation.)

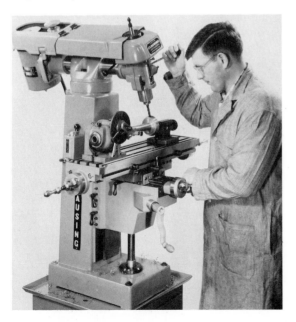

FIG. 1.12 Thomas A. Edison had essentially no formal education, but he gave the world some of its most creative designs, which have never been equaled by another single individual.

Designer

Designers are the people who have special talents for creating solutions to technological problems. The designer may be an engineer, an inventor, or a person who has special talents for devising creative solutions even though he or she may not have an engineering background. This is often the case in young areas of technology where little precedent has been established by previous experience. Thomas A. Edison (Fig. 1.12) had very little formal education, but he possessed an exceptional ability to design and perfect some of the world's most significant designs. Engineers may find that their formal backgrounds inhibit their design abilities. They may be too quick to label as impossible a particular problem that could well be solved by a designer who did not know it was "impossible."

Stylist

The person who is concerned with the outward appearance of a product rather than the development of a functional design is referred to as a stylist (Fig. 1.13). Stylists may be responsible for the design of an automobile body or the configuration of an electric iron, but they are not concerned with the design of the internal details of the product. The stylist must have a high degree of aesthetic awareness plus a feel for the consumer's acceptance of designs.

A stylist who designs an automobile body considers the functional requirements of the body, including driver vision, enclosure of passengers, space for the

FIG. 1.13 The stylist is more concerned with the outward appearance of a product than with the functional aspects of its design. (Courtesy of Ford Motor Corporation.)

power unit, etc. However, the stylist is not involved with the design of such items as the engine or the steering linkage.

1.4
The engineering fields

The following sections briefly outline the more prominent areas of engineering and describe the duties of engineers employed in these fields of the profession. The addresses of the professional societies associated with each field are given at the end of this chapter as sources for additional information concerning the opportunities and challenges of each branch.

Significant changes in engineering during recent years include the emergence of the technologist and technician and the growing number of women who are pursuing engineering careers. More than 3 percent of the present freshman engineering class is comprised of women, which indicates a growing trend in this career field.

Aerospace engineering

Aerospace engineers in the space exploration branch of this profession work on all types of aircraft and spacecraft, including missiles, rockets, and conventional propeller-driven and jet-powered planes. Their responsibilities include the development of aerospace products from initial planning and design to the final manufacture and testing (Fig. 1.14).

Second only to the auto industry in sales, the aerospace industry contributes immeasurably to national defense and the national economy. The future for aeronautical engineers appears to offer rewarding and challenging opportunities in the years ahead.

Aerospace engineering deals with flight in all aspects—at all speeds and all altitudes. Aerospace engineering assignments range from complex vehicles traveling 350 million miles to Mars to hovering aircraft used in deep sea exploration. This broad field of engineering encompasses many specialized disciplines. As a result, most aerospace engineers usually specialize in a particular area of work. Specialized areas are (1) aerodynamics, (2) structural design, (3) instrumentation, (4) propulsion systems, (5) materials, (6) reliability testing, and (7) production methods. They may also specialize in a particular product, such as conventional-powered planes, jet-powered military aircraft, rockets, satellites, or manned space capsules.

In the broadest sense, aerospace engineering can be divided into two major areas: research engineering and design engineering. The research engineer is concerned with the exploration of known principles in search of new ideas and concepts. The design engineer translates the new concepts developed by the researcher into workable applications for the improve-

FIG. 1.14 Aerospace engineers are responsible for the design and testing of aircraft to produce the most efficient craft possible. (Courtesy of Bell Helicopter Corporation.)

FIG. 1.15 The space shuttle *Columbia* was launched in April 1981 to introduce a new era for aerospace engineering—the era of the reusable spaceship. (Courtesy of NASA.)

ment of the existing state of the art. Engineers in each of these two areas must approach the unknowns of their fields with creativity, skill, and knowledge. This approach has elevated the field of aerospace engineering from the Wright brothers' first flight at Kitty Hawk, N.C., in 1903 to the penetration of outer space with an unlimited future (Fig. 1.15).

Related engineering specialists are vital members of the aerospace engineering team. A few of the specialty areas are: (1) avionics—the study of electronic, computerized communication and flight-control systems; (2) equipment—the design and installation of equipment for navigation, hydraulic, armament, survival, electrical, and comfort control related to the functional operation of the aircraft; (3) materials and processes—the development, testing, and evaluation of new materials to determine their applications to aircraft designs; and (4) metallurgy—the evaluation and testing of new and specially treated metals to select the most efficient metals for aerospace use.

The professional society of aeronautical engineers is the American Institute of Aeronautics and Astronautics (AIAA). Student branches of this association are located on most college campuses that offer degrees in aeronautical engineering.

Agricultural engineering

Agricultural engineers are trained to serve the world's largest industry—agriculture. Engineering problems of agriculture deal with the production, processing, and handling of food and fiber. The four major areas of specialization in agricultural engineering are: mechanical power and machinery, farm structures, electrical power and processing equipment, and soil and water control and conservation.

MECHANICAL POWER The agricultural engineer who works with one of the more than 800 manufacturers of farm equipment will be concerned with gasoline and diesel engine equipment such as pumps, irrigation machinery, and tractors. The use of farm machinery designed by agricultural engineers has been responsible for the increased production of agriculture at a higher degree of efficiency. Machinery must be designed for the electrical curing of hay, milk processing and pasteurizing, fruit processing, and artificially heating environments in which to raise livestock and poultry (Fig. 1.16).

FIG. 1.16 This walking sprinkler system is an example of the work of agricultural engineers. It distributes 1000 gallons per minute of water to irrigate 140 acres of farm land. (Courtesy of the Bureau of Reclamation.)

FARM STRUCTURES The construction of barns, shelters, silos, granaries, processing centers, and other agricultural buildings requires specialists in agricultural engineering. The design of buildings of this type requires that the engineer understand heating, ventilation, and chemical changes that affect the storage of crops. Engineers may also supervise research that will improve methods of design and construction of farm structures.

ELECTRICAL POWER A high percentage of the equipment used for agricultural purposes is driven by electrical power. Agricultural engineers design electrical systems and select equipment that will provide efficient operation to meet the requirements of the specific situation. They may serve as consultants or designers for manufacturers or large processors of agricultural products. Their knowledge improves living and working conditions connected with rural activities.

SOIL AND WATER CONTROL The agricultural engineer is responsible for devising systems for improving drainage and irrigation systems, resurfacing fields, and constructing water reservoirs (Fig. 1.17). These activities may be performed in association with the U.S. Department of Agriculture or the Department of the Interior, with state agricultural universities, with consulting engineering firms, or with irrigation companies.

FIG. 1.17 Agricultural engineers improve production by designing irrigation systems. This irrigation chute is 3¾ miles long. (Courtesy of the Bureau of Reclamation, U.S. Department of the Interior.)

Most agricultural engineers are employed in private industry, especially by manufacturers of heavy farm equipment and specialized lines of field, barnyard, and household equipment; electrical service companies; and distributors of farm equipment and supplies. Agricultural engineering is not limited to work in rural areas. Most agricultural engineers live in cities where the leading farm-equipment manufacturing centers are located. Although they need not live on a farm or be involved in agriculture, it is helpful if they have a clear understanding of agricultural problems, farmwork, crops, animals, and people involved in farming.

Agricultural engineers will be involved in the problems of conservation of resources and the introduction of new agricultural products in the future. Agricultural engineering has been instrumental in raising the farmer's efficiency so much that one farmer today raises enough food for thirty-two people, whereas he was capable of supplying only four other people 100 years ago. The future offers many opportunities in agricultural engineering.

The professional society of the agricultural engineer is the American Society for Agricultural Engineers.

Chemical engineering

Chemical engineering involves the design and selection of equipment that will facilitate the processing and manufacturing of chemicals in large quantities. These designs are closely related to the principles of chemistry and those of other engineering fields that aid in the economic and efficient production of chemical products (Fig. 1.18).

Chemical engineers design unit operations, including fluid transportation through ducts or pipelines, solid material transportation through pipes or conveyors, heat transfer from one fluid or substance to another fluid or substance through plate or tube walls, absorption of gases by bubbling them through liquids, evaporation of liquids to increase concentration of solutions, distillation under carefully controlled temperatures to separate mixed liquids, and many other similar chemical processes. The engineers may employ chemical reactions of raw products, such as oxidation, hydrogenation, reduction, chlorination, nitration, sulfonation, pyrolysis, and polymerization (Fig. 1.19). From these reactions come new materials and products.

FIG. 1.18 The chemical engineer conducts experiments on new refinery methods to improve the quality of products. (Courtesy of Exxon Corporation.)

Process control and instrumentation have developed as important specialties in chemical engineering. The handling and control of large quantities of materials must be done with a high degree of accuracy and precision. Process control is designed for operation with the measurement of quality and quantity by fully automatic instrumentation.

Chemical engineers develop and process chemicals such as acids, alkalies, salts, coal-tar products, dyes, synthetic chemicals, plastics, insecticides, fungicides, and many others for industrial and domestic

FIG. 1.19 The chemical engineer's efforts, in conjunction with efforts of engineers from other fields, are finalized in the construction and operation of a completed refinery producing vital products for our economy and needs. (Courtesy of Houston Oil and Refining Company.)

uses. Chemical engineers are associated with the development of drugs and medicine, cosmetics, explosives, ceramics, cements, paints, petroleum products, lubricants, synthetic fibers, rubber, and detergents. They also design equipment for food preparation and canning plants.

Chemical engineers work with metallurgical and mining engineers in designing processing equipment and in designing and laying out complete plants. They must be well versed in chemistry and must be able to discuss plant design and operation with plant operators. This combination requires that chemical engineers be well rounded in a number of disciplines, which enables them to select several specialties in which to work.

Approximately 80 percent of chemical engineers work in the manufacturing industries—primarily in the chemical industry. The remainder work for government agencies and independent research institutes, and as independent consulting engineers. New fields that require chemical engineers are nuclear sciences, rocket fuels, and environmental pollution areas. The development of new drugs, fertilizers, paints, and chemicals is expected to increase the demand for chemical engineers in the coming years.

The professional society for chemical engineers is the American Institute of Chemical Engineers (AIChE).

Civil engineering

Civil engineering, the oldest branch of engineering, is closely related to practically all our daily activities. The buildings we live in and work in, the transportation facilities we use, the water we drink, and the drainage and sewerage systems we use are the results of civil engineering. Civil engineers design and supervise the construction of roads, harbors, airfields, tunnels, bridges, water supply and sewerage systems, and many other types of structures. Civil engineers can specialize in a number of areas within the field, with the following being the most prominent: construction, city planning, structural engineering, hydraulic engineering, transportation, highways, and sanitation.

CONSTRUCTION ENGINEERS are responsible for the management of resources, manpower, finances, and materials necessary for construction projects. These projects may vary from the erection of skyscrapers to the movement of concrete and earth.

CITY PLANNERS develop plans for the future growth of cities and the various systems related to their operation. Street planning, zoning, and industrial-site development are problems encountered in the field of city planning.

STRUCTURAL ENGINEERS are responsible for the design and supervision of the erection of structural systems, including buildings, dams, powerhouses, stadiums, bridges, and numerous other structures. Strength and appearance are considered in designing structures of this type to serve the required needs economically.

HYDRAULIC ENGINEERS work with the behavior of water from its conservation to its transportation. They design wells, canals, dams, pipelines, drainage systems, and other methods of controlling and utilizing water and fluids (Fig. 1.20).

TRANSPORTATION ENGINEERS work with the development and improvement of railroads and airlines in all phases of their operations. Railroads are built, modified, and maintained under the supervision of civil engineers. Design and construction of airport runways, control towers, passenger and freight stations, and aircraft hangars are done by civil engineers who specialize in the field of transportation. (Fig. 1.21).

HIGHWAY ENGINEERS develop the complex network of highways and interchanges for moving automobile traffic. These systems require the design of tunnels, culverts, and traffic control systems.

SANITARY ENGINEERS assist in maintaining public health through the purification of water, control of water pollution, and sewage control. Such systems involve the design of pipelines, treatment plants, dams, and related systems.

The activities of the civil engineer are very diversified, with opportunities in a variety of locations from city centers to remote construction sites. Due to their experience in management and the solution of environmental problems, many civil engineers find positions in administration and municipal management. The majority of civil engineers are associated with federal, state, and local government agencies and the construction industry. Many work as consulting engineers for architectural firms and independent consulting engineering firms. The remainder work for public utilities, railroads, educational institutions, and manufacturing industries.

The professional society for civil engineering is the American Society of Civil Engineers (ASCE), founded in 1852, the oldest engineering society in the United States.

Electrical engineering

Electrical engineers are concerned with the utilization and distribution of electrical energy for the improve-

FIG. 1.20 Civil engineers design and supervise the construction of such structures as this third power plant of the Grand Coulee Dam of Washington. (Courtesy of the Bureau of Reclamation.)

FIG. 1.21 Civil engineers designed this gravity railroad yard in Houston, Texas, to aid in the switching of railroad cars. (Courtesy of Southern Pacific Railroad.)

ment of industrial and domestic functions. The two main divisions of electrical engineering are (1) power and (2) electronics. Power deals with the control of large amounts of energy used by cities and large industries, whereas electronics deals with small amounts of power used for communications and automated operations that have become integral parts of our everyday lives.

These two major divisions of electrical engineering have many areas of specialization. A few of these are given below.

POWER GENERATION poses many electrical engineering problems, from the development of transmission equipment to the design of generators for producing electricity. Modern methods of power transmission and generation have made it possible for electrical power to be an economical source of industrial energy (Fig. 1.22).

POWER APPLICATIONS are numerous in typical homes where toasters, washers, dryers, vacuum cleaners, and lights are used on a continuing basis. But only about one-quarter of the total energy consumed is used in the home. Industry uses about half of all energy for metal refining, heating, motor drives, welding, machinery controls, chemical processes, plating, and electrolysis.

TRANSPORTATION industries require electrical engineers to develop electrical systems for automobiles, aircraft, and other forms of transportation. These systems are used for starting ignition, lighting, and instrumentation. Locomotives and ships may power their own generators, which supply electrical power to turn their driving wheels or propellers. The sophisticated signal systems necessary for all forms of transportation require the services of electrical engineers.

FIG. 1.22 The design of power transmission systems is the responsibility of the electrical engineer. A helicopter is used to patrol this 1800-mile long Parker-Davis project in Arizona. (Courtesy of the Bureau of Reclamation, U.S. Department of the Interior.)

FIG. 1.23 Electrical engineers have made components smaller, lighter, and more efficient. Microminiature components, printed wiring, and advanced encapsulation techniques all combine to reduce complete circuits to a fraction of their former size. (Courtesy of General Dynamics Corporation.)

ILLUMINATION is required at all levels of human activities and environment. The improvement of illumination systems and the economy of illumination energy are challenging areas of study for the electrical engineer.

COMPUTERS have become a giant industry that is the domain of the electrical engineer. Computers used in conjunction with industrial electronics have changed manufacturing and production processes of industry, requiring fewer employees and resulting in greater precision. The trend in the production of electronic devices has been toward larger capacities and capabilities, while miniaturizing the circuits and components (Fig. 1.23).

COMMUNICATIONS is the field devoted to the improvement of radio, telephone, telegraph, and television systems, which are the nerve centers of most industrial operations. Communications is vital in the dispatching of a taxicab, in the control of ships and aircraft, and in many other everyday personal and industrial applications.

INSTRUMENTATION is the study of systems of electronic instruments used in the precise measuring of industrial processes. Extensive use has been made of the cathode-ray tube and the electronic amplifier in industrial applications and atomic power reactors. Instrumentation has been increasingly applied to medicine for diagnosis and therapy.

MILITARY ELECTRONICS is utilized in practically all areas of military weapons and tactical systems, from the walkie-talkie to the distant radar networks for detecting enemy aircraft. Remote-controlled electronic systems are used for navigation and interception of guided missiles. Many revolutionary advancements are expected to continue in fields of military applications of electrical engineering.

More electrical engineers are employed than any other type of engineer. Electrical engineers are employed in the United States by manufacturers of electrical and electronic equipment, aircraft and parts, business machines, and professional and scientific equipment. The increased need for electrical equipment, for automation, and for computerized systems is expected to contribute to the very rapid growth of this field during the coming years.

The professional society for electrical engineers is the Institute of Electrical and Electronic Engineers (IEEE). This is the world's largest technical society, and it was founded in 1884.

Industrial engineering

Industrial engineering, one of the newer branches of the engineering profession, is defined by the National Professional Society of Industrial Engineers as follows:

> Industrial engineering is concerned with the design, improvement, and installation of integrated systems of men, materials, and equipment. It draws upon specialized knowledge and skill in the mathematical, physical, and social sciences together with the principles and methods of engineering analysis and design to specify, predict, and evaluate the results to be obtained from such systems.

Industrial engineering is related to all areas of engineering and business. This field differs from other areas of engineering in that it is more closely related to people and their performance and working conditions (Fig. 1.24). Consequently, in many instances the industrial engineer is a manager of people, machines, materials, methods, money, and markets.

The industrial engineer may be assigned the responsibility of plant layout, the development of plant processes, or the determination of operating standards that will improve the efficiency of a plant operation. He or she may be responsible for quality control and cost analysis, two operations that are essential to a profitable manufacturing industry.

Several specific areas of industrial engineering are: management, plant design and engineering, electronic data processing, systems analysis and design, control of production and quality, performance standards and measurements, and research. In order for industrial engineers to work in these areas, they must act as members of a team composed of engineers from other branches of the profession. They must view the overall operations of industry and the factors affecting its efficiency rather than be concerned with isolated areas within the total structure.

People-oriented areas include the development of wage incentive systems, job evaluation, work measurement, and the design of environmental systems. Industrial engineers are often involved in management-labor agreements that affect the operation and production of an industry. They design and supervise systems for the improved safety and production of the working forces employed in an industry.

More than two-thirds of industrial engineers are employed in manufacturing industries. Others work for insurance companies, construction and mining firms, public utilities, large businesses, and governmental agencies. With the increasing complexities of industrial operations and the expansion of automated processes, the field of industrial engineering is expected to grow very rapidly.

The professional society of industrial engineers is the American Institute of Industrial Engineers (AIIE), which was organized in 1948.

FIG. 1.24 Industrial engineers lay out and design complex industrial facilities to provide efficient production. This plant produces and assembles automobiles. (Courtesy of Jervis B. Webb Company.)

FIG. 1.25 This hydroelectric power plant at Hoover Dam was completed by mechanical engineers. (Courtesy of the Bureau of Reclamation.)

Mechanical engineering

The major areas of specialization of the mechanical engineer are: power generation, transportation, aeronautics, marine vessels, manufacturing, power services, and atomic energy. Activities within each of these areas are outlined below.

POWER GENERATION requires that prime movers be developed to power electrical generators that will produce electrical energy in stationary power plants. Mechanical engineers design and supervise the operation of steam engines, turbines, internal combustion engines, and other prime movers required in power generation (Fig. 1.25). The storage and handling of fuel used in these systems is also a mechanical engineering problem.

TRANSPORTATION conveyances are designed and manufactured by mechanical engineers. Automobiles, trucks, buses, locomotives, marine vessels, and aircraft are designed and produced through the efforts of mechanical engineers, who must design the power systems of transportation vehicles as well as their structural and fuel systems (Fig. 1.26).

AERONAUTICS is a specialized field requiring the mechanical engineer to develop engines to power aircraft. The controls and environmental systems of air-

FIG. 1.26 The mechanical engineer is responsible for the design and production of the automobile's power system, suspension system, and much of its entire design. (Courtesy of Chrysler Corporation.)

craft are also problems solved by mechanical engineers. The fabrication of aircraft requires a close coordination between mechanical and aerospace engineers.

MARINE VESSELS that are powered by steam, diesel, or gas-generated engines are designed by mechanical engineers. They are also responsible for supplying power services throughout the vessel, including light, water, refrigeration, and ventilation.

MANUFACTURING is a challenging field that requires the mechanical engineer to design new products and new factories in which to build them. Mechanical engineers work closely with industrial engineers in the management of a wide variety of machines. Economy of manufacturing and the achievement of a uniform level of product quality are major functions of this area of mechanical engineering.

POWER SERVICES include the movement of liquids and gases through pipelines, refrigeration systems, elevators, and escalators. In applying the principles of mechanical engineering, the mechanical engineer must have a knowledge of pumps, ventilation equipment, fans, and compressors.

ATOMIC ENERGY development has used mechanical engineers for the development and handling of protective equipment and materials. Nuclear reactors, which provide nuclear power, are constructed as a joint effort, with the mechanical engineer playing an important role.

The professional society of the mechanical engineering field is the American Society of Mechanical Engineers (ASME).

Mining and metallurgical engineering

Mining and metallurgical engineers are often grouped together as a common profession, although their functions are somewhat separate.

Mining engineers are responsible for the extraction of minerals from the earth and for the preparation of minerals for use by manufacturing industries. They work with geologists to locate ore deposits, which must be exploited through the construction of extensive tunnels and underground operations, necessitating an understanding of safety, ventilation, water supply, and communications.

The metallurgical engineer develops methods of processing and converting metals into useful products. Two main areas of metallurgical engineering are extractive and physical. Extractive metallurgy is concerned with the extraction of metal from raw ores to form pure metals. Physical metallurgy is the development of new products and alloys.

Many metallurgical engineers work on development of machinery for electrical equipment and in the aircraft and aircraft parts industries. With the development of new materials for space flight vehicles, jet aircraft, missiles, and satellites, new lightweight, high-strength materials will increase the need for additional metallurgical engineers at a very rapid rate (Fig. 1.27). They will also be needed to develop economical meth-

FIG. 1.27 A metallograph is used to show the structure of an alloy that may be used in the construction of a refinery unit. Materials are specially developed for specific applications by the metallurgic engineer. (Courtesy of Exxon Corporation.)

ods for extracting metal from low-grade ore when the high-grade ore has been depleted.

Mining engineers who work at mining sites are usually employed near small communities or in out-of-the-way places, while those in research and consulting are often located in metropolitan areas. The development of new alloys is expected to increase the need for mining engineers for the recovery of relatively little-used ores.

The professional society of this field of engineering is the American Institute of Mining, Metallurgical, and Petroleum Engineering (AIME).

Nuclear engineering

The field of nuclear engineering has been confined to graduate studies until recent years, when undergraduate programs were developed to provide a more continuous curriculum. This branch of engineering promises to be a spectacular contributor to our future way of life.

The earliest work in the nuclear field has been for military applications and defense purposes. However, the utilization of nuclear power for domestic needs is being developed for a number of areas. Present applications of nuclear energy can be seen in the medical profession and in various other fields as a power source.

Although nuclear engineering degrees are now offered at the bachelor's level, advanced degrees are recommended for this area of engineering. Advanced degrees can provide the additional specialization especially important in the field. Peaceful applications are divided into two major areas: radiation and nuclear power reactors. Radiation is the propagation of energy through matter or space in the form of waves. In atomic physics this term has been extended to include fast-moving particles (alpha and beta rays, free neutrons, etc.), gamma rays, and x-rays. Nuclear science is closely associated with botany, chemistry, medicine, and biology.

The production of nuclear power in the form of mechanical or electrical power has become a major area of the peaceful utilization of nuclear energy (Fig. 1.28). For the production of electrical power, nuclear energy is used as the fuel for producing steam that will drive a turbine generator in the conventional manner. This source of power is expected to reduce the expenditure of our diminishing supply of coal, oil, and gas presently being used in great quantities for power production.

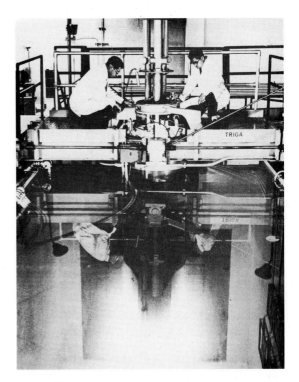

FIG. 1.28 This pulsing reactor is being operated by a nuclear engineer to test the effect of extremely high radiation on delicate equipment. (Courtesy of General Dynamics Corporation.)

Most training of the nuclear engineer is centered around the design, construction, and operation of nuclear reactors. Other areas include the processing of nuclear fuels, thermonuclear engineering, and the utilization of various nuclear by-products. The American Nuclear Society is the professional society of nuclear engineers.

Petroleum engineering

Petroleum engineering is the application of engineering to the development and recovery of petroleum resources. The primary concern of petroleum engineers is the recovery of petroleum and gases; however, they must also develop methods for the transportation and separation of various products. They are responsible for the improvement of drilling equipment and its economy of operation when drilling is being carried out at excessive depths, sometimes as great as four miles (Fig. 1.29).

FIG. 1.29 The petroleum engineer supervises the operation of offshore platforms used in the exploration for oil. (Courtesy of Marathon Oil Company.)

Conservation of petroleum reservoirs has gained increased emphasis because of greater consumption each year. New processes have been developed in recent years for recovering increasing amounts of petroleum from dormant oil reservoirs that had previously been abandoned (Fig. 1.30).

The production phase of petroleum engineering requires the cooperation of practically all branches of engineering, due to the wide variety of knowledge and

FIG. 1.30 Downtown Kilgore, Texas, is a reminder of one of the world's richest oil strikes in 1930. Buildings were torn down to make room for more wells.

skills demanded. Many new industries have emerged from new petroleum products that have been developed through modern production methods and research.

The petroleum engineer is assisted by the geologist in the exploration stages of searching for petroleum. Geologists use devices such as the airborne magnetometer, which gives readings on the earth's subsurfaces, indicating where uplifts that could hold oil or gas are located. If these findings are favorable, seismograph crews measure the depths of particular layers of the subsurface. The only sure way of determining whether oil or gas does exist after preliminary surveys have been made is to drill a well.

Oil well drilling is supervised by the petroleum engineer, who also develops the drilling equipment with members of other branches of engineering to offer the most efficient method of oil removal. When oil is found, the petroleum engineer must design piping systems to remove the oil and transport it to its next point of processing. The processing itself is a joint project in which chemical engineers work with petroleum engineers.

The future of petroleum engineering appears to be very promising, with the use of petroleum increasing yearly. The daily consumption of petroleum in the United States is 17 million barrels per day, and it is estimated that we will consume over 28 million barrels by the year 2000.

The Society of Petroleum Engineers is a branch of AIME, which includes mining and metallurgical engineers and geologists.

1.5

Technologists and technicians

As technology has expanded to include more disciplines, materials, and systems, it has become more impractical for an individual to perform all the steps required in the implementation of a design. Instead, it is more efficient to utilize trained specialists who are responsible for specific areas of technology.

With these changes in technology have emerged two new members of the technological team: the technologist and the technician. Both have as their primary mission the assistance of the engineer at a technical level below that of the engineer and above that of the craftsman.

The technologist

The technologist works under the supervision of an engineer but is called on to exercise considerable responsibility to free the engineer for more advanced applications. Most technologists have a four-year college background in a specialty area of engineering technology that enables them to perform semiprofessional jobs with a high degree of skill. They are trained in the basic engineering disciplines, which prepares them to use advanced equipment, from computers to testing equipment (Fig. 1.31).

Their interest in the practical aspects of engineering qualifies them to offer valuable advice to the engineer concerning production specifications and on-site procedures when new projects are being designed. They may work as designers responsible for converting the preliminary work of the engineer into finished designs that can be efficiently produced.

The technician

The separation between the technologist and the technician is not always clearly defined, and often these titles are mistakenly interchanged. The technician is a two-year graduate of a technical program who performs tasks that are less technical than those performed by the technologist. Technicians may be repairers, inspectors, production specialists, surveyors

FIG. 1.31 This technologist is gathering test data that might detect structural deficiencies in a newly designed tractor. (Courtesy of Ford Motor Company.)

(Fig. 1.32), or similar specialists. They may work under the supervision of an engineer, but ideally they are supervised by a technologist.

In turn, the technician will coordinate activities between the technologist and the craftsman. These levels of responsibility make it possible to ensure that

FIG. 1.32 These technicians are making a cartographic survey. Data from the tellurometer are being recorded in a notebook. (Courtesy of the U.S. Forest Service.)

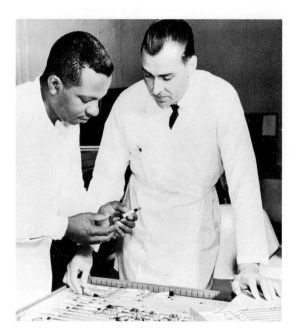

FIG. 1.33 Technicians check parts to within thousandths of an inch to ensure that the part conforms to the specifications. (Courtesy of DoAll Corporation.)

the proper skills and qualifications are available throughout the various levels of the chain of command, thereby creating a more efficient and productive use of personnel.

General

Some well-established areas of technology are aerospace technology, chemical technology, civil engineering technology, electronic technology, and design drafting technology. Approximately 700,000 engineering and science technologists are employed in all industries (Fig. 1.33). Twelve percent of these are women. Almost 475,000 technicians are employed in private industry.

The technologist/technician field has been one of the fastest-growing occupational groups during recent decades, and a continued increase is expected with the expansion of industry.

Additional information concerning technicians can be obtained from the American Society of Certified Engineering Technicians.

1.6
Drafters

Whereas the drafters of the past spent much of their time in the preparation of ink drawings from which prints could be made, the drafters of today carry a greater responsibility in assisting the engineer and the designer. The experienced drafter may be involved in preparing complex drawings, selecting materials, detailing designs, and writing specifications.

DESIGN AND CONSTRUCTION DRAWINGS are made by drafters to explain how to fabricate, build, or erect a project or product in fields such as aerospace engineering, architecture, machine design, mechanical engineering, and electrical/electronic engineering. Drawings must be prepared with sufficient details and specifications to permit them to be executed by the supervisors of construction and manufacturing with the minimum of supplementary information.

TECHNICAL ILLUSTRATION is a type of graphics that is usually prepared as three-dimensional pictorials to illustrate a project or product as realistically as possible. Technical illustration is the most artistic area of engineering design graphics.

MAPS, GEOLOGICAL SECTIONS, AND HIGHWAY PLATS are used for locating property lines, physical features, strata, rights-of-way, building sites, bridges, dams, mines, utility lines, and so forth. Drawings of this type are usually prepared as permanent ink drawings.

Career opportunities in the area of engineering design graphics are based on education, experience, and competence. The three levels of certification for drafters are: drafters, design drafters, and engineering designers.

DRAFTERS are graduates of a two-year post–high-school curriculum in the area of engineering design graphics.

DESIGN DRAFTERS complete two-year programs beyond the high-school level in an approved junior college or technical institute. Drafters with this background are technicians.

ENGINEERING DESIGNERS are drafters who have completed a four-year course at a college offering specialty training in the area of engineering design graphics, which leads to a degree in this area. Graduates of these programs can become certified as technologists.

Professional drafters have opportunities for advancement to positions with titles such as senior drafter, technical illustrator, chief drafter, designer, drafting supervisor, and manager of drafting. With each advancement, the drafter will be given more responsibility of a technical and supervisory nature.

Approximately 485,000 drafters and designers were employed in 1985, and the need for them is expected to increase in the future. Computerized systems are being adopted by industry to improve the drafter's productivity. Computer graphics systems do not lessen the need for the knowledge of the fundamental graphical principles by the drafter or engineer, but computer graphics does offer a different medium of expression (Fig. 1.34).

The professional society of this field is the American Institute for Design and Drafting (AIDD). Student chapters of AIDD are located on most campuses where programs in engineering design graphics are offered.

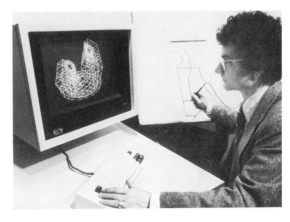

FIG. 1.34 This drafter has constructed a 3-D finite element model of a gun mount on the Applicon display by digitizing a multiview engineering drawing. (Courtesy of Applicon.)

Addresses of Professional Societies

Publications and information from these societies were used as the basis for the content of this chapter:

Alliance for Engineering in Medicine and Biology, 3900 Wisconsin Avenue, N.W., Suite 300, Washington, D.C. 20016

American Ceramic Society, 65 Ceramic Drive, Columbus, Ohio 43214

The American Institute of Aeronautics and Astronautics, 1290 Avenue of the Americas, New York, N.Y. 10019

American Institute of Chemical Engineers, 345 East 47th Street, New York, N.Y. 10017

American Institute for Design and Drafting, 3119 Price Road, Bartlesville, Okla. 74003

The American Institute of Industrial Engineers, 345 East 47th Street, New York, N.Y. 10017

American Institute of Mining, Metallurgical, and Petroleum Engineering, 345 East 47th Street, New York, N.Y. 10017

American Nuclear Society, 244A East Ogden Avenue, Hinsdale, Ill. 60521

American Society of Agricultural Engineers, 2950 Niles Road, St. Joseph, Mich. 49085

American Society of Civil Engineers, 345 East 47th Street, New York, N.Y. 10017

American Society for Engineering Education, Technical Institute Division, 11 DuPont Circle, Suite 200, Washington, D.C. 20036

American Society of Mechanical Engineers, 345 East 47th Street, New York, N.Y. 10017

The Institute of Electrical and Electronic Engineers, 345 East 47th Street, New York, N.Y. 10017

Society of Petroleum Engineers (AIME), 6300 North Central Expressway, Dallas, Tex. 75206

Engineer's Council for Professional Development, 345 East 47th Street, New York, N.Y. 10017

National Society of Professional Engineers, 2029 K Street, N.W., Washington, D.C. 20006

Society of Women Engineers, United Engineering Center, Room 305, 345 East 47th Street, New York, N.Y. 10017

2
The Design Process

2.1
Introduction

A knowledge of scientific and engineering principles is of little value if these disciplines cannot be harnessed to obtain a tangible end that will fulfill the needs of a given situation. For engineers to function to their fullest extent, they must exercise imagination combined with knowledge and curiosity. Engineers must be designers.

Engineering graphics and descriptive geometry, along with other engineering courses, provide methods of solving technical problems. An engineer who is developing a design solution must make many sketches and drawings to develop preliminary ideas before he or she can communicate with associates. Graphical methods used in this manner are creative tools.

2.2
Types of design problems

Design problems are numerous and take many forms; however, most fall into one of two categories: *systems design* and *product design*. The distinct separation of these types of problems is often difficult, due to an overlap of certain characteristics.

Systems design

Systems design deals with the incorporation of available products and components into a unique arrangement that yields a desired result. A residential building is a complex system made up of components and products. For example, the typical residence has a heating-cooling system, a utility system, a plumbing system, a gas system, an electrical system, and many other systems that form the overall composite system (Fig. 2.1). These component systems are also called systems because they are composed of a number of individual parts that can be used for other applications. The electrical system involves wiring, insulation,

FIG. 2.1 The typical residence is a system composed of many component systems.

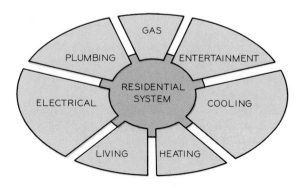

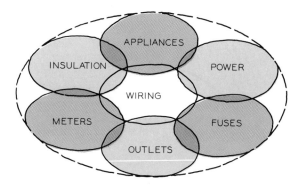

FIG. 2.2 The electrical system of a residence is a composite of related components.

electrical components, light bulbs, meters, controls, switches, and related items (Fig. 2.2).

An engineering project that requires the development of a traffic system for a particular need will overlap into other disciplines (Fig. 2.3). The engineering function will be the primary area that will support the project; however, the project will also involve legal problems, economic principles, historical data, human factors, social considerations, scientific principles, and political limitations. The engineer could, of course, design a suitable driving surface, a drainage system, overpasses, and other components of the traffic system by applying engineering principles without considering other areas and the limitations imposed by them. However, engineers must adhere to a specific budget in essentially all projects in which they are involved, and the budget is closely related to legal and political constraints. For example, traffic laws, zoning ordi-

FIG. 2.3 An engineering project may involve the complex interaction of many professions, with engineering as the primary function. An example of a project of this type is a traffic design problem.

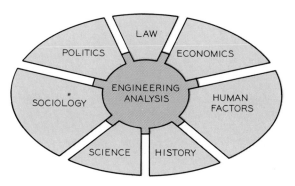

nances, right-of-way possession, and liability clearances are legal areas that must be considered.

Planning for the future is based on past needs and trends, which introduces historical data as a design consideration. Human factors include driver characteristics, safety features, and other factors that affect the function of traffic systems. Social problems are associated with traffic systems. Heavily traveled highways will attract commercial establishments, shopping centers, and filling stations, which will affect the adjacent property. Scientific principles developed through laboratory research can be applied by the engineer in building more durable roads, cheaper bridges, and a more functional system.

EXAMPLE SYSTEMS PROBLEM The following problem is given as an example of a simple systems design that can be used to illustrate the various steps of the design process.

PARKING AREA Select a building on your campus that needs an improved parking lot to accommodate the people housed in the building. This may be a dormitory, an office building, or a classroom building. Design a combination traffic and parking system that will be adequate for the requirements of the building. The solution of this problem must adhere to existing limitations, regulations, and policies of your campus in order for the problem to be as realistic as possible.

Product design

Product design is concerned with the design, testing, manufacture, and sale of an item that will be mass-produced and will perform a specific function. Such a product can be an appliance, a tool, a system component, a toy, or a similar item that can be purchased as a unit. Because of its more limited function, product development is considerably more specific than systems design (Fig. 2.4).

The distinction between a system and a product is not always clearly apparent. An automotive system has a primary function of providing transportation. However, the automobile must also furnish its passengers with communications, illumination, comfort, and safety, and these would classify it as a system. Nevertheless, the automobile is thought of as a product since it is mass-produced for a large consumer market. On the other hand, a petroleum refinery is definitely a system, composed of many interrelated functions and components. All refineries have certain processes in

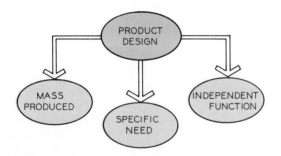

FIG. 2.4   Product design is more limited in scope than systems design and is mass produced.

common, but no two can be considered alike in all respects.

Product design is related to current market needs, cost of production, function, sales, method of distribution, and profit predictions (Fig. 2.5). This concept may be broadened to encompass an entire system that will have sweeping economic and social changes. An example of this transition from product to system is the automobile, whose function has had a significant effect on our way of life. This product has expanded to a system of highways, service stations, repair shops, parking lots, drive-in businesses, residential garages, traffic enforcement, and endless other related components.

EXAMPLE PRODUCT DESIGN PROBLEM   The product problem given below is an example of the type of problem that may be assigned as a class project.

FIG. 2.5   Areas associated with product design are related to the manufacture and sale of completed products.

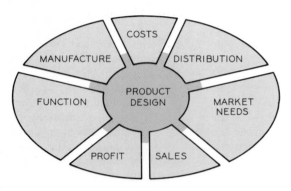

HUNTING SEAT   Many hunters, especially deer hunters, hunt from trees to obtain a vantage point. Sitting in a tree for several hours can be uncomfortable and hazardous to the hunter, which indicates a need for a hunting seat that could be used to improve this situation. Design a seat that would provide the hunter with comfort and safety while he was hunting from a tree, and would meet the general requirements of economy and hunting limitations.

2.3
The design process

Design is the act of devising an original solution to a problem by a combination of principles, resources, and products. Design is the most distinguishing responsibility that separates the engineer from the scientist and the technician. The engineer's solutions may involve a combination of existing components in a different arrangement to provide a more efficient result, or they may involve the development of an entirely new product; but in either case this work is referred to as the act of designing.

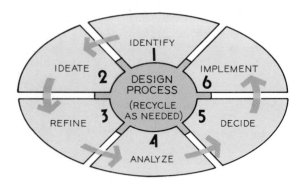

FIG. 2.6   The steps of the design process. Each step can be recycled when needed.

This book emphasizes a six-step design process that is a composite of the most commonly employed steps in solving problems. The six steps are (1) problem identification, (2) preliminary ideas, (3) problem refinement, (4) analysis, (5) decision, and (6) implementation (Fig. 2.6). Although designers work sequentially from step to step, they may recycle to previous steps as they progress.

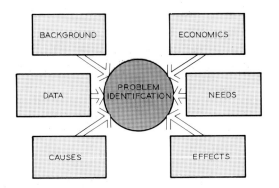

FIG. 2.7 Problem identification requires the accumulation of as much information as possible before a solution is attempted by the designer.

Problem identification

Most engineering problems are not clearly defined at the outset; consequently they must be identified before an attempt is made to solve the problem (Fig. 2.7). For example, a prominent concern today is air pollution. Before this problem can be solved, you must identify what air pollution is and what causes it. Is pollution caused by automobiles, by factories, by atmospheric conditions that harbor impurities, or by geographic features that contain impure atmospheres?

When you enter a bad street intersection where traffic is unusually congested, do you identify the reasons for the congestion? Are there too many cars? Are the signals poorly synchronized? Or are there visual obstructions resulting in congested traffic?

Problem identification requires considerable study beyond a simple problem statement like: "Solve air pollution." You will need to gather data of several types, including field data, opinion surveys, historical records, personal observations, experimental data, and physical measurements and characteristics (Fig. 2.7).

Preliminary ideas

Once the problem has been identified, the next step is to accumulate as many solutions as possible (Fig. 2.8). Preliminary ideas should be sufficiently broad to allow for unique solutions that could revolutionize present methods. All ideas should be recorded in written form. Many rough sketches of preliminary ideas should be made and retained as a means of generating original ideas and stimulating the design process. Ideas and comments should be noted on the sketches as a basis for further preliminary designs.

Problem refinement

Several of the better preliminary ideas are selected for further refinement to determine their true merits. Rough sketches are converted to scale drawings that will permit space analysis, critical measurements, and calculation of areas and volumes affecting the design (Fig. 2.9). Consideration is given to spatial relation-

FIG. 2.8 Preliminary ideas are developed after the identification process has been completed. All possibilities should be listed and sketched to give the designer a broad selection of ideas from which to work.

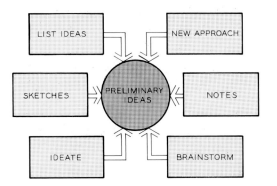

FIG. 2.9 Refinement begins with the construction of scale drawings of the better preliminary ideas. Descriptive geometry and graphical methods are used to find the necessary geometric characteristics.

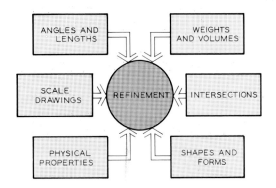

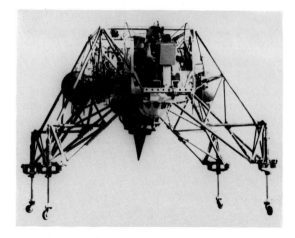

FIG. 2.10 The refinement of the lunar vehicle required the use of descriptive geometry and other graphical methods. (Courtesy of Ryan Aeronautics, Inc.)

ships, angles between planes, lengths of structural members, intersections of surfaces and planes.

Descriptive geometry is a valuable tool for determining information of this type, and it precludes the necessity for tedious mathematical and analytical methods.

An example of a problem of this nature is illustrated in the landing gear of the lunar vehicle shown in Fig. 2.10. The configuration of the landing gear was drawn to scale in the descriptive views of the landing craft. It was necessary, at this point, to determine certain fundamental lengths, angles, and specifications related to the fabrication of the gear. The length of each leg of the landing apparatus and the angles be-

FIG. 2.11 The analysis phase of the design process is the application of all available technological methods from science to graphics in evaluating the refined designs.

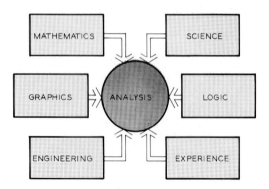

tween the members at the point of junction had to be found to design a connector, and the angles the legs made with the body of the spacecraft had to be known in order to design these joints. All this information was determined with the use of descriptive geometry.

Analysis

Analysis is the step of the design process where engineering and scientific principles are used the most (Fig. 2.11). Analysis involves evaluation of the best designs to determine the comparative merits of each with respect to cost, strength, function, and market appeal. Graphical principles can also be applied to analysis. The determination of forces is somewhat simpler with graphical vectors than with the analytical method. Functional relationships between moving parts will also provide data that can be obtained by graphical methods more easily than by analytical methods.

Graphical methods can also be applied to the conversion of functions of mechanisms to a format that will permit the designer to convert this action into an equation form that will be easy to utilize. Data that would otherwise be difficult to interpret by mathematical means can be gathered and graphically analyzed.

Models constructed at reduced scales are valuable to the analysis of a design to establish relationships of moving parts and outward appearances and to evaluate other design characteristics. Full-scale prototypes are often constructed after the scale models have been studied for function.

Decision

A decision must be made at this stage to select a single design that will be accepted as the solution of the design problem (Fig. 2.12). Each of the several designs that have been refined and analyzed will offer unique features, and it will probably not be possible to include all of these in a single, final solution. In many cases, the final design is a compromise that offers as many of the best features as possible.

The decision may be made by the designer on an independent, unassisted basis, or it may be made by a group of associates. Regardless of the size of the group making the decision as to which design will be accepted, graphics is a primary means of presenting the

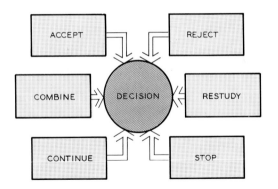

FIG. 2.12 Decision is the selection of the best design or design features to be implemented.

proposed designs for a decision. The outstanding aspects of each design usually lend themselves to presentation in the form of graphs that compare costs of manufacturing, weights, operational characteristics, and other data that will be considered in arriving at the final decision.

Implementation

The final design concept must be presented in a workable form. This type of presentation refers primarily to the working drawings and specifications that are used as the actual instruments for fabrication of a product, whether it is a small piece of hardware or a bridge (Fig. 2.13). Engineering graphics fundamentals must be used to convert all preliminary designs and data

FIG. 2.13 Implementation is the final step of the design process, where drawings and specifications are prepared from which the final product can be constructed.

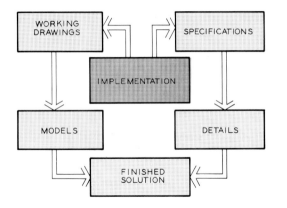

into the language of the manufacturer, who will be responsible for the conversion of the ideas into reality. Workers must have complete detailed instructions for the manufacture of each single part, measured to a thousandth of an inch to facilitate its proper manufacture. Working drawings must be sufficiently detailed and explicit to provide the legal basis for a contract that will be the document for the contractor's bid on the job.

2.4
Application of the design process to a simple problem

To illustrate the steps of the design process as they would be applied to a simple design problem, the following example is given.

Swing-set anchor problem

A child's swing set has been found unstable during the peak of the swing. The momentum of the swing causes the A-frame to tilt with a possibility of overturning and causing injury. The swing set has swings attached to accommodate three children at a time. Design a device that will eliminate this hazard and have market appeal for owners of swing sets of this type.

PROBLEM IDENTIFICATION As a first step, the designer writes down the problem statement (Fig. 2.14) and a statement of need. The limitations and desirable features are listed, along with necessary sketches, to enable the designer to develop a better understanding of the problem requirements (Figs. 2.14 and 2.15). Much of the information and notes in the problem identification step may be obvious to the designer, but the act of writing statements about the problem and making freehand sketches helps to get off "dead center," which is a common weakness at the beginning of the creative process.

PRELIMINARY IDEAS Work sheets are used to list brainstorming ideas (Fig. 2.16). The better ideas and design features are summarized on another work sheet (Fig. 2.17). These verbal ideas are then translated into rapidly drawn freehand sketches (Fig. 2.18)

PROBLEM IDENTIFICATION

1. Project title

   SWING SET ANCHOR

2. Problem statement

   SWING SETS (SEE SKETCH TEND TO OVERTURN WHEN IN USE WHICH CAUSES AN UNSAFE CONDITION.

3. Requirements and limitations

   A. SEE SKETCH FOR DIMEN-SIONS. 3 SWINGS ATTACHED.

   B. SALES PRICE: $5 - $10 RANGE

   C. EASY TO ATTACH BY HOUSEWIFE

   D. ALLOWS SWING SET TO BE MOVED

   E. SAFE FOR CHILDREN

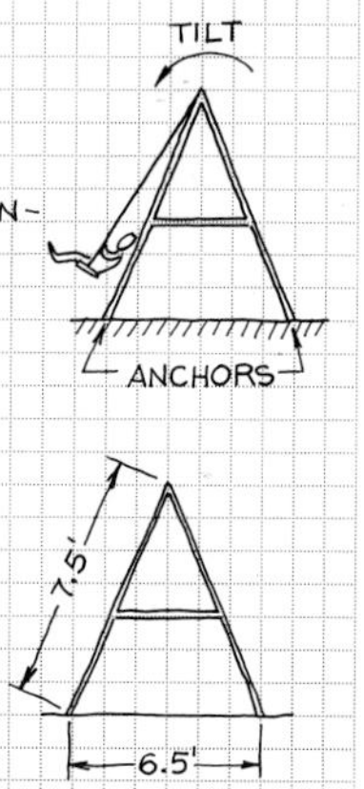

FIG. 2.14 A work sheet showing a portion of problem identification.

FIG. 2.15 The information needed to identify the problem is listed. This information would need to be gathered to complete this portion of the design process.

4. Needed information

   A. NUMBER OF SWING SETS SOLD
   – WRITE MANUFACTURERS
   – WRITE RETAILERS

   B. SIZES OF VARIOUS SWING SETS
   – WRITE MANUFACTURERS

   C. ARE THERE COMPETING PRODUCTS
   – REVIEW PRODUCT CATALOGS

5. Market Considerations

   A. SALES PRICES OF SWING SETS
   – SURVEY CATALOGS & LOCAL RETAILERS

   B. WOULD AN ANCHOR SELL – SURVEY FAMILIES WITH SWING SETS

   C. EFFECTS OF GEOGRAPHY WITHIN USA ON SWING SET SALES – WRITE MANUFACTUR-ERS OF SETS

   D. WHERE ARE SWING SETS SOLD?
   CHECK YELLOW PAGES
   VISIT STORES

PRELIMINARY IDEAS

1. Brainstorming ideas

   A. SAND BAGS

   B. STAKE IN GROUND – METAL OR WOOD

   C. SET IN CONCRETE

   D. WIDEN BASE

   E. FOOT ATTACHED TO LEGS

   F. WEIGHTS ON LEGS

   G. REDESIGN SWING SET A-FRAME

   H. STAKE WITH NYLON CORD

   I. STAKE WITH CHAIN

   J. STAKE WITH ROPE

   K. SUCTION CUPS ON PATIO

   L. WATER-FILLED WEIGHTS

   M. BRICKS AT BASE

Fig. 2.16 After holding a brainstorming session, the better ideas are listed as the first attempt to develop preliminary ideas.

FIG. 2.17 Several ideas are selected from the brainstorming list to be developed as preliminary ideas.

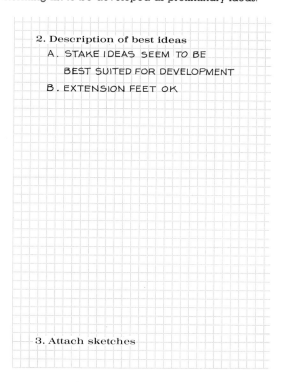

2. Description of best ideas

   A. STAKE IDEAS SEEM TO BE

      BEST SUITED FOR DEVELOPMENT

   B. EXTENSION FEET OK

3. Attach sketches

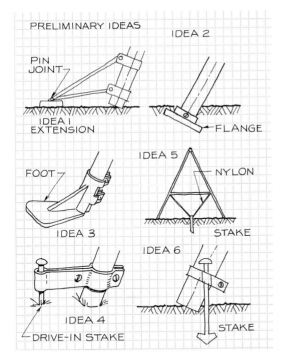

FIG. 2.18 Preliminary ideas are sketched. This is the most creative step of the design process.

FIG. 2.19 The refinement of several preliminary designs is begun by giving verbal descriptions of the selected designs.

REFINEMENT

1. Description of design

A. STAKE

   1. METAL OR WOOD STAKE - 6 IN - 10 IN

   2. ATTACHED WITH CHAIN, CORD, OR CABLE - ABOUT 10 IN LONG

   3. ATTACH TO A-FRAME WITH COLLAR

   4. NEED 4

B. EXTENSION FOOT

   1. ATTACH TO A-FRAME LEGS

   2. MUST DETERMINE LENGTH OF FOOT NECESSARY TO PREVENT TILTING

   3. MUST BE EASY TO ATTACH TO A-FRAME

   4. NEED 4 FEET

   5. NO HEAVIER THAN 1 POUND EACH

2. Attach scale drawings

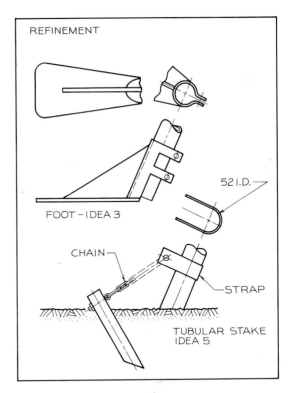

FIG. 2.20 Scaled refinement drawings are used to develop two or more of the designs. Only a few dimensions are needed.

as means of developing and presenting possible solutions to the design problem. The designer should attempt to develop as many ideas as possible. This is the most creative step of the design process.

PROBLEM REFINEMENT A verbal description of the design features of one or more of the the preliminary ideas is listed on a work sheet (Fig. 2.19) for comparison. The better designs are drawn to scale in preparation for their analysis (Fig. 2.20). Only a few dimensions need to be given at this stage.

Orthographic projection, working-drawing principles, and descriptive geometry may be used, depending on the particular problem being refined. In this example (Fig. 2.20), simple orthographic views with auxiliary views depict the two designs.

ANALYSIS Work sheets are used to analyze the tubular stake design (Figs. 2.21, 2.22, and 2.23). The major headings on these work sheets are suggested topics that should be analyzed. This process of analysis should be completed for each design if several solutions are being considered.

ANALYSIS

1. Function

  A. ANCHORS SWING SET

  B. ATTACHES TO GROUND - NOT CONC. SLAB

  C. PREVENTS OVERTUNING

2. Human engineering

  A. EASY TO INSTALL

  B. REQUIRES NO SPECIAL TOOLS

  C. PROVIDES SAFETY FOR CHILDREN

3. Market & consumer acceptance

  A. KEEP PRICE UNDER $15 FOR SWING SET

  B. SHOULD BE SOLD WITH SWING SET

  C. CAN BE SOLD AS AN ACCESSORY TO THE SWING SET.

  D. NEED ESTIMATE OF SWING SETS SOLD PER YEAR.

FIG. 2.21 An analysis work sheet.

FIG. 2.22 A work sheet that continues the analysis step of the design process.

4. Physical description

  A. STAKE: 14" LONG - CRIMPED AT ONE END FOR EASE OF DRIVING.

  B. CHAIN ATTACHED TO STAKE WITH EYE-BOLT THROUGH HOLE IN STAKE.

  C. WEIGHT - ABOUT 8 ONCES

  D. METAL COLLAR WITH NUT & BOLT TO FIT AROUND SWING SET LEG

5. Strength

  A. WITHSTANDS A TENSION AT EACH LEG OF 70 LEG.

6. Production procedures

  A. USE 15 GAGE GALVANIZED IRON PIPE (∅ 40 OD) FOR STAKE. CUT ON DIAGONAL TO CRIMP & POINT STAKE

  B. 15 GAGE METAL COLLAR TO BE FORMED INTO U-SHAPE & DRILLED FOR ∅5 BOLT

  C. STAKE DRILLED ∅8 FOR ∅5 BOLT FOR ATTACHING CHAIN

A. COSTS
1. CHAIN	$0.30
2. STAKE	.20
3. COLLAR	.20
4. EYE BOLT & NUT	.10
5. COLLAR BOLT	.10
	.90
B. LABOR	.60
C. PACKAGING	.30
D. PROFIT	1.15
E. WHOLESALE PRICE	2.95
F. RETAIL PRICE	$4.95

FIG. 2.23 The continuation of the analysis step of the design process.

The force F at the critical angle can be measured or estimated (Fig. 2.24) and used as the basis of a vector polygon to determine the magnitude of R at the base of the swing that must be resisted by the anchor. Again, graphics is used as a design tool.

DECISION Two designs, the stake and chain and the extension foot, are compared in the decision table (Fig. 2.25). Factors for analysis are assigned points in order for the total value to be 10 points. The factors for each design are evaluated to determine which has the best overall score.

The final conclusions are given in Fig. 2.26. A general summary of the recommended design is given, along with the profit outlook for the product in the marketplace. A decision is made to implement the stake and chain solution.

IMPLEMENTATION The tubular stake design is presented in the form of a working drawing, where each individual part is detailed and dimensioned. All principles of graphical presentation are used, including a freehand sketch illustrating how the parts will be assembled (Fig. 2.27). Note that changes have been made since the initial refinement of this design.

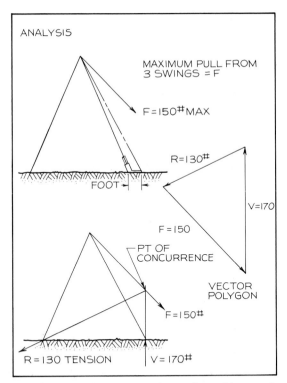

ANALYSIS

MAXIMUM PULL FROM 3 SWINGS = F

F=150# MAX

R=130#

FOOT

V=170

F=150

PT OF CONCURRENCE

VECTOR POLYGON

F=150#

R=130 TENSION   V=170#

FIG. 2.24 Descriptive geometry and graphics can be used to analyze a solution. In this case, the force (R) is calculated graphically.

FIG. 2.25 A work sheet with a decision table that is used to evaluate the final designs.

DECISION

1. Decision table for evaluation

Design 1: STAKE & CHAIN

Design 2: EXTENSION FOOT

Design 3:

Design 4:

Design 5:

Maximum value	Factors for analysis	Designs						
		1	2	3	4	5	6	7
2	Function	2	1					
2	Human Factors	1.5	1.5					
1	Market analysis	.5	.2					
1	Strength	1	1					
1	Production procedures	1	.5					
1	Cost	.5	.2					
2	Profitability	1.5	1.0					
0	Appearance	0	0					
10	TOTALS	8	5.9					

CONCLUSIONS

IMPLEMENT THE STAKE AND CHAIN.
THIS IS BEST DESIGN WITH THE BEST
MARKET POTENTIAL.
MARKET WITH SWING SETS.

SALES PRICE	$4.95
SHIPPING EXPENSES	.50
NUMBER TO SELL TO TO BREAK EVEN	1000
MANUFACTURED BY CONTRACTOR	
ESTIMATED PROFIT PER SET OF 4	$1.15

FIG. 2.26 The final decision and other conclusions are summarized on this work sheet.

FIG. 2.27 A working drawing is made of the final design that is to be implemented.

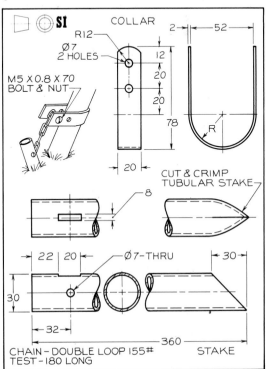

SI

COLLAR

R12

Ø7 2 HOLES

M5 X 0.8 X 70 BOLT & NUT

2   52

12
20
20

78

R

20

CUT & CRIMP TUBULAR STAKE

8

22 20   Ø7–THRU   30

30

32

360

CHAIN – DOUBLE LOOP 155#
TEST – 180 LONG

STAKE

FIG. 2.28 The completed swing-set anchor.

Standard parts, such as nuts, bolts, and the chain, should be merely noted. They should not be drawn, since they are parts that will not be specially fabricated. With this working drawing, the designer has implemented the design as far as can be done without

actually building a prototype, a model, or the actual part.

A photograph of the actual part is shown in Fig. 2.28. It is shown attached to the swing set in Fig. 2.29 with the stake driven into the ground.

FIG. 2.29 The anchor attached to a swing set.

Short Design Problems

The following problems can be completed as individual assignments or by the team approach, in which several students work on the same problem together. Design work sheets should be completed as shown in Fig. 2.14 through Fig. 2.27. Additional supplementary sheets can be used as desired.

1. **Lamp bracket.** Design a simple bracket to attach a desk lamp to a vertical wall for reading in bed. The lamp should be easily removable so that it can be used as a conventional desk lamp.

2. **Towel bar.** Design a towel bar for a kitchen or bathroom. Determine optimum size and consider styling, ease of use, and method of attachment. This design will be a modification of those already available on the market.

3. **Pipe aligner for welding.** Pipes are often welded together in sections. The initial problem of joining pipes with a butt weld is the alignment of the pipes in the desired position. Design a device with which to align pipes for on-the-job welding. For ease of operation, a hand-held device would be desirable. Assume that the pipes will vary in diameter from 2″ to 4″.

4. **Side-view mirror.** In most cars, rear-view mirrors are attached to the side of the automobile to improve the driver's view of the road. Design a

side-view mirror that is an improvement over those with which you are familiar. Consider the aerodynamics of your design, protection from inclement weather, and other factors that would affect the function of the mirror.

5. **Nail feeder.** Workers lose time in covering a roof with shingles if they have to fumble for nails. Design a device that can be attached to a worker's chest and will hold nails in such a way that these will be fed in a lined-up position ready for driving. Determine the number of nails that should be held in this device at any one time.

6. **Cupboard door closer.** Kitchen cupboard and cabinet doors are usually not self-closing and thus are safety hazards and can be unsightly. Design a device that will close doors left partially open. It would be advantageous to provide a means for disengaging the closer when desired.

7. **Paint-can holder.** Paint cans are designed with a simple wire bail that, when held, makes it difficult to get a paint brush in the can. Wire bails are also painful to hold for any length of time. Design a holding device that can be easily attached and removed from a gallon-size paint can (6.5″ diameter × 7.5″ height). Consider human factors such as comfort, grip, balance, and function.

8. **Self-closing or -opening hinge.** Interior doors of residential dwellings tend to remain partially open instead of staying in a completely open position against the wall. This introduces a problem when a door opens into the end of a hall, since the edge of the door that is ajar in the middle of the hall can be a hazard to the occupants.

Design a hinge that will hold the door in a completely open position. This can be used to replace door stops and other devices used to hold doors.

9. **Tape cartridge storage unit.** Tape players are used frequently in automobiles. Design a storage unit that will hold a number of these cartridges in an orderly fashion so that the driver can select and insert them with a minimum of motion and distraction. Determine the best location for this unit and the method of attachment to an automobile. Provision should be made to protect the cartridges from theft.

10. **Slide projector elevator.** Most commercial slide and movie projectors have adjustment feet used to raise the projector to the proper position for casting an image on a screen. The range of variation of this adjustment is usually less than 2″. Study a slide projector available to you to determine the specific needs and the limitations of an adjustment that would provide greater variation in elevation. Design a device to serve this purpose as part of the original design or as an accessory that could be used on existing projectors.

11. **Book holder for reading in bed.** As a student, you may often want to read while lying in bed. Design a holder that can be used for supporting a book in the desired position while providing comfortable conditions.

12. **Table leg design.** Do-it-yourselfers build a variety of tables using slab doors, plywood, and commercially available legs. Table tops come in a number of sizes, but table heights are fairly standard. Determine what the standard table heights are and design a family of legs that can be attached to table tops with screws. For improved appearance, the legs should not be vertical but should slant from the table top. Design the legs and attachments, indicating the method of manufacture, size, cost, and method of attachment.

13. **Canoe-mounting system.** Canoes and light boats are often transported on the top of automobiles on a luggage rack or similar attachment. Design an accessory that will enable a person to remove and load a boat on top of an automobile without additional help. This attachment should accommodate aluminum boats varying from 14′ to 17′ in length and weighing 100 lb to 200 lb. Give specifications for a method of securing the boat after it is in its final position on top of the automobile.

14. **Toothbrush holder.** Design a toothbrush holder that can be attached to a bathroom wall and can hold a drinking cup and two toothbrushes.

15. **Napkin holder.** Design a device that will hold 25 paper napkins on a dining room table for easy access.

16. **Book holder.** Design a holder that will support your textbook on your drawing table in a position that will make it more readable and accessible.

17. **Clothes hook.** Design a clothes hook that can be attached to a closet door for hanging clothes. It should be easy to manufacture and simple to install.

18. **Automobile ashtray.** Design an ashtray that can be attached to the dashboard of any automobile. It should be easy to attach and to remove for emptying.

19. **Door stop.** Design a door stop that can be attached to a vertical wall or to the floor to prevent the door knob from bumping the wall.

20. **Drawer handle.** Design a handle that would be satisfactory for a standard file cabinet drawer.

21. **Paper dispenser.** Design a dispenser that will hold a 6″-diameter-by-24″-wide roll of wrapping paper. The paper will be used on a table top for wrapping packages.

22. **Handrail bracket.** Design a bracket that will support a tubular handrail that will be used on a staircase (Fig. 2.30). Consider the weight that the handrail must support.

23. **Latch-pole hanger.** Design a hanger that can be used to support a latch-pole from a vertical wall. It should be easy to install and use (Fig. 2.31).

24. **Pipe clamp.** A pipe with a 4″ diameter must be supported by angles (Fig. 2.32) that are spaced 8′ apart. Design a clamp that will support the pipe without drilling holes in the angles.

25. **TV yoke.** Design a yoke that can support a TV set from the ceiling of a classroom and permit it to be adjusted to the best position for viewing.

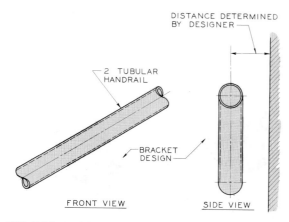

FRONT VIEW   SIDE VIEW

FIG. 2.30 Problem 22. A handrail bracket.

26. Flagpole socket. Design a socket for a flag that is to be attached to a vertical wall. Determine the best angle of inclination for the flagpole.

27. Crutches. Design a portable crutch that could be used by a person with a temporary leg injury.

28. Cup holder. Design a holder that will support a soft-drink can or bottle on the interior of an automobile.

29. Gate hinge. Design a hinge that could be attached to a 3″-diameter tubular post to support a 3′-wide gate.

30. Safety lock. Design a safety lock that will hold a high-voltage power switch in either the "off" or the "on" position to prevent an accident (Fig. 2.33).

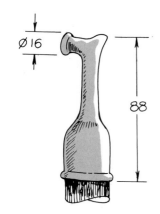

FIG. 2.31 Problem 23. A latch-pole hanger.

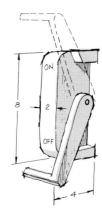

FIG. 2.33 Problem 30. A safety lock.

FIG. 2.32 Problem 24. A pipe clamp.

FIG. 2.34 Problem 31. A tubular hinge.

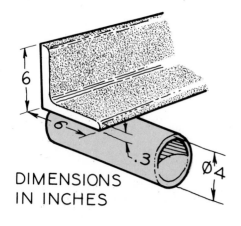

DIMENSIONS
IN INCHES

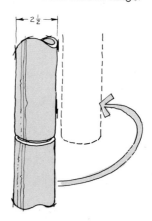

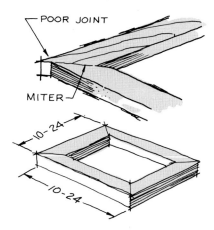

FIG. 2.35 Problem 32. A miter jig.

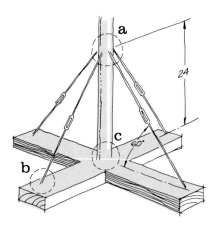

FIG. 2.36 Problem 33. Base hardware for a volleyball net.

31. **Tubular hinge.** Design a hinge that can be used to hinge 2.5″ OD high-strength aluminum pipe in the manner shown in Fig. 2.34. A hinge of this type is needed for portable scaffolding.

32. **Miter jig.** Design a jig that can be used for assembling wooden frames at 90° angles. The stock for the frames is to be rectangular in cross sections that vary from 0.75″ × 1.5″ to 1.60″ × 3.60″. Outside dimensions vary from 10″ to 24″ (Fig. 2.35).

33. **Base hardware.** Design the hardware needed at the points indicated for a standard volleyball net. The 7′ pipes are supported by crossing two-by-fours. Design the hardware needed at points *a, b,* and *c* (Fig. 2.36).

34. **Conduit connector hanger.** Design a support that will attach to a 0.75″ conduit that will support a channel that is used as an adjustable raceway for electrical wiring. Your design should permit ease of adjustment. (Fig. 2.37).

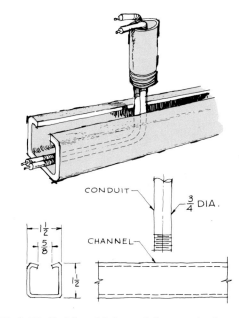

FIG. 2.37 Problem 34. A conduit connector hanger.

Product Design Problems

A product design involves the development of a device that will perform a specific function and will be mass-produced and sold to a broad market.

35. **Hunting blind.** Hunters of geese and ducks must remain concealed while hunting. Design a portable hunting blind to house two hunters. This blind should be completely portable so that it can be carried in separate sections by each of the hunters. Specify its details and how it is to be assembled and used.

36. **Convertible drafting table.** Many laboratories are equipped with drawing tables that have draft-

FIG. 2.38 Problem 37. Folding chair with a writing-tablet arm attached.

ing machines attached to them. This arrangement clutters the table top surface when the tables are used for classes that do not require drafting machines. Design a drafting table that can enclose a drafting machine, thereby concealing it and protecting it when it is not needed.

37. Writing-tablet for a folding chair. Design a writing-tablet arm for a folding chair that could be used in an emergency or when a class needs more seating. To allow easy storage, the tablet arm must fold with the chair (Fig. 2.38). Use a folding chair available to you for dimensions and specifications.

38. Portable truck ramp. Delivery trucks need ramps to load and unload supplies and materials at their destinations (Fig. 2.39). Assume that the bed of the truck is 20″ from ground level. Design a portable ramp that would permit the unloading

FIG. 2.39 Problem 38. Delivery ramp for unloading goods.

of goods with the use of a hand dolly and would reduce manual lifting.

39. Sensor-retaining device. The Instrumentation Department of the Naval Oceanographic Office uses underwater sensors to learn more about the ocean. These sensors are submerged on cables from a boat on the surface. The winch used to retrieve the sensor frequently overruns (continues pulling when it has been retrieved), causing the cable to break and the sensor to be lost. Design a safety device that will retain the sensor if the cable is broken when the sensor reaches a pulley. The problem is illustrated in Fig. 2.40. The sensor weighs 75 lb.

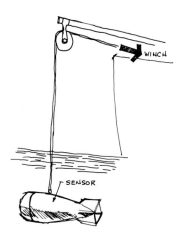

FIG. 2.40 Problem 39. Underwater sensor.

40. Flexible trailer hitch. In combat zones, vehicles must tow trailers where terrain may be very uneven and hazardous. Design a trailer hitch that will provide for the most extreme conditions possible. Study the problem requirements and limitations to identify the parameters within which your design must function.

41. Worker's stilts—human engineering. Workers who apply gypsum board and other types of wallboard to the interiors of buildings and homes must work on scaffolds or wear some type of stilts to be able to reach the ceiling to nail the 4′ × 8′ boards into position.

   Design stilts that will provide the worker with access to a ceiling 8′ high while permitting him or her to perform the job of nailing ceiling panels with comfort. The stilts should be adjustable to accommodate workers of various weights, sizes, and heights. Analyze the needs of the workers and the requirements for stilts that will provide comfort, traction, and safety and will adapt to the human body.

42. **Pole vault uprights.** Many pole vaulters are exceeding the 18′ height in track meets, which introduces a problem for the officials of this event. The pole vault uprights must be adjusted for each vaulter by moving them forward or backward plus or minus 18″. Also, the crossbar must be replaced with great difficulty at these heights by using forked sticks and ladders, which are crude and inefficient. Develop a more efficient set of uprights that can be easily repositioned and will allow the crossbar to be replaced with greater ease.

43. **Sportman's chair.** Analyze the need for a sportman's chair that could be used for camping, for fishing from a bank or a boat, for watching sporting events, and for as many other purposes as you can think of. The need is not for a special-purpose chair, but for a chair that is suitable for a wide variety of uses to fully justify it as a marketable item. Make a list of as many applications as possible and use this as a basis for your design.

44. **Child carrier for a bicycle.** Design a seat that can be used to carry a small child as a passenger on a bicycle. Assume that the bicycle will be ridden by an older youth or an adult. Determine the age of the child who would probably be carried as a passenger. Design the seat for safety and comfort.

45. **Power lawn-fertilizer attachment.** The rotary-power lawn mower emits a force through its outlet caused by the air pressure from the rotating blades. This force might be used to distribute fertilizer while the lawn is being mowed. Design an attachment for a power mower that could spread fertilizer while the mower is performing its usual cutting operation.

46. **Projector cabinet.** Many homes have slide projectors, but each showing of the family slides must be preceded by the time-consuming effort of setting up the equipment. Design a cabinet that could serve as an end table or for some other function while also housing a slide projector ready for use at any time. The cabinet might also serve as storage for slide trays. It should have electrical power for the projector. Evaluate the market for a multipurpose cabinet of this type.

47. **Heavy-appliance mover.** Design a device that can be used for moving large appliances, such as stoves, refrigerators, and washers, about the house. This product would not be used often—only for rearranging, cleaning, and servicing equipment.

48. **Bicycle-for-two adapter.** Design the parts and assembly required to convert the typical bicycle into a bicycle built for two (tandem) when mated with another bicycle of the same make and size. Work from an existing bicycle, and consider, among other things, how each rider can equally share in the pedaling. Determine the cost of your assembly and its method of attachment to the average bicycle. Use existing stock parts when possible to reduce special machining.

49. **Stump remover.** Assume that a number of tree stumps must be removed from the ground to clear land for construction. The stumps are dead with partially deteriorated root systems and require a force of approximately 2000 pounds to remove them. Design an apparatus that could be attached to the bumper of a car and could be used to remove the stumps by either pushing or pulling. The device should be easy to attach to the stump and to the car.

50. **Gate opener.** An aggravation to farmers and ranchers is the necessity of opening and closing gates when driving from one fenced area to the next. It would be desirable to design a gate that could be opened and closed without getting out of the vehicle. Design a manually operated gate that would appeal to this market.

    Study all aspects of the problem to determine the features that would make your design as marketable as possible.

51. **Mounting for an outboard motor on a canoe.** Unlike a square-end boat, the pointed-end canoe does not provide a suitable surface for attaching an outboard motor. Design an attachment that will adapt an outboard motor to a canoe. Indicate how the motor will be controlled by the operator in the canoe.

52. **Baby seat (cantilever).** Design a child's chair that can be attached to a standard table top and will support the child at the required height. The chair should be designed to ensure that the child cannot crawl out of or detach it from the table top. A possible solution could be a design that would cantilever from the table top, using the child's body as a means of applying the force necessary to grip the table top. The design would be further improved if the chair were collapsible or

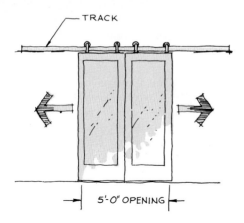

FIG. 2.41 Problem 56. Door dimensions.

suitable for other purposes. Determine the age group that would be most in need of the chair and base your design on the dimensions of a child of this age.

53. **Automobile controls.** Design driving controls that can be attached to the standard automobile and will permit an injured person to drive the car without the use of his or her legs. This device should be easy to attach and to operate with the maximum of safety.

54. **Adjustable TV base.** Design a TV base to support full-sized TV sets that would allow the maximum of adjustment: up and down, and rotation about a vertical axis and about a horizontal axis. Design the base to be as versatile as possible.

55. **Projector cabinet.** Design a portable cabinet that could be left permanently in a classroom and would house a slide projector and a movie projector. The cabinet should provide both convenience and security from vandalism and theft.

56. **Door opener.** Design a method whereby a trucker at a loading dock could open the warehouse door without having to get out of the

truck. The doors are dimensioned in Fig. 2.41, and the dock extends 8′ from the doors.

57. **Projector eraser.** Design a device that would erase grease pencil markings from the acetate roll of an overhead projector as the acetate is cranked past the stage of the projector.

58. **Boat trailer.** Design a trailer that supports the boat under the trailer rather than on top of it. Develop this design so that a boat can be launched in water that is shallower than is required now.

59. **Washing machine.** Design a manually operated washing machine that can be used by less-developed countries without power. This could be considered as an "undesign" of a powered washing machine.

60. **Pickup truck hoist.** Design a tail gate that can be attached to the tail gate of a pickup truck and can be used for raising and lowering loads from the ground to the floor of the truck (Fig. 2.42). Design it to be operated without a motor.

FIG. 2.42 Problem 60. Pickup truck hoist.

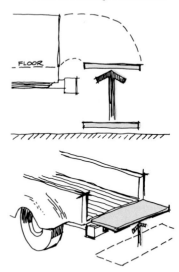

3

Drawing Instruments

3.1
Introduction

The preparation of technical drawings is possible only through knowledge of and skill in the use of a variety of drafting instruments. Drawing instruments are designed to help the drafter prepare many types of drawings in a productive manner.

You will find that your skill and productivity will increase with practice and as you become more familiar with the use of the tools that are available to you. People with little artistic ability can produce technical drawings of a professional quality when they learn to use drawing instruments and tools properly.

3.2
Pencil

Since any drawing begins with the pencil, the proper pencil must be selected and sharpened correctly to yield the desired results. Pencils may be the conventional wood pencil or the lead holder, which is a mechanical pencil (Fig. 3.1). Both types are identified by a number and/or letter at the end. Sharpen the end opposite these markings so you will not sharpen away the identity of the grade of lead.

FIG. 3.1 The mechanical pencil (lead holder) or the wood pencil can be used for mechanical drawing. The ends of the lead and the wood pencil are labeled to indicate the grade of the pencil lead.

Pencil grades are shown graphically in Fig. 3.2, ranging from the hardest, 9H, to the softest, 7B. The pencils in the medium-grade range of 3H–B are the pencils most often used for drafting work.

The wood pencil can be sharpened with a small knife or a drafter's pencil sharpener, which removes the wood and leaves approximately ⅜ inch of lead exposed (Fig. 3.3). The point can then be sharpened with a sandpaper pad to a conical point by stroking the sandpaper with the pencil point while the pencil is being revolved between the fingers (Fig. 3.4). The excess graphite is wiped from the point with a cloth or tissue.

A pencil pointer that is used by professional drafters (Fig. 3.5) can be used to sharpen either wood or mechanical pencils. Insert the pencil in the hole and revolve it to sharpen the lead to a conical point. Other types of small hand-held point sharpeners that work on the same principle are available.

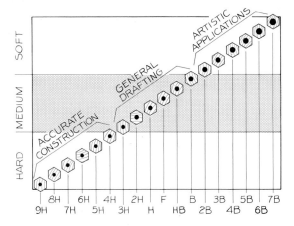

FIG. 3.2 The hardest pencil lead is 9H and the softest is 7B. Note the diameter of the hard leads is smaller than that of the soft leads.

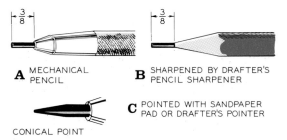

FIG. 3.3 Approximately ⅜″ of lead should project from the end of your pencil as shown at A and B. The point should be sharpened to a tapered conical point with a sandpaper pad or a drafter's pointer.

FIG. 3.4 The drafting pencil is revolved about its axis as you stroke the sandpaper pad to form a conical point. The graphite is wiped from the sharpened point with a tissue or a cloth.

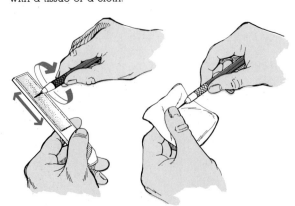

FIG. 3.5 The professional drafter will often use a pencil pointer of this type to sharpen pencils.

3.3
Papers and drafting media

SIZES The surface on which a drawing is made must be carefully selected to yield the best results for a given application. The drafter usually begins by selecting the sheet size for making a set of drawings. Sheet sizes are specified by letters such as Size A, Size B, and so forth. These sizes are multiples of either the standard 8½″ × 11″ sheet or the 9″ × 12 ″ sheet. These sizes are listed below:

Size A	8½″ × 11″	9″ × 12″
Size B	11″ × 17″	12″ × 18″
Size C	17″ × 22″	18″ × 24″
Size D	22″ × 34″	24″ × 36″
Size E	34″ × 44″	36″ × 48″

DETAIL PAPER When drawings are not to be reproduced by the diazo process (which is a blue-line print), an opaque paper, called *detail paper,* can be used as the drawing surface.

The higher the rag content (cotton additive) of the paper, the better will be its quality and durability because it will contain cotton rather than just wood pulp.

Preliminary layouts can be drawn on detail paper and then traced onto the final tracing surface.

TRACING PAPER A thin, translucent paper that is used for making detail drawings is *tracing paper* or *tracing vellum*. These papers are translucent to permit the passage of light through them so drawings can be reproduced by the diazo process (blue-line process). The highest-quality tracing papers are very translucent and yield the best reproductions.

Vellum is tracing paper that has been treated chemically to improve its translucency. Vellum does not retain its original quality as long as does high-quality, untreated tracing paper.

It is advantageous to lay out a drawing on detail paper, overlay it with tracing paper, and trace the final reproducible drawing.

TRACING CLOTH *Tracing cloth* is a permanent drafting medium that is available for both ink and pencil drawings. It is made of cotton fabric that has been covered with a compound of starch to provide a tough, erasable drafting surface that yields excellent blue-line reproductions. This material is more stable than paper, which means that it does not change its shape with variations in temperature and humidity as much as does tracing paper.

Erasures can be made on tracing cloth repeatedly without damaging the surface. This is especially important when drawing with ink.

POLYESTER FILM An excellent drafting surface is polyester film. It is available under several trade names, such as *Mylar film*. This material is highly transparent (much more so than paper and cloth), very stable, and is the toughest medium available. It is waterproof and very difficult to tear.

Mylar film is used for both pencil and ink drawings. The drawing is made on the matte side of the film; the other side is glossy and will not take pencil or ink lines. Some films specify that a plastic-lead pencil be used, whereas others adapt well to standard lead pencils. India ink and inks especially made for Mylar film can be used for ink drawings.

Ink lines will not wash off with water and will not erase with a dry eraser, but erasures can be made with a dampened hand-held eraser. An electric eraser is not recommended for use with this medium unless it is equipped with an eraser of the type recommended by the manufacturer of the film.

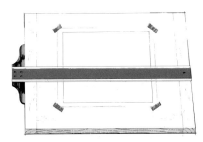

FIG. 3.6 The T-square and drafting board are the basic tools used by the student drafter. The drawing is taped to the board with drafting tape.

3.4
T-square and board

The T-square and drafting board are the basic pieces of equipment used by the beginning drafter (Fig. 3.6). The T-square can be moved with its head in contact with the edge of the board for drawing parallel horizontal lines. The drawing paper should be attached with drafting tape to the drawing board parallel to the blade of the T-square (Fig. 3.7).

The drafting board is made of basswood, which is lightweight but strong. Standard sizes of boards are 12″ × 14″, 15″ × 20″, and 21″ × 26″. The working edge of a drawing board is the edge where the T-square head is held firmly in position when each horizontal line is drawn. Some drawing boards have built-in steel working edges for a higher degree of accuracy.

FIG. 3.7 The drafting tape placed at each corner of a drawing should be square and be cut prior to taping the drawing.

FIG. 3.8 The drafting machine is often used instead of the T-square and drafting board in both industry and school laboratories. (Courtesy of Keuffel & Esser Co., Morristown, N.J.)

FIG. 3.10 The professional who uses a drafting station may work in an environment similar to the one shown here. (Courtesy of Martin Instrument Company.)

3.5
Drafting machines

Although the T-square is used in industry and in the classroom, most professional drafters prefer the mechanical *drafting machine,* which is attached to the drawing board or table top (Fig. 3.8). These machines have fingertip controls for drawing lines at any angle and can be easily returned to their original position.

Drafting machines provide a considerable degree of convenience to the drafter, but the same drawing can be made with the T-square and triangle in the hands of a skilled drafter.

For large drawings such as those made by architects, a parallel blade is available (Fig. 3.9). The blade is attached to a cable at each end that keeps it parallel in any position. The angle of the blade can be changed by adjusting the cables.

The professional who must have access to a drafting work station will use equipment that is usually

FIG. 3.9 The parallel blade is advantageous to the drafter who makes very large drawings. It is guided by the cables attached to the board. (Courtesy of Keuffel & Esser Co., Morristown, N.J.)

FIG. 3.11 More and more professional work stations are being equipped with computer graphics equipment. (Courtesy of Bausch & Lomb.)

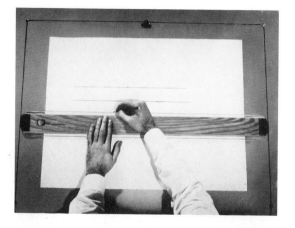

more sophisticated and advanced than that used by the student. A modern, fully equipped drafting station is shown in Fig. 3.10. Also, most offices of the future will be equipped with computer graphics equipment to supplement manual equipment and techniques (Fig. 3.11).

3.6
Alphabet of lines

The type of line produced by a pencil depends on the hardness of the lead, the drawing surface, and the technique of the drafter. Examples of the standard lines, or the *alphabet of lines,* is shown in Fig. 3.12, along with the recommended pencils for drawing the

FIG. 3.12 The alphabet of lines varies in width to produce a finished mechanical drawing of an object. The full-size lines are shown in the right column along with the recommended pencil grades for drawing the lines.

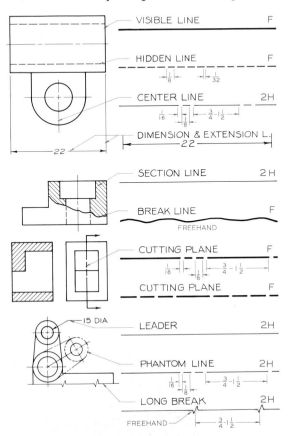

lines. These pencil grades may vary greatly with the drawing surface being used.

Guidelines that are used to aid in lettering and in laying out a drawing are very light lines, just dark enough to be seen. A 4H pencil is recommended for drawing most guidelines.

Except for guidelines, all other lines should be drawn dark and black so they will reproduce well. The important characteristic of pencil lines is their relative widths, as shown in Fig. 3.12. When drawing these lines, assume that you are drawing them in ink where their blackness is uniform and the only variable is their widths.

3.7
Horizontal lines

A horizontal line is drawn by using the upper edge of your horizontal straightedge and drawing the line from left to right, for the right-handed person (Fig. 3.13). Your pencil should be held in a vertical plane to make a 60° angle with the drawing surface and the line being drawn.

FIG. 3.13 Horizontal lines are drawn with a pencil held in a plane perpendicular to the paper and at 60° to the surface. These lines are drawn left to right along the upper edge of the T-square.

As horizontal lines are drawn, the pencil should be rotated about its axis to allow its point to wear evenly (Fig. 3.14). If necessary, lines can be darkened by drawing over them one or more times. A small space should be left between the straightedge and the pencil point for the best line (Fig. 3.15).

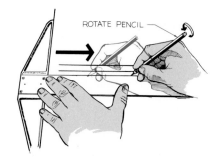

FIG. 3.14 As the horizontal lines are drawn, the pencil should be rotated about its axis so that the point will wear down evenly.

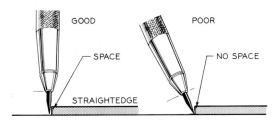

FIG. 3.15 The pencil point should be held in a vertical plane and inclined 60° to leave a space between the point and the straightedge being used.

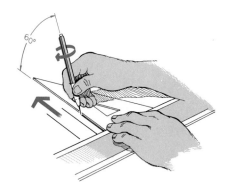

FIG. 3.16 Vertical lines are drawn along the left side of a triangle in an upward direction with the pencil held in a vertical plane at 60° to the surface.

3.8
Vertical lines

A triangle is used in conjunction with the blade of a T-square for drawing vertical lines, since all drawing triangles have one 90° angle. While the T-square is held firmly with one hand, the triangle can be positioned where needed and the vertical lines drawn (Fig. 3.16).

Vertical lines are drawn upward along the left side of the triangle while holding the pencil in a vertical plane at 60° to the drawing surface.

3.9
Drafting triangles

The two most often used triangles are the 45° triangle and the 30°–60° triangle. The 30°–60° triangle is specified by the longer of the two sides adjacent to the 90° angle (Fig. 3.17). Standard sizes of 30°–60° triangles range in 2-inch intervals from 4″ to 24″. The variety of lines that can be drawn with this triangle and a straightedge are shown in Fig. 3.17.

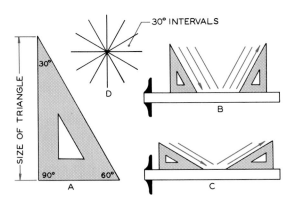

FIG. 3.17 The 30°–60° triangle can be used to construct lines at 60° and 30° to the horizontal. Lines can be spaced at 30° intervals throughout 360°.

FIG. 3.18 The 45° triangle can be used to draw lines at 45° intervals throughout 360°.

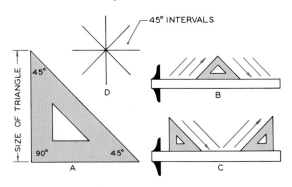

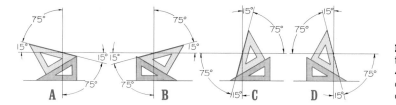

FIG. 3.19 By using the 30°–60° triangle in combination with the 45° triangle, angles can be drawn at 15° intervals without the use of a protractor.

The 45° triangle is specified by the length of the sides adjacent to the 90° angle. These range in size from 4″ to 24″ at 2-inch intervals, but the 6″ and 10″ sizes are adequate for most classroom applications. The various angles that can be drawn with this triangle are shown in Fig. 3.18.

By using the 45° and 30°–60° triangles in combination, angles can be drawn at 15° intervals throughout 360° (Fig. 3.19).

3.10
Protractor

When lines must be drawn or measured at angles other than at multiples of 15°, a protractor is used (Fig. 3.20). Protractors are available as semicircles (180°) or as circles (360°).

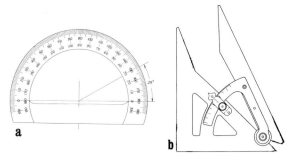

FIG. 3.20 The semicircular protractor can be used to measure angles. The adjustable triangle can be used as a drawing edge in addition to being used to measure angles.

An adjustable triangle also serves as a protractor and a drawing edge at the same time (Fig. 3.20b). Most drafting machines have built-in protractors that are easier to use and more convenient than traditional protractors.

3.11
Parallel lines

A series of lines can be drawn parallel to a given line by using a triangle and a straightedge (Fig. 3.21).

The 45° triangle is placed parallel to a given line and is held in contact with the straightedge (which may be another triangle). By holding the straightedge in one position, the triangle can be moved to various positions for drawing series of parallel lines.

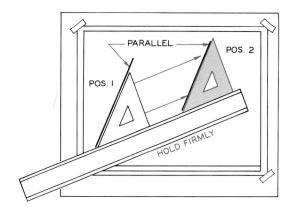

FIG. 3.21 A straightedge and a 45° triangle can be used for drawing a series of parallel lines. The T-square is held firmly in position, and the triangle is moved from position 1 to position 2.

3.12
Perpendicular lines

Perpendicular lines can be constructed by using either of the standard triangles. A 30°–60° triangle is used with a T-square or another triangle to draw line 3–4

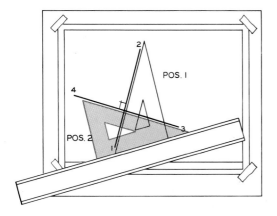

FIG. 3.22 A 30°–60° triangle and a straightedge can be used to construct a line perpendicular to line 1–2. The triangle is aligned with 1–2 in position 1 and is then rotated to position 2 to construct line 3–4.

3.13
Irregular curves

Curves that are not arcs must be drawn with an instrument called an *irregular curve*. These plastic curves come in a variety of sizes and shapes, but the one shown in Fig. 3.23 is typical of those that are used.

The use of the irregular curve is shown in this figure where a series of points is connected to form a smooth curve. Note that the plastic curve must be repositioned several times to draw the complete curve.

The *flexible spline* is an instrument used for drawing long, irregular curves (Fig. 3.24). The spline is held in position by weights while the curve is drawn.

3.14
Erasing

Erasers are manufactured to match the lines and papers used for drafting. Surfaces may vary from soft paper to polyester film, and the lines may be drawn in ink or in pencil.

Erasing should be done with the softest eraser that will serve the purpose. For example, ink erasers should not be used to erase pencil lines, because ink

perpendicular to line 1–2 (Fig. 3.22). One edge of the triangle is placed parallel to line 1–2 in position 1 with the straightedge in contact with the triangle. By holding the straightedge in place, the triangle is rotated and moved to position 2 to draw the perpendicular line.

FIG. 3.23 Use of the irregular curve

Step 1 To connect points with a smooth curve, the irregular curve is positioned to pass through as many points as possible, and a portion of the curve is drawn.

Step 2 The irregular curve is positioned for drawing another portion of the connecting curve.

Step 3 The last portion is drawn to complete the smooth curve. Most irregular curves must be drawn in separate steps, in this manner.

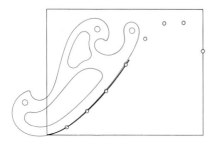

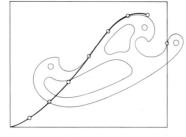

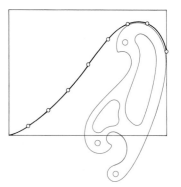

FIG. 3.24 A flexible spline can be used for drawing long, irregular curves. The spline is held in position by weights.

FIG. 3.26 This cordless electric eraser is typical of those used by professional drafters.

erasers are coarse and may damage the surface of the paper.

When it is necessary to erase in small areas, an erasing shield can be used to prevent accidental erasing of adjacent lines (Fig. 3.25). Place the shield over the line, hold it firmly, and erase the desired line. All erasing should be followed by brushing away the ''crumbs'' from the drawing with a brush. Do not use

your hands for this purpose; this may smudge your drawing.

Electric erasers are available. A cordless model is shown in Fig. 3.26. This eraser is recharged by its desk stand, which is left plugged into a wall outlet. The erasers that are used in these machines are available in several grades for erasing ink and pencil lines.

FIG. 3.25 The erasing shield is used for erasing in tight spots without removing the wrong lines. The dusting brush is used to remove the erased material when finished. Do not brush with the palm of your hand; this will smear the drawing.

3.15
Scales

All engineering drawings require the use of scales to measure lengths, sizes, and other measurements. Triangular engineers' and architects' scales are shown in Fig. 3.27.

FIG. 3.27 The architects' scale is used to measure in feet and inches, whereas the engineers' scale measures in decimal units.

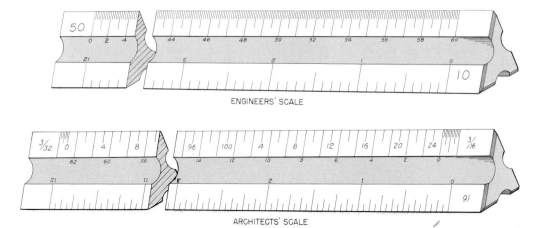

ENGINEERS' SCALE

ARCHITECTS' SCALE

ARCHITECTS' SCALE

BASIC FORM $SCALE: \frac{X}{X} = 1'-0$ ⟵ *FROM END OF SCALE*

TYPICAL SCALES

SCALE: FULL SIZE (USE 16-SCALE)

SCALE: HALF SIZE (USE 16-SCALE)

SCALE: 3 = 1'-0   *SCALE: $1\frac{1}{2}$ = 1'-0*

SCALE: $1\frac{1}{2}$ = 1'-0   *SCALE: $\frac{3}{4}$ = 1'-0*

SCALE: $\frac{1}{2}$ = 1'-0   *SCALE: $\frac{3}{8}$ = 1'-0*

SCALE: $\frac{3}{16}$ = 1'-0   *SCALE: $\frac{1}{8}$ = 1'-0*

SCALE: $\frac{3}{32}$ = 1'-0

FIG. 3.28 The basic form of indicating the scale when the architects' scale is used, and the variety of scales available on this scale.

FIG. 3.29 Examples of lines measured by using an architects' scale.

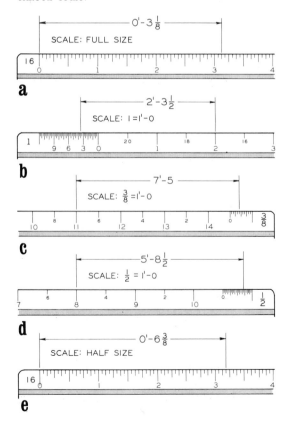

a

b

c

d

e

Most scales are either 6″ or 12″ long. The scales covered in this section are the architects', engineers', mechanical engineers', and metric scales.

Architects' scale

The architects' scale is used to dimension and scale features encountered by the architect, such as cabinets, plumbing, and electrical layouts. Most indoor measurements are made in feet and inches with an architects' scale.

The basic form of indicating on a drawing the scale that is being used is shown in Fig. 3.28. This form should be placed on the drawing in the title block or in some prominent location.

Since the dimensions made with the architects' scale are in feet and inches, it is very difficult to handle the arithmetic associated with these dimensions. It is necessary to convert all dimensions to decimal equivalents (all feet or all inches) before the simplest arithmetic can be performed.

SCALE: FULL SIZE The 16 scale is used for measuring full-size lines (Fig. 3.29a). An inch on the 16 scale is divided into sixteenths to match the ruler used by the carpenter. This example is measured to be $3\frac{1}{8}$″. Note that when the measurement is less than one foot, a zero may be used to precede the inch measurements, and the inch marks are omitted in all cases.

SCALE: 1 = 1′-0 In Fig. 3.29b, a line is measured to its nearest whole feet (2′ in this case) and the remainder is measured in inches at the end of the scale ($3\frac{1}{2}$″) for a total of 2′-$3\frac{1}{2}$. At the end of each architects' scale, a foot has been divided into inches for measuring dimensions that are less than a foot.

FIG. 3.30 When marking off measurements along a scale, be sure to hold your pencil vertically for the most accurate measurement.

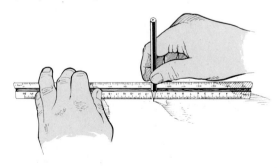

The scale 1 = 1'–0 is the same as saying 1" is equal to 12", or a ¹⁄₁₂th size (inch marks omitted).

SCALE: ³⁄₈ = 1'–0 When this scale is used, ³⁄₈" is used to represent 12" on a drawing. Figure 3.29c is measured to be 7'–5 (inch marks omitted).

SCALE: ½ = 1'–0 A line is measured to be 5'–8½ in Fig. 3.29d (inch marks omitted).

SCALE: HALF SIZE The 16 scale is used to measure or draw a line that is half size. This is sometimes specified as Scale: 6 = 12 (inch marks omitted). The line in Fig. 3.29e is measured to be 0'–6³⁄₈.

When locating dimensions using any scale, hold your pencil in a vertical position for the greatest accuracy when marking measurements (Fig. 3.30).

When indicating dimensions in feet and inches, they should be in the form shown in Fig. 3.31. Notice that fractions are twice as tall as whole numerals.

FIG. 3.31 Inch marks are omitted according to current standards, but foot marks are shown. A leading zero is used when the inch measurements are less than a whole inch. When representing feet, a zero is optional if the measurement is less than a foot.

Engineers' scale

The engineers' scale is a decimal scale on which each division is a multiple of 10 units. It is used for making drawings of engineering projects that are located outdoors, such as streets, structures, land measurements, and other large dimensions associated with topography. For this reason it is sometimes called the civil engineers' scale.

Since the measurements are in decimal form, it is easy to perform arithmetic operations without the need of converting feet and inches, as when the architects' scale is used.

The form of specifying scales on the engineers' scale is shown in Fig. 3.32, such as Scale: 1 = 10'. Each end of the scale is labeled 10, 20, 30, etc. This indicates the number of units per inch on the scale.

ENGINEERS' SCALES

BASIC FORM SCALE: 1 = XX (FROM END OF ENGR. SCALE)

EXAMPLE SCALES

10	SCALE: 1=10';	SCALE: 1 = 1,000'
20	SCALE: 1=200';	SCALE: 1=20LB
30	SCALE: 1=0.3;	SCALE: 1=3,000'
40	SCALE: 1=4';	SCALE: 1 = 40'
50	SCALE: 1=50';	SCALE: 1 = 500'
60	SCALE: 1=6;	SCALE: 1 = 0.6'

FIG. 3.32 The basic form for indicating the scale when the engineers' scale is used, and the variety of scales that are available on this scale.

Many combinations may be obtained by moving the decimal places of a given scale, as indicated in Fig. 3.32.

10 SCALE In Fig. 3.33a, the 10 scale is used to measure a line at the scale of 1 = 10'. The line is 32.0 feet long.

20 SCALE In Fig. 3.33b, the 20 scale is used to measure a line drawn at a scale of 1 = 200.0'. The line is 540.0 feet long.

30 SCALE A line of 10.6 (inch marks omitted) is measured using the scale of 1 = 3.0 in Fig. 3.33c.

FIG. 3.33 Examples of lines measured with the engineers' scale.

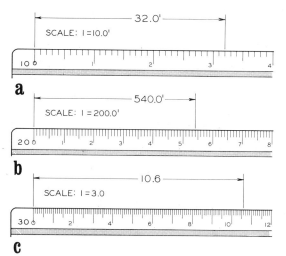

FIG. 3.34 When using English units (inches), decimal fractions do not have leading zeros and inch marks are omitted. Be sure to provide adequate space for decimal points between the numbers. Foot marks are shown.

The format for indicating measurements in feet and inches is shown in Fig. 3.34. It is customary to omit zeros in front of decimal points when dimensioning an object using English (Imperial) units, and inch marks are always omitted if the dimensions are given in inches.

Mechanical engineers' scale

The mechanical engineers' scale is used to draw small parts (Fig. 3.35). This scale is used to represent drawings in inches using common fractions. These scales are available in ratios of half size, one-quarter size, and one-eighth size. For example, on the half-size scale, 1 inch is used to represent 2 inches. On a quarter-size scale, 1 inch would represent 4 inches.

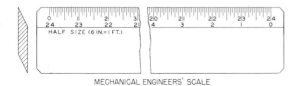

MECHANICAL ENGINEERS' SCALE

FIG. 3.35 The mechanical engineers' scales are used for measuring small parts at scales of half size, quarter size, and one-eighth size. These units are in inches with common fractions.

Metric system—SI units

The English system (Imperial system) of measurement has been used in the United States, Britain, and Canada since these countries were established. At present a movement is under way to convert to the more universal metric system.

The English system was based on arbitrary units of the inch, foot, cubit, yard, and mile (Fig. 3.36). There is no common relationship between these units of measurement; consequently the system is cumbersome to use when simple arithmetic is performed. For example, finding the area of a rectangle that is 25 inches by 6¾ yards using the English system is a complex problem.

The metric system was proposed by France in the fifteenth century. In 1793 the French National Assembly agreed that the meter (m) would be one ten-millionth of the meridian quadrant of the earth (Fig. 3.37). Fractions of the meter were expressed as decimal fractions. Debate continued until an international commission officially adopted the metric system in 1875. Since a slight error in the first measurement of the meter was found, the meter was later established as equal to 1,650,763.73 wavelengths of the orange-red light given off by krypton-86 (Fig. 3.37).

The international organization charged with the establishment and the promotion of the metric system is called the *Internationale Standards Organization (ISO)*. The system they have endorsed is called *Système International d'Unités* (International System of Units) and is abbreviated SI. The basic SI units are shown in Fig. 3.38 with their abbreviations. It is important that lowercase and uppercase abbreviations be used properly as shown.

Several practical units of measurement have been derived (Fig. 3.39) to make them easier to use in many applications. These unofficial SI units are widely used. Note that degrees Celsius (centigrade) is recommended over the official temperature measurement,

FIG. 3.36 The units of the English system were based on arbitrary dimensions.

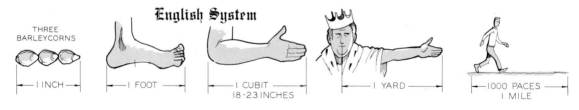

English System

THREE BARLEYCORNS — 1 INCH — 1 FOOT — 1 CUBIT 18-23 INCHES — 1 YARD — 1000 PACES 1 MILE

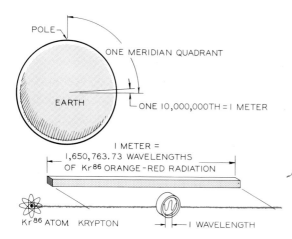

FIG. 3.37 The origin of the meter was based on the dimensions of the earth, but it has since been based on the wavelength of krypton-86. A meter is 39.37 inches.

SI UNITS			DERIVED UNITS		
LENGTH	METER	m	AREA	SQ METER	m²
MASS	KILOGRAM	kg	VOLUME	CU METER	m³
TIME	SECOND	s	DENSITY	KILOGRAM/CU MET	kg/m³
ELECTRICAL			PRESSURE	NEWTON/SQ MET	N/m²
CURRENT	AMPERE	A			
TEMPERATURE	KELVIN	K			
LUMINOUS					
INTENSITY	CANDELA	cd			

FIG. 3.38 The basic SI units and their abbreviations. The derived units are units that have come into common usage.

PARAMETER	PRACTICAL UNITS		SI EQUIVALENT
TEMPERATURE	DEGREES CELSIUS	°C	0°C = 273.15 K
LIQUID VOLUME	LITER	l	l = dm³
PRESSURE	BAR	BAR	BAR = 0.1 MPa
MASS WEIGHT	METRIC TON	t	t = 10³ kg
LAND MEASURE	HECTARE	ha	ha = 10⁴ m²
PLANE ANGLE	DEGREE	°	1° = π/180 RAD

FIG. 3.39 These practical metric units are a few of those that are widely used because they are easier to deal with than the official SI units.

FIG. 3.40 The prefixes and abbreviations used to indicate the decimal placement for SI measurements.

VALUE		PREFIX	SYMBOL
1,000,000	= 10⁶ =	MEGA	M
1,000	= 10³ =	KILO	k
100	= 10² =	HECTO	h
10	= 10¹ =	DEKA	da
1	= 10⁰ =		
.1	= 10⁻¹ =	DECI	d
.01	= 10⁻² =	CENTI	c
.001	= 10⁻³ =	MILLI	m
.000 001	= 10⁻⁶ =	MICRO	μ

Kelvin. When using Kelvin, the freezing and boiling temperatures are 273.15°K and 373.15°K, respectively. Pressure is measured in bars, where one bar is equal to 0.1 megapascal or 100,000 pascals.

Many SI units have prefixes to indicate placement of the decimal. The more common prefixes and their abbreviations are shown in Fig. 3.40.

Several comparisons of English and SI units are given in Fig. 3.41. Other conversion factors are given in Appendix 2.

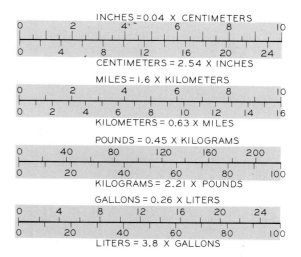

FIG. 3.41 A comparison of metric units with those used in the English system of measurement.

3.16
Metric scales

The basic unit of measurement on an engineering drawing is the millimeter (mm), which is one-thousandth of a meter, or one-tenth of a centimeter. These units are understood unless otherwise specified on a drawing. The width of the fingernail of your index finger can serve as a convenient gage to approximate the dimension of a centimeter, or ten millimeters (Fig. 3.42).

Metric scales are indicated on a drawing in the form shown in Fig. 3.43. A colon is placed between the numeral 1 and the ratio of the drawing size. The units are not specified, since the millimeter is understood to be the unit of measurement.

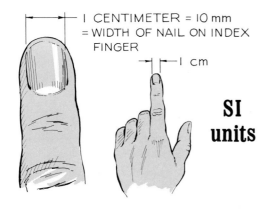

1 CENTIMETER = 10 mm
= WIDTH OF NAIL ON INDEX FINGER
1 cm
SI units

FIG. 3.42 The width of the nail on your index finger is approximately equal to a centimeter or ten millimeters.

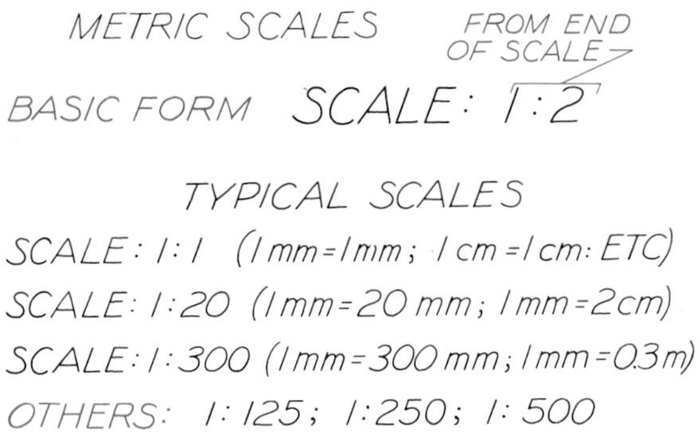

METRIC SCALES
FROM END OF SCALE
BASIC FORM SCALE: 1:2
TYPICAL SCALES
SCALE: 1:1 (1 mm=1 mm; 1 cm=1 cm: ETC)
SCALE: 1:20 (1 mm=20 mm; 1 mm=2 cm)
SCALE: 1:300 (1 mm=300 mm; 1 mm=0.3 m)
OTHERS: 1:125; 1:250; 1:500

FIG. 3.43 The basic form for indicating scales when the metric scale is used, and the variety of metric scales that are available.

Decimal units are unnecessary on most metrically dimensioned drawings; consequently the dimensions are usually rounded off to whole numbers except for those measurements that are dimensioned with specified tolerances. For metric units less than 1, a leading zero is placed in front of the decimal. In the English system, the zero is omitted from inch measurements (Fig. 3.44).

SCALE 1:1 The full-size metric scale (Fig. 3.45) shows the relationship between the metric units of the dekameter, the centimeter, the millimeter, and the micrometer. There are 10 dekameters in a meter; 100 centimeters in a meter; 1000 millimeters in a meter; and 1,000,000 micrometers in a meter. A line of 59 mm is measured in Fig. 3.46.

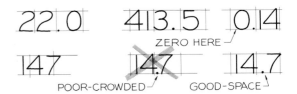

22.0
413.5
0.14
ZERO HERE
147
14.7
14.7
POOR-CROWDED
GOOD-SPACE

FIG. 3.44 When decimal fractions are shown in metric units, a zero is used to precede the decimal. Be sure to allow adequate space for the decimal point when numbers with decimals are lettered.

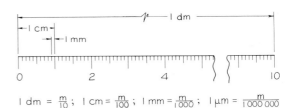

1 dm
1 cm
1 mm
0
2
4
10
1 dm = m/10; 1 cm = m/100; 1 mm = m/1000; 1 μm = m/1 000 000

FIG. 3.45 The dekameter is one-tenth of a meter; the centimeter is one-hundredth of a meter; a millimeter is one-thousandth of a meter; and a micrometer is one-millionth of a meter.

a
59 mm
1 cm = 10 mm
1:1
0 1 2 3 4 5 6
b
106 mm
1 cm = 20 mm
1:2
0 5 10
c
165 mm
1 cm = 30 mm
1:3
0 5 10 15

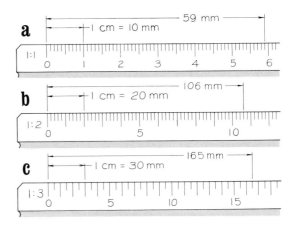

FIG. 3.46 Examples of lines measured with metric scales.

SCALE 1:2 This scale is used when 1 mm is equal to 2 mm, 20 mm, 200 mm, etc. The line in Fig. 3.46b is 106 mm long.

SCALE 1:3 A line of 165 mm is measured in Fig. 3.46c, where 1 mm is used to represent 30 mm.

Other scales

Many other metric (SI) scales are used: 1:250, 1:400, 1:500, and so on. The scale ratios mean that one unit represents the number of units on the right of the colon. For example, 1:20 means that one millimeter equals 20 mm, or one centimeter equals 20 cm, or one meter represents 20 m.

Metric symbols

When drawings are made in metric units, this fact can be noted in the title block or elsewhere using the SI symbol (Fig. 3.47). The large SI indicates Système Internationale. The two views of the partial cone are used to denote whether the orthographic views were drawn in the U.S. system (third-angle projection) or the European system (first-angle projection).

Scale conversion

Tables for converting inches to millimeters are given in Appendix 2; however, this conversion can be performed by multiplying decimal inches by 25.4 to obtain millimeters. For example, 1.5 inches would be 1.5 × 25.4 = 38.1 mm.

To convert an architects' scale to an approximate metric scale, the scale must be multipled by 12. For example, Scale: ⅛ = 1′−0 is the same as ⅛ inch = 12 inches, or 1 inch = 96 inches. This scale closely approximates the metric scale of 1:100. Many of the scales used in the metric system cannot be converted to exact English scales. The scale of 1 = 5′ converts exactly to the metric scale of 1:60.

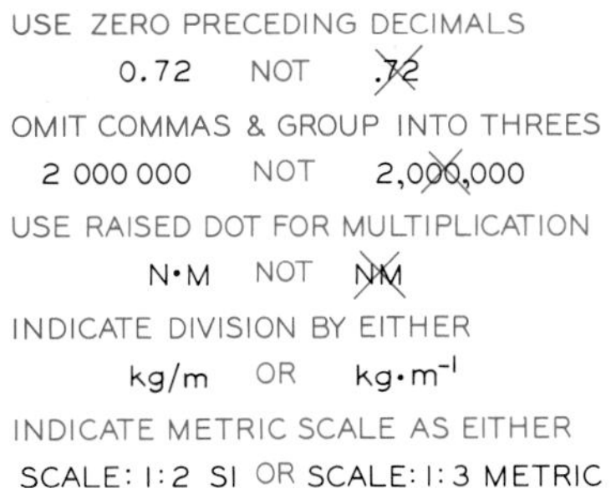

USE ZERO PRECEDING DECIMALS
0.72   NOT   .72
OMIT COMMAS & GROUP INTO THREES
2 000 000   NOT   2,000,000
USE RAISED DOT FOR MULTIPLICATION
N·M   NOT   NM
INDICATE DIVISION BY EITHER
kg/m   OR   kg·m⁻¹
INDICATE METRIC SCALE AS EITHER
SCALE: 1:2 SI OR SCALE:1:3 METRIC

FIG. 3.48 General rules to be used with the SI system.

Expression of metric units

The general rules for expressing SI units are given in Fig. 3.48. Commas are not used between sets of zeros, but instead a space is left between them.

3.17
The instrument set

A basic set of drawing instruments is shown in Fig. 3.49, and the name of each part is given. These can be purchased separately, but they are available assembled as a set in a case similar to the one shown in Fig. 3.50. A more elaborate set of instruments is shown in Fig. 3.51.

Compass

The *compass* is used to draw circles and arcs in ink and in pencil (Fig. 3.52). In order to obtain good results with the compass, its pencil point must be sharpened

FIG. 3.47 The large letters SI indicate that measurements are in metric units. The partial cones indicate that the views are arranged using third-angle projection (the U.S. system) or first-angle projection (the European system).

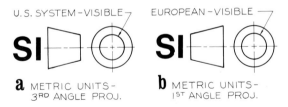

a METRIC UNITS −
3ᴿᴰ ANGLE PROJ.

b METRIC UNITS −
1ˢᵀ ANGLE PROJ.

FIG. 3.49 The parts of a set of drafting instruments.

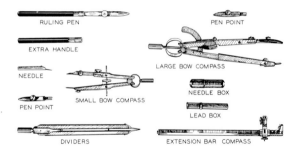

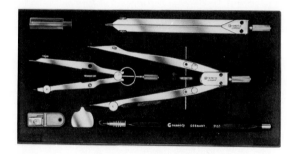

FIG. 3.50 Instruments usually come as a cased set. (Courtesy of Gramercy Guild.)

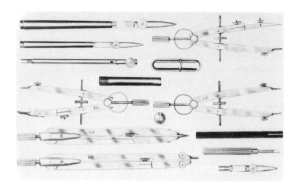

FIG. 3.51 A set of instruments with three bows for the advanced student. (Courtesy of Keuffel & Esser Co., Morristown, N.J.)

FIG. 3.52 The compass is used for drawing circles.

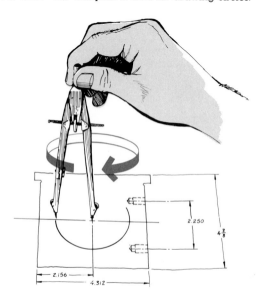

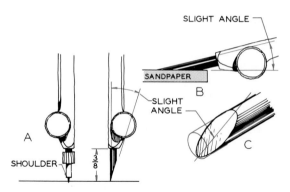

FIG. 3.53 The compass lead should be sharpened from the outside on a sandpaper pad at B, to a wedge point (C). The pencil point should be about the same length as the compass point at A.

on its outside with a sandpaper board (Fig. 3.53). A bevel cut of this type gives the best all-round point for drawing a circle. Note that the needle point of the compass must be adjusted to match the length of the lead. When the compass point is set in the drawing surface, it should not be inserted to the shoulder of the point but just far enough for a firm set.

When the table top has a hard covering, several sheets of paper should be placed under the drawing to provide a seat for the compass point. A center tack (the circular part shown in Fig. 3.51) can be placed over the center point and used for setting the compass point for drawing circles and for preventing the enlargement of the center hole when used repetitively.

FIG. 3.54 Two types of small bow compasses for drawing circles of about one-inch radius.

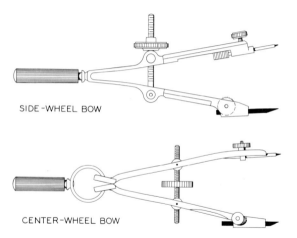

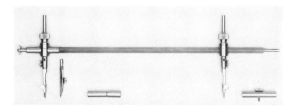

FIG. 3.55 Large circles can be drawn with the beam compass. Note that ink attachments are available also.

Bow compasses are provided in some sets (Fig. 3.54) for drawing small circles with ink and pencil. For large circles, extension bars are provided to extend the range of the large bow compass. If the circles are much larger, a beam compass can be used (Fig. 3.55).

Small circles and ellipses can be effectively drawn with a circle template that is aligned with the perpendicular center lines of the circle. The circle or ellipse is drawn with a pencil to match the other lines of the drawing (Fig. 3.56).

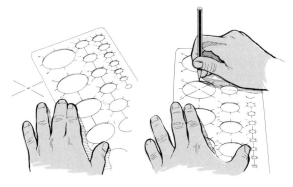

FIG. 3.56 The circle template can be used for drawing circles without the use of a compass. The circle or ellipse template is aligned with the center lines.

FIG. 3.57 Dividers are used to step off measurements and to transfer dimensions on a drawing.

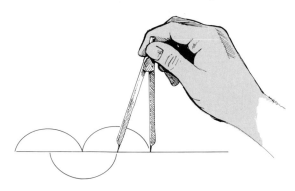

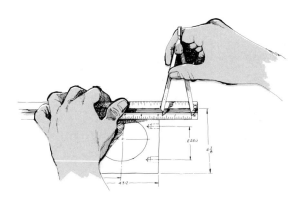

FIG. 3.58 Dividers are used to transfer dimensions from a scale to a drawing.

Dividers

Dividers look much like a compass but are used for laying off and transferring dimensions onto a drawing. For example, equal divisions can be stepped off rapidly along a line (Fig. 3.57). A slight impression is made in the drawing surface with the points as each measurement is made.

Dividers can be used to transfer dimensions from a scale to a drawing (Fig. 3.58). Another use for dividers is the division of a line into a number of equal parts. This is done by trial and error. You begin by estimating the spacing and stepping off the space until the correct spacing is found.

Small bow dividers can be used for transferring smaller dimensions, such as the spacing between the guidelines for lettering. Two types of bow dividers are shown in Fig. 3.59.

FIG. 3.59 Two types of bow dividers for transferring small dimensions, such as the spacing between guidelines for lettering.

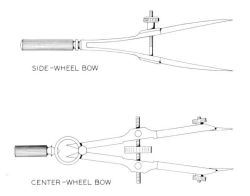

SIDE-WHEEL BOW

CENTER-WHEEL BOW

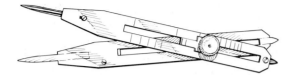

FIG. 3.60 Proportional dividers can be used for making measurements that are proportional to other dimensions. The pivot can be set to give the desired ratio.

Proportional dividers

Dimensions can be transferred from one scale to another by using a special type of dividers, the *proportional dividers*. The central pivot point can be moved from position to position to vary the ratio of the spacing at one end of the dividers to the ratio at the other end (Fig. 3.60). This instrument is very helpful in enlarging and reducing dimensions on a drawing.

3.18
Ink drawing

Although the majority of drafting and design work is done in pencil, inking is required for many applications, especially for drawings that will be reproduced in publications and reports. Pencil drawings have a tendency to lose their sharpness as instruments are moved about the drawing surface during preparation. Ink drawings remain dark and distinct without danger of losing their quality.

FIG. 3.61 The ruling pen has two nibs that are adjusted by a set screw to vary the width of the ink lines drawn with the pen.

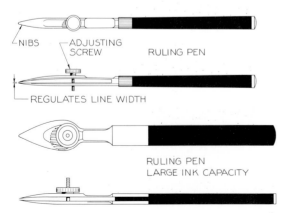

NIBS   ADJUSTING
       SCREW   RULING PEN

REGULATES LINE WIDTH

RULING PEN
LARGE INK CAPACITY

Ink lines give the best reproductions when reproduced by the printing press, diazo process, or microfilming.

Materials for ink drawing

A good grade of tracing paper can be used for ink drawings, but erasing errors may result in holes in the paper and the loss of your time. Therefore, film or tracing cloth should be used, which will withstand many erasings and corrections.

When using drafting film, the drawing should be made on the matte surface in accordance with the manufacturer's directions. A cleaning solution is available to prepare the surface for ink and to remove spots that might not take the ink properly. These films can also be cleaned with a damp cloth.

Tracing cloths need to be prepared for inking by applying a coating of powder or pounce to absorb oily spots that will otherwise repel an ink line. These oily spots might be left on a drawing by fingerprints.

The drawing ink used for engineering drawings is called India ink and is available under numerous trade names. This dense black carbon ink is much thicker and faster-drying than regular fountain pen ink. Some drafters prefer to "season" their ink by leaving the top of the bottle open for several days to thicken it.

Ink should be removed from instruments before drying to prevent clogging, which restricts an easy flow of ink from the pen to the paper. Ultrasonic pen cleaners are available to aid in the cleaning of pen points that have been used for inking. Dirty points are immersed in a cup of pen cleaner solution that is vibrated at about 80,000 cycles per second to free the dried particles of ink.

FIG. 3.62 The pen is inked between the nibs with the spout on the ink bottle cap.

Ruling pen

Two types of ruling pens are shown in Fig. 3.61. Both have set screws for varying the widths of the lines that are drawn.

The ruling pen should be inked with the spout on the cap of the ink bottle (Fig. 3.62). Do not overload the pen with ink, since this will cause the lines to be wet and perhaps run on the drawing. Experiment with your particular pen to learn the proper amount of ink to apply to the nibs.

When ruling horizontal lines, the ruling pen is held in the same position as a pencil (Fig. 3.63), maintaining a space between the nibs and the straightedge.

An extra margin of safety can be obtained by placing a triangle or template under the straightedge, as shown in Fig. 3.64. This prevents wet ink from coming in contact with the straightedge and avoids smearing.

An alternative pen that can be used is the technical ink fountain pen as illustrated in Fig. 3.65. These

FIG. 3.65 An India ink technical pen that is commonly used to make an ink drawing. (Courtesy of Koh-I-Noor Rapidograph, Inc.)

FIG. 3.66 A set of technical inking pens in a humidifier that attaches to the drafting table. (Courtesy of Koh-I-Noor Rapidograph, Inc.).

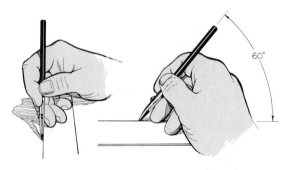

FIG. 3.63 The ruling pen is held in a vertical plane at 60° to the drawing surface for drawing ink lines.

FIG. 3.64 The ruling pen should be held in a vertical plane so that there will be a space between the T-square and the nibs. A triangle or template can be placed under the straightedge for a greater margin of safety.

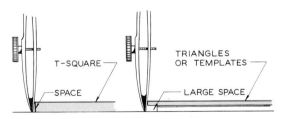

pens come in sets (Fig. 3.66) with various-sized pen points that are used for the alphabet of lines (Fig. 3.67). Lines drawn with fountain pens of this type dry much faster than those drawn with ruling pens because the ink is applied in a thinner layer.

Examples of poorly drawn ink lines are shown in Fig. 3.68. These mistakes can be avoided by learning the technical faults that cause them.

Inking compass

The inking compass is usually the same compass used for the circles drawn by pencil, with the inking attachment inserted in place of the pencil attachment. The inking compass may have an elbow in its legs to allow the points to be approximately perpendicular to the

Available in 18 "Kolor-Koded" line widths*

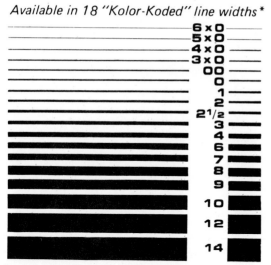

6x0
5x0
4x0
3x0
00
0
1
2
2½
3
4
6
7
8
9
10
12
14

*Approximate only. (Line widths will vary, depending on type of surface, type of ink, speed at which line is drawn, etc.)

FIG. 3.67 This chart shows the variety of technical pens available for drawing lines of graduated widths. (Courtesy of Koh-I-Noor Rapidograph, Inc.)

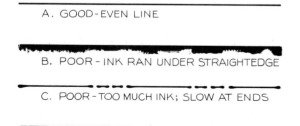

A. GOOD-EVEN LINE

B. POOR-INK RAN UNDER STRAIGHTEDGE

C. POOR-TOO MUCH INK; SLOW AT ENDS

D. POOR-NIBS TOUCH IMPROPERLY

FIG. 3.68 Examples of poor ink lines are shown at B, C, and D.

drawing surface. The circle can be drawn with one continuous line, as shown in Fig. 3.69.

Two types of compasses for drawing smaller circles are shown in Fig. 3.70. These can be used to draw circles up to a radius of one inch. The spring bow compass (Fig. 3.71) can be used to draw pencil and ink circles that are very small, usually about one-eighth inch in radius.

Larger circles can be drawn using the extension bar with the large compass (Fig. 3.72). Much larger circles can be drawn with the beam compass (as

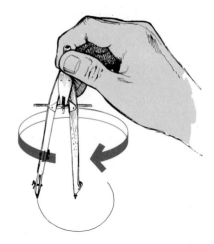

FIG. 3.69 An inking attachment can be used with a large bow compass for drawing circles and arcs in ink.

shown in Fig. 3.55) using the inking attachment. Special compasses are available for drawing circles with a technical fountain pen (Fig. 3.73).

Order of inking

When a drawing is to be inked, begin by locating the center lines and tangent points associated with the arcs as shown in Step 1 of Fig. 3.74. This construction should be laid out in pencil prior to inking.

Ink the arcs and circles first, being careful to stop all arcs at their points of tangency, Step 2. Connect the

FIG. 3.70 Two types of small bow compasses for drawing arcs in ink up to a radius of about one inch.

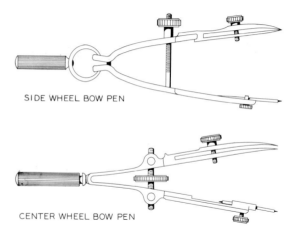

SIDE WHEEL BOW PEN

CENTER WHEEL BOW PEN

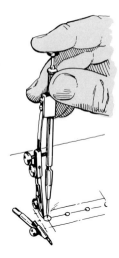

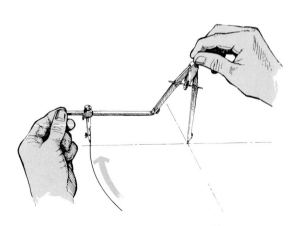

FIG. 3.71 The spring bow compass is used for drawing small circles of about one-quarter inch diameter. A pencil attachment is available for pencil circles.

FIG. 3.72 An extension bar can be used with a bow compass for drawing large arcs.

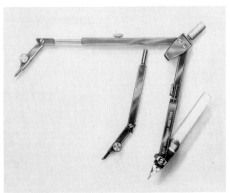

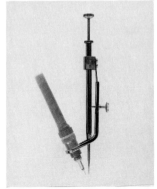

FIG. 3.73 Special compasses with adapters are available for using technical ink pens for drawing arcs. (Courtesy of Koh-I-Noor Rapidograph, Inc.)

FIG. 3.74 Order of inking

Step 1 The drawing is laid out with light pencil construction lines. All centers and tangent points are accurately located.

Step 2 The arcs and circles are always inked first. Arcs should stop at their points of tangency.

Step 3 Straight lines are drawn to match the ends of the arcs. Center lines are shown to complete the drawing.

CONSTRUCTION LAYOUT

INK ARCS

INK STRAIGHT LINES

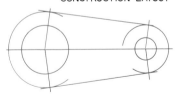

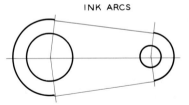

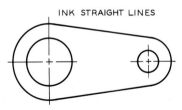

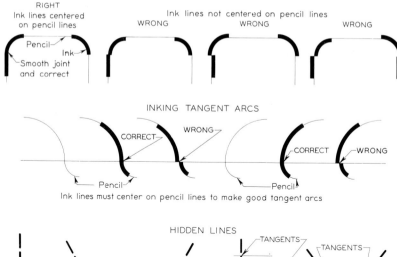

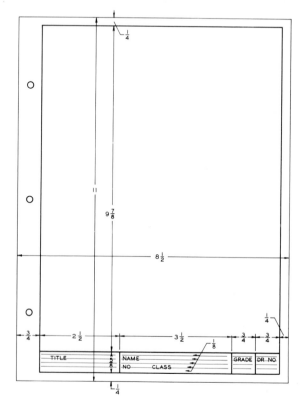

FIG. 3.75 Techniques of matching and joining ink lines.

arcs with straight lines at the points of tangency to complete the drawing.

When a drawing is composed mostly of straight lines, begin at the top of the drawing and draw all the horizontal lines as you progress from the top of the sheet to the bottom. In this manner, you are moving away from the wet lines, which are left to dry as you progress. After allowing the last horizontal line to dry, the right-handed person should ink the vertical lines by beginning with the far left and moving across the drawing to the right, away from the wet lines. The direction may be reversed for the left-handed drafter.

FIG. 3.76 Many types of templates are available to aid drafters in their work. (Courtesy of Rapidesign.)

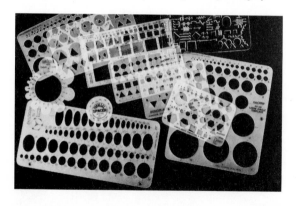

FIG. 3.77 The format and title strip for a Size A sheet (8½″ × 11″) suggested for solving the problems at the end of each chapter. When the sheet is in the vertical format, it will be called a Size AV.

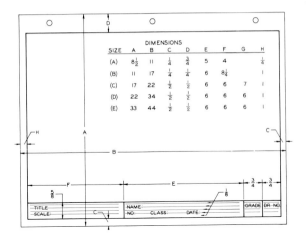

FIG. 3.78 The format for Size AH (an 8½″ × 11″ sheet in a horizontal position) and the sizes of other sheets. The dimensions under columns A through H give the various layouts.

The proper methods of connecting ink lines are shown in Fig. 3.75. Ink lines should make good junctions with their connecting lines. Ink lines should be centered over the previously drawn pencil guidelines.

Templates

A wide variety of templates are available for preparing drawings in both pencil and ink. These are available for drawing nuts and bolts, circles and ellipses, architectural symbols, and many other applications. They work best when used with technical fountain pens rather than with the traditional ruling pen. The wide range of templates is shown in Fig. 3.76.

3.19
Solutions of problems

The following formats are suggested for the layout of problem sheets. Most problems will be drawn on 8½″ × 11″ sheets, as shown in Fig. 3.77. A title strip is suggested in this figure, with a border as shown. The 8½″ × 11″ sheet in the vertical format is called Size AV throughout the remainder of this textbook. When this size sheet is in the horizontal format, as shown in Fig. 3.78, it will be called Size AH.

The standard sizes of sheets from Size A through Size E are shown in Fig. 3.78. An alternative title strip for Sizes B, C, D, and E is shown in Fig. 3.79. Guidelines should always be used for lettering title strips.

Another title block and parts list is given in Fig. 3.80. These are placed in the lower right-hand corner of the sheet against the borders. When both are used on the same drawing, the parts list is placed directly above and in contact with the title block or the title strip, as the case may be.

FIG. 3.80 A title block and parts list that can be used on some problem sheets if needed.

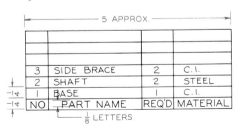

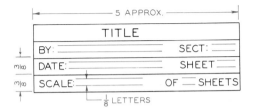

FIG. 3.79 An alternative title strip that can be used on sheet Sizes B, C, D, and E instead of the one given in FIG. 3.78.

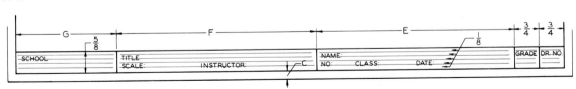

Problems

The problems (Figs. 3.81–3.85) are to be constructed on Size AH (8½″ × 11″) paper, plain or with a printed grid, using the format shown in Fig. 3.77. Two prob-lems can be constructed per sheet. Use pencil or ink as assigned by your instructor.

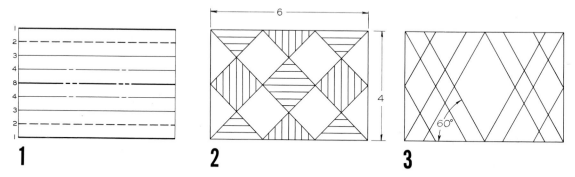

FIG. 3.81 Each of these problems is to be drawn on a Size AV sheet. Begin by lightly laying out a 4″ × 6″ rectangle. Problem 1: Construct the following lines: 1—visible line, 2—hidden line, 3—dimension line, 4—center line, and 8—cutting plane line. Problems 2 and 3: Construct the patterns shown using line weights equal to those of visible lines.

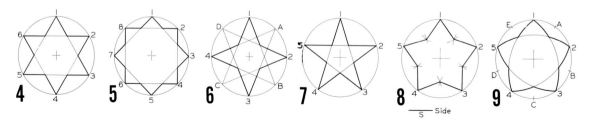

FIG. 3.82 Study the figures closely and draw one per sheet of Size AV paper. The circles are 4″ in diameter. The dimension *S* in Problem 8 is 1.2 inches.

FIG. 3.83 Draw the problems on a Size AV sheet, one per sheet, using the given dimensions and your instruments.

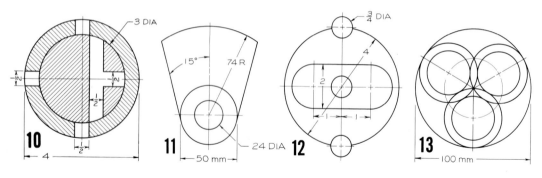

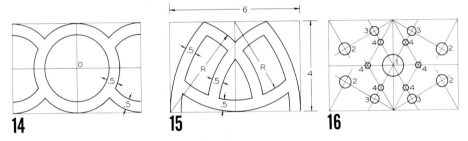

FIG. 3.84 Construct the problems inside of the 4″ × 6″ rectangles, using the given dimensions. Problem 16: The hole diameters are as follows: No. 1—1″, No. 2—0.5″, No. 3—0.4″, and No. 4—0.25″. The holes are located along lines drawn at 30° and 60°.

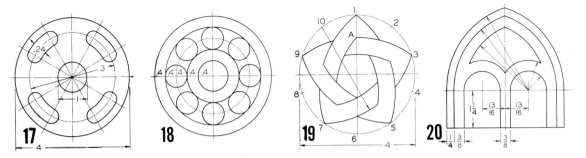

FIG. 3.85 Construct the drawings in Problems 17–20 using the given dimensions (the large circles are 4″ in diameter), one problem per sheet. Use a Size AV sheet.

4
Lettering

4.1
Lettering

All drawings are supplemented with notes, dimensions, and specifications that must be lettered. Consequently the ability to construct legible freehand letters is a very important skill to develop, since it affects the usage and interpretation of a drawing.

A measure of the professionalism of drafters, technologists, and engineers is their technique of lettering. Good lettering results from a good attitude and the willingness to put forth one's best efforts.

FIG. 4.1 Good lettering begins with a properly sharpened pencil point. The point should be slightly rounded, not a needle point. The F pencil is usually the best grade for lettering.

FIG. 4.2 When lettering a drawing, use a protective sheet under your hand to prevent smudges. Your lettering will be best when you are working from a comfortable position; you may wish to turn your paper for the most natural strokes.

4.2
Tools of lettering

The best-grade pencil for lettering on most surfaces is a pencil in the H–HB grade range, with an F pencil being the most commonly used grade. Some papers

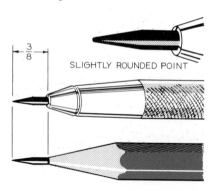

SLIGHTLY ROUNDED POINT

$\frac{3}{8}$

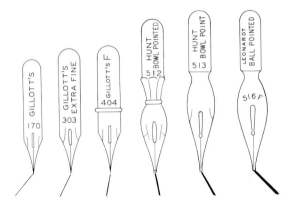

FIG. 4.3 Inking pens are available to give a variety of line widths for freehand lettering. These are used with pen staffs and India ink.

and films are coarser than others and may require a harder pencil lead. The point of the pencil should be slightly rounded to give the desired line width (Fig. 4.1). A needle point will break off when pressure is applied.

When lettering, the pencil should be revolved about its axis between your fingers as the strokes are being made so the lead will wear down gradually and evenly. Bear down firmly to make the letters black and bright for good reproduction. To prevent smudging your drawing during the lettering process, it is

FIG. 4.4 Ink fountain pens for the drafter have been widely used by professionals who work with ink. The line widths vary with each pen, each of which has its own number. (Courtesy of J. S. Staedtler, Inc.)

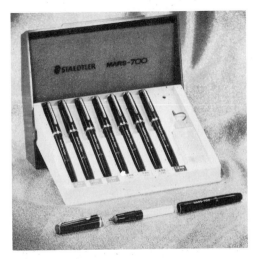

helpful to place a sheet of paper under your hand to protect the drawing (Fig. 4.2).

For freehand lettering with ink, a number of pen points are available to use with a pen staff. You will notice in Fig. 4.3 that points come in various graduations to give lines of varying widths. These pens should be used with India ink for best results.

Another type of inking pen that is widely used is the India ink technical pen (Fig. 4.4). These pens have tubular points that are kept clear by gently shaking the pen up and down to activate the plunger inside the point. They are available in a range of tubular point sizes for making lines of varying widths. They can hold a supply of ink sufficient for hours of use without refilling.

4.3
Gothic lettering

The standard type of lettering that is recommended for engineering drawings is *single-stroke Gothic lettering*. This form of lettering is given this name because the letters are made with a series of single strokes and the letter form is a variation of Gothic lettering.

Two general categories of Gothic lettering are *vertical* and *inclined* lettering (Fig. 4.5). Each is equally acceptable; however, these types should not be mixed on the same drawing.

FIG. 4.5 Two types of Gothic lettering recommended by engineering standards are vertical and inclined lettering.

4.4
Guidelines

The most important rule of lettering is: *Use guidelines at all times.* This applies whether you are lettering a paragraph or a single letter or numeral. The method of constructing and using guidelines can be seen in Fig. 4.6. Use a sharp pencil in the 3H–5H grade range

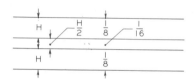

FIG. 4.6 Lettering guidelines

Step 1 Letter heights, *H*, are laid off, and thin construction lines are drawn with a 4H pencil. The spacing between the lines should be no closer than *H*/2, or ¹⁄₁₆″, when ⅛″ letters are used.

Step 2 Vertical guidelines are drawn as light, thin lines. These are randomly spaced to serve as visual guides for lettering.

Step 3 The letters are drawn with single strokes using a medium-grade pencil, H-HB. The guidelines need not be erased since they are drawn lightly.

and draw these lines very lightly, just dark enough for them to be seen.

Most lettering is done with the capital letters ⅛″ high (3 mm high). The spacing between the lines of lettering should be no closer than half the height of the capital letters, ¹⁄₁₆″ in this case.

Vertical guidelines should be drawn at random to serve as a visual guide in addition to the horizontal lines. These guidelines will be slanted for inclined lettering.

Lettering guides

The two most-used instruments for drawing guidelines for lettering are the *Braddock-Rowe lettering triangle* and the *Ames lettering instrument*.

The Braddock-Rowe triangle is pierced with sets of holes for spacing guidelines (Fig. 4.7). The numbers under each set of holes represent thirty-seconds of an inch. For example, the numeral 4 represents ⁴⁄₃₂″, or guidelines that are placed ⅛″ apart for making uppercase (capital) letters. Some triangles are marked for metric lettering in millimeters. Note in Fig. 4.7 that intermediate holes are provided for guidelines for lowercase letters, which are not as tall as the capital letters.

The Braddock-Rowe triangle is used in conjunction with a horizontal straightedge held firmly in position with the triangle placed against its edge. A sharp 4H pencil is placed in one hole of the desired set of holes to contact the drawing surface, and the pencil point is guided across the paper to draw the guideline while the triangle slides against the straightedge. This

FIG. 4.7 A. The Ames lettering guide can be used for drawing guidelines for uppercase and lowercase letters, vertical or inclined. The dial is set to the desired number of thirty-seconds of an inch for the height of uppercase letters. **B.** The Braddock-Rowe triangle can be used as a 45° triangle as well as an instrument for constructing guidelines. The numbers designating the guidelines represent thirty-seconds of an inch. For example, the number 4 represents ⁴⁄₃₂, or ⅛ inch, for the height of uppercase letters.

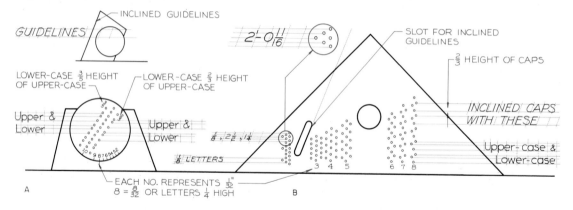

process is repeated as the pencil point is moved successively to each hole until the desired number of guidelines are drawn.

A slanted slot for drawing guidelines for inclined lettering is cut in the triangle. These slanting guidelines are spaced randomly by eye.

The Ames lettering guide is a very similar device with a circular dial for selecting the proper spacing of guidelines. Again, the numbers around the dial represent thirty-seconds of an inch. The number 8 represents ⅜₂″, or guidelines for drawing capital letters that are ¼″ tall. Metric guides are labeled in millimeters.

This instrument is used with a pencil and straightedge, as previously explained for using the Braddock-Rowe triangle.

4.5
Vertical letters

Vertical capital letters

Capital letters (uppercase letters) are commonly used on working drawings. They are very legible and result in a word or phrase that is easy to read.

The capital letters for the *single-stroke Gothic* alphabet are shown in Fig. 4.8. Each letter is drawn inside a square box of guidelines to help you learn their correct proportions. Each straight-line stroke should be drawn as a single stroke without stopping. For example, the letter A is drawn with three single strokes.

FIG. 4.8 The uppercase letters used in single-stroke Gothic lettering. Each is drawn inside a square to help you learn the proportions of each letter.

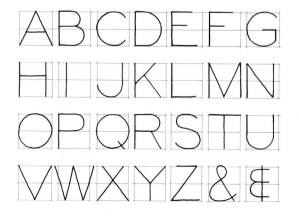

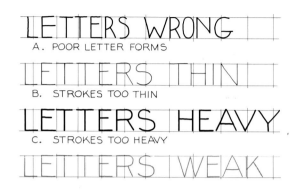

FIG. 4.9 There are many ways to letter poorly. A few of them, and the reasons why the lettering is inferior, are shown here. *Do not* make these mistakes.

Letters composed of curves can best be drawn in segments. The letter O can be drawn by joining two semicircles to form the full circle.

The shape and proportion of letters is important to good lettering. Memorize the shape of each letter given in this alphabet. Small wiggles in your strokes will not detract from your lettering if the letter forms are correct.

Examples of poor lettering are shown in Fig. 4.9. Observe the reason given for the lettering being poor in each example.

Vertical lowercase letters

An alphabet of lowercase letters is shown in Fig. 4.10. Lowercase letters are either two-thirds or three-fifths as tall as the uppercase letters they are used with. Both of these ratios are labeled on the Ames guide. Only the two-thirds ratio is available on the Braddock-Rowe triangle.

Some lowercase letters have ascenders that extend above the body of the letter, such as the letter *b*; and some have descenders that extend below the body of the letters, such as the letter *y*. The ascenders are the same length as the descenders.

The guidelines in Fig. 4.10 form perfect squares about the body of each letter to illustrate the proportions. A number of these letters have bodies that are perfect circles that touch all sides of the squares.

Capital and lowercase letters are used together in Fig. 4.11, as in a title. You can see the difference between the lowercase letters that are two-thirds the height of capitals and those that are three-fifths the height of capitals.

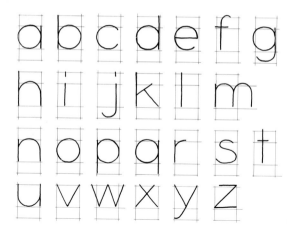

FIG. 4.10 The lowercase alphabet used in single-stroke Gothic lettering. The body of each letter is drawn inside a square to help you learn the proportions.

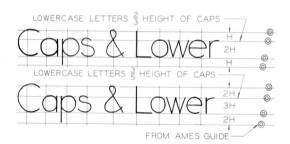

FIG. 4.11 Uppercase and lowercase letters are sometimes used together. The ratio of the lowercase letters to the uppercase letters will be either two-thirds or three-fifths. The Ames guide has both, and the Braddock-Rowe triangle has only the three-fifths ratio.

FIG. 4.12 The numerals for single-stroke Gothic lettering. Each is drawn inside a square to help you learn the proportions.

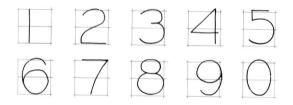

FIG. 4.13 The uppercase alphabet for single-stroke inclined Gothic lettering.

Vertical numerals

Vertical numerals are shown in Fig. 4.12, where each number is enclosed in a square box of guidelines. As with letters, you must learn the proportions of the numerals in order to use them properly. Each number is made the same height as the capital letters being used, usually ⅛″ high. The numeral zero is an oval and the letter *O* is a perfect circle in vertical lettering.

4.6
Inclined letters

Inclined capital letters

Inclined uppercase letters (capitals) have the same heights and proportions as vertical letters, the only difference being their inclination of 68° to the horizontal. The inclined alphabet is shown in Fig. 4.13.

Inclined guidelines should be drawn using the Braddock-Rowe triangle or the Ames guide, as illustrated in Fig. 4.7.

Inclined lowercase letters

Inclined lowercase letters are drawn in the same manner as the vertical lowercase letters. This alphabet is shown in Fig. 4.14. Ovals (ellipses) are used instead of the circles used in vertical lettering. The angle of inclination is 68°, the same as for uppercase letters.

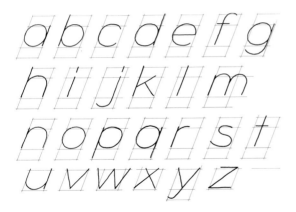

FIG. 4.14 The lowercase, single-stroke alphabet for inclined Gothic lettering. The body of each letter is drawn inside a rhombus to help you learn the proportions.

FIG. 4.15 The numerals for single-stroke inclined Gothic lettering. Each number is drawn inside a rhombus to help you learn the proportions.

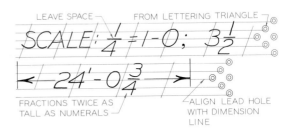

FIG. 4.16 Inclined common fractions are twice as tall as single numerals. Inch marks are omitted when numerals are used to show dimensions.

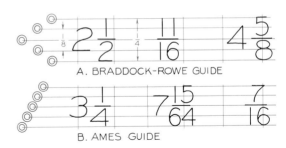

FIG. 4.17 Common fractions are twice as tall as single numerals. Guidelines for these can be drawn by using the Ames guide or the Braddock-Rowe triangle.

FIG. 4.18 Examples of poor spacing of numerals that results in poor lettering.

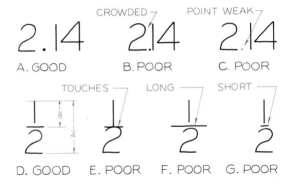

Inclined numerals

The inclined numerals that should be used in conjunction with inclined lettering are shown in Fig. 4.15. Except for the inclination of 68° to the horizontal, they are drawn the same as vertical numbers.

The use of inclined letters and numbers in combination is seen in Fig. 4.16. The guidelines in this example were constructed using the Braddock-Rowe triangle (Fig. 4.7).

4.7
Spacing numerals and letters

Common fractions are twice as tall as single numerals (Fig. 4.17). The fractions will be ¼″ tall when they are used with ⅛″ lettering. A separate set of holes for common fractions is given on the Braddock-Rowe triangle and on the Ames guide. These are equally spaced ¹⁄₁₆″ apart, with the center line being used for the fraction's crossbar. Examples of these holes can be seen in Figs. 4.17 and 4.7.

When numbers are used with decimals, space should be provided for the decimal point (Fig. 4.18). Common mistakes of spacing decimal fractions are shown at B and C. The correct method of drawing common fractions is illustrated at D, and several of the often-encountered errors are shown at E, F, and G.

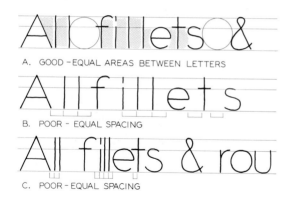

FIG. 4.19 Proper spacing of letters is necessary for good lettering and good appearance. The areas between letters should be approximately equal.

FIG. 4.20 Always leave space between lines of lettering. After constructing guidelines, *use them.* Use vertical guidelines to improve your vertical strokes.

When letters are grouped together to spell words, the areas between the letters should be approximately equal for the most pleasing result (Fig. 4.19). The incorrect use of guidelines and other violations of good lettering practice are shown in Fig. 4.20. Avoid making these errors when you are lettering.

4.8
Mechanical lettering

Drawings and illustrations that are to be reproduced by a printing process are usually drawn in India ink, and the lettering must also be in ink. Several mechanical aides for ink lettering are available.

FIG. 4.21 A Wrico lettering template can be used for mechanical lettering. (Courtesy of Wood-Reagan Instrument Company.)

The Wrico lettering template (Fig. 4.21) can be placed against a fixed straightedge for aligning the letters. You move the template from position to position while drawing each letter (with a lettering pen) through the raised portion of the template where the holes form the letters.

A slightly different template and pen are used by the Rapid-O-Graph system, as illustrated in Fig. 4.22. The method of lettering with this system is the same as with the Wrico system. A variety of pen sizes are available from each manufacturer for drawing different-sized letters and numbers with thin or bold lines.

Another system of mechanical lettering makes use of a grooved template used in conjunction with a scriber that follows the grooves and inks the letters on the drawing surface. Note that a standard India ink technical pen can be unscrewed from its barrel and at-

FIG. 4.22 A typical India ink fountain pen and template that can be used for mechanical lettering. (Courtesy of Koh-I-Noor Rapidograph, Inc.)

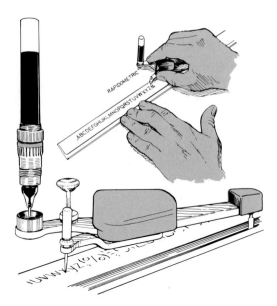

FIG. 4.23 Templates and scribes of this type are available for mechanical lettering.

FIG. 4.24 This portable typewriter can be placed on the drawing surface so that notes and numerals can be typed on a drawing. (Courtesy of Grintzner, Inc.)

FIG. 4.25 Transfer lettering can be transferred from film to the drawing surface. Transfer lettering comes in many sizes and styles. (Courtesy of Artype, Incorporated.)

tached to the scriber (Fig. 4.23). Many templates for varying styles of lettering and symbols are available for this system of mechanical lettering.

4.9
Lettering by typing

Some drafting departments type many of their notes and specifications to reduce drafting time and to improve the readability of a drawing. Large typewriters are available with long carriages that will accept extremely large drawings, and the notes are typed on the drawings in the conventional manner.

A portable typewriter (Fig. 4.24) can be placed on top of a drawing for typing without a typewriter carriage. Numbers and letters are typed after the drawing has been made.

4.10
Transfer lettering

Transfer lettering comes printed on transparent sheets with adhesive backings on one side (Fig. 4.25). The letters are cut from the sheet, aligned with drawn guidelines, and then burnished down permanently.

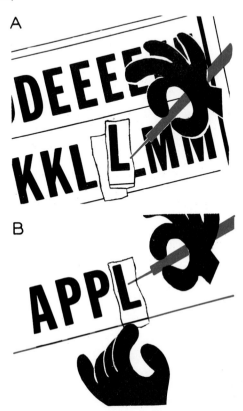

FIG. 4.26 A computer-based lettering machine that transcribes and stores lettering and symbols for engineering drawings. (Courtesy of AlphaMerics Corporation.)

FIG. 4.27 Computer-drawn symbols, and the program that was written to plot them.

ABCDEFGHI
JKLMNOPQR
STUVWXYZ
0123456789

```
100 REM ALPHABET OF CHARACTERS
110 REM INITIALIZE
120 IP% = 0: GOSUB 9000: GOSUB 9400
130 REM DEFINE CHARACTER STRING
140 REM DEFINE CHAR ORIGIN & HEIGHT
150 CHAR$ = "ABCDEFGHI"
160 REM DEFINE ORIGIN
170 CHAR.X = .5: CHAR.Y = 5
180 REM DEFINE ROTATION & HEIGHT
190 CHAR.A = 0: CHAR.H = .6
200 REM PLOT CHARACTER STRING
210 GOSUB 9300
220 REM PLOT 3 MORE STRINGS
230 CHAR.X = .5: CHAR.Y = 4
240 CHAR.A = 0: CHAR.H = .6
250 CHAR$ = "JKLMNOPQR"
260 GOSUB 9300
270 CHAR.X = .5: CHAR.Y = 3
280 CHAR.A = 0: CHAR.H = .6
290 CHAR$ ="STUVWXYZ"
300 GOSUB 9300
310 CHAR.X = .5: CHAR.Y = 2
320 CHAR.A = 0: CHAR.H = .6
330 CHAR$ = "0123456789"
340 GOSUB 9300
350 XP =0: YP = 0: IP%=3: GOSUB 9000
360 STOP
370 END
```

Some types of transfer lettering are burnished directly from the transparent sheet to the drawing surface. Each letter is rubbed firmly and evenly to make it transfer.

A multitude of letter forms and sizes are available from many manufacturers of transfer letters. It is possible to have custom transfer sheets produced for trademarks, title blocks, and other often-used applications.

4.11
Lettering by computer

Computer-based equipment may also be used for lettering on engineering drawings. The Datascribe IV (Fig. 4.26) is a portable machine that operates on a microprocessor with a typewriter keyboard. The operator can type text at the keyboard. The text is then transcribed onto the drawing with the selected pen and the selected type font, which ranges in letter size from 1/16 inch to 3.5 inches. Often-repeated notes can be stored on cassette tapes for recall when needed.

Most computer graphics systems have programs for plotting lettering and notes on drawings. A typical alphabet and its operating program are shown in Fig. 4.27. The letters must be specified by location, size, and angle in the program. Additional information on computer graphics lettering is given in Chapter 22.

Problems

Lettering problems are to be presented on Size AV (8½″ × 11″) paper, plain or grid, using the format shown in Fig. 4.28.

1. Practice lettering the vertical uppercase alphabet shown in Fig. 4.8. Construct each letter four times: four A's, four B's, etc. Use a medium-weight pencil—H, F, or HB.
2. Practice lettering vertical numerals and the lowercase alphabet as shown in Fig. 4.29. Construct each letter and numeral three times: three 1's, three 2's, etc. Use a medium-weight pencil—H, F, or HB.
3. Practice lettering the inclined uppercase alphabet shown in Fig. 4.13. Construct each letter four times: four A's, four B's, etc. Use a medium-weight pencil—H, F, or HB.
4. Practice lettering the vertical numerals and the lowercase alphabet shown in Figs. 4.10 and 4.12.

Construct each letter three times: three 1's, three 2's, etc. Use a medium-weight pencil—H, F, or HB.

5. Construct guidelines for ⅛″ capital letters starting ¼″ from the top border. Each should end ½″ from the left and right borders. Using these guidelines, letter the first paragraph of the text of this chapter. Use all vertical capitals. Spacing between the lines should be ⅛″.
6. Repeat Problem 5, but use all inclined capital letters. Use inclined guidelines to assist you in slanting your letters uniformly.
7. Repeat Problem 5, but use vertical capitals and lowercase letters in combination. Capitalize only the words that are capitalized in the text.
8. Repeat Problem 5, but use inclined capitals and lowercase letters in combination. Capitalize only the words that are capitalized in the text.

FIG. 4.28 Problem 1. Construct each vertical uppercase letter four times.

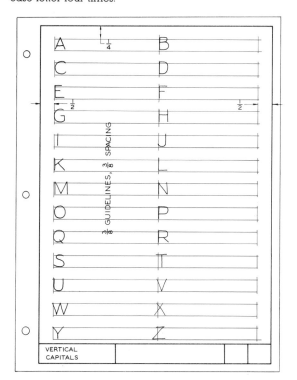

FIG. 4.29 Problem 2. Construct each vertical numeral and lowercase letter three times.

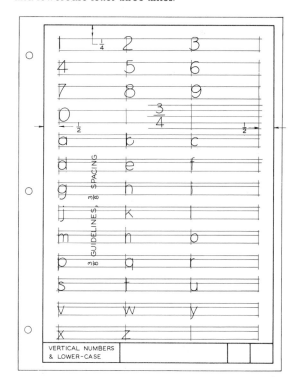

5

Geometric Construction

5.1
Introduction

Many problems in drafting and graphics can be solved only by the application of geometry and geometric construction. Engineering and technical problems are often solved by geometric construction during the design process.

Mathematics was an outgrowth of graphical construction; consequently there is a close relationship between the two areas. The proofs of many principles of plane geometry and trigonometry may be developed by using graphics. Graphical methods can be ap-

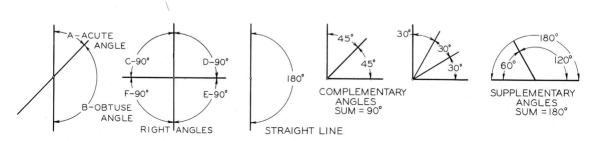

FIG. 5.1 Standard types of angles and their definitions.

FIG. 5.2 Types of triangles and their definitions. The hypotenuse of the right triangle is equal to the square root of the sum of the squares of the other two sides. This is the Pythagorean theorem.

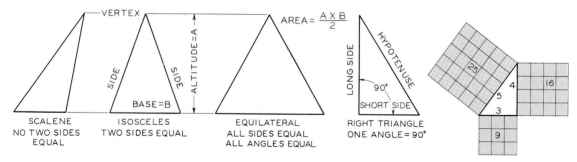

plied to algebra and arithmetic, and virtually all problems of analytical geometry can be solved graphically.

5.2
Angles

A fundamental requirement of geometric construction is the construction of lines that join at specified angles with each other. The definitions of various angles are given in Fig. 5.1.

The unit of angular measurement is the degree, and a circle has 360 degrees. A degree (°) can be divided into 60 parts called minutes ('), and a minute can be divided into 60 parts called seconds ("). An angle of 15°32'14" is an angle of 15 degrees, 32 minutes, and 14 seconds.

5.3
Triangles

The *triangle* is a three-sided polygon (or figure) that is named according to its shape. The four types of triangles are the *scalene, isosceles, equilateral,* and *right triangles* (Fig. 5.2). The sum of the angles inside a triangle will always be equal to 180°.

5.4
Quadrilaterals

A *quadrilateral* is a four-sided figure of any shape. The sum of the angles inside a quadrilateral is 360°.

The various types of quadrilaterals and their respective names are shown in Fig. 5.3, along with the equations for the areas of these figures.

5.5
Polygons

A *polygon* is a multisided plane figure of any number of sides. (The triangle and quadrilateral are polygons.) If the sides of the polygon are equal in length, the polygon is a *regular polygon.* Four types of regular

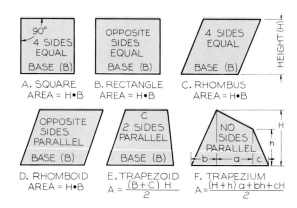

FIG. 5.3 Types of quadrilaterals (four-sided plane figures).

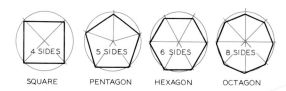

FIG. 5.4 Regular polygons inscribed in circles.

polygons are shown in Fig. 5.4. Note that a regular polygon can be inscribed in a circle, and all the corner points will lie on the circle.

Other regular polygons not pictured are: the *heptagon* with 7 sides, the *nonagon* with 9 sides, the *decagon* with 10 sides, and the *dodecagon* with 12 sides.

The sum of the angles inside any polygon can be found by the equation: Sum = $(n - 2) \times 180°$, where n is equal to the number of sides of the polygon.

5.6
Elements of circles

A circle of 360° can be divided into a number of parts, each of which has its own special name (Fig. 5.5). The equation for finding the area of a circle is $A = \pi r^2$, where r is the radius and π is equal to 3.14 (pi). The equation for finding the circumference is $C = \pi d$, where d is the diameter.

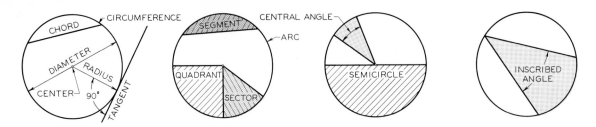

FIG. 5.5 Definitions of the elements of a circle.

5.7
Geometric solids

The various types of solid geometric shapes are shown in Fig. 5.6 along with their names and definitions.

POLYHEDRA A multisided solid formed by intersecting planes is called a *polyhedron*. If the faces of a

polyhedron are regular polygons, it is called a *regular polyhedron*. Five regular polyhedra are the *tetrahedron* with 4 sides, the *hexahedron* with 6 sides, the *octahedron* with 8 sides, the *dodecahedron* with 12 sides, and the *icosahedron* with 20 sides.

PRISMS A *prism* is a solid that has two parallel bases that are equal in shape. The bases are connected by sides that are parallelograms. The line from the

FIG. 5.6 Types of geometric solids and their elements and definitions.

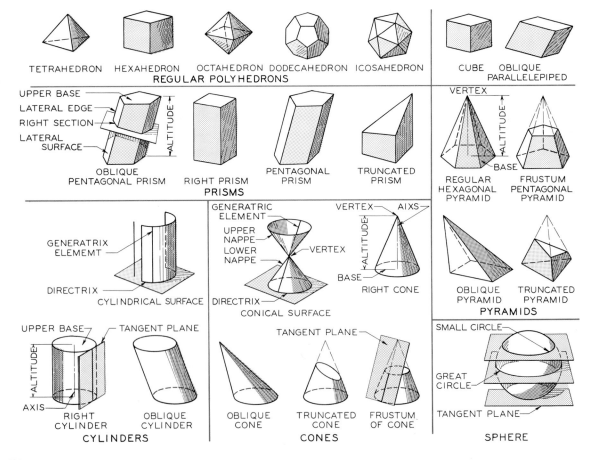

TETRAHEDRON HEXAHEDRON OCTAHEDRON DODECAHEDRON ICOSAHEDRON
REGULAR POLYHEDRONS

CUBE OBLIQUE PARALLELEPIPED

UPPER BASE
LATERAL EDGE
RIGHT SECTION
LATERAL SURFACE
ALTITUDE
OBLIQUE PENTAGONAL PRISM RIGHT PRISM PENTAGONAL PRISM TRUNCATED PRISM
PRISMS

VERTEX
ALTITUDE
BASE
REGULAR HEXAGONAL PYRAMID FRUSTUM PENTAGONAL PYRAMID

GENERATRIX ELEMEMT
DIRECTRIX
CYLINDRICAL SURFACE

GENERATRIC ELEMENT
UPPER NAPPE
LOWER NAPPE
VERTEX
DIRECTRIX
CONICAL SURFACE

VERTEX AIXS
ALTITUDE
BASE
RIGHT CONE

OBLIQUE PYRAMID TRUNCATED PYRAMID
PYRAMIDS

UPPER BASE TANGENT PLANE
ALTITUDE
AXIS
RIGHT CYLINDER OBLIQUE CYLINDER
CYLINDERS

TANGENT PLANE
OBLIQUE CONE TRUNCATED CONE FRUSTUM OF CONE
CONES

SMALL CIRCLE
GREAT CIRCLE
TANGENT PLANE
SPHERE

center of one base to the other is called the *axis*. If the axis is perpendicular to the bases, the axis is called the *altitude,* and the prism is a *right prism*. If the axis is not perpendicular to the base, the prism is an *oblique prism*.

A prism that has been cut off to form a base that is not parallel to the other is called a *truncated prism*.

A *parallelepiped* is a prism with a base that is either a rectangle or a parallelogram.

PYRAMIDS A *pyramid* is a solid with a polygon as a base and triangular faces that converge at a point called the *vertex*. The line from the vertex to the center of the base is called the *axis*. If the axis is perpendicular to the base, it is the *altitude* of the pyramid, and the pyramid is a *right pyramid*. If the axis is not perpendicular to the base, the pyramid is an *oblique pyramid*. A truncated pyramid is called a *frustum* of a pyramid.

CYLINDERS A *cylinder* is formed by a line or element, called a generatrix, that moves about the circle while remaining parallel to its axis.

The axis of a cylinder connects the centers of each end of the cylinder. If the axis is perpendicular to the bases, it is the *altitude* of the cylinder. When the axis is perpendicular to the bases, the cylinder is a *right cylinder*. When the axis does not make a 90° angle with the base, the cylinder is an *oblique cylinder*.

CONES A *cone* is formed by a line or element, called a generatrix, with one end that moves about the circular base while the other end remains at a fixed point called the *vertex*. The line from the center of the base to the vertex is called the *axis*. If the axis is perpendicular to the base, it is called the *altitude,* and the cone is a *right cone*. A truncated cone is called a *frustum* of a cone.

SPHERES A *sphere* is generated by the plane of a circle that is revolved about one of its diameters to form a solid. The ends of an axis through the center of the sphere are called *poles*.

The equations for finding the volumes of geometric solids are given in Fig. 5.7.

FIG. 5.7 Standard geometric solids and the equations for finding their volumes.

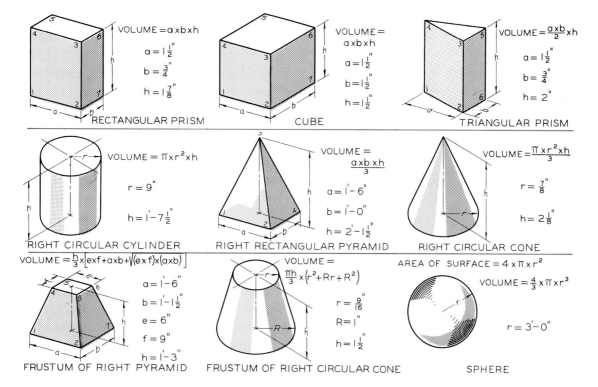

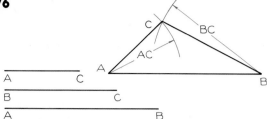

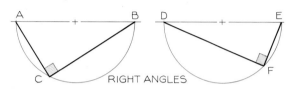

FIG. 5.8 When three sides are given, a triangle can be drawn with a compass.

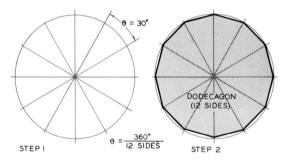

FIG. 5.9 Any angle that is inscribed in a semicircle will be a right angle.

FIG. 5.10 The regular polygon

Step 1 To construct a regular polygon, divide the circumference of a circle into the same number of divisions as the sides of the polygon, 12 in this case.

Step 2 Connect the divisions along the circumference with straight lines to form the sides of the polygon.

FIG. 5.11 A circle can be inscribed or circumscribed to form a hexagon by using a 30°–60° triangle.

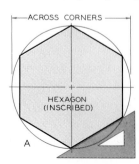

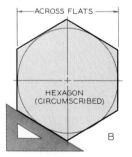

5.8
Constructing triangles

When three sides of a triangle are given, the triangle can be constructed by using a compass as shown in Fig. 5.8. Only one triangle can be found when the sides are given.

A right triangle can be constructed by inscribing it inside a semicircle as shown in Fig. 5.9. Any triangle inscribed in a semicircle will always be a right triangle.

5.9
Constructing polygons

A regular polygon (having equal sides) can be inscribed in a circle or circumscribed about a circle. When inscribed, all the corner points will lie along the circle. This makes it possible to divide the circle into the desired number of sectors to locate the points (Fig. 5.10).

For example, a 12-sided polygon is constructed by dividing the circle into 12 sectors and connecting the points to form the polygon.

5.10
Hexagons

Examples of inscribed and circumscribed hexagons are shown in Fig. 5.11. These are drawn with 30°–60° triangles either inside or outside the circles. Note that the circle represents the distance from corner to corner when inscribed, and from flat to flat when circumscribed.

5.11
Octagons

The octagon, an eight-sided regular polygon, can be inscribed in or circumscribed about a circle (Fig. 5.12) by using a 45° triangle. A second method inscribes the octagon inside a square (Fig. 5.13).

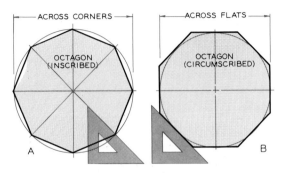

FIG. 5.12 A circle can be inscribed or circumscribed to form an octagon with a 45° triangle.

5.12
Pentagons

Since the pentagon is a five-sided regular polygon, it can be inscribed in or circumscribed about a circle, as previously covered. Another method of constructing a pentagon is shown in Fig. 5.14. This construction is performed with the use of a compass and a straight-edge.

5.13
Bisecting lines and angles

Finding the midpoint of a line, or the perpendicular bisector of a line, is a basic technique of geometric construction. Two methods are illustrated in Fig. 5.15.

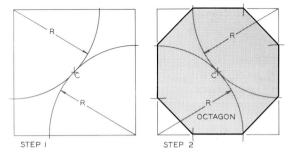

FIG. 5.13 Octagon in a square

Step 1 Construct a square and locate its center, *C*. Construct two arcs from opposite corners that will pass through *C*.

Step 2 Repeat this step by using the other corners. Connect the points located on the square to form the octagon.

FIG. 5.14 The pentagon

Step 1 Bisect radius *OP* to locate point *A*. With *A* as the center and *AC* as the radius *R1*, locate point *B* on the diameter.

Step 2 With point *C* as the center and *BC* as the radius *R2*, locate point *D* on the arc. Line *CD* is the chord that can be used to locate the other corners of the pentagon.

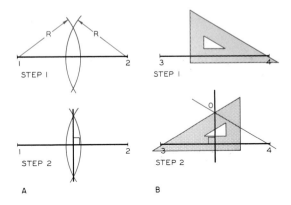

FIG. 5.15 Bisecting a line

A line can be bisected by using a compass and any radius or a standard triangle and a straightedge.

The first method involves the use of a compass to construct a perpendicular to a line. The second method uses a standard triangle. The compass method can be used to find the midpoint of an arc as well as a straight line.

The angle in Fig. 5.16 can be bisected with a compass by drawing three arcs.

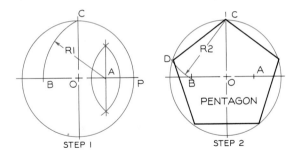

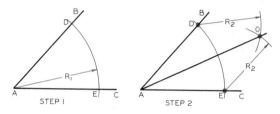

FIG. 5.16 Bisecting an angle

Step 1 Swing an arc of any radius to locate points *D* and *E*.

Step 2 Using the same radius, draw two arcs from *D* and *E* to locate point *O*. Line *AO* is the bisector of the angle.

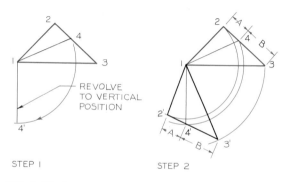

FIG. 5.17 Rotation of a figure

Step 1 A plane figure can be rotated about any point. Line 1–4 is rotated about point 1 to its desired position with a compass.

Step 2 Points 2′ and 3′ are located by measuring distances *A* and *B* from 4′.

5.14
Revolution of figures

Rotating a triangle about one of its points is demonstrated in Fig. 5.17. In this case, the triangle is rotated about point 1 of line 1–4. Point 4 is rotated to its desired position by using a compass. Points 2 and 3 are found by triangulation to complete the rotated view.

This same principle will work on any plane regardless of its shape.

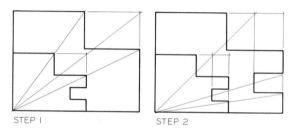

FIG. 5.18 Enlargement of a figure

Step 1 A proportional enlargement can be made by using a series of diagonals drawn through a single point, the lower left corner in this case.

Step 2 Additional diagonals are drawn to locate the other features of the object. This process can be reversed for reducing an object.

5.15
Enlargement and reduction of figures

In Fig. 5.18, the small figure is enlarged by using a series of radial lines from the lower left corner. The smaller figure is completed as a rectangle, and the larger rectangle is drawn proportional to the small one. The upper right notch is located by using the same technique.

The smaller notch is found by using three construction lines projected from the lower left corner. This is based on the principle of similar and proportional triangles.

FIG. 5.19 Division of a line

Step 1 Line *AB* is divided into 7 equal divisions by constructing a line through *A* and dividing it into 7 known units with your dividers. Point *C* is connected to point *B*.

Step 2 A series of lines are drawn parallel to *CB* to locate the divisions along line *AB*.

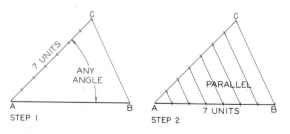

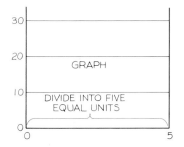
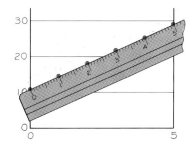
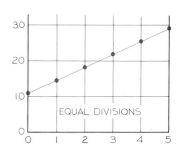

FIG. 5.20 Division of a space

Given It is desired to divide a graph into five equal divisions along the *x*-axis.

Step 1 A convenient scale with five units of measurement that approximate the horizontal distance is laid across the graph aligning the 0 and 5 markings with the lines.

Step 2 Construct vertical lines through the points that were found in Step 1. This method could have been used to calibrate the divisions along the *y*-axis.

This method could have been used to reduce the larger drawing to the smaller one.

5.16
Division of lines

It is often necessary to divide a line into a number of equal parts when a convenient scale is not available for this purpose. For example, suppose a one-inch line is to be divided into seven equal parts. No scale is available that divides an inch into sevenths, and mathematical units involve hard-to-measure decimals.

The method shown in Fig. 5.19 is an efficient way to solve this problem.

An application of this principle is used for locating lines on a graph that are equally spaced (Fig. 5.20). A scale with the desired number of units (0 to 5) is laid across from left to right on the graph. Vertical lines are drawn through these points to divide the graph into five equal divisions.

5.17
Circles through three points

An arc can be drawn through any three points by connecting the points with two lines (Fig. 5.21). Perpendicular bisectors are found for each line to locate the center at point *C*. The radius is drawn, and the lines *AB* and *BD* become chords of the circle.

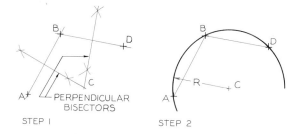

FIG. 5.21 An arc through three points

Step 1 Connect points *A* and *B* and points *B* and *D* with two lines and find their perpendicular bisectors. The bisectors will intersect at the center, *C*.

Step 2 Using the center, *C*, and the distance to the points as the radius, construct the arc through the points.

This system can be reversed to find the center of a given circle. Draw two chords that intersect at a point on the circumference and bisect them. The perpendicular bisectors will intersect at the center of the circle.

5.18
Parallel lines

A line can be drawn parallel to another by using either of the methods shown in Fig. 5.22.

The first method involves the use of a compass to draw two arcs to locate a parallel line that is the desired distance away.

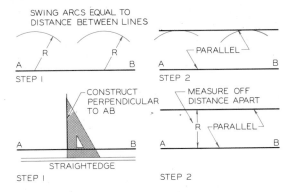

FIG. 5.22 Construction of parallel lines

Either of the above methods can be used for constructing one line parallel to another. The first method uses a compass and a straightedge; the other uses a triangle and a straightedge.

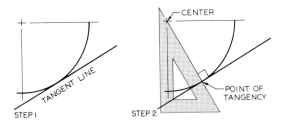

FIG. 5.23 Locating a tangent point

Step 1 Align your triangle with the tangent line while holding it firmly against a straightedge.

Step 2 Hold the straightedge in position, rotate the triangle, and construct a line through the center that is perpendicular to the line, to locate the point of tangency.

FIG. 5.24 Thin lines that extend beyond the arcs from their centers are used to mark points of tangency. These lines should always be shown in this manner.

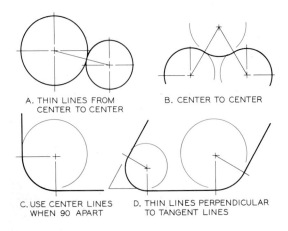

A. THIN LINES FROM CENTER TO CENTER

B. CENTER TO CENTER

C. USE CENTER LINES WHEN 90 APART

D. THIN LINES PERPENDICULAR TO TANGENT LINES

The second method requires the construction of a perpendicular from a given line and the measurement of the distance R to locate the parallel, which is drawn with a straightedge.

5.19
Points of tangency

A point of tangency is the theoretical point where a straight line joins an arc or where two arcs join, making a smooth transition. In Fig. 5.23, a line is tangent to an arc. The point of tangency is located by constructing a perpendicular to the line from the center of the arc. A thin line is drawn from the center through the point to mark the point of tangency.

The conventional methods of marking points of tangency are shown in Fig. 5.24. Thin lines are drawn through the points from the centers of the arcs. This method should be used to mark all points of tangency when solving tangency problems.

5.20
Line tangent to an arc

Although you can approximate the point of tangency between a line and an arc, the method of finding the exact point of tangency is shown in Fig. 5.25. Point A and the arc are given. Point A is connected to the center in Step 1, AC is bisected in Step 2, and T is located in Step 3. This is the exact point of tangency.

The point of tangency could also have been found by using a standard triangle, as shown in Fig. 5.26. One edge of the triangle is aligned with TA while the straightedge is held firmly. The triangle is rotated to construct a line through the center that is perpendicular to AT, locating the point of tangency, T.

5.21
Arc tangent to a line from a point

If an arc is to be constructed tangent to line CD at T (Fig. 5.27) and pass through point P, a perpendicular bisector of TP is drawn. A perpendicular to CD is drawn at T to locate the center at O.

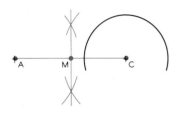
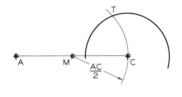
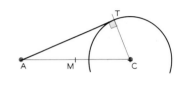

FIG. 5.25 A line from a point tangent to an arc

Step 1 Connect point *A* with center *C*. Locate point *M* by bisecting *AC*.

Step 2 Using point *M* as the center and *MC* as the radius, locate point *T* on the arc.

Step 3 Draw the line from *A* to *T* that is tangent to the arc of point *T*.

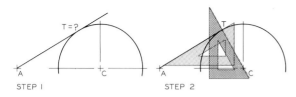

FIG. 5.26 A line from a point tangent to an arc

Step 1 A line can be drawn from point *A* tangent to the arc by eye.

Step 2 By rotating your triangle, the point of tangency (*T*) can be located at a 90° angle with lines *CT* and *AT*.

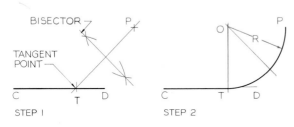

FIG. 5.27 An arc through two points

Step 1 If an arc must be tangent to a given line at a certain point and pass through *P*, find the perpendicular bisector of line *TP*.

Step 2 Construct a perpendicular to the line at *T* to intersect the bisector. The arc is drawn from center *O* with radius *OT*.

A similar problem in Fig. 5.28 requires you to draw an arc of a given radius that will be tangent to line *AB* and pass through point *P*. In this case the point of tangency on the line is not known until the problem has been solved.

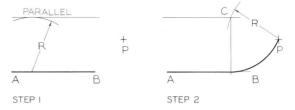

FIG. 5.28 An arc tangent to a line and a point

Step 1 When an arc of a given radius is to be drawn tangent to a line and through point *P*, draw a line parallel to *AB* and *R* from it.

Step 2 Draw an arc from *P* with radius *R* to locate the center at *C*. The arc is drawn with radius *R* and center *C*.

5.22
Arc tangent to two lines

An arc of a given radius can be constructed tangent to two nonparallel lines if the radius is given. This construction may be used to round a corner of a product or to design a curb at a traffic intersection. This method is shown in Fig. 5.29, where two lines form an acute angle.

The same steps are used to find an arc that is tangent to two lines that form an obtuse angle (Fig. 5.30). In both cases, the points of tangency are located with thin lines drawn from the centers through the points of tangency.

A different technique can be used to find an arc of a given radius that is tangent to perpendicular lines (Fig. 5.31). This method will work only for perpendicular lines.

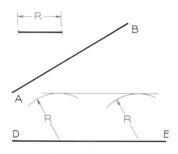

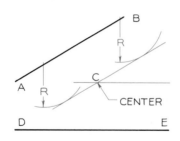

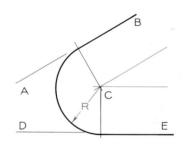

FIG. 5.29 An arc tangent to two lines

Step 1 Construct a line parallel to *DE* with the radius of the specified arc, *R*.

Step 2 Draw a second construction line parallel to *AB* to locate the center, *C*.

Step 3 Thin lines are drawn from *C* perpendicular to *AB* and *DE* to locate the points of tangency. The tangent arc is drawn using the center *C*.

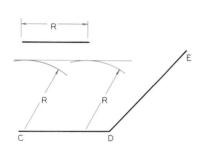

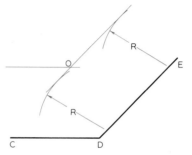

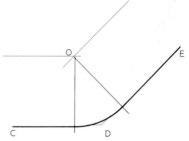

FIG. 5.30 An arc tangent to two lines

Step 1 Using the specified radius, *R*, construct a line parallel to *CD*.

Step 2 Construct a line parallel to *DE* that is distance *R* from it to locate center *O*.

Step 3 Construct thin lines from center *O* perpendicular to lines *CD* and *DE* to locate the points of tangency. Draw the arc using radius *R* and center *O*.

FIG. 5.31 An arc tangent to perpendicular lines

Step 1 Using the specified radius, *R*, locate points *D* and *E* by using center *A*.

Step 2 Swing two arcs using the radius *R* that was used in Step 1 to locate point *C*.

Step 3 Locate the tangent points with lines from the center, *C*. Draw the arc with radius *R* and center *C*.

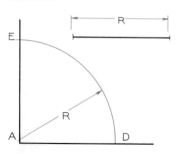

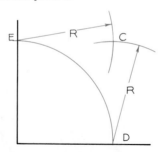

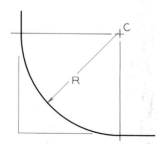

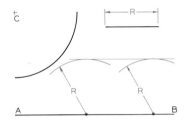 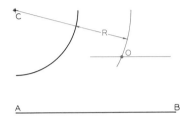 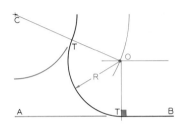

FIG. 5.32 An arc tangent to an arc and a line

Step 1 Construct a line parallel to *AB* that is *R* from it. Use thin construction lines.

Step 2 Add radius *R* to the extended radius through point *C*. Use this large radius to locate point *O*.

Step 3 Lines *OC* and *OT* are drawn to locate the tangency points. The arc with radius *R* and center *O* is drawn.

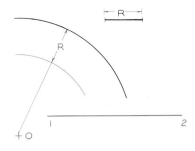 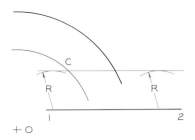 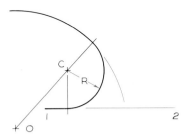

FIG. 5.33 An arc tangent to an arc and a line

Step 1 The specified arc, *R*, is subtracted from the extended radius through the arc's center at *O*. A concentric arc is drawn with the shortened radius.

Step 2 A line parallel to 1–2 is drawn a distance of *R* from it to locate the center, point *C*.

Step 3 The tangent points are located with lines from *O* through *C*, and through *C* perpendicular to 1–2. Draw the tangent arc with radius *R*.

5.23
Arc tangent to an arc and a line

When a radius is given, an arc can be drawn that is tangent to an arc and a line. These steps of construction are given in Fig. 5.32.

A variation of this principle of construction is shown in Fig. 5.33, where the arc is drawn tangent to an arc and a line with the arc in a reverse position.

5.24
Arc tangent to two arcs

A third arc is drawn tangent to two given arcs in Fig. 5.34. Thin lines are drawn from the centers to locate the points of tangency. This tangent arc is concave from the top.

A convex arc can be drawn tangent to the given arcs if the radius of the arc is greater than the radius of either of the given arcs (Fig. 5.35).

A variation of this problem is shown in Fig. 5.36, where an arc of a given radius is drawn tangent to the top of one arc and the bottom of the other. The two arcs are of different sizes.

A similar problem is shown in Fig. 5.37, where an arc is drawn tangent to a circle and a larger arc.

In all cases, the points of tangency are located and marked by thin lines drawn from center to center of the tangent arcs.

5.25
Ogee curves

The ogee curve can be thought of as a double curve formed by tangent arcs. The ogee curve in Fig. 5.38 was found by constructing two arcs tangent to three intersecting lines.

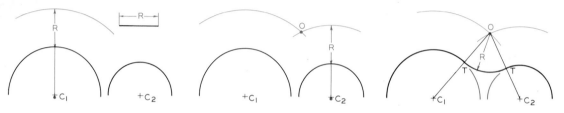

FIG. 5.34 An arc tangent to two arcs

Step 1 The radius of one circle is extended, and the radius R is added to it. The extended radius is used for drawing a concentric arc.

Step 2 The radius of the other circle is extended, and the radius R is added to it. The extended radius is used to construct an arc and to locate point O, the center.

Step 3 The centers are connected with point O to locate the points of tangency. The arc is drawn tangent to the two arcs with radius R.

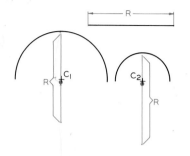
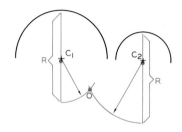
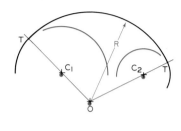

FIG. 5.35 An arc tangent to two arcs

Step 1 The radius of each arc is extended from the arc past its center, and the specified radius, R, is laid off from the arcs along these lines.

Step 2 The distance from each center to the ends of the extended radii are used for drawing two concentric arcs to locate the center O.

Step 3 Thin lines from O through centers C_1 and C_2 locate the points of tangency. The arc is drawn using point O as the center.

FIG. 5.36 An arc tangent to two circles

Step 1 The specified radius, R, is laid off from the arc along the extended radius to locate point D. Radius AD is used to construct a concentric arc.

Step 2 The radius through center B is extended, and the radius R is added to it from point F. Radius BE is used to locate the center, C.

Step 3 The tangent arc is drawn from center C with radius R. The points of tangency are located with thin lines from C through the given centers.

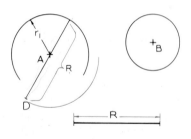
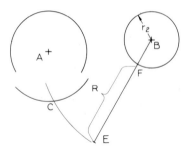
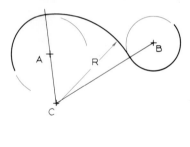

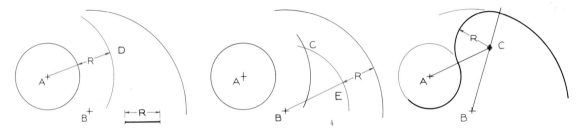

FIG. 5.37 An arc tangent to two arcs

Step 1 Radius R is added to the radius through center A. Radius AD is used to draw a concentric arc with center A.

Step 2 Radius R is subtracted from the radius through B. Radius BE is used to construct a concentric arc and locate point C.

Step 3 The points of tangency are located with thin lines BC and AC. The tangent arc is drawn with center C.

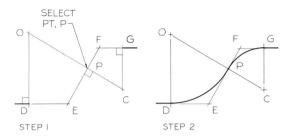

STEP I   STEP 2

FIG. 5.38 An ogee curve

Step 1 To draw an ogee curve between two parallel lines, draw line EF at any angle. Locate a point of your choosing along EF, P in this case. Find the tangent points by making FG equal to FP and DE equal to EP. Draw perpendiculars at G and D to intersect the perpendicular bisector of EF.

Step 2 Using radii CP and OP, draw the two tangent arcs to complete the ogee curve.

FIG. 5.39 An ogee curve

Step 1 To draw an ogee curve formed by two equal arcs passing through points B and C, draw a line between the points. Bisect the line BC and draw a line parallel to AB and CD to find the radius, R.

Step 2 Construct perpendiculars at B and C to locate the centers at both points O. Draw the arcs to complete the ogee curve.

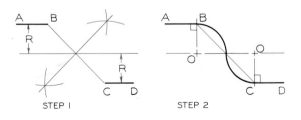

STEP I   STEP 2

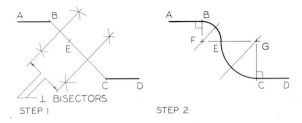

STEP I   STEP 2

FIG. 5.40 The unequal ogee curve

Step 1 Two parallel lines are to be connected by an ogee curve that passes through B and C. Draw line BC and select point E on the line. Bisect BE and EC.

Step 2 Construct perpendiculars at B and C to intersect the bisectors to locate centers F and G. Locate the points of tangency and draw the ogee curve using radii FB and GC.

An ogee curve can be drawn between two parallel lines (Fig. 5.39) from points B to C by geometric construction.

An alternative method of drawing an ogee curve that passes through points B, E, and C is illustrated in Fig. 5.40. The method of drawing an ogee curve through two nonparallel lines is shown in Fig. 5.41.

5.26
Curve of arcs

An irregular curve formed with tangent arcs can be constructed as shown in Fig. 5.42. In this case, the radii of the arcs are selected to give the desired curve by moving from one set of points to the next.

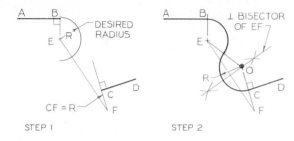

FIG. 5.41 Ogee curve between nonparallel lines

Step 1 Draw a perpendicular to *AB* at *B*, and draw the arc with the desired radius from center *E*. Draw a perpendicular to line *CD* at *C*, and make *CF* equal to the radius *R*. Connect *E* and *F*.

Step 2 Draw a perpendicular bisector of *EF* to locate point *O*. Use radius *OC* (which may be different from the first radius) to complete the curve. Mark points of tangency.

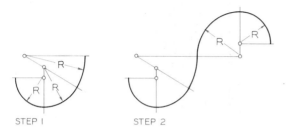

FIG. 5.42 A curve of arcs

Step 1 A series of arcs can be joined to form a smooth curve. Begin with the small arc, extend the radius through its center, draw the second and then the third.

Step 2 The curve can be reversed by extending the radius in the opposite direction and repeating the same process.

FIG. 5.43 To rectify an arc

Step 1 An arc has been rectified when its length has been laid out along a straight line. Construct a line tangent to the arc and divide the arc into a series of equal divisions from *A* to *B*.

Step 2 The chordal distances *D* along the arc are laid out along the straight line until point *B* is located.

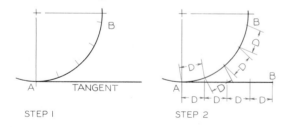

5.27
Rectifying arcs

An arc is rectified when its true length is laid out along a straight line. A method of rectifying an arc is illustrated in Fig. 5.43.

A second method of rectifying an arc is given in Fig. 5.44 by using a different form of geometric construction.

In addition to these methods, an arc can be rectified by using the mathematical equation for finding the circumference of the circle. Since a circle has 360°, the arc of a 30° sector is $\frac{1}{12}$ of the full circumference. Therefore, if the circumference is 12 inches, the arc of 30° is equal to 1 inch.

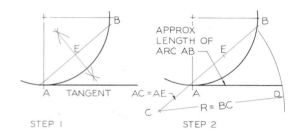

FIG. 5.44 To rectify an arc—compass method

Step 1 If you wish to rectify an arc from *A* to *B*, draw chord *AB* and find its midpoint, *E*.

Step 2 Extend *AB* to make *AC* equal to *AE*. The length of arc *AB* is approximated to be the distance from *A* to *D* found by swinging arc *BC*.

5.28
Conic sections

Conic sections are plane figures that can be described graphically as well as mathematically. They are formed by passing imaginary cutting planes through a right cone (Fig. 5.45).

5.29
Ellipses

The *ellipse* is a conic section formed by passing a plane through a right cone at an angle (Fig. 5.45). The el-

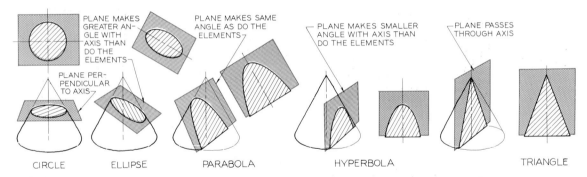

FIG. 5.45 Conic sections are formed by passing cutting planes at various angles through right cones. The conic sections are the circle, ellipse, parabola, hyperbola, and triangle.

lipse is mathematically defined as the path of a point that moves in such a way that the sum of the distances from two focal points is a constant.

The construction of an ellipse is found by revolving the edge view of a circle as shown in Fig. 5.46. These points are connected by a smooth curve. This ellipse could have been drawn by using the ellipse template as shown in Fig. 5.47. The angle between the line of sight and the edge view of the circle is the angle of the ellipse template that should be used (or the one closest to this size).

The largest diameter of an ellipse is always true length and is called the *major diameter*. The shortest diameter is perpendicular to the major diameter and is called the *minor diameter*. The crossing diameters are used to align the ellipse template.

FIG. 5.46 An ellipse by revolution

Step 1 When the edge view of a circle is perpendicular to the projectors between its adjacent view, the view will be a true circle. Mark equally spaced points along the arc and project them to the edge.

Step 2 Revolve the edge of the circle to the desired position and project the points to the circular view, which will now appear as an ellipse. Note that the points are projected vertically downward to their new positions.

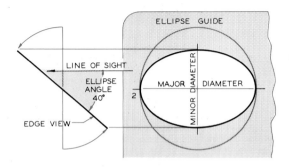

FIG. 5.47 The ellipse template

When the edge view of a circle is revolved so the line of sight between the two views is not perpendicular to the edge view, the circle will appear as an ellipse. The major diameter remains constant, but the minor diameter will vary. The angle between the line of sight and the edge view of the circle will give the angle of the ellipse template.

Ellipse templates are available in intervals of 5° and in variations in size of the major diameter of about ⅛″ (Fig. 5.48).

The mathematical equation of the ellipse is

$$\frac{x^2}{a^2} + \frac{y^2}{b^2} = 1, \qquad \text{where } a, b \neq 0.$$

Letters *a* and *b* are constants, and *x* and *y* are variables. This equation can be plotted on graph paper.

The ellipse can be constructed inside a rectangular box or a parallelogram, as illustrated in Fig. 5.49, where a series of points is plotted to form an elliptical curve.

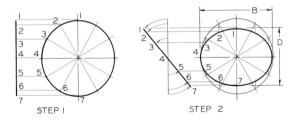

FIG. 5.48 Ellipse templates come in a variety of sizes. Most are calibrated at 5° intervals from 15° up to 60°. (Courtesy of Timely Products, Inc.)

Two circles can be used for constructing an ellipse by making the diameter of the large one equal to the major diameter and the diameter of the small one equal to the minor diameter (Fig. 5.50).

FIG. 5.49 Ellipse—parallelogram method

Step 1 An ellipse can be drawn inside a rectangle or a parallelogram by dividing the horizontal center line into the same number of equal divisions as the shorter sides, *AF* and *CD*.

Step 2 The construction of the curve in one quadrant is shown by using sets of rays from *E* and *B* to plot the points.

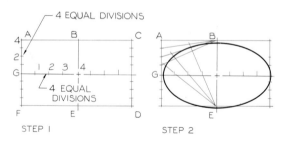

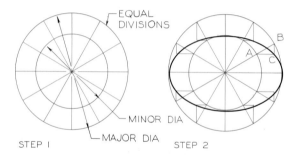

FIG. 5.50 Ellipse—circle method

Step 1 Two concentric circles are drawn, with the large one equal to the major diameter and the small one equal to the minor diameter. Divide them into equal sectors.

Step 2 Plot points on the ellipse by projecting downward from the large curve to intersect horizontal construction lines drawn from the intersections on the small circle.

The *conjugate diameters* of an ellipse are diameters that are parallel to the tangents at the ends of each, as illustrated in Fig. 5.51. A single ellipse has an infinite number of sets of conjugate diameters.

When the ellipse and a pair of conjugate diameters are given, the major and minor diameters of the ellipse can be found by using the method illustrated in Fig. 5.52. If the conjugate diameters are not given, they can be constructed as shown in Fig. 5.51, and the major and minor diameters found. The major and minor diameters are necessary for using the ellipse template.

FIG. 5.51 Conjugate diameters

Step 1 Many sets of conjugate diameters can be constructed for an ellipse. A conjugate diameter is one that is parallel to the tangents at the ends of the other conjugate diameter. A diameter is selected, and parallel tangents are drawn.

Step 2 The horizontal conjugate diameter is drawn parallel to the horizontal tangents, and the inclined tangents are drawn parallel to the conjugate diameter found in Step 1.

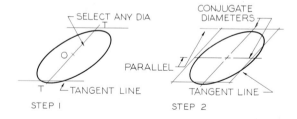

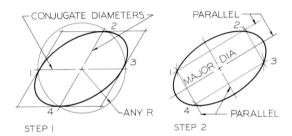

FIG. 5.52 Finding the axes of an ellipse

Step 1 When the conjugate diameters of an ellipse are given, you can find the major and minor diameters of the ellipse by drawing a circle of any radius with its center at the intersection of the diameters.

Step 2 The circle cuts four points along the ellipse. The points are connected to form a rectangle. The major and minor diameters are parallel to the sides of the rectangle.

5.30
Parabolas

The parabola is mathematically defined as a plane curve, each point of which is equidistant from a directrix (a straight line) and its focal point. The parabola

FIG. 5.53 The parabola—mathematical method

Step 1 Draw an axis perpendicular to a line called a directrix. Choose a point for the focus, F, on the axis.

Step 2 Locate points by using a series of selected radii to plot points on the curve. For example, draw a line parallel to the directrix and R_2 from it. Swing R_2 from F to intersect the line, and plot the point.

Step 3 Continue the process with a series of arcs of varying radii until an adequate number of points have been found to complete the curve.

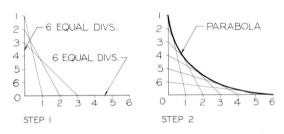

FIG. 5.54 The parabola—tangent method

Step 1 Construct two lines at a convenient angle and divide each of them into the same number of divisions. Connect the points with a series of diagonals.

Step 2 When finished, construct the parabolic curve to be tangent to the diagonals.

is formed when the cutting plane makes the same angle with the base of a cone as do the elements of the cone.

The construction of a parabola by using its mathematical definition is shown in Fig. 5.53.

A parabolic curve can be constructed by geometric construction, as shown in Fig. 5.54, by dividing the two perpendicular lines into the same number of divisions. The parabola is drawn through the plotted points with an irregular curve.

A third method of construction is illustrated in Fig. 5.55, the parallelogram method.

The mathematical equation of the parabola is

$$y = ax^2 + bx + c, \qquad \text{where } a \neq 0.$$

FIG. 5.55 The parabola—parallelogram method

Step 1 Construct a rectangle or a parallelogram to contain the parabola, and locate its axis parallel to the sides through O. Divide the sides into equal divisions. Connect the divisions with point O.

Step 2 Construct lines parallel to the sides (vertical in this case) to locate the points along the rays from O. Draw the parabola.

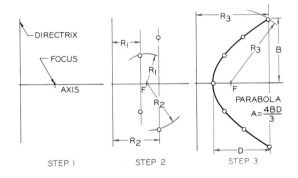

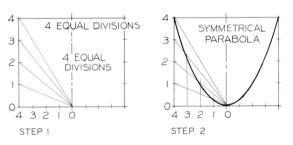

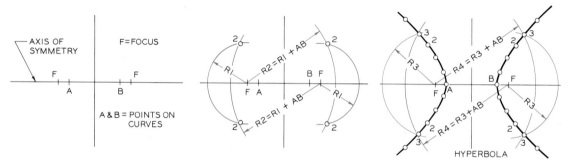

FIG. 5.56 The hyperbola

Step 1 A perpendicular is drawn through the axis of symmetry, and focal points *F* are located equidistant from it on both sides. Points on the curve, *A* and *B*, are located equidistant from the perpendicular at a location of your choice but between the focal points.

Step 2 Radius R_1 is selected to draw arcs using focal points *F* as the centers. R_1 is added to *AB* (the distance between the nearest points on the hyperbolas) to find R_2. Radius R_2 is used to draw arcs using the focal points as centers. The intersections of R_1 and R_2 establish points 2 on the hyperbola.

Step 3 Other radii are selected and added to distance *AB* to locate additional points in the same manner as described in Step 2. A smooth curve is drawn through the points to form the hyperbolic curves.

Letters *a*, *b*, and *c* are constants, and *x* and *y* are variables. This equation can be written by interchanging the *y*'s and *x*'s. The equation can be plotted on graph paper.

5.31
Hyperbolas

The hyperbola is a two-part conic section that is mathematically defined as the path of a point that moves in such a way that the difference of its distances from two focal points is a constant. This definition is used to construct the hyperbola in Fig. 5.56.

A second method of construction is shown in Fig. 5.57. Two perpendicular lines are drawn through point *B* as asymptotes. The hyperbolic curve becomes more nearly parallel and closer to the asymptotes as the hyperbola is extended, but the curve never merges with the asymptotes.

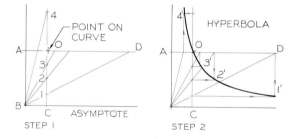

FIG. 5.57 The equilateral hyperbola

Step 1 Two perpendiculars are drawn through *B*, and any point *O* on the curve is located. Horizontal and vertical lines are drawn through *O*. Line *CO* is divided into equal divisions, and rays from *B* are drawn through them to the horizontal line.

Step 2 Horizontal construction lines are drawn from the divisions along line *OC*, and lines from *AD* are projected vertically to locate points 1' through 4' on the curve.

5.32
Spirals

The spiral is a coil that begins at a point and becomes larger as it travels around the origin. The spiral lies in a single plane. The steps of constructing a spiral are shown in Fig. 5.58.

5.33
Helixes

The helix is a curve that coils around a cylinder or a cone at a constant angle of inclination. Examples of helixes are corkscrews or threads on a screw. A helix is constructed about a cylinder in Fig. 5.59 and about a cone in Fig. 5.60.

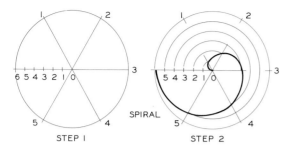

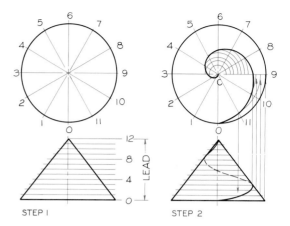

FIG. 5.58 The spiral

Step 1 Draw a circle and divide it into equal parts. The radius is divided into the same number of equal parts—six in this example.

Step 2 By beginning on the inside, draw arc 0–1 to intersect radius 0–1. Then swing arc 0–2 to radius 0–2 and continue until the last point is reached at 6, which lies on the original circle.

5.34
Involutes

The involute is the path of the end of a line as it unwinds from a line or a plane figure. In Fig. 5.61, an involute is formed by unwinding a line from a rectangle. Successively different radii that are equal in length to the sides are used to develop the involute.

FIG. 5.59 The helix

Divide the top view of the cylinder into equal divisions and project them to the front view. Lay out the circumference and the height of the cylinder, which is the lead. Divide the circumference into the same number of equal parts by taking the measurements from the top view. Project the points along the inclined rise to their respective elements to find the helix.

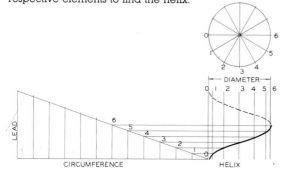

FIG. 5.60 A conical helix

Step 1 Divide the cone's base into equal parts. Pass a series of horizontal cutting planes through the front view of the cone. Use the same number as the divisions on the base, 12 in this case.

Step 2 Project all the divisions along the front view of the cone to line $C9$, and draw a series of arcs from center C to their respective radii in the top view to plot the points. Project the points to their respective cutting planes in the front view.

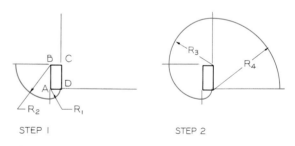

FIG. 5.61 The involute

Step 1 Side AD is used as a radius for drawing arc AD. AB is added to AD to form radius R_2. Draw a second arc using center B.

Step 2 The two remaining sides are used to unwind the involute back to its point of origin.

5.35
Cycloids

The cycloid is a plane curve formed by a point on a circle as the circle rolls along a straight line.

In Fig. 5.62, the distance from 1 to 9 must be equal to the circumference of the circle. The circle is

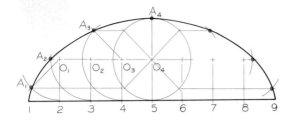

FIG. 5.62 The cycloid

Step 1 A circle is centered on a horizontal tangent line. It is divided into a number of equal divisions. The circumference of the circle is laid off along line 1–9, which is divided into the same number of parts as the circle.

Step 2 The center of the arc is moved from O_4 to O_3 to locate point A_3, then to O_2, etc. The points are connected to form the cycloidal curve.

located at the center point, 5, and it is rolled to the left to locate points A_4, A_3, A_2, and A_1 as the center moves from O_4 to O_1. These points are connected with a smooth curve to complete the left side of the cycloid.

The same construction is repeated as the circle moves to the right to complete the symmetrical curve.

5.36
Epicycloids and hypocycloids

The *epicycloid* is a curve formed by a point on a circle as the circle rolls along the convex side of a larger circle (Fig. 5.63). The circumference of the rolling circle is divided into equal divisions. These same units are measured along the circumference of the large arc. These will be the contact points as the circle rolls along the arc. The plotted points are connected to form the epicycloid.

The *hypocycloid* is a curve formed by a point on a circle as the circle rolls along the concave side of a larger arc. As in the epicycloid, the circumference of the small circle is divided into equal divisions that are laid off along the larger arc. The curve is plotted and drawn as a smooth curve as the circle is rolled left and right.

The elliptical-appearing curve in Step 3 of Fig. 5.63 is a curve formed by connecting the epicycloid and hypocycloid curves by rolling a circle of the same size on each side of a given arc.

FIG. 5.63 The hypocycloid and epicycloid

Step 1 Divide the circle into a number of equal parts. Measure the same lengths along the large arc. Locate the positions of center O along an arc drawn through center O.

Step 2 Move the circle to positions 1 through 6 to find the points along the epicycloid. Repeat this process at the right side and draw the curve.

Step 3 The hypocycloid is found in the same manner, but the circle rolls along the inside of the large arc. This figure shows both the epicycloid and the hypocycloid drawn together.

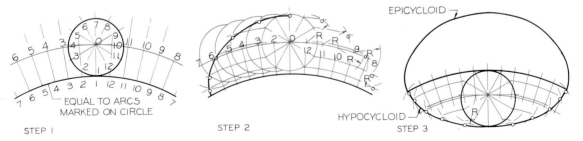

Problems

These problems are to be solved on Size AV paper similar to the one shown in Fig. 5.64, where Problems 1–5 are laid out. Each inch on the grid is equal to 0.20 inches; therefore, use your engineers' 10 scale to lay out the problems. By equating each grid to 5 mm, you can use your full-size metric scale to lay out and solve the problems.

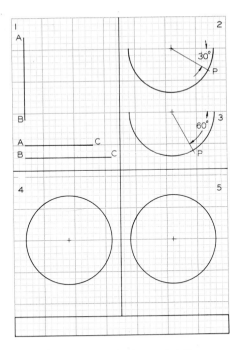

FIG. 5.64 Problems 1–5: Basic constructions.

17. Divide the two vertical lines into four equal divisions. Draw three equally spaced vertical lines at the divisions.

18. Construct an arc with radius R that is tangent to the line at J and passes through point P.

19. Construct an arc with radius R that is tangent to the line and passes through P.

20. Construct a line from P that is tangent to the semicircle. Locate the points of tangency. Use the compass method.

21–23. Construct arcs with the given radii tangent to the lines.

Show your construction and mark all points of tangency, as discussed in the chapter.

1. Draw triangle ABC using the given sides.

2–3. Inscribe an angle in the semicircles with the vertexes at point P.

4. Inscribe a nine-sided regular polygon inside the circle.

5. Circumscribe a ten-sided regular polygon about the circle.

6. Circumscribe a hexagon about the circle.

7. Inscribe an octagon in the circle.

8. Circumscribe an octagon about the circle.

9. Construct a pentagon inside the circle using the compass method.

10–11. Bisect the lines.

12–13. Bisect the angles.

14. Rotate the triangle 80° in a clockwise direction about point A.

15. Enlarge the given shape to the size indicated by the diagonal.

16. Divide AB into seven equal parts. Draw the construction line through B for your construction.

FIG. 5.65 Problems 6–9: Construction of regular polygons.

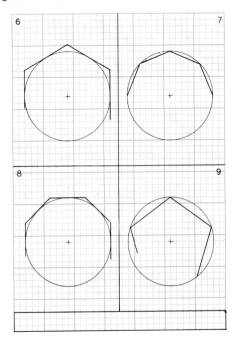

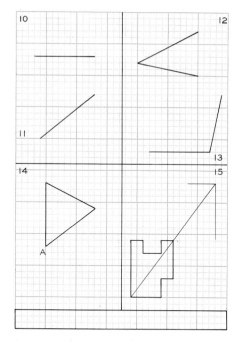

FIG. 5.66 Problems 10–15: Basic constructions.

FIG. 5.67 Problems 16–19: Tangency construction.

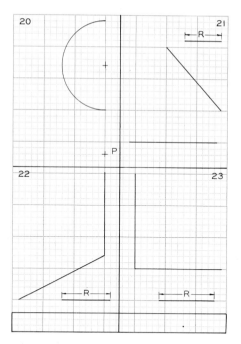

FIG. 5.68 Problems 20–23: Tangency construction.

FIG. 5.69 Problems 24–27: Tangency construction.

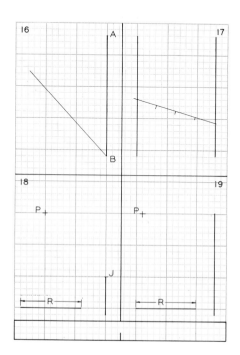

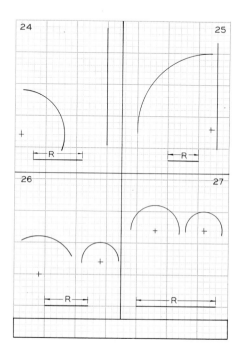

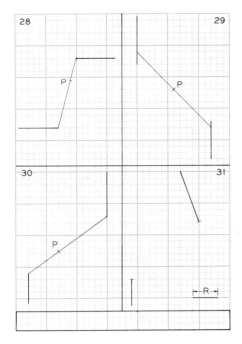

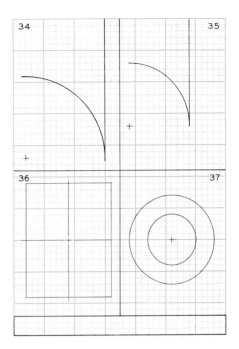

FIG. 5.70 Problems 28–31: Ogee curve construction.

FIG. 5.72 Problems 34–37: Rectifying an arc, ellipse construction.

FIG. 5.71 Problems 32–33: Tangency construction.

FIG. 5.73 Problems 38–40: Ellipse and parabola construction.

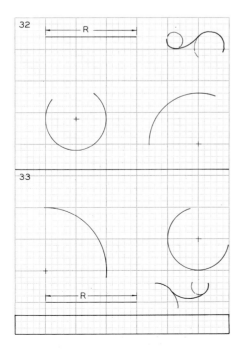

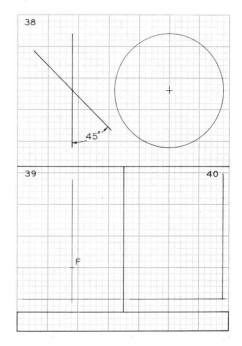

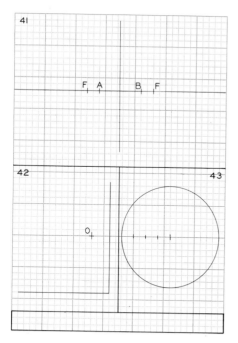

FIG. 5.74 Problems 41–43: Hyperbola and spiral construction.

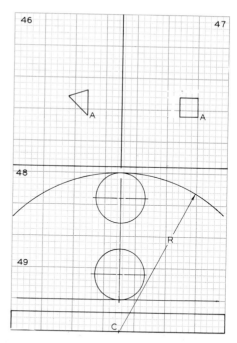

FIG. 5.76 Problems 46–49: Involute, hypocycloid, and epicycloid construction.

FIG. 5.75 Problems 44–45: Helix construction.

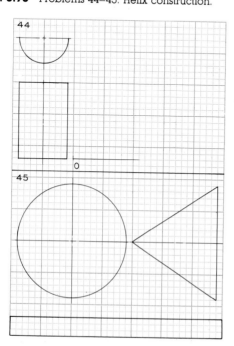

24–27. Construct arcs that are tangent to the arcs and/or lines. The radii are given for each problem.

28–31. Construct ogee curves that connect the ends of the given lines and pass through points P where given. In Problem 31, the radii for the arcs are given.

32–33. Using the given radii, connect the given arcs with a tangent arc as indicated in the freehand sketches.

34. Rectify the arc along the given line by dividing the circumference into equal divisions and laying them off with your dividers.

35. Rectify the arc by using the compass method as shown in Fig. 5.44.

36. Construct an ellipse inside the rectangular layout.

37. Construct an ellipse inside the large circle. The small circle represents the minor diameter.

38. Construct an ellipse inside the circle when the edge view has been rotated 45° as shown.

39. Using the focal point F and the directrix, plot and draw the parabola formed by these elements.

40. Construct a parabola using perpendicular lines by either of the methods shown in Figs. 5.54 and 5.55.

41. Using the focal points *F,* points *A* and *B* on the curve, and the axis of symmetry, construct the hyperbolic curve.

42. Construct a hyperbola that passes through *O*. The perpendicular lines are asymptotes.

43. Construct a spiral by using the four divisions that are marked along the radius.

44–45. Construct a helix that has a rise equal to the heights of the cylinder and cone. Show construction and the curve in all views.

46–47. Construct involutes by unwinding the triangle and rectangle in a clockwise direction, beginning with point *A* in each.

48. Construct an epicycloid by rolling the circle along the curve whose center is at point *C*.

49. Construct a cycloid by rolling the circle about the horizontal line.

50–61. Construct these problems (Figs. 5.77–5.88) on Size A sheets, one problem per sheet. Select the proper scale that will best fit the problem to the sheet. Mark all points of tangency and strive for good line quality.

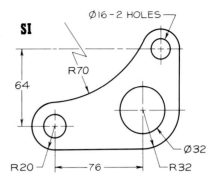

FIG. 5.78 Problem 51: Lever crank.

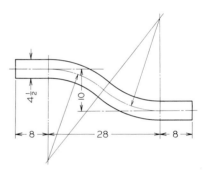

FIG. 5.79 Problem 52: Road tangency.

FIG. 5.80 Problem 53: Road tangency.

FIG. 5.77 Problem 50: Gasket.

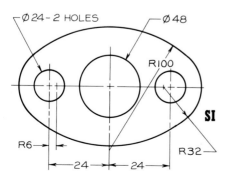

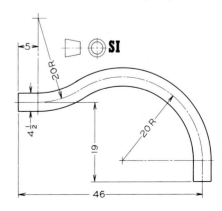

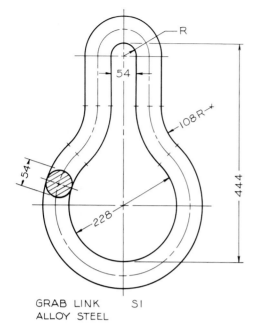

GRAB LINK   SI
ALLOY STEEL

FIG. 5.81 Problem 54: Grab link.

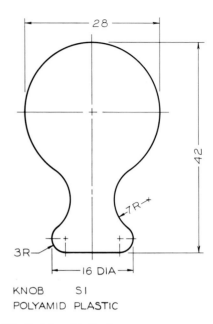

KNOB   SI
POLYAMID PLASTIC

FIG. 5.83 Problem 56: Knob.

FIG. 5.82 Problem 55: Three-lobe knob.

FIG. 5.84 Problem 57: Foundry hook.

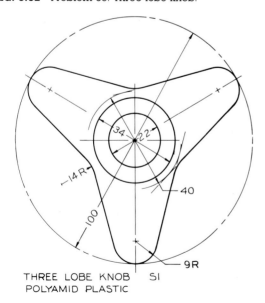

THREE LOBE KNOB   SI
POLYAMID PLASTIC

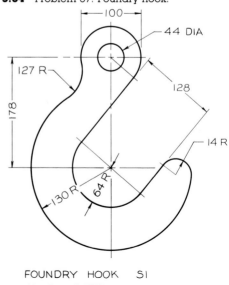

FOUNDRY HOOK   SI
ALLOY STEEL

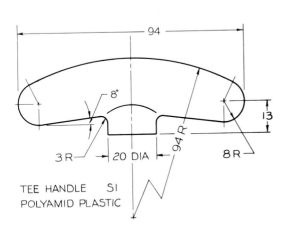

FIG. 5.85 Problem 58: Tee handle.

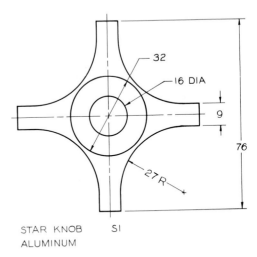

FIG. 5.87 Problem 60: Star knob.

FIG. 5.86 Problem 59: Five-lobe knob.

FIG. 5.88 Problem 61: Lug link plate.

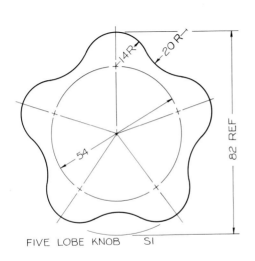

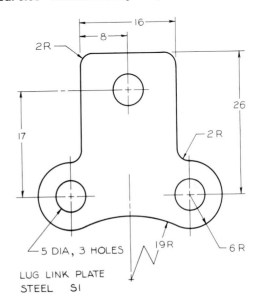

6

Multiview Sketching

6.1

The purpose of sketching

Sketching is a thinking process as well as a technique of communication. Designers must develop their ideas by making many sketches, revising them, and finally arriving at the desired solution. Later, these sketches are converted into instrument drawings.

Sketching should be a rapid method of drawing. If speed is not developed with a degree of skill in sketching, this technique will not be effective. If you

develop your sketching skills, you can assign drafting work to assistants who can then prepare the finished drawings by working from your sketches. If your sketches are not sufficiently clear to communicate your ideas to someone else, then it is likely that you have not thought out the solution well enough, even for your own understanding. The ability to communicate by any means is a great asset, and sketching is one of the more powerful techniques of communication.

6.2

Shape description

A pictorial of an object is shown in Fig. 6.1 with three arrows that indicate the directions of sight that will give top, front, and right-side views of the object. Each view is a two-dimensional view rather than a three-dimensional pictorial.

This is called *orthographic projection* or *multiview projection*. In multiview projection, it is important that the views be located as shown in Fig. 6.1. The top view is placed over the front view, since both views share the dimension of width. The side view is placed to the right of the front view, since these views share the dimension of height. The distance between the views can vary, but they must be positioned so that the views project from each other as shown here.

FIG. 6.1 Three views of an object can be found by looking at the object in this manner. The three views—top, front, and right side—describe the object.

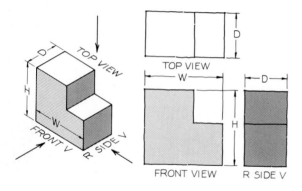

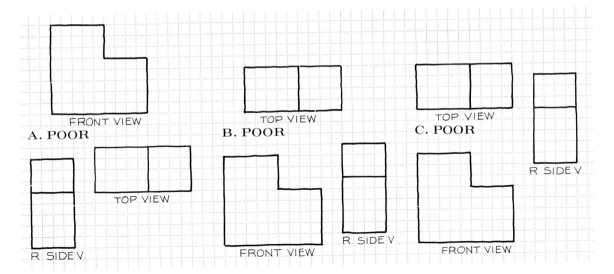

FIG. 6.2 Arrangement of views

A. These views are sketched in-correctly. The views are scrambled.

B. These views are nearly correct, but they do not project from view to view.

C. The top and front views are correctly positioned, but the right-side view is incorrectly positioned.

Several examples of poorly arranged views are shown in Fig. 6.2. You can see that although these views are correct, they are hard to interpret because the views are not placed in their standard positions.

6.3
Six-view drawings

Six principal views may be found for any object by using the rules of orthographic or multiview projection. This is the maximum number of principal views.

The directions of sight for the six views are shown in Fig. 6.3, where the views are drawn in their standard positions as illustrated. The width dimension is common to the top, front, and bottom views. Height is common to the right-side, front, left-side, and rear views.

Seldom will an object be so complex as to require six orthographic views, but if six views are needed, they should be arranged as shown in this example.

6.4
Sketching techniques

Sketching means freehand drawing without the use of instruments or straightedges. The best pencil grades for sketching are medium weight pencils, such as H, F, or HB grades. The standard lines used in multiview drawing and their respective line weights are shown in Fig. 6.4. Using the correct line weight improves the readability of a drawing.

You will be able to use the same grade of pencil for all lines when you are sketching by sharpening the pencil point to match the desired line width. The different point sizes are shown in Fig. 6.5. A line that is drawn freehand should have a freehand appearance; no attempt should be made to give the line the appearance of one drawn by instruments, since these two techniques of drawing are completely different.

Sketching technique can be improved by using a printed grid on sketching paper or by overlaying a printed grid with translucent tracing paper (Fig. 6.6) so the grid can be seen through the paper.

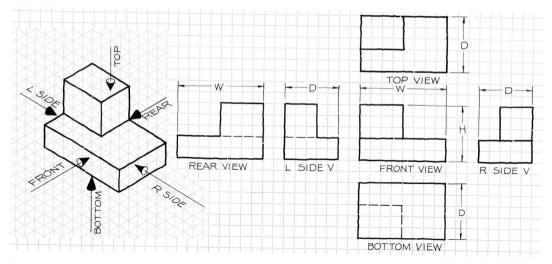

FIG. 6.3 Six principal views can be sketched by looking at the object in the directions indicated by the lines of sight. Note how the dimensions are placed on the views. Height (*H*) is shared by all four of the horizontally positioned views.

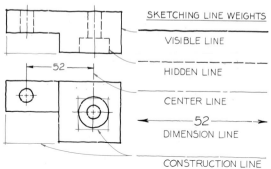

FIG. 6.4 These lines are examples of those that you should sketch with an F or an HB pencil when drawing views of an object. Note that some lines are thin and others are wider, but all are black except construction lines.

FIG. 6.5 The alphabet of lines that are sketched freehand and are all made with the same pencil grade (F or HB). The variation in the lines is achieved by varying the sharpness of the pencil point.

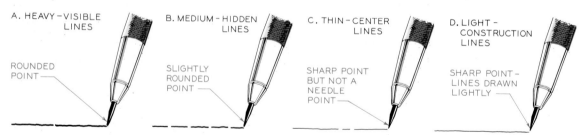

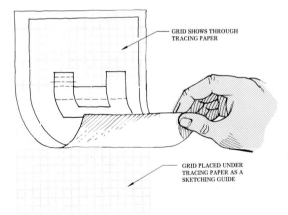

GRID SHOWS THROUGH
TRACING PAPER

GRID PLACED UNDER
TRACING PAPER AS A
SKETCHING GUIDE

FIG. 6.6 A grid can be placed under a sheet of tracing paper to aid you in freehand sketching. The grid can be used as guidelines for sketching.

A. VERTICAL STROKES

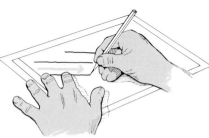

B. ANGULAR STROKES

C. HORIZONTAL STROKES

FIG. 6.7 Freehand sketching techniques

A. Freehand vertical lines should be sketched in a downward direction.

B. If you rotate your sheet slightly, angular strokes can be sketched left to right.

C. Horizontal strokes are made best in a left-to-right direction. Sketch from a comfortable position, and turn your paper if necessary.

FIG. 6.8 For good sketches, follow these examples of technique. Compare the good drawing (B) with the drawing made when these rules were not followed (A).

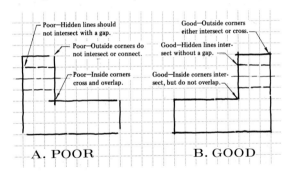

Poor—Hidden lines should not intersect with a gap.

Poor—Outside corners do not intersect or connect.

Poor—Inside corners cross and overlap.

Good—Outside corners either intersect or cross.

Good—Hidden lines intersect without a gap.

Good—Inside corners intersect, but do not overlap.

A. POOR   B. GOOD

When a freehand sketch is made, some lines will be vertical, others horizontal or angular. If you do not tape your drawing to the table top, you will be able to position the sheet for the most comfortable strokes, which are (for the right-handed drafter) from left to right (Fig. 6.7).

The lines that are sketched to form the various views should intersect, as indicated in Fig. 6.8, for the best effect.

6.5
Three-view sketch

The steps of drawing three orthographic views on a printed grid are shown in Fig. 6.9. The most commonly used combination of views is the front, top, and

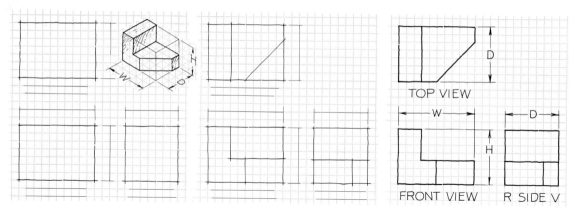

FIG. 6.9 Three-view sketching

Step 1 Block in the views by using the overall dimensions. Allow proper spacing for labeling and dimensioning the views.

Step 2 Remove the notches and project from view to view as shown.

Step 3 Check your layout for correctness; darken the lines and complete the labels and dimensions.

right-side views, as in this example. The overall dimensions of the object are sketched in Step 1. The slanted surface is drawn in the top view and projected to the other views. The final lines are darkened, the views are labeled, and the overall dimensions of height, width, and depth are applied to the views.

When surfaces are slanted, they will not appear true shape in the principal views of orthographic projection (Fig. 6.10). Surfaces that do not appear true size are either *foreshortened* or they appear as *edges*. In

Fig. 6.10C, two planes of the object are slanted; thus both appear foreshortened in the right-side view.

A good exercise for analyzing the given views is to find the missing view when two views are given (Fig. 6.11). In Step 1, the front view can be blocked in by projecting from the top and side views. In Step 2, the notch is located; and in Step 3, the final view is completed and darkened to match the other views.

The right-side view is found in Fig. 6.12 where the top and front views are given. The right-side view

FIG. 6.10 Views of planes

A. The plane appears as an edge in the front view, and it is foreshortened in the top and side views.

B. The plane is an edge in the top view, and it is foreshortened in the front and side views.

C. These planes appear foreshortened in the right-side view. Each appears as an edge in either the top or front views.

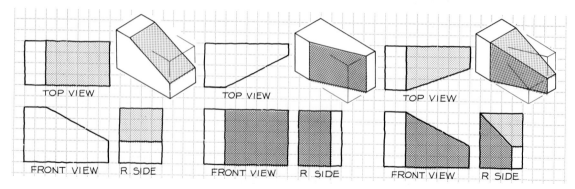

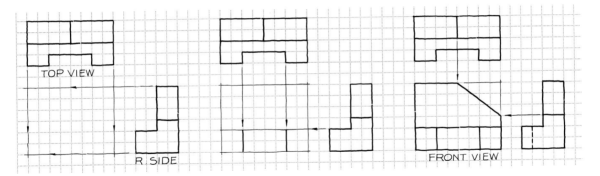

FIG. 6.11 Missing views

Step 1 When sketches of two views are given and the third is required, begin by projecting the overall dimensions from the top and right-side views.

Step 2 The various features of the object are sketched with the use of construction lines.

Step 3 The features are completed, the view is checked for correctness, and the lines are darkened to the proper line quality.

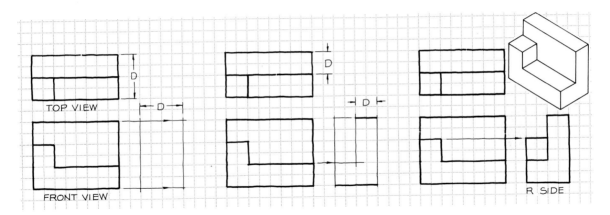

FIG. 6.12 Missing views

Step 1 To find the right-side view when the top and front views are given, block in the view with the overall dimensions.

Step 2 Develop the features of the view by analyzing the views together. Use light construction lines.

Step 3 Check the view for correctness and darken the lines to their proper line weight.

has the depth dimension in common with the top view, and height in common with the front view. Knowing this enables you to block in the side view in Step 1. The side view is developed in Step 2 and is completed in Step 3. A pictorial of the part is shown in Step 3 to help you visualize the object.

Another exercise is that of completing the views when some or all of them have missing lines (Fig. 6.13). A pictorial is provided to help you analyze the given views. Remember that depth is common to the top and side views, as shown in Step 2.

6.6
Circular features

The circle is often used in the design of a part. For example, the top view of a drilled hole will be a circle, which means that it is a cylinder.

The lines that are used with circles and cylinders to indicate that the features are true circles or cylinders are called *center lines*. Examples of these are shown in Fig. 6.14.

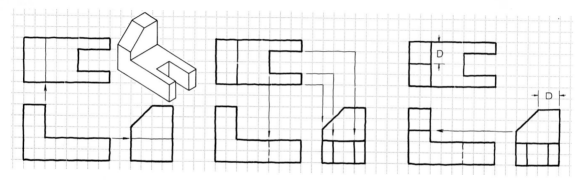

FIG. 6.13 Missing lines

Step 1 Lines may be missing in all views in this type of problem. The first missing line is found by projecting the edges of the planes from the front to the top and side views.

Step 2 The notch in the top view is projected to the front and side views. The line in the front view is a hidden line.

Step 3 The line formed by the beveled surface is found in the front view by projecting from the side view.

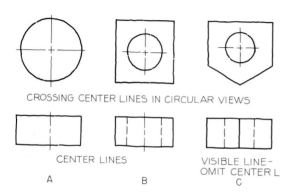

CROSSING CENTER LINES IN CIRCULAR VIEWS

CENTER LINES

A   B

VISIBLE LINE—
OMIT CENTER L
C

FIG. 6.14 Center lines are used to indicate the centers of circles and the axes of cylinders. These are drawn as very thin lines. When they coincide with visible or hidden lines, center lines are omitted.

FIG. 6.15 The center line should extend beyond the last arc that has the same center. When the arcs are not concentric, separate center lines should be drawn.

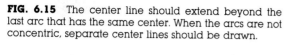

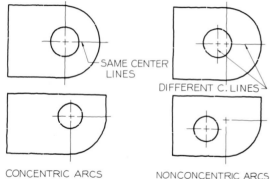

SAME CENTER LINES

DIFFERENT C. LINES

CONCENTRIC ARCS   NONCONCENTRIC ARCS

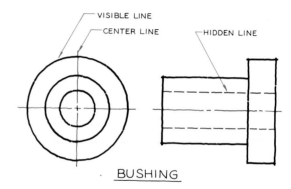

VISIBLE LINE

CENTER LINE

HIDDEN LINE

BUSHING

FIG. 6.16 Here you can see the application of center lines of concentric cylinders and the relative weight of hidden, visible, and center lines.

Center lines cross in the circular views to indicate the position of the center of the circle. Center lines are thin lines with short dashes spaced at intervals about every inch along the line. The short dashes should cross in the circular views. Refer to Fig. 6.4 for several examples of center lines applied to a drawing.

If a center line coincides with an object line—visible or hidden—the center line should be omitted, since the object lines are more important (Fig. 6.14C).

The application of center lines is shown in Fig. 6.15 where they indicate whether or not circles and arcs are concentric (share the same centers). The center line should extend beyond the arc by about one-eighth of an inch. The correct manner of applying center lines is shown in Fig. 6.16. The circular view clearly indicates that the cylinders are concentric, since each shares the same center lines.

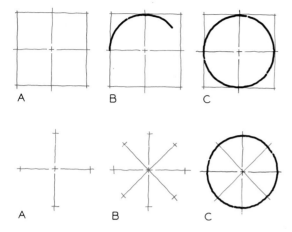

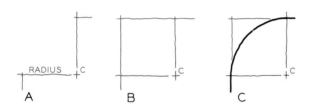

FIG. 6.17 Circles can be sketched by using either of the methods shown here. The use of guidelines is essential to freehand sketching of circles and arcs.

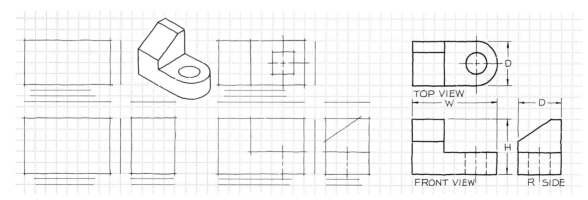

FIG. 6.18 Partial circles (arcs) should be drawn by using guidelines.

Sketching circles

Circles can be sketched by using light guidelines in conjunction with center lines (Fig. 6.17). The circles are drawn by using the guidelines and a series of short arcs. It is difficult to draw a freehand circle in one continuous line. Arcs of less than a full circle are drawn as shown in Fig. 6.18 by using light guidelines and center lines.

The construction of three orthographic views with circular features is shown in Fig. 6.19. Note that the circular features are located with center lines and light guidelines in Step 2; then they are sketched in and darkened in Step 3.

A similar example is given in Fig. 6.20, where the part is composed of circular features and arcs. The circles should be drawn first so that their corresponding rectangular views (such as the hidden hole in the top view) can be found by projecting from the circular view.

6.7
Isometric sketching

Another type of pictorial is the *isometric drawing*, which may be drawn on a specially printed grid. You will notice that the grid in Fig. 6.21 is composed of a

FIG. 6.19 Circular features in orthographic views

Step 1 To sketch orthographic views of the object shown in this pictorial, begin by blocking in the overall dimensions. Leave room for the labels and dimensions.

Step 2 Construct the center lines and the squares about the center lines in which the circles will be drawn. Show the slanted surface in the side view.

Step 3 Sketch the arcs and darken the final lines of the views. Label the views and show the dimensions of *W, D,* and *H.*

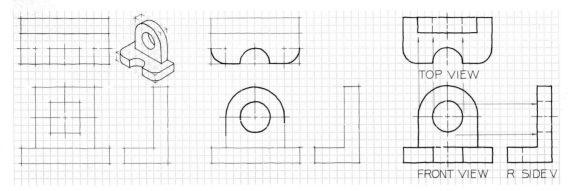

FIG. 6.20 Circular features in orthographic views

Step 1 When sketching orthographic views with circular features, you should begin by sketching the center lines and guidelines.

Step 2 Using the guidelines, sketch the circular features. These can be darkened as they are drawn if they will be final lines.

Step 3 The outlines of the circular features are found by projecting from the views found in Step 2. All the lines are darkened.

series of lines making 60° angles with each other to form the axes for drawing the pictorials.

The squares in the orthographic views can be laid off along the isometric grids as shown in Step 1. The notch is located in the same manner in Steps 2 and 3 to complete the isometric pictorial.

Isometric pictorials are helpful in communicating an idea in three dimensions rather than by a series of two-dimensional orthographic views.

Angles cannot be measured with a protractor in

isometric pictorials; they must be drawn by measuring coordinates along the three axes of the printed grid. In Fig. 6.22, the sloping surface is located by measuring in the direction of width and height to locate the ends of the angular slope.

When an object has two sloping planes that intersect (as shown in Fig. 6.23), it is necessary to draw the sloping planes one at a time to find point *B*. The line from *A* to *B* is the line of intersection between the two planes.

FIG. 6.21 Isometric pictorial sketching

Step 1 When orthographic views of a part are given, an isometric pictorial can be sketched on an isometric grid. Begin by constructing a box using the overall dimensions from the given views.

Step 2 The notch can be located by measuring over 5 squares as shown in the orthographic views. The notch is measured 4 squares downward by counting the units.

Step 3 The pictorial is completed and the lines are darkened. This is a three-dimensional pictorial, whereas the orthographic views are each two-dimensional views.

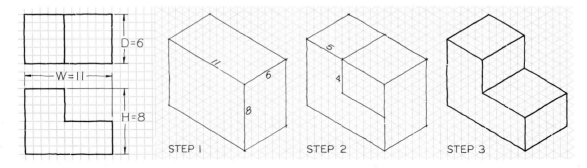

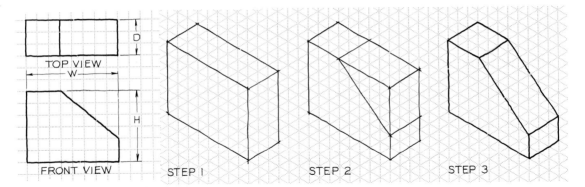

FIG. 6.22 Angles in isometric pictorials

Step 1 Begin by drawing a box using the overall dimensions given in the orthographic views. Count the squares and transfer them to the isometric grid.

Step 2 Angles cannot be measured with a protractor. Angles will be either larger or smaller than their true measurement. Find each end of the angle by measuring along the axes.

Step 3 The ends of the angles are connected to give the pictorial view of the object. Note that dimensions can be measured only in directions parallel to the three axes.

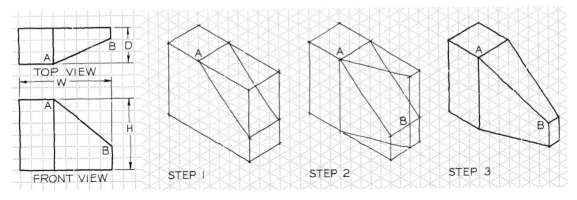

FIG. 6.23 Double angles in isometric pictorials

Step 1 When part of an object has a double angle, begin by constructing the overall box and finding one of the angles.

Step 2 Find the second angle that locates point B. Point A will connect to point B to give the intersection line.

Step 3 The final lines are darkened. Line AB is the line of intersection between the two sloping planes.

Circles in isometric pictorials

Circles will appear as ellipses in isometric pictorials. These can be sketched by locating their center lines as shown in Step 1 of Fig. 6.24. The center must be located equidistant from the top, bottom, and end of the front view (Step 1).

The circles are blocked in with guidelines, which will appear as rhombuses. The approximate elliptical views of the circular features will pass through the points where the center lines intersect the guidelines (Step 2). The rear of the object is found in the same manner to complete the pictorial in Step 3.

The steps of constructing elliptical views of circles by two methods are shown in Fig. 6.25. The first

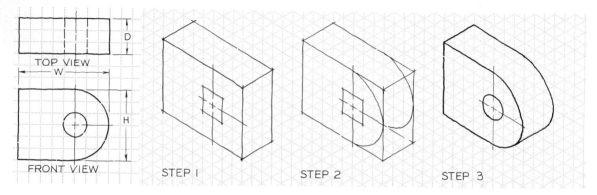

FIG. 6.24 Circles in isometric pictorials

Step 1 Begin by constructing a box using the overall orthographic dimensions, omitting arcs and circles. Draw the center lines and a rhombus of guidelines around the hole.

Step 2 Draw the pictorial views of the arcs tangent to the boxes formed by the guidelines. These arcs will appear as ellipses instead of circular arcs.

Step 3 Construct the small hole and darken the lines. Hidden lines are normally omitted in pictorial sketches.

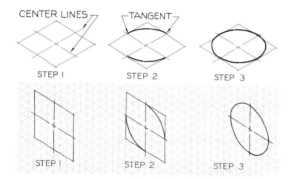

FIG. 6.25 Sketching ellipses

Two methods of sketching ellipses are shown: one without a grid, and one with a grid. When a grid is not used, the center lines are drawn at 30° to the horizontal (for a horizontal circle). A rhombus is drawn about them, and the ellipse is sketched inside it. When there is a grid, the same technique is used, except that the lines of the grid become the guidelines.

method is sketched without the use of guidelines, and the second method utilizes guidelines. The ellipses are sketched tangent to the sides of the rhombus formed by the guidelines or the grid.

Techniques of sketching circular and cylindrical shapes are shown in Fig. 6.26. When drawing cylinders, the outside elements are drawn parallel to the axis of the cylinder and tangent to the elliptical ends of the cylinder.

A part that is composed of a number of cylindrical forms is shown in Fig. 6.27a. These three views are used as the basis for an isometric pictorial shown in the steps of construction. When an isometric grid is not used, the axes of the isometric sketch are positioned 120° apart. In other words, the height dimension is vertical, and the width and depth dimensions make 30° angles with the horizontal direction on your paper.

FIG. 6.26 Sketching circular features

Several examples of sketching circles and cylinders are shown here. Note that in all cases, guidelines are used to aid in proportional sketching.

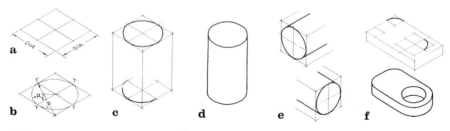

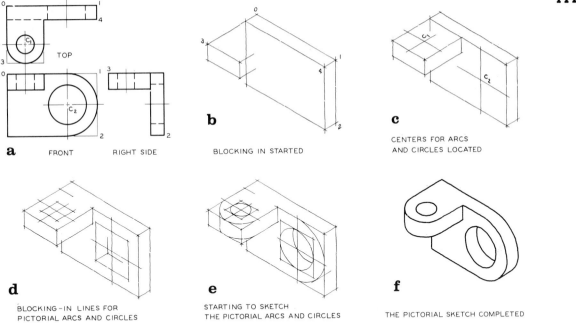

a FRONT RIGHT SIDE

b BLOCKING IN STARTED

c CENTERS FOR ARCS AND CIRCLES LOCATED

d BLOCKING-IN LINES FOR PICTORIAL ARCS AND CIRCLES

e STARTING TO SKETCH THE PICTORIAL ARCS AND CIRCLES

f THE PICTORIAL SKETCH COMPLETED

FIG. 6.27 Sketching an isometric pictorial

The steps of constructing an isometric pictorial of a part are shown here. Since a grid is not given, the guidelines are drawn vertical and at 30° to the horizontal. Center lines are used for locating the circular features at c. The object is developed at d and e and darkened at f.

Problems

These sketching problems should be drawn on Size A (8½″ × 11″) paper with or without a printed grid. A typical format for this size of sheet is shown in Fig. 6.28, where a one-quarter-inch grid is given. (This grid can be converted to an approximate metric grid by equating each square to five millimeters.) All sketches and lettering should be neatly executed by applying the principles covered in this chapter. Figures 6.29–6.31 contain the problems and instructions.

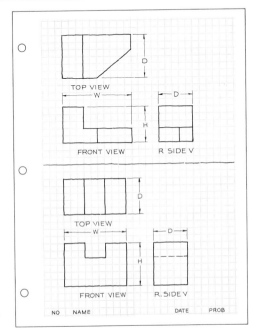

FIG. 6.28 The layout of a Size A sheet for sketching problems.

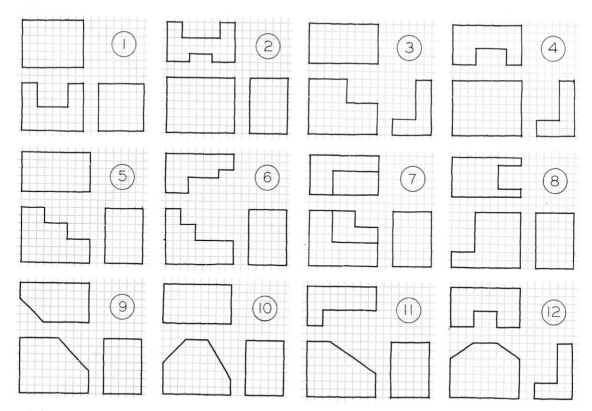

FIG. 6.29 On Size A paper, sketch top, front, and right-side views of the problems assigned. Two problems can be drawn on each sheet. Give the overall dimensions of *W, D,* and *H* and label each view.

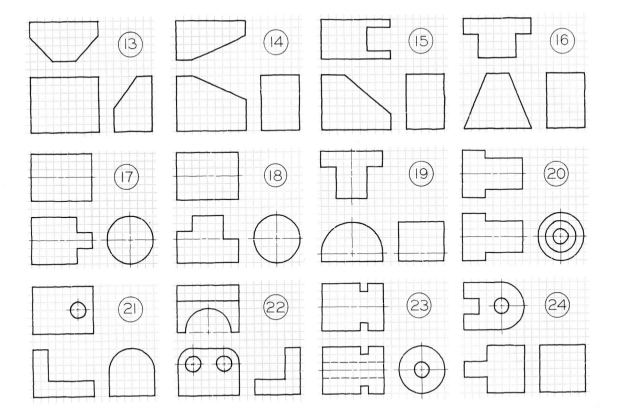

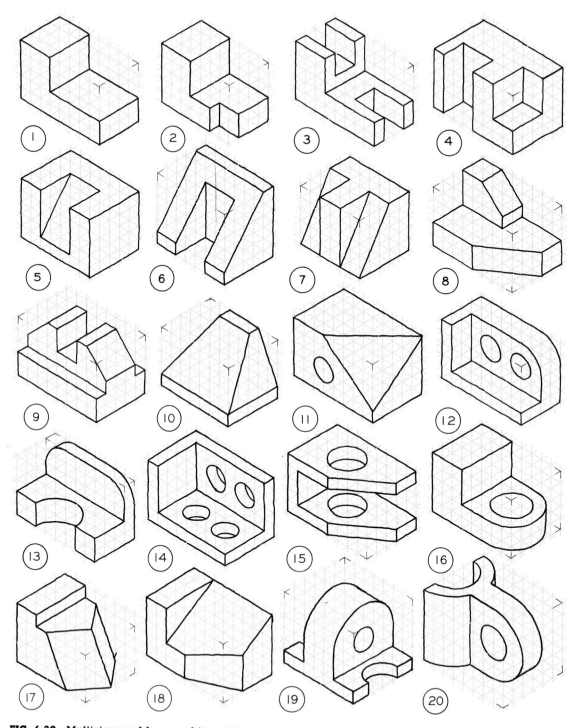

FIG. 6.30 Multiview problems and isometric sketching

On Size A paper, sketch the top, front, and right-side views of the problems assigned. Supply the lines that may be missing from all views. Then sketch isometric pictorials of the object assigned, two per sheet.

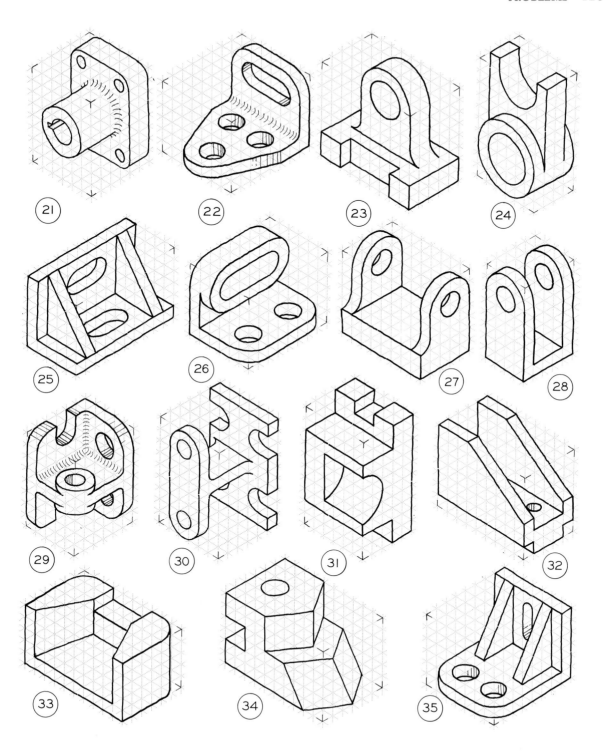

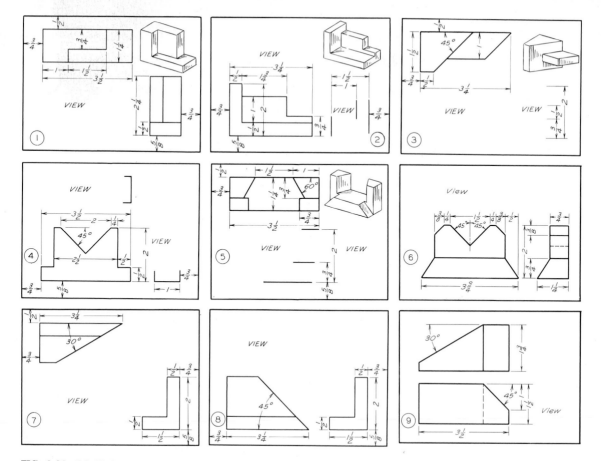

FIG. 6.31 Multiview problems and isometric pictorials

Sketch the problems assigned by your instructor, two per sheet, on Size A paper. Use the given dimensions to locate and draw the views. Draw the missing views. Then sketch isometric pictorials of the objects assigned, two per sheet, on Size A paper.

7

Multiview Drawing with Instruments

7.1
Introduction

Multiview drawing is the system of representing three-dimensional objects on a sheet of paper by separate views arranged in a standard manner that is familiar to members of the engineering and technological fields. These drawings are usually executed with the instruments and drafting aids that were discussed in Chapter 3; consequently these drawings are often referred to as *mechanical drawings*. When dimensions and notes have been added to complete the specifications of the parts that have been drawn, they are called *working drawings* or *detail drawings*.

This system of constructing multiview drawings is called *orthographic projection*. This method has evolved over the years into a system that, when the rules of the system are followed, is readily understood by the technical community. Orthographic drawing is truly the language of the engineer.

7.2
Orthographic projection

Whereas the artist is likely to draw pictorials to represent objects in an impressionistic manner, drafters must be precise and give more attention to detail in order for their drawings to be understood clearly. The method of preparing this type of drawing is *orthographic projection* or *multiview drawing*.

In this system, the views of an object are projected perpendicularly onto projection planes with parallel projectors. The process of finding one orthographic view, the front view, is illustrated in Fig. 7.1. The resulting front view is two dimensional since it has no depth and lies in a single plane described by two dimensions, width and height.

The top view of the same object is projected onto a horizontal projection plane that is perpendicular to the frontal projection plane in Fig. 7.2A. The right-side view is projected onto a vertical profile plane that is perpendicular to both the horizontal and frontal planes (Fig. 7.2B).

Imagine that the same object has been enclosed in a glass box composed of the frontal, horizontal, and profile projection planes. While in the glass box, the views are projected onto the projection planes (Fig. 7.3), and then the box is opened into the plane of the drawing surface. This gives the standard positions for the three orthographic views.

A similar example of the same principle is the object in the projection box in Fig. 7.4. Again, the three views are positioned in the same manner: the top view over the front, and the right-side view to the right of the front view.

> The three principal projection planes used in orthographic projection are referred to as the *horizontal (H)*, *frontal (F)*, and *profile (P)* planes.

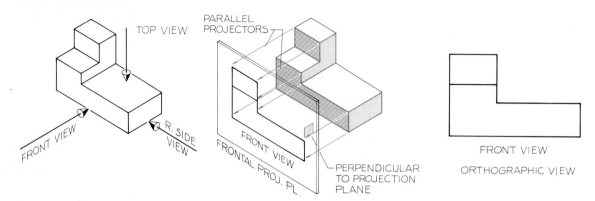

FIG. 7.1 Orthographic projection

Step 1 Three mutually perpendicular lines of sight are drawn to obtain three views of the object.

Step 2 The frontal plane is a vertical plane on which the front view is projected with parallel projectors that are perpendicular to the frontal plane.

Step 3 The resulting view is the front view of the object. This is a two-dimensional orthographic view.

Any view that is projected onto one of these principal planes is called a *principal view.* The three dimensions of an object that are used to show its three-dimensional form are *height, width,* and *depth.*

FIG. 7.2 The top view is projected onto a horizontal projection plane. The right-side view is projected onto a vertical profile plane that is perpendicular to the horizontal and frontal planes.

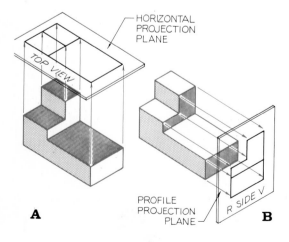

7.3
Alphabet of lines

The use of proper line weights in a drawing will greatly improve the drawing's readability and appearance. All lines should be drawn dark and dense as if drawn with ink; the lines should be varied only by their width. The only lines that are exceptions are guidelines and construction lines, which are drawn very lightly for lettering and laying out a drawing.

The lines of an orthographic view are labeled along with the suggested pencil grades for drawing them in Fig. 7.5. The lengths of dashes in hidden lines and centerlines are drawn longer as the size of a drawing increases. Additional specifications for these lines are given in Fig. 7.6.

7.4
Six-view drawings

A view that is projected onto a principal plane—the horizontal, frontal, or profile plane—is called a principal view. Again, if you visualize an object placed in-

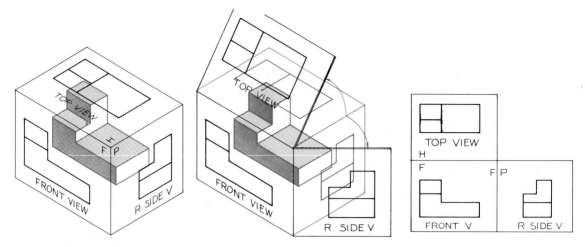

FIG. 7.3 Glass box theory

Step 1 Imagine that the object has been placed inside a box formed by the horizontal, frontal, and profile planes on which the top, front, and right-side views have been projected.

Step 2 The three projection planes are then opened into the plane of the drawing surface.

Step 3 The three views are positioned with the top view over the front view and the right-side view to the right. The planes are labeled H, F, and P at the fold lines.

FIG. 7.4 Principal projection planes

A. The three principal projection planes of orthographic projection can be thought of as planes of a glass box.

B. The views of an object are projected onto the projection planes that are opened into the plane of the drawing surface.

C. The outlines of the planes are omitted. The fold lines are drawn and labeled.

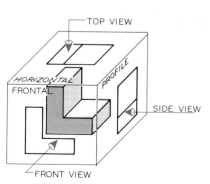
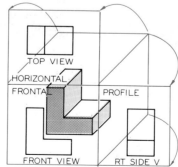
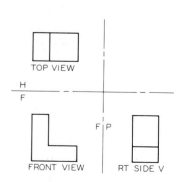

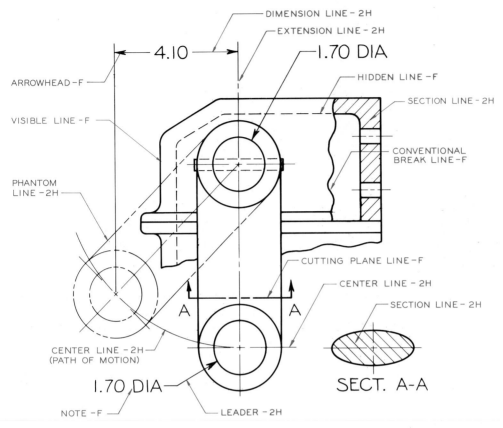

FIG. 7.5 The line weights and suggested pencil grades recommended for orthographic views.

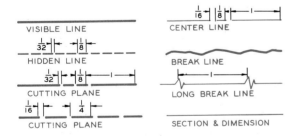

FIG. 7.6 A comparison of the line weights for orthographic views. These dimensions will vary for different sizes of drawings and should be approximated by eye.

FIG. 7.7 Six principal views of an object can be drawn in orthographic projection. You can imagine that the object is in a glass box with the views projected onto the six planes of projection.

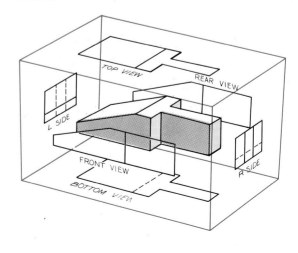

side a glass box, you will see that there are two horizontal planes, two frontal planes, and two profile planes (Fig. 7.7). Consequently the maximum number of principal views that can be used to represent an object is six.

The top and bottom views are projected onto horizontal planes, the front and rear views are projected

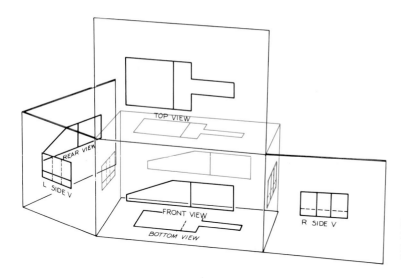

FIG. 7.8 The glass box is opened into the plane of the drawing surface, which locates the views in their standard positions.

onto frontal planes, and the right- and left-side views are projected onto profile planes.

To draw the views on a sheet of paper, the glass box is imagined to be opened up into the plane of the drawing paper (Fig. 7.8). When in a fully opened position, the views will appear as shown in Fig. 7.9.

Note the order and arrangement of the six views, and you will see the logic behind the system of

orthographic projection. The top view is placed over the front view, the bottom view under the front view, the right-side view to the right, the left-side view to the left, and the rear view to the left of the left-side view.

The three dimensions of an object that are necessary to give its size are *height, width,* and *depth.* The standard arrangement of the six views allows some of the views to share dimensions by projection. For ex-

FIG. 7.9 Once the box is completely opened into a single plane, the six views are arranged to describe the object. The outlines of the planes are usually omitted. They are shown here to assist you in relating this figure to the previous one.

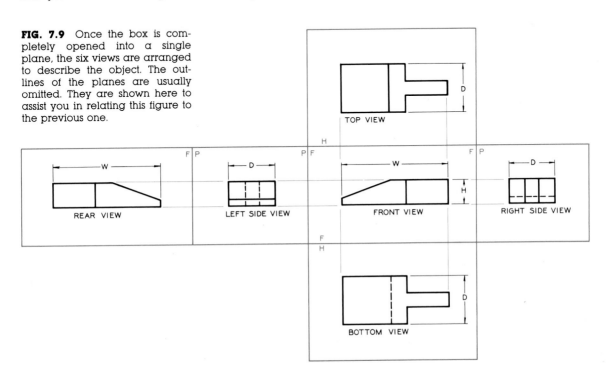

ample, the height dimension applies to the four views that are arranged horizontally, and this dimension is shown only once between the front and right-side views. The width dimension is placed between the top and front views, but it also applies to the bottom view, which is located under the front view.

Note that the projectors align the views both horizontally and vertically about the front view in Fig.

7.9. This is one of the reasons why this system of drawing is referred to as a system of projection.

Each side of the fold lines of the glass box is labeled with the letters *H, F,* or *P.* These letters identify the projection planes on a given side of the fold lines.

7.5
Three-view drawings

The most commonly used orthographic arrangement is the three-view drawing composed of the front, top, and right-side views. This is because three views are usually adequate to describe an object.

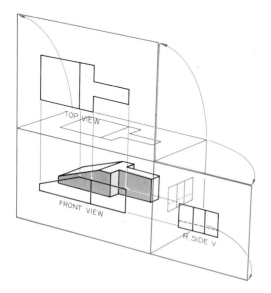

FIG. 7.10 Three-view drawings are commonly used for describing machine parts and small parts. The glass box is used to illustrate how the views are projected onto their projection planes.

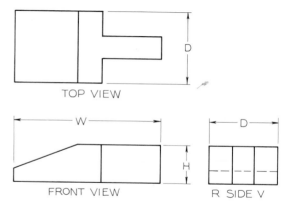

FIG. 7.11 The resulting three-view drawing of the object from the previous figure.

FIG. 7.12 Positioning orthographic views

A. This is a correct arrangement of views, labels, and dimensions. The views project from each other in proper alignment.

B. These views have been scrambled into unconventional positions, making it hard to interpret them. Dimensions have been unnecessarily repeated.

C. These views have been misaligned where they do not project from one to the other. This is an incorrect arrangement.

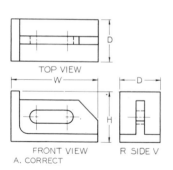

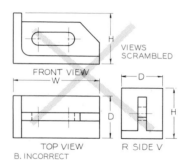

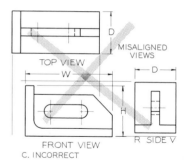

The same object used in the previous example is shown placed in a glass box in Fig. 7.10, which is opened into the plane of the drawing surface. The resulting three-view arrangement is shown in Fig. 7.11, where the views are labeled and dimensioned.

Since the views are understood to be projections from one view to the next, the dimensions are placed between the adjacent views and are not repeated. A single dimension will apply to all views that project in a single direction.

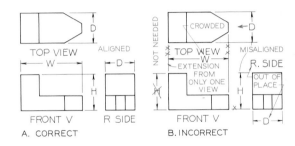

FIG. 7.13 **A**. These views are properly aligned, and the dimensions are correctly located. **B**. A number of errors are indicated in this incorrect arrangement.

7.6
Arrangement of views

The proper arrangement of orthographic views is essential for understanding and interpreting them. The standard positions for a three-view drawing consisting of the top, front, and right-side views are shown in Fig. 7.12A. The top and side views are projected directly from the front view. The views are properly labeled and dimensioned.

You can see the problems that you would cause by rearranging the views in a nonstandard sequence as shown in Fig. 7.12B. Similarly, views that do not project from view to view are improperly drawn, as in the example in Fig. 7.12C. These rules of arrangement are emphasized in Fig. 7.13.

7.7
Selection of views

When drawing an object by orthographic projection, you should select the views with the fewest hidden lines. That is why the right-side view is preferred over the left-side view in Fig. 7.14A.

Although the three-view arrangement of top, front, and right-side views is the most commonly used one, the arrangement of the top, front, and left-side views is equally acceptable (Fig. 7.14B) if this view has fewer hidden lines than the right-side view.

FIG. 7.14 Selection of views

A. In orthographic projection, you should select the sequence of views with the fewest hidden lines.

B. The left-side view has fewer hidden lines; therefore this view is selected over the right-side view.

C. When both views have an equal number of hidden lines, the right-side view is traditionally selected.

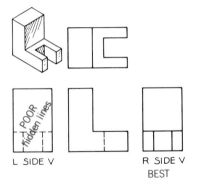

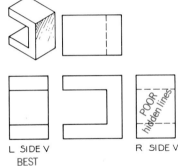

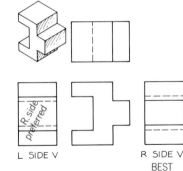

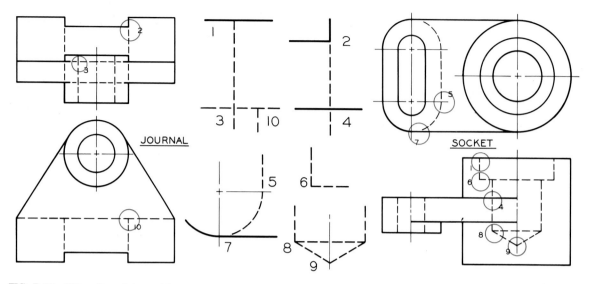

FIG. 7.15 When lines intersect in orthographic projection, they should intersect as shown here.

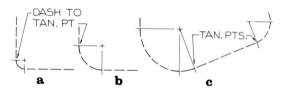

FIG. 7.16 Hidden lines in orthographic projection that are composed of curves should be drawn in this manner.

FIG. 7.17 The order of importance (precedence) of lines is: visible, hidden, and center lines, part A. The symbol made of the letters *C* and *L* in part B is used to label a center line when needed on symmetrical parts.

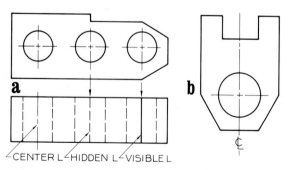

Some objects have standard views that are considered to be the front view, top view, and so forth. For example, a chair has front and top views that are recognized as such by everyone; therefore the accepted front view should be used as the orthographic front view.

7.8
Line techniques

As drawings become more complex, you will encounter more instances where lines overlap and intersect in a variety of ways similar to those shown in Fig. 7.15. This illustration shows the standard techniques of handling intersecting lines of most types.

The methods of constructing hidden lines composed of straight lines and curved segments is shown in Fig. 7.16.

You should become familiar with the order of precedence (priority) of lines (Fig. 7.17). The most important line that dominates all others is the visible object line. It will be shown regardless of any other type of line that lies behind it. The hidden object line is of next importance, and it takes precedence over the center line.

7.9
Point numbering

It will be helpful to you in constructing orthographic views if you become familiar with the method of numbering points and lines of an object. The rules of point numbering are introduced in Fig. 7.18.

An object is shown in Fig. 7.19 that has been numbered to aid in the construction of the missing front view when the top and side views are given. By projecting a selected point from the top and side views, the front view of the point can be found.

This method is recommended when you are having difficulty in interpreting given views.

7.10
Line and planes

An orthographic view of a line can appear true length, foreshortened, or as a point (Fig. 7.20). When a line appears true length, it must be parallel to the reference line in the previous view.

A plane in orthographic projection can appear true size, foreshortened, or as an edge (Fig. 7.20).

7.11
Alternate arrangement of views

Although the right-side view is usually placed to the right of the front view as shown in Fig. 7.21A, the side view can be projected from the top view as shown in part B. This arrangement is advisable when the object has a large depth dimension as compared with its height.

Different methods of positioning and arranging views are shown in Fig. 7.22. All these arrangements are acceptable.

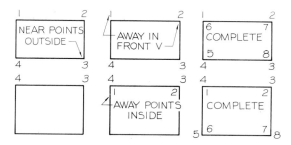

FIG. 7.18 When numbering points, the near points are labeled outside the view, and away points are labeled inside the view.

FIG. 7.19 Point numbering

Step 1 When a missing orthographic view is to be drawn, it is helpful to number the points in the given views.

Step 2 Points 1, 5, 6, and 7 are found by projecting from the given views of these points.

Step 3 The plotted points are connected to form the missing front view.

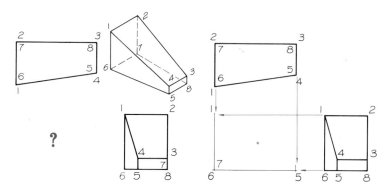
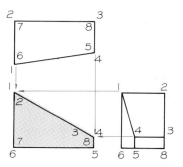

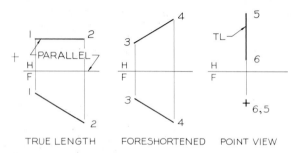

FIG. 7.20 Lines and planes

A line will project in orthographic projection as true length, foreshortened, or a point.

A plane in orthographic projection will appear as true size, foreshortened, or an edge.

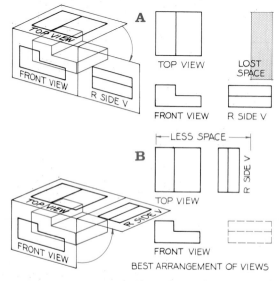

FIG. 7.21 The right-side view can be projected from the top view and positioned as shown in part B to save space when an object has a much larger depth than height.

7.12
Laying out three-view drawings

The depth dimension applies to both the top and the side views, but these views are usually positioned where this dimension does not project between them (Fig. 7.23). The depth dimension can be graphically transferred to the two views by using a 45° line, an arc, or a pair of dividers. It is preferable that you learn to use your dividers for this purpose.

These and similar methods of transferring dimensions from view to view are illustrated in Fig. 7.24.

The method of laying out a three-view drawing in a 10″ × 15″ space is shown in Fig. 7.25. The views are blocked in with light construction lines using the overall dimensions of the views. Center lines and tangent points are drawn next, and then the straight lines. Once the layout has been checked, the lines are darkened to their proper weight.

FIG. 7.22 Arrangement of views

A. This is the most standard arrangement of views for a three-view drawing.

B. The right-side view is projected off the top view in this arrangement.

C. This is an unconventional but acceptable arrangement of three views.

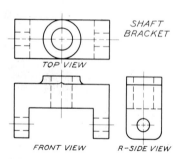

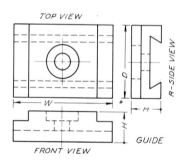

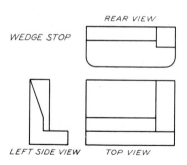

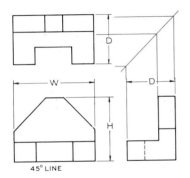
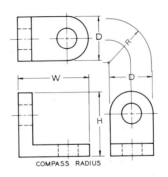
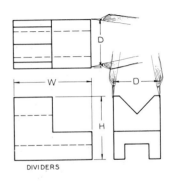

FIG. 7.23 Transferring depth dimensions

A. The depth dimension can be projected from the top view to the right-side view by constructing a 45° line positioned as shown.

B. The depth dimension can be projected from the top view to the side view by using a compass and a center point.

C. The depth dimension can be transferred from the top view to the side view by using dividers, which is the most desirable method.

FIG. 7.24 Several methods of transferring dimensions from the views are shown here.

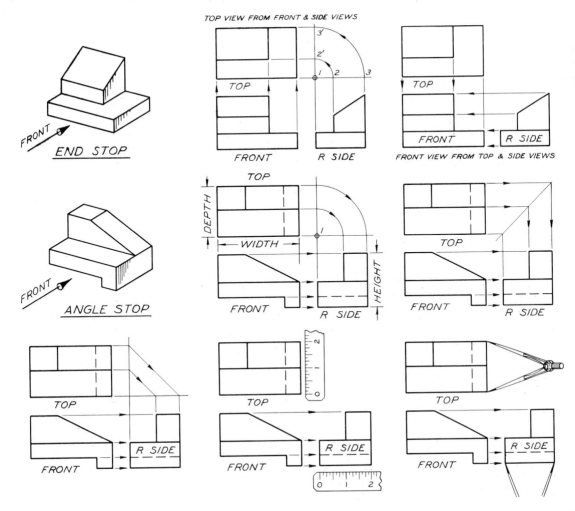

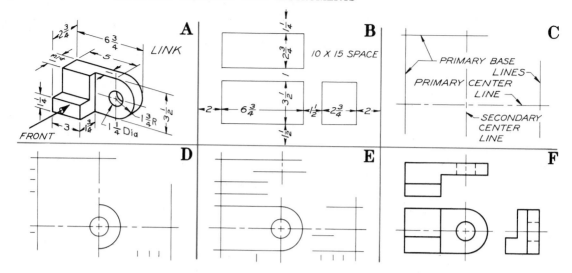

FIG. 7.25 Lay out a three-view drawing in this order: Center the views, draw the center lines, draw the arcs, draw the straight lines, and darken your layout.

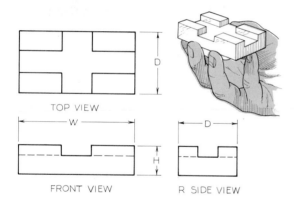

FIG. 7.26 A three-view drawing of an object.

FIG. 7.27 A three-view drawing of an object.

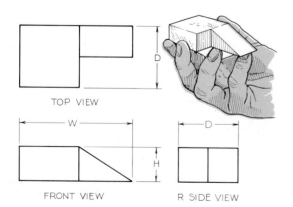

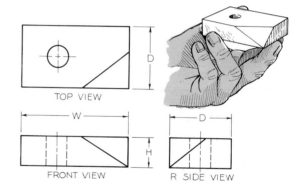

FIG. 7.28 A three-view drawing of an object.

When overall dimensions and labels are required, the three-view sketch should be positioned as shown in Figs. 7.25 through 7.28.

7.13

Two-view drawings

Some objects can be adequately explained in two views, and it is good economy of time and space to use only the views that are necessary to depict an object. Two parts that require only two views are shown in Fig. 7.29.

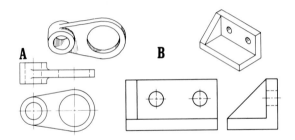

FIG. 7.29 These objects can be adequately described with two orthographic views.

Cylindrical parts need only two views, as shown in Fig. 7.30. It is preferable to select the views with the fewest hidden lines; consequently the right-side view is the best view in this case.

The layout of a two-view drawing is shown in Fig. 7.31 in a 10″ × 15″ space. First, the overall dimensions are used to block in and center the views. Second, the center lines are located and drawn by us-

ing light construction lines with a 2H–4H pencil. When the views have been completed, they are checked, and then the lines are darkened to complete the two-view drawing.

Views that involve arcs and tangent lines should be laid out by locating the centers and center lines of the arcs as the first step (Fig. 7.32). The tangent points

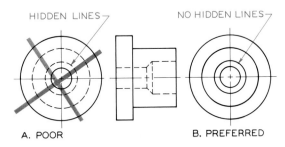

FIG. 7.30 Cylindrical objects can be depicted with two views. Always select views with the fewest hidden lines.

FIG. 7.31 Lay out two-view drawings in this order: Position the views, locate the center lines, block in the views, locate the centers, draw the arcs, draw the straight lines, and darken the lines of the finished views.

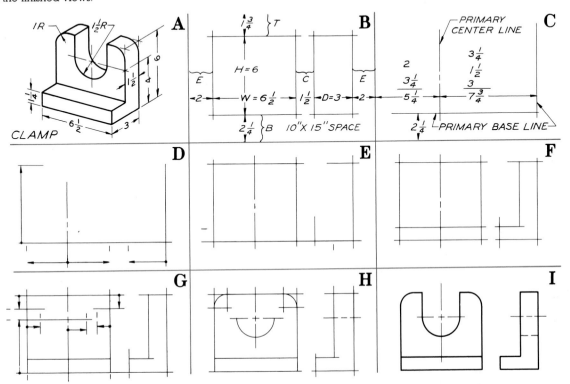

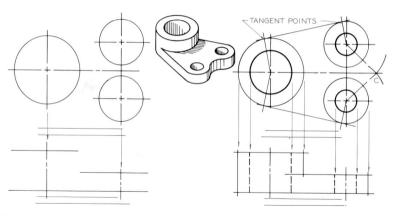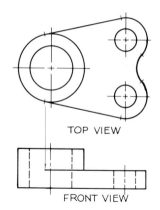

FIG. 7.32 Views with circular features

Step 1 Begin the layout by locating the centers and drawing the center lines and arcs. Draw these lines lightly.

Step 2 Draw the tangent lines and locate the points of tangency. Block in the front view.

Step 3 After checking the views for correctness, darken the lines to their proper widths. Draw the arcs first, being careful to stop them at the tangent points.

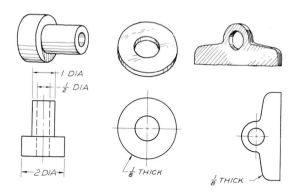

FIG. 7.33 Objects of this type can be described with only one orthographic view and supplementary notes.

FIG. 7.34 Unnecessary and confusing hidden lines have been omitted in the side views to improve their clarity.

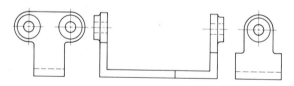

are found with the use of light construction lines. Begin by drawing the arcs first. Then check the completed views and darken the lines to their proper weights. It is important to draw the arcs so as to stop at the points of tangency. This is especially necessary when the drawing is being inked.

7.14
One-view drawings

Cylindrical parts and those with a uniform thickness can be described in a single view (Fig. 7.33).

In both cases, notes are used to explain the missing feature or dimension. The note DIA is placed after the diameter dimension for the cylindrical part. The thickness of the washer and the shim is specified by a note.

7.15
Incomplete and removed views

The right- and left-side views of the part in Fig. 7.34 would be very complex and hard to interpret if all hidden lines were shown as specified by the rules of or-

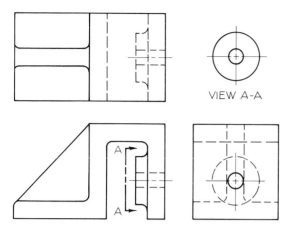

VIEW A-A

FIG. 7.35 A removed view, indicated by the directional arrows, can be used to draw views in hard-to-view locations.

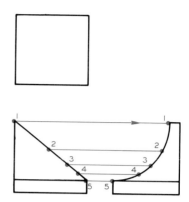

Step 1 Curves in orthographic projection can be plotted by locating and numbering the points in two views.

thographic projection. Therefore it is best to omit lines that confuse a clear understanding of the views. The two side views that are shown are easier to understand than the views drawn in their entirety.

In many cases, it is difficult to show a feature because of its location. Standard views can be confusing when lines overlap from other features. The view in Fig. 7.35 is more clearly shown when the view indicated by the lines of sight is removed to an isolated position.

7.16
Curve plotting

An irregular curve can be drawn by following the rules of orthographic projection. Such an example is shown in Fig. 7.36.

The process of plotting points is begun by locating a series of points along the curve in two given views. By following the rules of projection, these points are projected to the top view where each point is located, and the points are then connected by a smooth curve.

Figure 7.37 is a similar example where an ellipse is plotted in the top view by projecting from the front and side views.

You will find it helpful to number points that are to be plotted when curves are being located by projection.

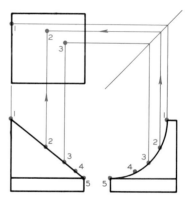

Step 2 Two views of each point are projected to the third view, where the projectors intersect.

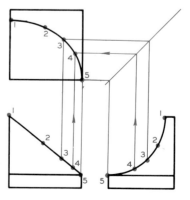

Step 3 All points are projected in this manner. The points are connected with an irregular curve.

FIG. 7.36 Plotting curved lines

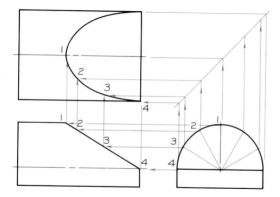

FIG. 7.37 The ellipse in the top view was found by numbering points in the front and side views and then projecting them to the top view.

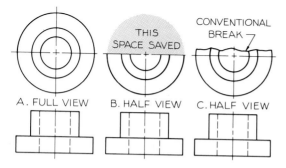

FIG. 7.38 To save space and drawing time, the top view of a cylindrical part can be drawn as a partial view by using either of these methods.

Fig 7.39 Revolving holes

A. A true projection of equally spaced holes gives a misleading impression that the center hole passes through the center of the plate.

B. A conventional view is used to show the true radial distances of the holes from the center by revolution. The third hole is omitted.

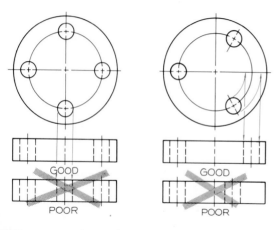

7.17
Partial views

A partial view can be used to save time and space when the parts are symmetrical or cylindrical. By omitting the rear of the circular top view in Fig. 7.38, space can be saved without sacrificing clarity.

A partial break may be used to make it more apparent that a portion of the view has been omitted. Either method is correct and acceptable.

7.18
Conventional revolutions

The readability of an orthographic view may be improved in some cases if the rules of projection are violated.

Established violations of rules that are customarily made for the sake of clarity are called *conventional practices*.

When holes are symmetrically spaced in a circular arrangement as shown in Fig. 7.39, it is conventional practice to show them at their true radial distance from the center of their circle of centers. This requires an imagined revolution of the holes in the top view.

This same principle of revolution applies to symmetrically positioned features such as the three lugs on the outside of the part in Fig. 7.40. The conventional view is better than the true orthographic projection.

FIG. 7.40 Symmetrically positioned external features, such as these lugs, are revolved to their true-size positions for the best conventional views.

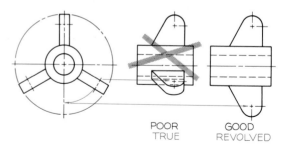

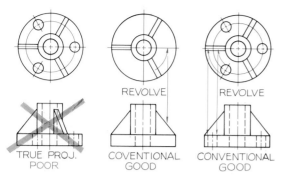

REVOLVE   REVOLVE

TRUE PROJ. POOR   COVENTIONAL GOOD   CONVENTIONAL GOOD

FIG. 7.41 The conventional methods of revolving holes and ribs in combination for improved clarity.

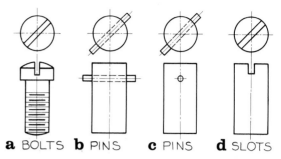

a BOLTS   **b** PINS   **c** PINS   **d** SLOTS

FIG. 7.44 Parts of this type are drawn at 45° angles in the top views, but the front views are drawn to show the details as revolved views.

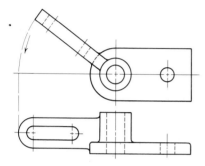

FIG. 7.42 It is conventional practice to imagine that features of this type have been revolved in order for them to appear true shape in the front view.

FIG. 7.43 The arm in the front view is imagined to be revolved so its true length can be drawn in the top view. This is an accepted conventional practice.

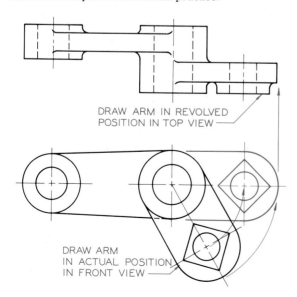

DRAW ARM IN REVOLVED POSITION IN TOP VIEW

DRAW ARM IN ACTUAL POSITION IN FRONT VIEW

The conventional and desired method of drawing holes and ribs in combination in the same view is shown in Fig. 7.41.

Another conventional practice is illustrated in Fig. 7.42, where an inclined feature is revolved to a frontal position in the top view so it can be drawn as true size in the front view. The revolution of the part is not drawn, since it is an imagined revolution. A similar object with a revolved feature is shown in Fig. 7.43.

Other parts whose views are improved by revolution are those shown in Fig. 7.44. Unless there is a reason for them to be positioned otherwise, it is desirable to show the top view features at 45° so they will not coincide with the center lines. The front views are drawn by imagining that the features have been revolved.

A closely related type of conventional view is the developed view where a bent piece of material is drawn as if it were flattened out (Fig. 7.45).

FIG. 7.45 Objects that have been shaped by bending thin stock can be shown as developed views as if the views have been flattened out.

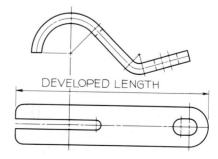

DEVELOPED LENGTH

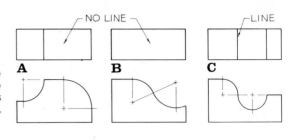

FIG. 7.46 Object lines are drawn only where there are sharp intersections or where arcs are tangent at their center lines, as in part C.

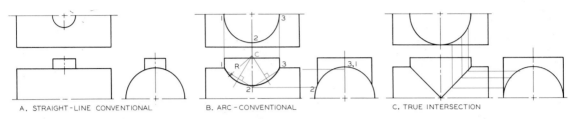

A. STRAIGHT-LINE CONVENTIONAL   B. ARC-CONVENTIONAL   C. TRUE INTERSECTION

FIG. 7.47 The conventional methods of showing intersections between cylinders. These are approximations except for part C.

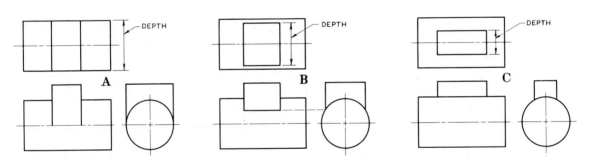

FIG. 7.48 Conventional intersections between cylinders and rectangular shapes. The intersection at C is an approximate intersection.

FIG. 7.49 Conventional intersections between cylinders and holes piercing them.

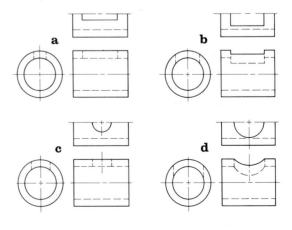

7.19

Intersections

In orthographic projection, the intersection between planes results in a line that describes the object. In Fig. 7.46, examples of views are shown where lines may or may not be required.

The standard types of intersections between cylinders are shown in Fig. 7.47. Those at A and B are conventional intersections, which means that they are approximations, and they are drawn in this manner for ease of construction. The example at C is a true intersection, where cylinders of equal diameters intersect. Similar intersections are shown in Fig. 7.48.

FIG. 7.50 The edges of this Collet Index Fixture are rounded to form fillets and rounds. Note that the surface of the casting is rough except where it has been machined. (Courtesy of Hardinge Brothers Inc.)

The types of intersections that are formed by holes in cylinders are shown in Fig. 7.49. These are conventional intersections that are sufficient for depicting these features.

7.20
Fillets and rounds

Fillets and *rounds* are rounded corners that are used on castings such as the body of the Collet Index Fixture shown in Fig. 7.50. A fillet is an inside rounding, and a round is an external rounding. The radii of fillets and rounds may be many sizes, but they are usually about ¼″ in radius. They are used on castings for added strength and improved appearance.

The orthographic views of a part with fillets and rounds must be drawn in such a manner that these detailed features can be seen (Fig. 7.51).

A casting will have square corners only when its surfaces have been finished, which is the process of machining away a portion of the surface to a smooth finish (Fig. 7.51B).

FIG. 7.51 Fillets and rounds

A. When a surface has been finished by machining, rounds are removed and the corners are squared. A finish mark is a V that is placed on the edge view of the finished surfaces.

B. A fillet is a rounded inside corner. The rounds are removed when the outside surfaces are finished. The fillets can be seen only in the front view.

C. The views of an object with fillets and rounds must be drawn in such a way as to call attention to them.

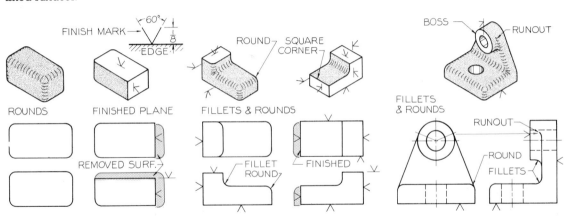

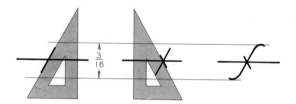

FIG. 7.52 An alternate type of finish mark is the traditional **f** that is applied to the edge views of the finished surface, whether hidden or finished.

> The finished surface is indicated on a view by placing a *finish mark* (V) on the outside edge views of the surface that is to be finished in all views, whether the edges are *visible* or *hidden*.

You can see in Fig. 7.51 how fillets and rounds are shown, as well as the square corners without rounded corners due to finishing. Note that a *boss* is a raised cylindrical feature that is thickened to receive a shaft or to be threaded.

An alternate finish mark that can be used is the one shown in Fig. 7.52.

The techniques of showing fillets and rounds on orthographic views are given in Fig. 7.53.

The curve formed by a fillet at a point of tangency where features of an object intersect is called a *runout*.

A comparison of intersections and runouts of parts with and without fillets and rounds is shown in Fig. 7.54. Large runouts are constructed as an eighth of a circle with a compass, as illustrated in Fig. 7.55. When the runouts are small, they can be drawn with a circle template.

When properly drawn, the runouts on orthographic views will tell much about the details of an object (Fig. 7.56). For example, the runouts in the top views tell us that the ribs at A have rounded corners, and the rib at B is completely round.

Methods of showing intersections of other types are shown in Fig. 7.57.

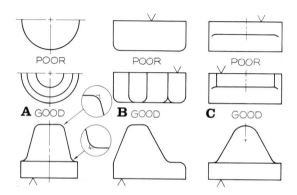

FIG. 7.53 Examples of conventionally drawn fillets and rounds.

FIG. 7.54 Intersections between features of objects. Curves formed by filleted intersections are called runouts.

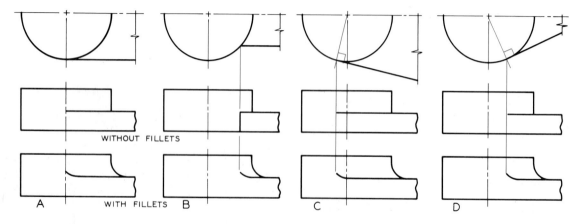

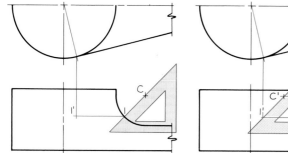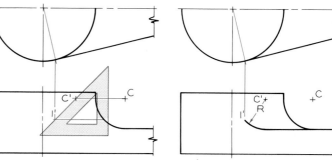

FIG. 7.55 Plotting runouts

Step 1 Find the point of tangency in the top view and project it to the front view. A 45° triangle is used to find point 1, which is projected to locate point 1'.

Step 2 A 45° triangle is used to locate point C', which is on the horizontal projector from the center of the fillet, C.

Step 3 The radius of the fillet is used to draw the runout with C' as the center. The runout arc is equal to one-eighth of a circle.

7.21
Left-hand and right-hand views

Two parts are often required that are very similar to each other, but one part is actually a "mirror image" of the other. Two parts of this type are shown in Fig. 7.58. Your first impression may be that the parts are interchangeable, but the parts are actually as different as a pair of shoes.

The drafter can reduce drawing time by drawing views of only one of the parts and labeling these views as shown in Fig. 7.58B. A note can be added to indicate that the other matching part has the same dimensions.

FIG. 7.56 Runouts are shown for differently shaped ribs. One has fillets, and the one at B is a rib with rounded edges.

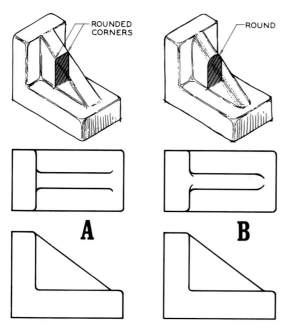

ROUNDED CORNERS

ROUND

A

B

7.22
First-angle projection

The problems of this chapter have been presented as third-angle projections where the top view is placed over the front view, and the right-side view to the right of the front view. This method is used extensively in the United States, Britain, and Canada. Most of the rest of the industrial world uses the *first-angle* of projection.

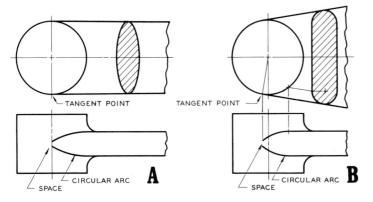

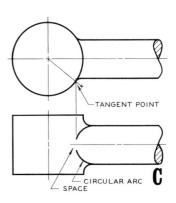

FIG. 7.57 Conventional runouts involving parts of different cross sections.

The first-angle system is illustrated in Fig. 7.59, where an object is placed above the horizontal plane and in front of the frontal plane. When these projection planes are opened onto the surface of the drawing paper, the front view is projected over the top view, and the left-side view is placed to the right of the front view.

FIG. 7.58 Left-hand and right-hand parts

A. Some parts are required to be similar except that one is a left-hand part and the other is a right-hand part.

B. One of the parts can be drawn and labeled, the right-hand one in this case. The other part need not be drawn, but merely indicated by a note.

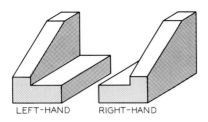

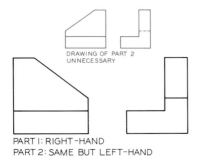

FIG. 7.59 First-angle projection

A. The first angle of projection is used by many of the countries that use the metric system. You imagine that the object is placed above the horizontal and in front of the frontal plane.

B. The views are drawn in this location, which is different from the third angle of projection that is used in the United States.

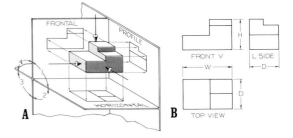

FIG. 7.60 The angle of projection that is used to prepare a set of drawings is indicated by this truncated cone. It is placed in or near the title block of a drawing.

It is important that the angle of projection be indicated on a drawing to aid in the interpretation of the views. This is done by placing a truncated cone in or near the title block (Fig. 7.60). When metric units of measurements are used, the cone and the symbol SI are placed together on the drawing.

Problems

The following problems are to be drawn as orthographic views on Size A or Size B paper as assigned by your instructor. (Refer to Section 3.19 for instructions on laying out a sheet.)

1–11 (Figs. 7.61–7.71) Draw the given views using the dimensions provided, and then construct the missing view—either the top, front, or right-side view. Use Size A sheets and draw one or two problems per sheet as assigned by your instructor.

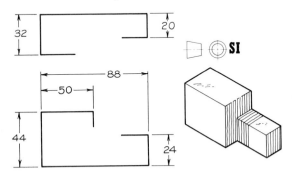

FIG. 7.61 Problem 1: Guide block.

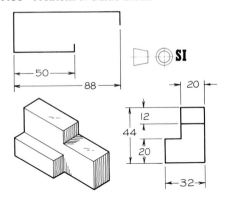

FIG. 7.62 Problem 2: Double step.

FIG. 7.63 Problem 3: Adjustable stop.

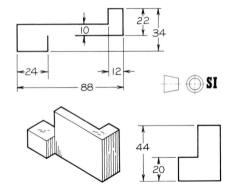

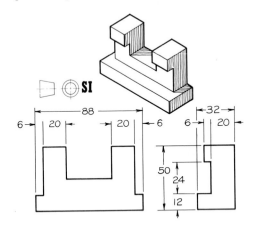

FIG. 7.64 Problem 4: Lock catch.

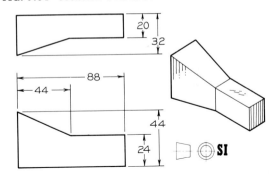

FIG. 7.65 Problem 5: Two-way adjuster.

FIG. 7.66 Problem 6: Vee block.

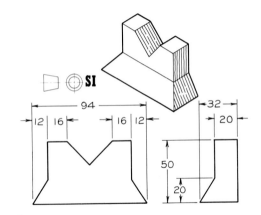

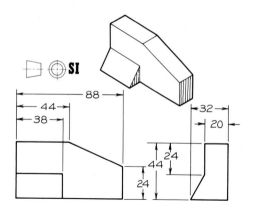

FIG. 7.67 Problem 7: Filler.

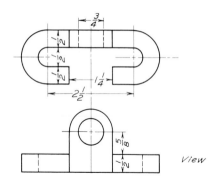

FIG. 7.68 Problem 8: Slide stop.

FIG. 7.69 Problem 9: Shaft support.

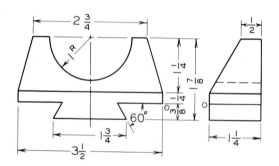

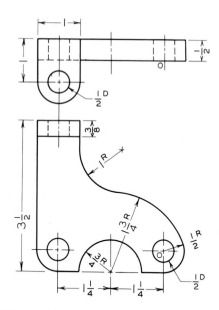

FIG. 7.70 Problem 10: Support brace.

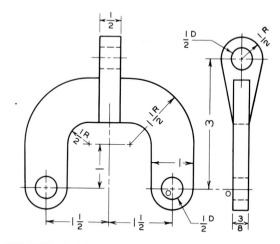

FIG. 7.71 Problem 11: Yoke.

12–52 Construct the necessary orthographic views to describe the objects from Fig. 7.72 through Fig. 7.112. Draw these on Size A or Size B sheets. Label the views and show the overall dimensions of *W, D,* and *H* on the appropriate views.

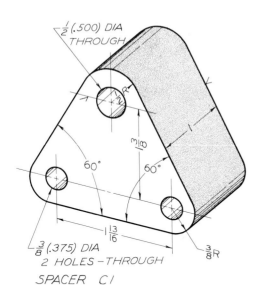

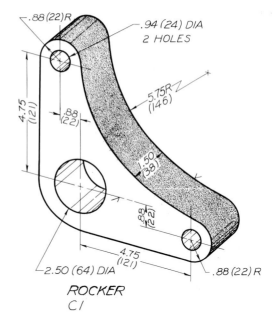

$\frac{1}{2}$ (.500) DIA
THROUGH

.88 (22)R

.94 (24) DIA
2 HOLES

5.75R
(146)

4.75
(121)

.88
(22)

1.50
(38)

60°   60°

1 $\frac{13}{16}$

.88
(22)

4.75
(121)

$\frac{3}{8}$ (.375) DIA
2 HOLES – THROUGH

$\frac{3}{8}$R

2.50 (64) DIA

.88 (22) R

SPACER C1

ROCKER
C1

FIG. 7.72 Problem 12: Spacer.

FIG. 7.74 Problem 14: Rocker.

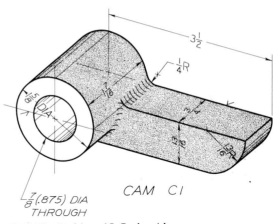

CAM C1

FIG. 7.75 Problem 15: Rod guide.

$\frac{7}{8}$ (.875) DIA
THROUGH

$3\frac{1}{2}$

$\frac{1}{4}$R

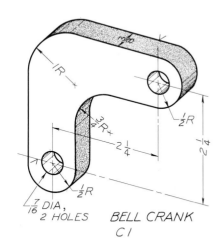

1R

$\frac{3}{4}$R

$2\frac{1}{4}$

$\frac{1}{2}$R

$2\frac{1}{4}$

$\frac{1}{2}$R

$\frac{7}{16}$ DIA,
2 HOLES

BELL CRANK
C1

FIG. 7.73 Problem 13: Bell crank.

FIG. 7.76 Problem 16: Cam.

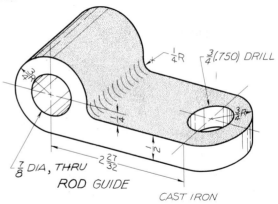

$\frac{1}{4}$R

$\frac{3}{4}$ (.750) DRILL

$\frac{3}{4}$R

$\frac{7}{8}$ DIA, THRU

$2\frac{27}{32}$

ROD GUIDE

CAST IRON

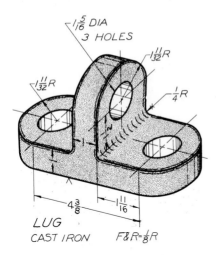

LUG
CAST IRON   F&R=⅛R

FIG. 7.77 Problem 17: Lug.

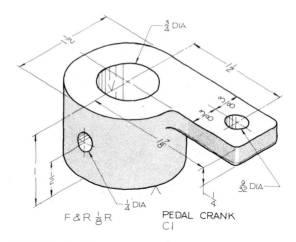

F&R ⅛R   PEDAL CRANK
C1

FIG. 7.80 Problem 20: Pedal crank.

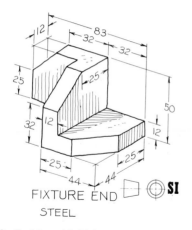

FIXTURE END   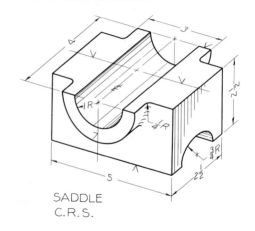 **SI**
STEEL

FIG. 7.78 Problem 18: Fixture end.

FIG. 7.79 Problem 19: Adjusting slide.

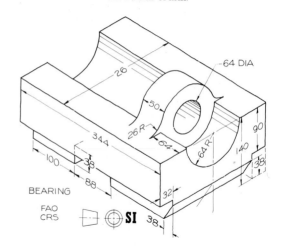

BEARING
FAO
CRS   **SI**

FIG. 7.81 Problem 21: Bearing.

FIG. 7.82 Problem 22: Saddle.

ADJUSTING SLIDE
CAST IRON   **SI**

SADDLE
C.R.S.

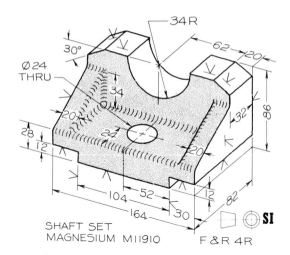

FIG. 7.83 Problem 23: Shaft set.

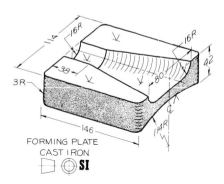

FIG. 7.86 Problem 26: Forming plate.

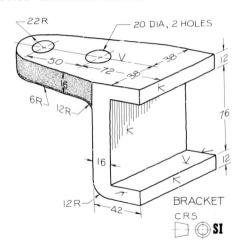

FIG. 7.84 Problem 24: Bracket.

FIG. 7.85 Problem 25: Lifting block.

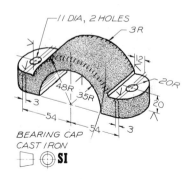

FIG. 7.87 Problem 27: Bearing cap.

FIG. 7.88 Problem 28: Stop plate.

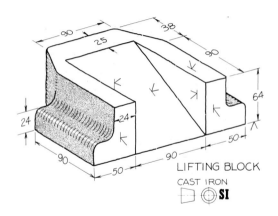

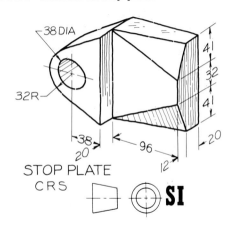

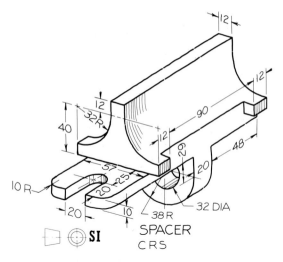

FIG. 7.89 Problem 29: Spacer.

SPACER
C R S

10 R

12
32R
40
12
90
12
12
29
48
20
57
20 25
20
20
10
38 R
32 DIA

SI

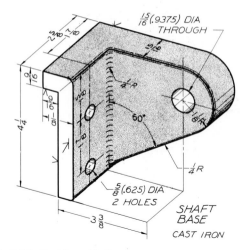

FIG. 7.92 Problem 32: Shaft base.

$\frac{15}{16}$ (.9375) DIA
THROUGH

SHAFT
BASE
CAST IRON

$\frac{5}{8}$ (.625) DIA
2 HOLES

$4\frac{1}{4}$

$3\frac{3}{8}$

$\frac{9}{16}$
$\frac{1}{8}$
$\frac{1}{8}$
60°
$\frac{1}{2}$R
$\frac{1}{4}$R

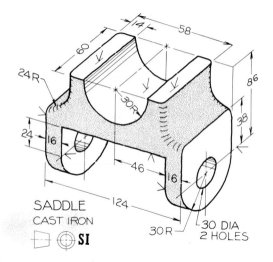

SADDLE
CAST IRON

SI

24R
24
16
60
14
58
30R
86
38
46
16
124
30 R
30 DIA
2 HOLES

FIG. 7.90 Problem 30: Saddle.
FIG. 7.91 Problem 31: Linkage arm.

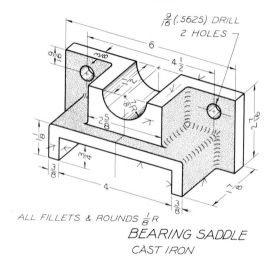

FIG. 7.93 Problem 33: Bearing saddle.

$\frac{9}{16}$ (.5625) DRILL
2 HOLES

6
$4\frac{1}{2}$
$2\frac{1}{16}$
$\frac{9}{16}$
$2\frac{5}{8}$
$\frac{1}{8}$
$\frac{3}{4}$
$\frac{3}{8}$
4
$\frac{3}{8}$
$\frac{7}{8}$

ALL FILLETS & ROUNDS $\frac{1}{8}$R

BEARING SADDLE
CAST IRON

FIG. 7.94 Problem 34: Clamp.

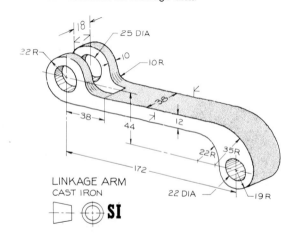

18
25 DIA
22 R
10
10 R
38
38
44
12
172
22R
35R
22 DIA
19 R

LINKAGE ARM
CAST IRON

SI

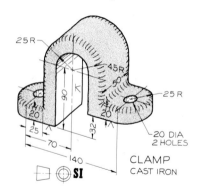

25R
45R
60
90
25
70
20
32
20
140
25R
20 DIA
2 HOLES

CLAMP
CAST IRON

SI

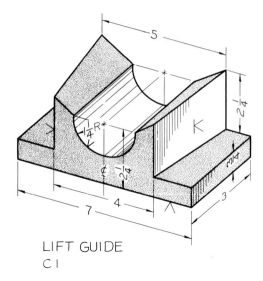

LIFT GUIDE
C1

FIG. 7.95 Problem 35: Lift guide.

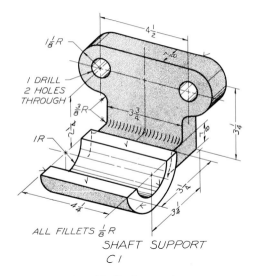

SHAFT SUPPORT
C1

ALL FILLETS $\frac{1}{8}$ R

FIG. 7.98 Problem 38: Shaft support.

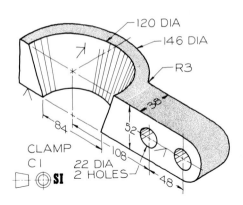

CLAMP
C1

FIG. 7.96 Problem 36: Clamp.
FIG. 7.97 Problem 37: Clamp jaw.

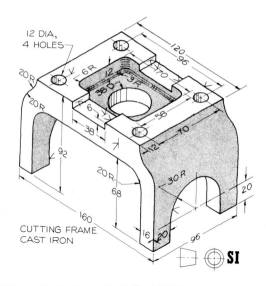

CUTTING FRAME
CAST IRON

FIG. 7.99 Problem 39: Cutting frame.

FIG. 7.100 Problem 40: Link.

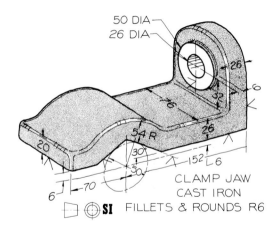

CLAMP JAW
CAST IRON
FILLETS & ROUNDS R6

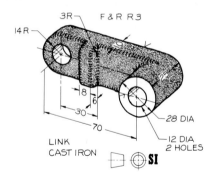

LINK
CAST IRON

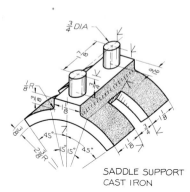

SADDLE SUPPORT
CAST IRON

FIG. 7.101   Problem 41: Saddle support.

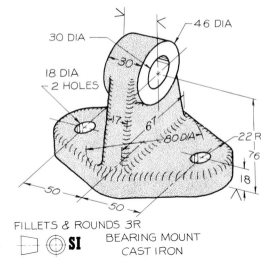

FILLETS & ROUNDS 3R

BEARING MOUNT
CAST IRON

FIG. 7.104   Problem 44: Bearing mount.

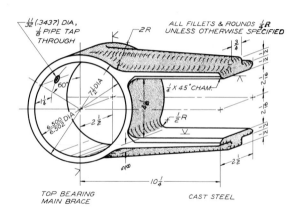

TOP BEARING
MAIN BRACE   CAST STEEL

FIG. 7.102   Problem 42: Top bearing.

FIG. 7.103   Problem 43: Hold down.

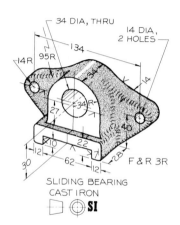

SLIDING BEARING
CAST IRON

FIG. 7.105   Problem 45: Sliding bearing.

FIG. 7.106   Problem 46: Rocker arm.

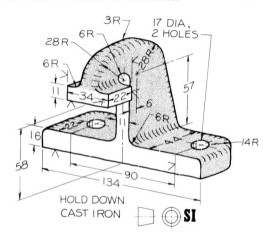

HOLD DOWN
CAST IRON   SI

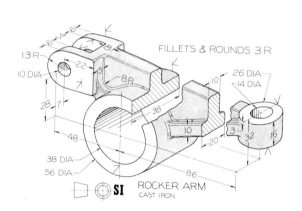

FILLETS & ROUNDS 3 R

SI   ROCKER ARM
CAST IRON

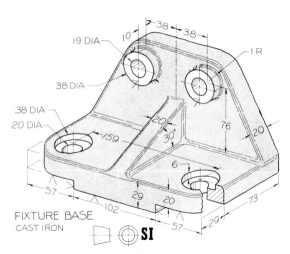

FIG. 7.107 Problem 47: Fixture base.

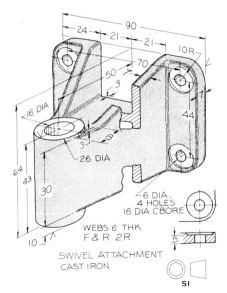

FIG. 7.110 Problem 50: Swivel attachment.

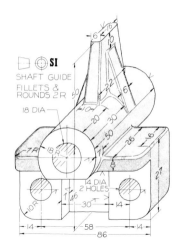

FIG. 7.108 Problem 48: Shaft guide.

FIG. 7.109 Problem 49: Yoke.

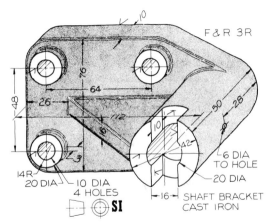

FIG. 7.111 Problem 51: Shaft bracket.

FIG. 7.112 Problem 52: Pivot base.

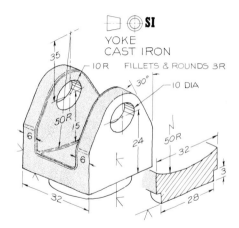

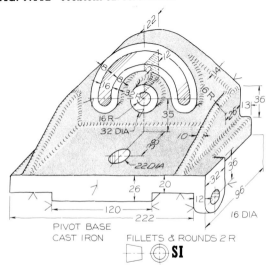

8

Auxiliary
Views

8.1
Introduction

The auxiliary view is a type of orthographic view, but it is not a *principal* view as previously covered. The principal views, such as the top, front, and side views, are usually drawn so that their surfaces are parallel to the principal projection planes. In these views, the surfaces will appear true size.

However, if the surfaces of an object are not mutually perpendicular, all planes cannot be parallel to the projection planes. This means that a nonprincipal

plane, an *auxiliary plane,* must be used on which the auxiliary view can be projected.

Such an example is shown in Fig. 8.1. The inclined plane is not true shape but is foreshortened in the top view. The edge view of the surface is not horizontal in the front view; therefore it cannot appear true shape in the top view. This surface is true shape in the auxiliary view that is projected from the front view.

This chapter will cover the auxiliary view and its usage in the preparation of engineering drawings.

8.2
Folding-line approach

The three principal orthographic planes are the *frontal, horizontal,* and *profile* planes. These are the names of the planes of an imaginary glass box in which an object is placed for drawing its various views.

> A primary auxiliary plane is a plane that is perpendicular to one of the principal planes but is oblique to the other two planes. In other words, a *primary auxiliary* view is an auxiliary view that is projected from a *primary orthographic view.*

FIG. 8.1 When a surface appears as an inclined edge in a principal view, it can be found true size by an auxiliary view. The top view at (a) is foreshortened, but this plane is true size in an auxiliary view at (b).

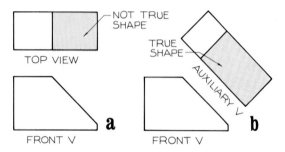

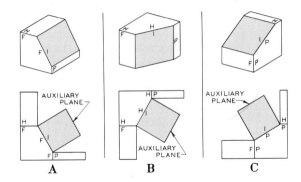

FIG. 8.2 A primary auxiliary plane can be folded from the frontal, horizontal, or profile planes. The fold lines are labeled F-1, H-1, and P-1, respectively.

Auxiliary planes can be thought of as planes that fold from principal planes as shown in Fig. 8.2. The plane at A folds from the frontal plane to make a 90° angle with it. The fold line between the two planes is labeled F-1. The F is an abbreviation for frontal, and the numeral 1 represents *first*, or *primary*, auxiliary plane.

The examples at B and C illustrate the positions for auxiliary planes that fold from the horizontal (H-1) and profile planes (P-1).

8.3
Auxiliaries projected from the top view

The inclined plane in Fig. 8.3 is an edge in the top view, and it is perpendicular to the top projection plane, the horizontal plane. An auxiliary plane can be drawn that is parallel to the inclined surface of the object, and the view projected onto it will be a true-size view of the inclined plane.

> It is necessary that a surface appear as an edge in a principal view before it can be found as true size in a primary auxiliary view.

Fold line H-1 is drawn parallel to the edge view of the inclined plane in Step 1. The line of sight is drawn perpendicular to the edge view. Each corner of the inclined plane is projected perpendicularly to the auxiliary plane and is located by using the dimension of height (*H*) from the side or front view.

A similar example is the object shown in the glass box in Fig. 8.4. Since this object has an inclined surface that appears as an edge in the top view, it can be

FIG. 8.3 Auxiliary from the top—folding-line method

Given An object with an inclined surface shown pictorially.

Required Find the inclined surface true size by an auxiliary view.

Step 1 Construct a line of sight perpendicular to the edge view of the inclined surface. Draw the H-1 fold line parallel to the edge, and draw the H-F fold line between the top and front views.

Step 2 Project the four corners of the inclined edge parallel to the line of sight. Locate the corners by measuring perpendicularly from the horizontal plane with height (*H*) dimensions.

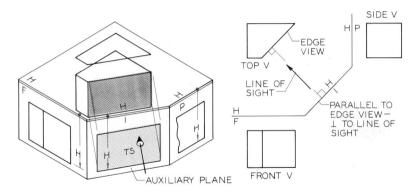
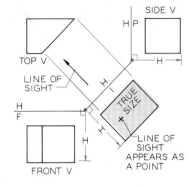

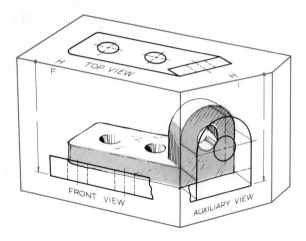

FIG. 8.4 A pictorial showing the relationship of the projection planes used to find the true-size view of the inclined surface.

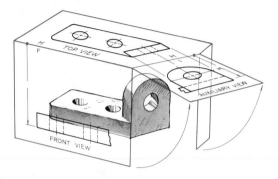

FIG. 8.5 The auxiliary plane is opened into the plane of the top view by revolving it about the H-1 fold line. This is how the views would be positioned on your drawing surface.

FIG. 8.6 When the object is drawn on a sheet of paper, it would be laid out in this manner. The front view is drawn as a partial view since the omitted part is shown true size in the auxiliary view.

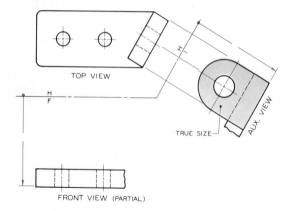

found true size in a primary auxiliary view. The fold line between the horizontal and auxiliary planes is labeled H-1. The height dimension *(H)* is transferred from the front view to the auxiliary view, since both views are located from the same horizontal plane.

The auxiliary plane is rotated about the H-1 fold line into the plane of the top view in Fig. 8.5.

When drawn on a sheet of paper, the drawing of this part would appear as shown in Fig. 8.6. The front view is shown as a partial view because the omitted portion would have been hard to draw, and it would not have been true size. The auxiliary view shows the true-size view of the inclined surface.

8.4

Constructing an auxiliary from the top view—
folding-line method

The steps of constructing an auxiliary view that is projected from the top view are shown in Fig. 8.7.

This inclined surface can be found true size in a primary auxiliary view since it appears as an edge in the top view. The line of sight is drawn perpendicular to the edge view, and the fold line is drawn parallel to the edge. The dimension that must be transferred from the front view is height *(H)*. Height is perpendicular to the horizontal plane; consequently it cannot be found in the top view. It must be taken from an adjacent view, such as the front or side view.

The fold line (H-1) is drawn as a thin but black line. It is also helpful to number or letter the points on the views. The projectors are construction lines, and they should be drawn with a hard pencil (3H–4H) just dark enough to be seen.

8.5

Auxiliaries from the top view—reference-line method

A similar method of locating an auxiliary view is a method that uses a *reference plane* instead of the folding-line technique. An example of this type of auxiliary view is shown in Fig. 8.8, where an auxiliary view is projected from the top view.

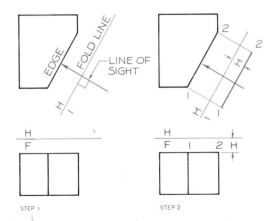

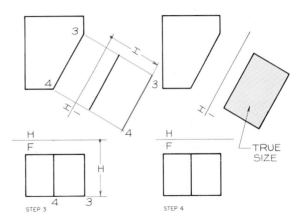

FIG. 8.7 Construction of an auxiliary view

Step 1 The line of sight is drawn perpendicular to the edge view of the inclined surface. The H-1 fold line is drawn parallel to the edge view of the inclined surface. An H-F fold line is drawn between the given views.

Step 2 Points 1 and 2 are found by transferring the height (H) dimensions from the front view to the auxiliary view to locate line 1–2.

Step 3 Points 3 and 4 are found in the same manner by using the dimensions of height (H) to locate line 3–4.

Step 4 The corner points are connected to complete the true-size view of the inclined plane.

Instead of placing a fold line between the top and front views, a reference plane is passed through the bottom of the front view, shown pictorially in Fig. 8.8a. The height dimensions (H) are measured upward from the reference plane instead of downward from a fold line.

In Fig. 8.8b, the reference plane is shown as a horizontal edge in the front view, where it is labeled HRP (horizontal reference plane). This plane will appear as an edge that is parallel to the edge view of the inclined surface from which the auxiliary view is projected. The auxiliary view will lie between the HRP and the top view.

When using the folding-line method, the fold line H-1 is located between the top view and the auxiliary view.

FIG. 8.8 A horizontal reference plane can be used instead of the folding-line technique to construct an auxiliary view. Instead of placing the reference plane between the top and auxiliary views, it is placed outside the auxiliary view (b).

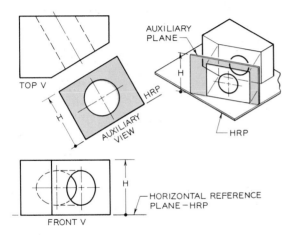

8.6
Auxiliaries from the front view—folding-line method

A plane that appears as an edge in the front view (Fig. 8.9) can be found true size in a primary auxiliary view projected from the front view. Fold line F-1 is drawn parallel to the edge view of the inclined plane in the front view at a convenient location.

The line of sight is drawn perpendicular to the edge view of the inclined plane in the front view. The frontal plane appears as an edge in the auxiliary view; consequently the measurements that are perpendicu-

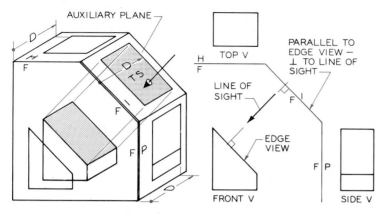

FIG. 8.9 Auxiliary from the front—folding-line method

Given An object with an inclined surface shown pictorially.

Required Find the inclined surface true size by an auxiliary view.

Step 1 Draw a line of sight perpendicular to the edge view of the inclined surface. Draw the F-1 fold line parallel to the edge, and draw the H-F and/or F-P fold lines between the given views.

Step 2 Project the corners of the edge view parallel to the line of sight. Locate the corners by measuring perpendicularly from the frontal plane with depth (D) dimensions to find the true-size view.

lar to the frontal plane will be seen true length. These dimensions are depth *(D)* dimensions. Depth dimensions are transferred from the top view to the auxiliary view by using dividers.

The inclined surface will be true size in Step 2. Note that the *line of sight* will appear as a *point* in this view.

FIG. 8.10 A pictorial showing the relationship of the projection planes used to find the true-size view of the inclined surface of the object.

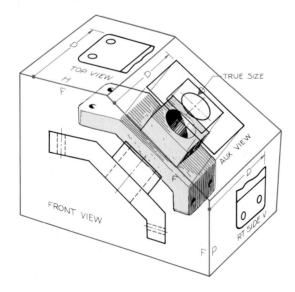

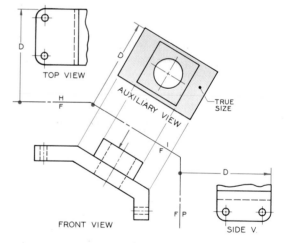

FIG. 8.11 The layout and construction of an auxiliary view of the object in Fig. 8.10 as it would appear on your drawing paper.

A practical application of this type of auxiliary view is the part shown in Fig. 8.10. It is enclosed in a glass box where an auxiliary plane is constructed parallel to the inclined plane of the part.

When this arrangement is drawn on a sheet of paper, it will appear as shown in Fig. 8.11. The top and side views are drawn as partial views, since the auxiliary view eliminates a need for more than is shown of them. The auxiliary view is located by using

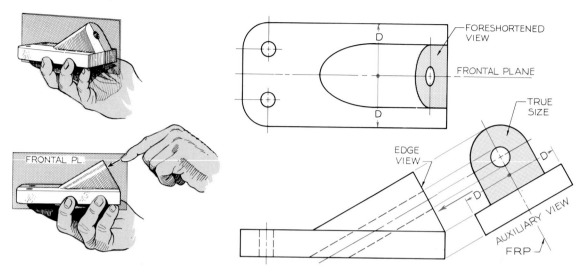

FIG. 8.12 Since the inclined surface of this part is symmetrical, it would be advantageous to use a frontal reference plane (FRP) that is passed through the object. The auxiliary view is projected perpendicularly from the edge view of the inclined plane in the front view. The FRP appears as an edge in the auxiliary view, and depth (D) dimensions are made on each side of it to locate points on the true-size view of the inclined surface.

the depth dimension measured from the edge view of the frontal projection plane. The auxiliary view shows the surface's true size.

8.7
Auxiliaries from the front
view—reference-plane method

The object in Fig. 8.12 has an inclined surface that appears as an edge in the front view. Consequently this plane can be found true size in a primary auxiliary view.

Since it is symmetrical, there is an advantage to using a reference plane that passes through the center of the symmetrical top view. The reference plane is a frontal plane; consequently this will be called a *frontal reference plane* (FRP) in the auxiliary view.

The FRP is located parallel to the edge view of the inclined plane. In the auxiliary view, the reference plane will pass through the center of the view (Fig. 8.13). The auxiliary fold line, P-1, is drawn parallel to the edge view of the inclined surface.

Whereas the folding line would be placed between the top and auxiliary views, the FRP is located through the center of the auxiliary view.

8.8
Auxiliaries from the profile
view—folding-line method

Since the inclined surface in Fig. 8.13 appears as an edge in the profile plane, it can be found true size in a primary auxiliary view projected from the profile view. The auxiliary fold line, P-1, is drawn parallel to the edge view of the inclined surface.

A line of sight that is perpendicular to the auxiliary plane will see the profile plane as an edge. Consequently dimensions of width *(W)* will appear true length in the auxiliary view. Width dimensions are transferred from the front view to the auxiliary view.

8.9
Auxiliaries from the profile—
reference-plane method

The object in Fig. 8.14 has two inclined surfaces that appear as edges in the right-side view, the profile view. These are found true size by using a *profile ref-*

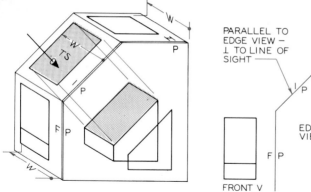

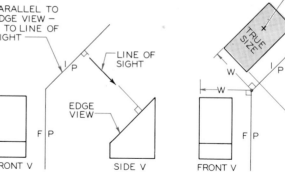

FIG. 8.13 Auxiliary from the side view—folding-line method

Given An object with an inclined surface shown pictorially.

Required Find the inclined surface true size by an auxiliary view.

Step 1 Draw a line of sight perpendicular to the edge view of the inclined surface. Draw the P-1 fold line parallel to the edge view, and draw the F-P fold line between the given views.

Step 2 Project the corners of the edge parallel to the line of sight. Locate the corners by measuring perpendicularly from the P-1 fold line with width (*W*) dimensions.

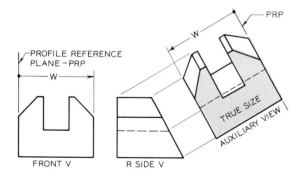

FIG. 8.14 An auxiliary view is projected from the right-side view by using a profile reference plane (PRP). The auxiliary view shows the true-size view of the inclined surfaces.

FIG. 8.15 This auxiliary view requires that a series of points be plotted. Since the object is symmetrical, the reference plane is positioned through its center.

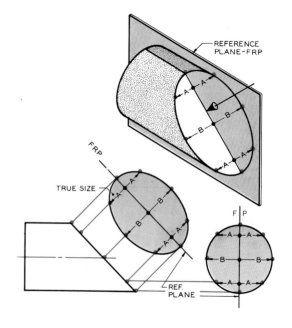

erence plane (PRP) that is a vertical edge in the front view.

The two auxiliary views are drawn to find the inclined surfaces' true size. The profile reference planes are positioned at the far outsides of the auxiliary views instead of between the profile and auxiliary views as in the folding-line method. The views are found by transferring the width *(W)* dimensions from the edge view of the PRP in the front view to the auxiliary view.

8.10
Auxiliaries of curved shapes

When an auxiliary view is drawn to show a curve that is not a true arc, a series of points must be plotted by using the principles previously covered. The cylinder in Fig. 8.15 has a beveled surface that appears as an edge in the front view. When found true size, this surface will be elliptical in shape.

Since the cylinder is symmetrical, it is beneficial to use an FRP through the object so that dimensions can be measured on both sides of it. Points are located about the circular right-side view, and these are projected to the edge view of the surface in the front view.

The FRP is located parallel to the edge view of the plane, and the points are projected perpendicularly from the edge view of the plane. Dimensions *A* and *B* are shown as examples for plotting points in the auxiliary view. More points are needed than are shown to construct a smooth curve.

An irregular curve appears as an edge in the front view of Fig. 8.16. Points are located on the curve in the top view and are projected to the front view. These points are found in the auxiliary view by locating each point using the depth dimension *(D)* transferred from the top view.

FIG. 8.16 The auxiliary view of this surface required that a series of points be located in the given view and then projected to the auxiliary view. The curve is made by drawing an irregular curve through the plotted points.

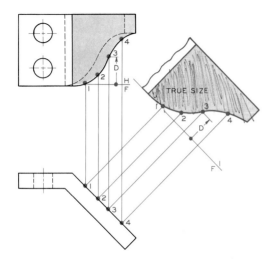

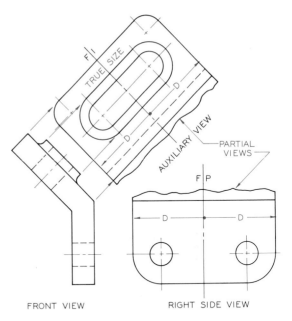

FRONT VIEW   RIGHT SIDE VIEW

FIG. 8.17 This auxiliary view was drawn as a partial view since it is supplemented by the true-size partial view shown in the right-side view. Note that the reference planes are labeled the same as a folding line, which is permissible.

8.11
Partial views

Since an auxiliary view is a supplementary view, some views of an orthographic arrangement can be drawn as partial views. Omitted features in one view will be shown in another view, perhaps in the auxiliary view.

The object in Fig. 8.17 is shown by a complete front view and partial auxiliary and side views. These views are easier to draw, and they are more functional without sacrificing clarity.

A similar example is given in Fig. 8.18, where the front view is a complete view and the other two views are partial views. A frontal reference plane is passed through the center of the side view. Note that even though it is a reference plane, it is labeled F-1, which is a permissible practice. The guide bracket that was drawn in Fig. 8.18 is shown in Fig. 8.19.

You can see by comparison of the views in Fig. 8.20 that the partial views at A are correctly drawn but are hard to read. The arrangement at B is much better, since the views have been shown as partial views supplemented by an auxiliary view. Note that a cross section is shown in part B to describe the part.

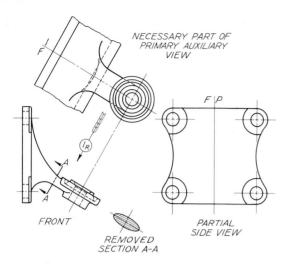

NECESSARY PART OF
PRIMARY AUXILIARY
VIEW

FRONT

REMOVED
SECTION A-A

$F \mid P$

PARTIAL
SIDE VIEW

FIG. 8.18 This object is represented by a series of partial views. The hub, which would appear elliptical, is not shown in the side view at all.

FIG. 8.19 A photograph of the guide bracket that was drawn orthographically in the previous figure.

FIG. 8.20 The three view at A are completely drawn orthographic views with all hidden lines shown. The better method of showing this part is the method at B where an auxiliary view is used to supplement the partial primary views. Hidden lines have been omitted to make the views easier to interpret.

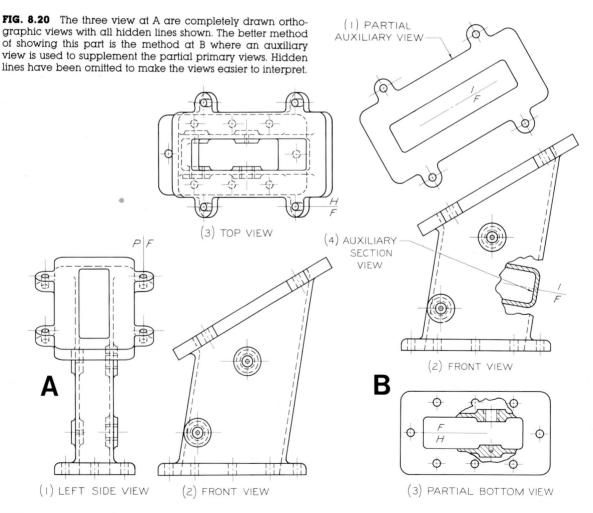

(1) PARTIAL
AUXILIARY VIEW

(3) TOP VIEW

(4) AUXILIARY
SECTION
VIEW

(2) FRONT VIEW

A

(1) LEFT SIDE VIEW

(2) FRONT VIEW

B

(3) PARTIAL BOTTOM VIEW

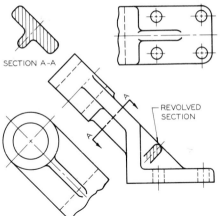

FIG. 8.21 Auxiliary Section A-A is projected from the cutting plane that is labeled A-A to show the cross section of the object.

8.12
Auxiliary sections

Auxiliary sections can be drawn to show the cross sections through a part to describe it better. A section through a part that is projected as an auxiliary view is shown in Fig. 8.21.

The section is labeled as section A-A, and a cutting plane was passed through the object and labeled A-A. This cutting plane was used to show where the sectional view was projected from and where the object was cut to show the section.

The auxiliary section gives a good description of the part that cannot be as readily understood from the given principal views.

8.13
Secondary auxiliary views

> A *secondary auxiliary view* is an auxiliary view that has been projected from a primary auxiliary view. An example of a secondary auxiliary view is shown in Fig. 8.22.

The inclined plane is found as an edge view in the primary auxiliary view, and then a line of sight is established that is perpendicular to the edge. This second auxiliary view shows the inclined surface true size.

You will remember that it has been necessary in all cases to project from an edge view of a plane before it can be found true size.

The problem in Fig. 8.23 is a secondary auxiliary projection in which the point view of a diagonal of a

FIG. 8.22 Secondary auxiliary views

Step 1 A line of sight is drawn parallel to the true-length view of a line on the oblique surface. The folding line F-1 is drawn perpendicular to the line of sight.

Step 2 The primary auxiliary view of the oblique surface is an edge view. Dimension depth (*D*) is used to locate two points in the primary auxiliary view.

Step 3 A line of sight is drawn perpendicular to the edge view, and a secondary auxiliary view is projected in this direction. The dimension *L* is used to locate one of the points of the true-size view.

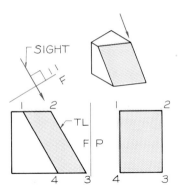

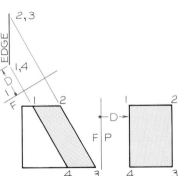

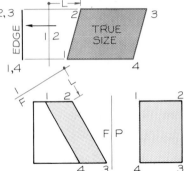

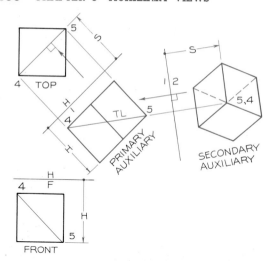

FIG. 8.23   A point view of a diagonal of a cube is found in the secondary auxiliary view. The diagonal is drawn from 4 to 5 in the given views. This line is found true length in the primary auxiliary view and then as a point in the secondary auxiliary view. The secondary auxiliary view is an isometric projection of the cube.

cube is found. When this is found, the three surfaces of the cube are equally foreshortened. This view is an *isometric projection*, and it is the basis for isometric pictorial drawing.

Since auxiliary views are supplementary views, the views can be partial views if all features are sufficiently shown. The object in Fig. 8.24 is shown as a series of orthographic views, some of which are partial views.

FIG. 8.24   An example of a secondary auxiliary view projected from a partial auxiliary view.

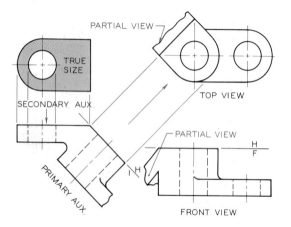

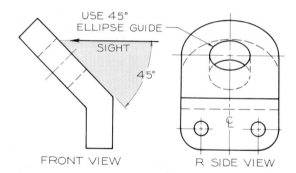

FIG. 8.25   The ellipse guide angle is the angle that the line of sight makes with the edge view of the circular feature. The ellipse guide for the right-side view is 45°.

8.14
Elliptical features

There are occasions when circular shapes will project as ellipses, and these views may need to be shown for clarity. Ellipses of this type can be drawn by using any of the techniques introduced in Chapter 5 once the necessary points have been plotted.

FIG. 8.26   Ellipses in secondary auxiliary views

A. The inclined surface is found true size in a secondary auxiliary view. The edge view of the plane in the primary auxiliary view makes a 30° angle with the line of sight from the top view. This is the angle of the ellipse guide for drawing the circular features in the top view.

B. The elliptical features in the front view can be completed by constructing a box about the circular features as shown in the top view. When transferred to the front view, these boxes are used to find the conjugate diameters, and then the ellipses are constructed.

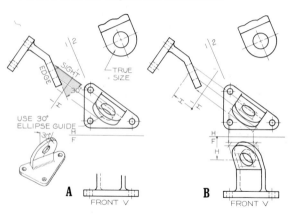

The most convenient method of drawing ellipses is with the use of an ellipse guide (template). The angle of the ellipse guide is the angle the line of sight makes with the edge view of the circular feature. In Fig. 8.25, the angle is found to be 45°. The right-side view is drawn as a 45° ellipse.

A more complex problem that involves a secondary auxiliary view is shown in Fig. 8.26. The inclined surface is found true size in the secondary auxiliary view by projecting from the edge view of the surface found in the primary auxiliary view.

The edge view of the circular feature found in the primary auxiliary view can be used to complete the top view of the object. Since the line of sight makes a 30° angle with the edge, this is the angle of the ellipse template to be used for drawing the top view.

The elliptical features in the front view can be found by locating a box around the ellipses in the top view and projecting them to the front view. The front view of these points is located by transferring dimensions of height *(H)* from the primary auxiliary view to the front view. This gives the conjugate diameters that can be used for constructing an ellipse as covered in Chapter 5.

The auxiliary views in this problem could not have been drawn unless you knew what the dimensions and specifications of the object were before you started drawing the views. This can be thought of as an object that was drawn by a designer who was translating his or her ideas to the paper while working.

Problems

The following problems are to be solved on Size A or Size B sheets, as assigned by your instructor.

1–13. (Fig. 8.27) Using the example layout, change the top and front views by substituting the top views given at the right in place of the one given in the example. The angle of inclination in the front view is 45° for all problems, and the height is 38 mm in the front view. Construct auxiliary views that show the inclined surface true size. Draw four problems per Size A sheet if drawn full size or one per sheet if drawn double size.

FIG. 8.27 Problems 1–13: Primary auxiliary views.

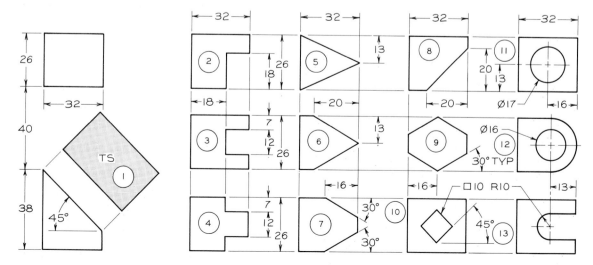

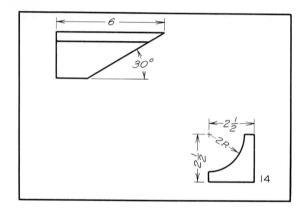

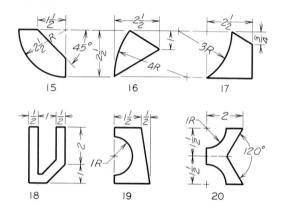

FIG. 8.28 Problems 14–20: Primary auxiliary views.

14–20. (Fig. 8.28) Using the example layout, change the top and right-side views by using the side views given at the right of the example problem. Complete the front view and the auxiliary view. Draw two problems per Size AV sheet, showing each half size.

21–43. (Figs. 8.29–8.51) Draw the necessary primary views and auxiliary views to describe the parts shown. Draw one per Size A or Size B sheet, as assigned. Adjust the scale of each to fit the space on the sheet.

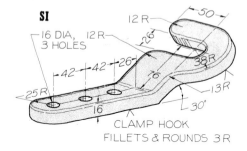

SI

CLAMP HOOK
FILLETS & ROUNDS 3R

FIG. 8.29 Problem 21: Clamp hook.

FIG. 8.30 Problem 22: Shaft mount.

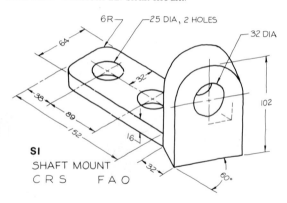

SI
SHAFT MOUNT
C R S   F A O

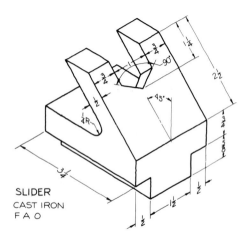

SLIDER
CAST IRON
F A O

FIG. 8.31 Problem 23: Slider.

FIG. 8.32 Problem 24: Eye fixture.

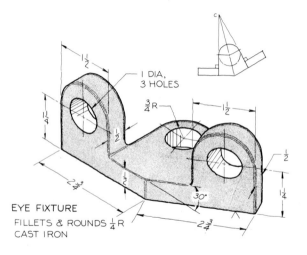

EYE FIXTURE
FILLETS & ROUNDS ¼ R
CAST IRON

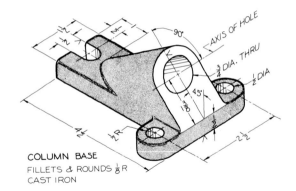

COLUMN BASE
FILLETS & ROUNDS $\frac{1}{8}$ R
CAST IRON

FIG. 8.33 Problem 25: Column base.

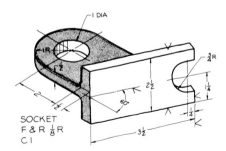

SOCKET
F & R $\frac{1}{8}$ R
C I

FIG. 8.36 Problem 28: Socket.

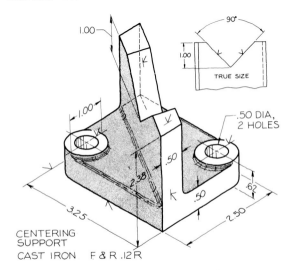

90°

1.00

TRUE SIZE

.50 DIA,
2 HOLES

CENTERING
SUPPORT
CAST IRON   F & R .12 R

FIG. 8.34 Problem 26: Centering support.

FIG. 8.35 Problem 27: Crank arm.

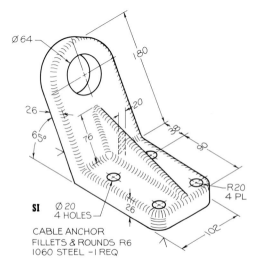

Ø 64

180

26

65°

76

SI   Ø 20
4 HOLES

R20
4 PL

CABLE ANCHOR
FILLETS & ROUNDS R6
1060 STEEL -1 REQ

FIG. 8.37 Problem 29: Cable anchor.

FIG. 8.38 Problem 30: Wedge lift.

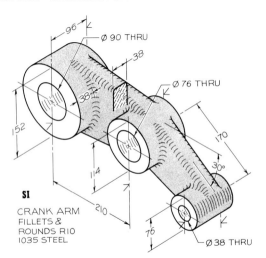

96

Ø 90 THRU

38

Ø 76 THRU

152

38

170

30°

114

SI

210

CRANK ARM
FILLETS &
ROUNDS R10
1035 STEEL

76

Ø 38 THRU

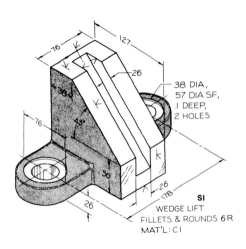

127

76

26

38 DIA,
57 DIA SF,
1 DEEP,
2 HOLES

76

26

26

18

SI

WEDGE LIFT
FILLETS & ROUNDS 6R
MAT'L: C I

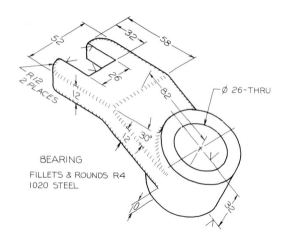

FIG. 8.39 Problem 31: Bearing.

BEARING

FILLETS & ROUNDS R4
1020 STEEL

Ø 26-THRU

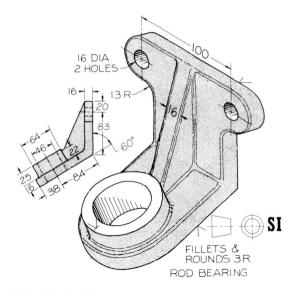

16 DIA
2 HOLES

FILLETS &
ROUNDS 3R
ROD BEARING

FIG. 8.42 Problem 34: Rod bearing.

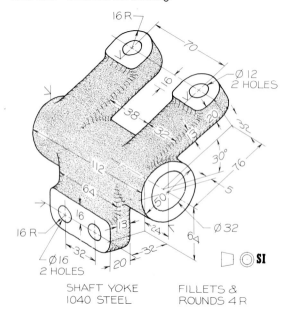

16 R

Ø 12
2 HOLES

Ø 32

16 R

Ø 16
2 HOLES

SHAFT YOKE
1040 STEEL

FILLETS &
ROUNDS 4 R

FIG. 8.40 Problem 32: Shaft yoke.

FIG. 8.41 Problem 33: Lever arm.

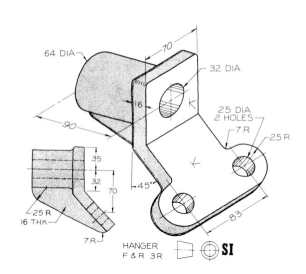

64 DIA

32 DIA

25 DIA
2 HOLES

7 R

25 R

35

32

70

25 R

16 THK

7 R

HANGER
F & R 3R

FIG. 8.43 Problem 35: Hanger.

FIG. 8.44 Problem 36: Floating lever.

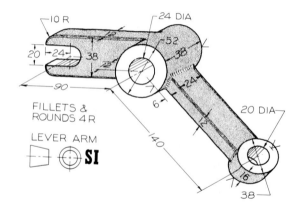

10 R

24 DIA

52

38

20 DIA

FILLETS &
ROUNDS 4 R

LEVER ARM

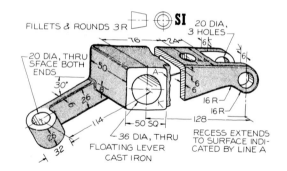

FILLETS & ROUNDS 3R

20 DIA,
3 HOLES

20 DIA, THRU
SFACE BOTH
ENDS

50 SQ

36 DIA, THRU

FLOATING LEVER
CAST IRON

16 R

16 R

RECESS EXTENDS
TO SURFACE INDI-
CATED BY LINE A

162

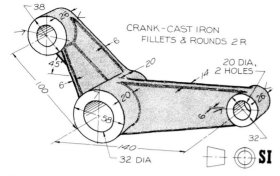

FIG. 8.45 Problem 37: Crank.

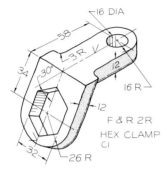

FIG. 8.46 Problem 38: Hexagon angle.

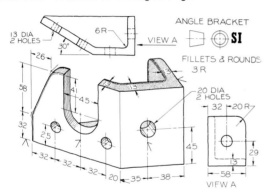

FIG. 8.47 Problem 39: Angle bracket.

FIG. 8.48 Problem 40: Double bearing.

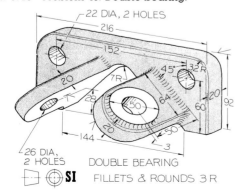

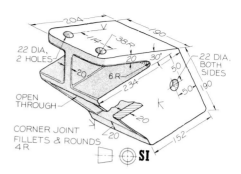

FIG. 8.49 Problem 41: Corner joint.

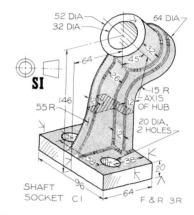

FIG. 8.50 Problem 42: Shaft socket.

FIG. 8.51 Problem 43: Dovetail bracket.

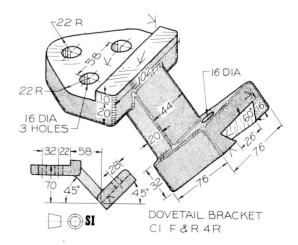

44–45. (Figs. 8.52–8.53) Lay out these orthographic views on Size B sheets and complete the auxiliary and primary views.

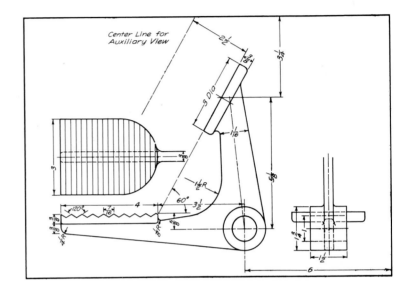

FIG. 8.52   Problem 44: Clutch pedal.

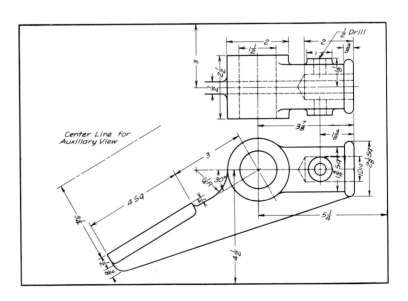

FIG. 8.53   Problem 45: Adjustment.

46–48. (Figs. 8.54–8.56) Construct orthographic views of the given objects and draw auxiliary views that give the true-size views of the inclined surfaces by using secondary auxiliary views. Draw one per Size B sheet.

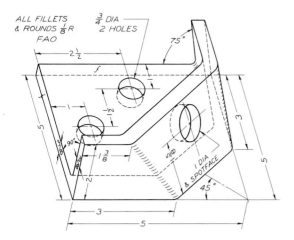

FIG. 8.54 Problem 46: Corner connector.

FIG. 8.56 Problem 48: Oblique support.

FIG. 8.55 Problem 47: Shaft bearing.

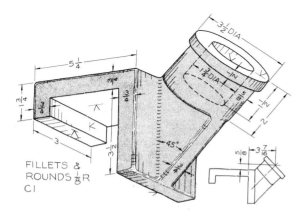

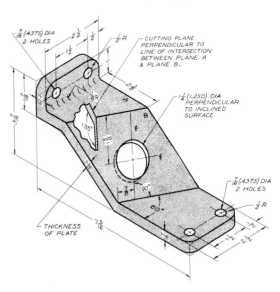

9 Sections

9.1
Sections

The standard orthographic views that show all hidden lines may not effectively reveal the true details of an object. This shortcoming can often be improved by using a technique of cutting away part of the object and looking at the cross-sectional view. Such a cutaway view is called a *section*.

A section is shown pictorially in Fig. 9.1A, where an imaginary cutting plane is passed through the object in order to show its internal features. The standard top and front orthographic views are shown at B. The method of drawing a section is shown at C, where the front view has been converted to a *full section,* and the cut portion is cross-hatched or section-lined. Hidden lines have been omitted since they are not needed. The cutting plane is drawn in the top view as a heavy line with short dashes at intervals; this can be thought of as a knife-edge cutting through the object.

By referring to the top view and the front sectional view, you have no hidden lines to interpret, and you can understand the cross-sectional view of the object more easily.

Two types of cutting planes are shown in Fig. 9.2. Either is acceptable, although the upper example is

FIG. 9.1 A comparison of a regular orthographic view with a full-section view of the same object to show the internal features as well as the external features.

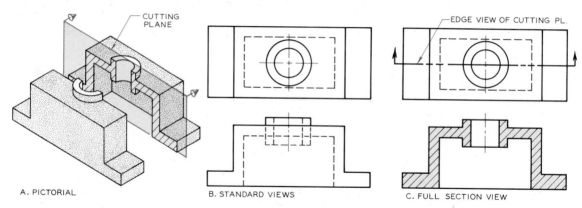

A. PICTORIAL

B. STANDARD VIEWS

C. FULL SECTION VIEW

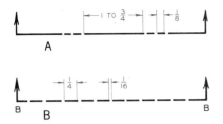

FIG. 9.2 Typical cutting-plane lines used to represent sections. The cutting planes marked B-B will produce a section that will be labeled section B-B.

more often used. The spacing of the dashes depends on the size of the drawing. The weight of the cutting plane is the same as that of a visible object line. Letters can be placed at each end of the cutting plane to label the sectional view, such as Section B-B, wherever it is drawn. Think of the cutting-plane line as a knife-edge cutting through a part.

The three basic views that may appear as sections are shown in Fig. 9.3 with their respective cutting planes. Each cutting plane has perpendicular arrows at the ends pointing in the direction of the line of sight for the section. For example, the cutting plane at A passes through the top view, the front of the top view is removed, and the line of sight is toward the remaining portion of the top view. The top view will appear as a section when the cutting plane passes through the front view and the line of sight is downward (B). When the cutting plane passes through the front view (C), the right-side view will be a section.

9.2
Sectioning symbols

The symbols that are used to distinguish between different materials in section are shown in Fig. 9.4. Although the symbols can be used to indicate the mate-

FIG. 9.3 The three standard positions of cutting planes that pass through given views to result in sectional views in the front, top, and side views. The arrows point in the direction of the line of sight for each section.

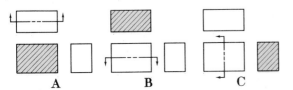

rials within a section, it is good practice to provide supplementary notes specifying the materials to avoid misinterpretation.

> The *cast-iron* symbol of evenly spaced section lines can be used to represent *any material* and is the most often used sectioning symbol.

Cast-iron symbols are usually drawn with a 2H pencil with lines that are slanted at 45° or at any other standard angle, such as 30° or 60°, and spaced about $\frac{1}{16}$" apart by eye.

The proper spacing of section lines for cast iron is shown at the top left of Fig. 9.5, where the lines are evenly spaced. Common errors of section-lining are shown in the remainder of the figure.

Extremely thin parts, such as sheet metal, washers, or gaskets (Fig. 9.6), are sectioned by blacking in the areas completely rather than using section lines. Large parts are sectioned with an *outline section* to save time and effort. Section lines in small parts are drawn closer together than in larger parts.

> Sectioned areas should be section-lined with line symbols that are neither parallel nor perpendicular to the outlines of the parts (Fig. 9.7) but are drawn at some other angle with the outline of the part (A).

Parallel and perpendicular section lines may be confused as serrations or other machining treatments of the surface.

9.3
Sectioning assemblies

When an assembly of several parts is sectioned it is important that the section lines be drawn at varying angles to distinguish the parts (Fig. 9.8). The use of different material symbols, when the assembly is composed of parts made from different materials, is helpful in distinguishing the parts from one another. The same part is cross-hatched at the same angle and with

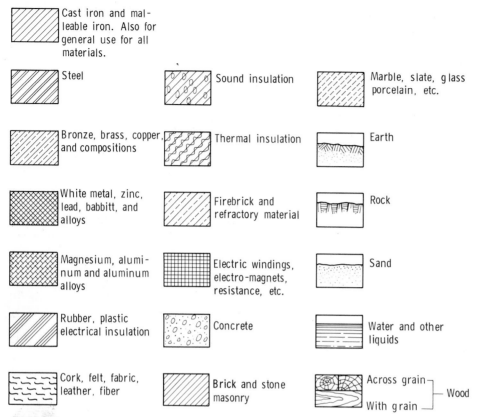

FIG. 9.4 The symbols used for section-lining parts in section. Note that the cast-iron symbol can be used for all materials.

FIG. 9.5 Good section lines are drawn as thin lines and are spaced about 1/16" to 1/8" apart. Some typical section-lining errors are shown here.

FIG. 9.6 Thin parts are blacked in, and large areas are section-lined around their outlines to save time and effort.

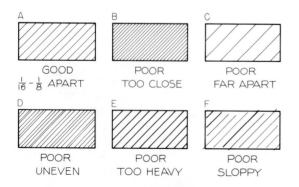

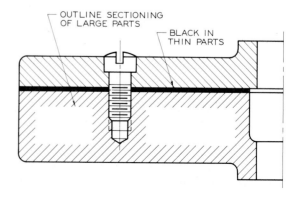

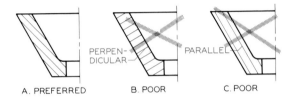

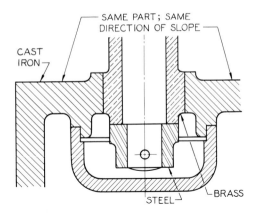

FIG. 9.7 Section lines should be drawn so that they are not parallel or perpendicular to the outline of a part.

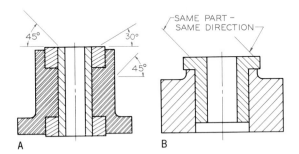

FIG. 9.8 The section lines of the same part should be drawn in the same direction. Section lines of different parts should be drawn at varying angles to separate the parts.

FIG. 9.9 A typical assembly in section with well-defined parts and correctly drawn section lines.

the same symbol even though the part may be separated into different areas, as shown in Fig. 9.8B.

An assembly is illustrated in Fig. 9.9 where section lines are effectively used to identify the parts of the assembly.

9.4
Full sections

A **full section** is a sectional view formed by passing a cutting plane fully through an object and removing half of it to give a view of its internal features.

The object shown in Fig. 9.10A is drawn as three orthographic views in which a number of hidden lines are shown. The front view can be drawn as a full section by passing a cutting plane fully through the top view (B) and removing the front portion as shown at C. The standard way of showing this section can be seen at D, where the cutting plane is drawn through

the top view with arrows indicating the direction of sight. The front view is then section-lined to give the full section.

A full section through a cylindrical part is shown in Fig. 9.11, where half of the object is removed. A common mistake in constructing sectional views is the

FIG. 9.10 A full section is a section formed by a cutting plane that passes fully through it. The cutting plane is shown passing through the top view, and the direction of sight is indicated by the arrows at each end. The front view is converted to a sectional view to give a clear understanding of the internal features.

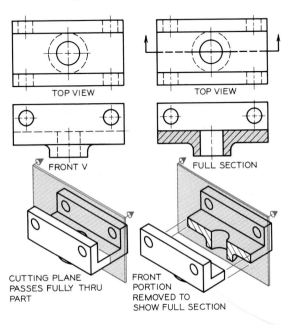

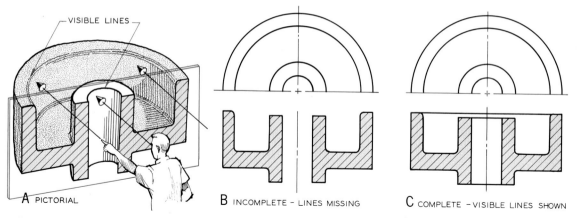

FIG. 9.11 Full section–cylindrical part

A. When a full section is passed through an object, you will see the lines behind the sectioned area.

B. If only the sectioned area were shown, the view would be incomplete.

C. Visible lines behind the sectioned area must be shown also.

omission of the visible lines behind the cutting plane (B); the correctly drawn sectional view is shown at C. Hidden lines are omitted in all sectional views unless they are believed to be necessary to provide a clear understanding of the view. Visible lines behind the sectional view are omitted also if they confuse the view.

Figure 9.12 is an example of a part whose right-side view is shown as a full section. Likewise, the part in Fig. 9.13 illustrates a front view that appears as a full section. Note that the lines behind the cutting plane are shown as visible lines.

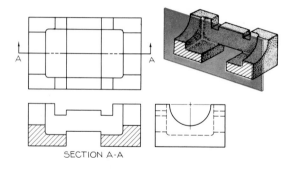

SECTION A-A

FIG. 9.13 A full Section, Section A-A, is used to supplement the given views of the object.

FIG. 9.12 A full section with the cutting plane shown.

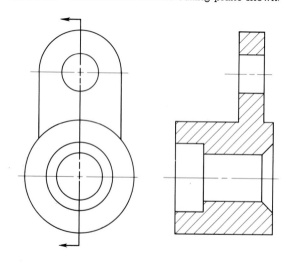

FIG. 9.14 These parts are not cross-sectioned even though the cutting plane passes through them.

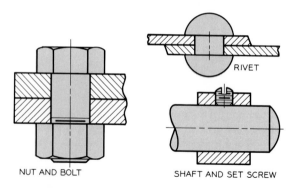

RIVET

NUT AND BOLT

SHAFT AND SET SCREW

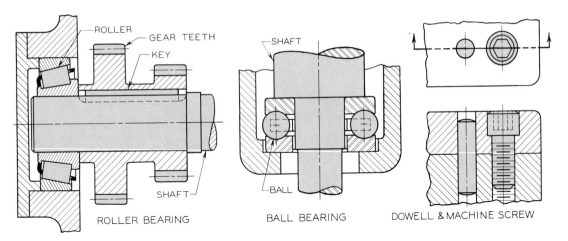

FIG. 9.15 These parts are not section-lined even though cutting planes pass through them.

9.5
Parts not section-lined

Many standard parts are not section-lined even though the cutting plane passes through them. Examples of such parts in Fig. 9.14 are nuts and bolts, rivets, shafts, and set screws. These parts have no internal features, and sections through them would not be of value.

Other parts not section-lined are roller bearings, ball bearings, gear teeth, shafts, dowels, pins, set screws, and washers (Fig. 9.15).

9.6
Ribs in section

Ribs are not section-lined when the cutting plane passes flatwise through them, as shown in Fig. 9.16a, since this would give a misleading impression of the rib. A similar example is the section shown in Fig. 9.16c. However, a rib is section-lined when the cutting plane passes through it and shows its true thickness (Fig. 9.16b).

An alternative method of section-lining webs and ribs is shown in Fig. 9.17. At (a) the ribs are not sec-

FIG. 9.16 A rib that is cut in a flatwise direction by a cutting plane is not section-lined. Ribs are section-lined when cutting planes pass perpendicularly through them, as shown at (b) and (c).

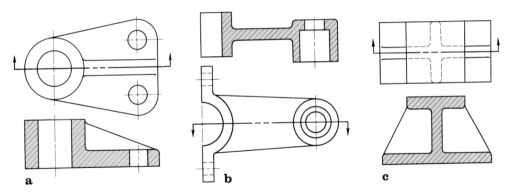

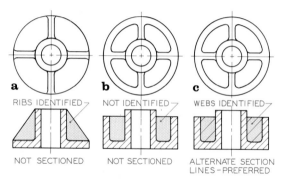

RIBS IDENTIFIED⌐   NOT IDENTIFIED⌐   WEBS IDENTIFIED⌐

NOT SECTIONED   NOT SECTIONED   ALTERNATE SECTION
                                     LINES – PREFERRED

FIG. 9.17 Outside ribs in section are not section-lined. Webs that are poorly identified, as at (b), should be identified by cross-hatching them with alternate section lines as shown at (c). This calls attention to the fact that webs are present.

tion-lined since the cutting plane passes through them in a flatwise direction. The webs at (b) are symmetrically spaced about the hub. As a rule, webs are not cross-hatched, but this would leave the webs unidentified. Therefore it is better to use the *alternate sectioning* technique and extend every other section line through the webs to call attention to them as shown at (c).

The ribs in Fig. 9.18 are not section-lined and thus afford a more descriptive view of the part. If the ribs had been section-lined, the section would have given the impression that the part was solid and conical in shape (b). Note that the top views are partial views, and the portion nearest the sectional view is the portion that has been omitted.

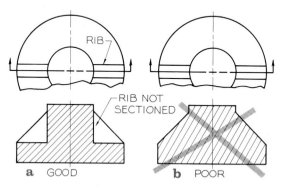

RIB⌐

⌐RIB NOT
SECTIONED

a   GOOD   b   POOR

FIG. 9.18 When ribs are not section-lined, the view is more descriptive of the part. Partial views are used in the top views to save space. The front part of the top view is removed when the front view is a section.

FIG. 9.19 The cutting plane of a half-section passes halfway through the object, which results in a sectional view that shows half of the outside and half of the inside of the object. Hidden lines are omitted unless they are necessary to clarify the view.

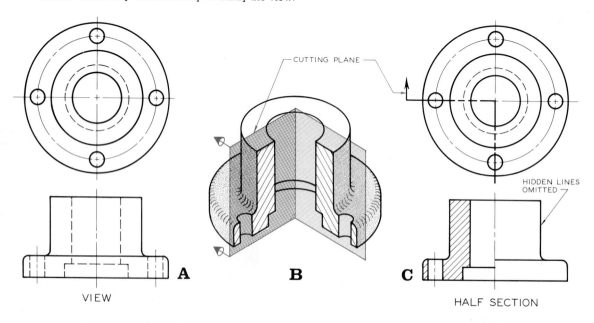

VIEW   A   B   C   HALF SECTION

⌐CUTTING PLANE⌐

HIDDEN LINES
OMITTED⌐

9.7
Half-sections

A *half-section* is a view that results from passing a cutting plane halfway through an object, thereby removing a quarter of it.

Half-sections are most often used with symmetrical parts, cylinders in particular. A cylindrical part in Fig. 9.19A is shown as a pictorial half-section at B. The method of drawing the orthographic half-section is shown at C. In this view, both internal and external features can be seen; the hidden lines are omitted in the sectional view.

A similar half-section is given in Fig. 9.20. Hidden lines are omitted unless they are believed to be essential for understanding or dimensioning the object. The half-section in Fig. 9.21 has been drawn without showing the cutting plane. This is permissible if it is obvious where the cutting plane was passed through the part.

Two half-sections are drawn in Fig. 9.22 to describe the object. The front and right-side views are constructed as half-sections, as if the quarter cut away by the cutting plane had been removed. A few, but

FIG. 9.20 A typical half-section drawing.

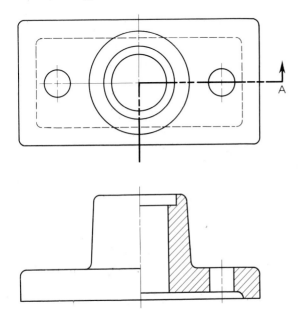

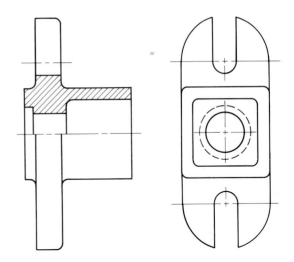

FIG. 9.21 When it is obvious where the cutting plane is located, it is unnecessary to show it, as in this example.

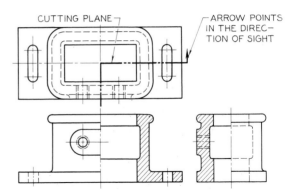

FIG. 9.22 Two separate half-sections are found in this example, using the same cutting plane for both.

not all, of the hidden lines have been given in each view to clarify the sections. Note that instead of using an object line, center lines are given to separate the sectional half from the half that appears as an external view.

9.8
Partial views

A conventional method of representing symmetrical views is shown in Fig. 9.23. A half-view is sufficient when it is drawn adjacent to the sectional view, as at

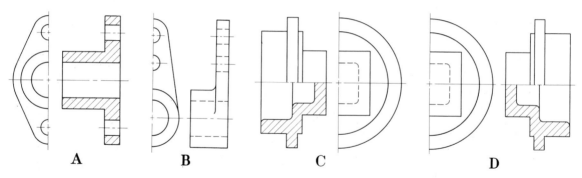

FIG. 9.23 Half-views can be used for symmetrical parts to conserve space and time. A. The omitted portion of the view is toward the full section. B. The omitted portion of the view is away from the adjacent view when it is not sectioned. C and D. The omitted half can be toward or away from the section in the case of a half-section.

A. In full sections, the removed half is the portion nearest the section (part A). When drawing half-views that are associated with nonsectional views, the removed half of the partial view will be the half that is away from the adjacent view (Fig. 9.23B).

When partial views are drawn in conjunction with half-sections, either the near or the far halves of the partial views can be omitted, as shown in Fig. 9.23C and D. Partial views are faster to draw and require less space.

9.9
Offset sections

An *offset section* is a type of full section in which the cutting plane is offset to pass through important features that would be missed by the usual full section formed by a flat plane.

FIG. 9.24 Offset section

A. An offset section may be necessary to show typical features of a sectioned part.

B. When the front portion is removed, the internal features can be seen. Note that the cutting plane has been offset.

C. In an offset section, the offset cut is not shown. The section is shown as if it were a typical full section.

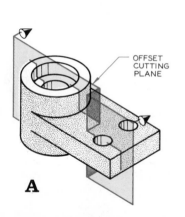

OFFSET CUTTING PLANE

A

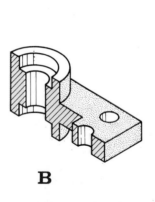

B

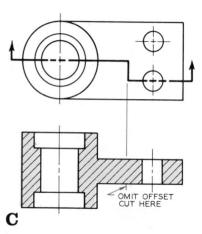

OMIT OFFSET CUT HERE

C

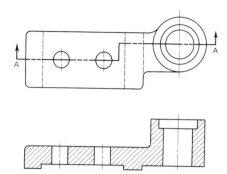

FIG. 9.25 An offset section.

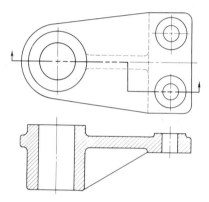

FIG. 9.26 An offset section.

An offset section is shown pictorially in Fig. 9.24A, where the plane is offset to pass through the large hole and one of the small holes. The cut object is shown at B. The method of drawing an offset section orthographically is given in part C. The cut formed by the offset is not shown in the section since this is an imaginary cut.

The part in Fig. 9.25 is one that lends itself to representation by an offset section. Another offset section is used to describe a part in Fig. 9.26. As previously discussed, the ribs are not section-lined in this case, whether the plane passes through them or not.

FIG. 9.27 Examples of revolved sections with and without conventional breaks.

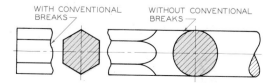

WITH CONVENTIONAL BREAKS

WITHOUT CONVENTIONAL BREAKS

9.10
Revolved sections

> A *revolved section* is used to describe a cross section of a part to eliminate the need for drawing an entirely separate view.

For example, revolved sections are used to indicate cross sections of the parts in Fig. 9.27. One of the revolved sections is positioned within the view, and conventional breaks are drawn on each side of the hexagonal section. The circular cross section is drawn superimposed on the cylindrical portion of the part without using conventional breaks. Either of these methods can be used when drawing revolved sections.

A more advanced type of revolved section is illustrated in Fig. 9.28, where a cutting plane is passed through the object (Step 1). The plane is imagined to be revolved in the top view to give a true-size revolved section in the front view (Step 2). Note that the object lines do not pass through the revolved section in the front view. It would have been permissible to use conventional breaks on each side of the revolved section, as used in Fig. 9.27.

Typical revolved sections are shown in Fig. 9.29 at A and B. These sections provide a method of giving a part's cross section without drawing another complete orthographic view.

9.11
Removed sections

> A *removed section* is a revolved section that has been removed from the view where it was revolved, as shown in Fig. 9.30.

Center lines are used as axes of rotation to show where the sections were taken from. Removed sections may be necessary where room does not permit revolution on the given view (Fig. 9.31A); instead, the cross section must be removed from the view, as shown at B.

Removed sections do not have to be positioned directly along an axis of revolution adjacent to the view from where the sections were taken. Instead,

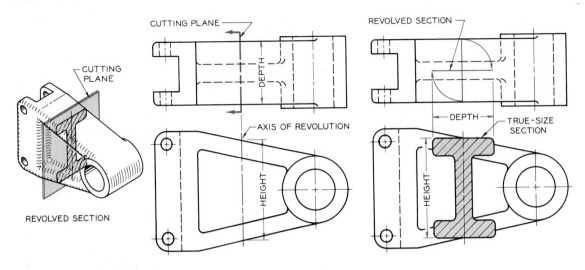

FIG. 9.28 Revolved section

Step 1 An axis of revolution is shown in the front view. The cutting plane would appear as an edge in the top view if it were shown.

Step 2 The vertical section in the top view is revolved so that the section can be seen true size in the front view. Object lines are not drawn through the section.

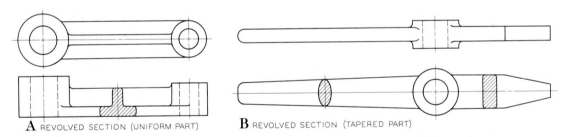

A REVOLVED SECTION (UNIFORM PART)   **B** REVOLVED SECTION (TAPERED PART)

FIG. 9.29 The revolved sections given here are helpful in describing the cross-sectional characteristics of the two parts. This is much more effective than using additional orthographic views.

FIG. 9.30 Removed sections are similar to revolved sections, but they have been removed outside the object along an axis of revolution.

FIG. 9.31 Removed sections can be used to good advantage where space does not permit the use of a revolved section.

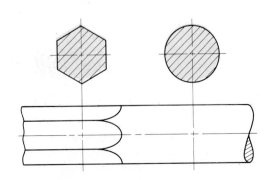

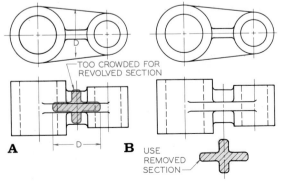

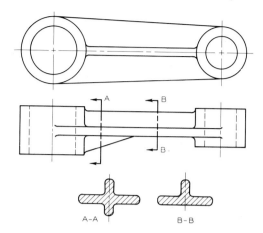

FIG. 9.34 It may be necessary to remove a section to another page in a set of drawings. In this case, each end of the cutting plane can be labeled with a letter and a number. The letters refer to section A-A, and the numbers mean that this section is located on page 7.

FIG. 9.32 Sections can be lettered at each end of a cutting plane, such as A-A. This removed section can then be shown elsewhere on the drawing, where it is designated as section A-A.

cutting planes can be labeled at each end, as shown in Fig. 9.32, to specify the sections. For example, the plane labeled with an A at each end is used to label section A-A. Section B-B is found in a similar manner.

A series of removed sections is shown in Fig. 9.33. Removed sections may be put on different sheets when a set of drawings is composed of many pages. In this case, a cutting plane may be labeled as shown in Fig. 9.34. The A at each end identifies the section as section A-A, and the numerals indicate on which page of the set of drawings the section is located.

FIG. 9.33 A series of removed sections, each labeled according to the cutting plane that was used to find them.

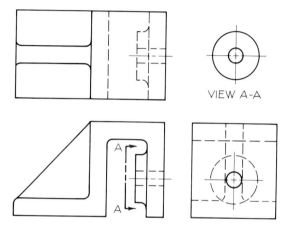

FIG. 9.35 A removed view (not a section) can also be used to view a part from an unconventional direction that would be more effective than an entire orthographic view.

Removed views can also be used to provide inaccessible orthographic views (nonsectional views) when these are needed, as shown in Fig. 9.35.

9.12
Broken-out sections

> A *broken-out section* is a convenient method used to show interior features that would improve the understanding of a drawing without drawing a separate view as a section.

A portion of the part in Fig. 9.36 is broken out to reveal details of the wall thickness to better explain the drawing. This reduces the need for hidden lines; therefore they may be omitted if desired.

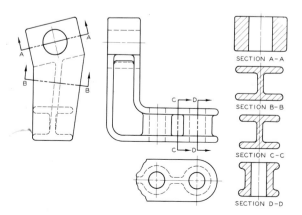

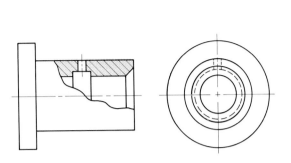

FIG. 9.36 A broken-out section is used to show an internal feature by using a conventional break.

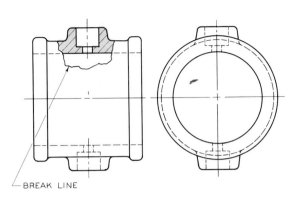

— BREAK LINE

FIG. 9.37 A broken-out section that shows a section through the boss at the top of a collar.

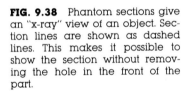
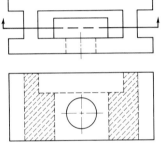

FIG. 9.38 Phantom sections give an "x-ray" view of an object. Section lines are shown as dashed lines. This makes it possible to show the section without removing the hole in the front of the part.

FIG. 9.39 These are often-used conventional breaks that are used to remove a portion of an object so it can be drawn at a larger scale.

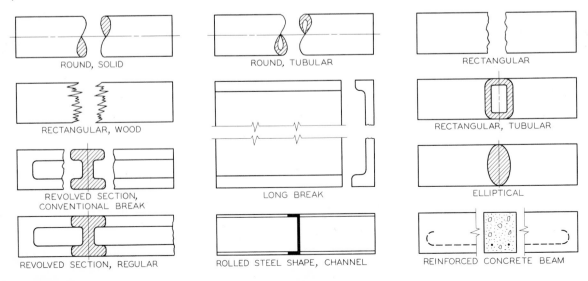

ROUND, SOLID

ROUND, TUBULAR

RECTANGULAR

RECTANGULAR, WOOD

RECTANGULAR, TUBULAR

REVOLVED SECTION, CONVENTIONAL BREAK

LONG BREAK

ELLIPTICAL

REVOLVED SECTION, REGULAR

ROLLED STEEL SHAPE, CHANNEL

REINFORCED CONCRETE BEAM

Figure 9.37 is a similar example of a broken-out section where hidden lines are given. The irregular lines that are used to represent the break are called *conventional breaks;* these are discussed in Section 9.14.

9.13
Phantom (ghost) sections

> A *phantom section,* or *ghost section,* is used occasionally to depict parts as if they were viewed by an x-ray.

An example of a phantom section is shown in Fig. 9.38. The cutting plane is drawn in the usual manner, but the section lines are drawn as dashed lines. You can see that if the object had been shown as a regular full section, the circular hole through the front surface could not have been shown in the same view.

9.14
Conventional breaks

Some of the previous examples have used conventional breaks to indicate the removal of parts of an object. Examples of conventional breaks are illustrated in Fig. 9.39. These examples should be reviewed and

FIG. 9.40 Conventional breaks in cylindrical and tubular sections can be drawn freehand with the aid of the guidelines shown. The radius, R, is used to establish the width of both "figure 8's."

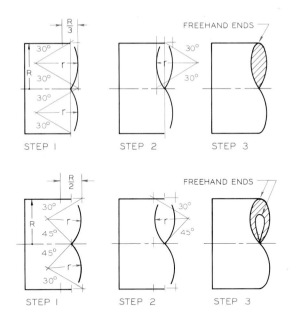

FIG. 9.41 The steps of constructing conventional breaks of solid and tubular shapes with instruments.

learned so these short-cut methods can be used in drawing views. The "figure 8" breaks that are used for cylindrical and tubular parts can be drawn freehand (Fig. 9.40), or they can be drawn by using a compass, as shown in Fig. 9.41, when they are drawn large.

One use of conventional breaks is to shorten a long piece that has a uniform cross section. The long part in Fig. 9.42 has been shortened and drawn at a larger scale for more clarity by using the conventional

FIG. 9.42 By using conventional breaks and a revolved section, this part can be drawn on a larger scale that is easier to read.

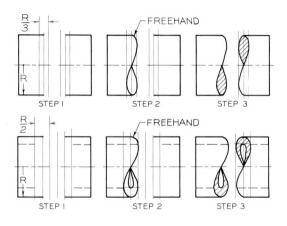

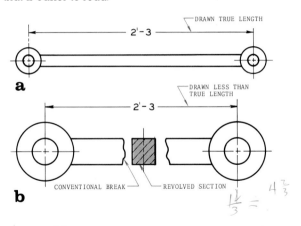

breaks shown at (b). The dimension specifies the true length of the part, and the breaks indicate that a portion of the length has been removed.

9.15
Conventional revolutions

Three conventional sections are shown in Fig. 9.43. The center hole is omitted at (a) since it does not pass through the center of the circular plate. However, the hole at (b) does pass through the plate's center and is sectioned accordingly. At (c), the cutting plane does not pass through one of the symmetrically spaced holes in the top view, but the hole is revolved to the cutting plane in order to show the recommended full section.

When ribs are symmetrically spaced about a hub (Fig. 9.44), it is conventional practice to revolve them to where they will appear true size in either a view (B) or a section (D). A full section that shows both ribs and holes revolved to their true-size locations can be seen in Fig. 9.45.

The cutting plane can be positioned as shown in Fig. 9.46. Even though the cutting plane does not pass through the ribs and holes at (a), the sectional view should be drawn as at (b), where the cutting plane is shown passing through the hole. The cutting plane

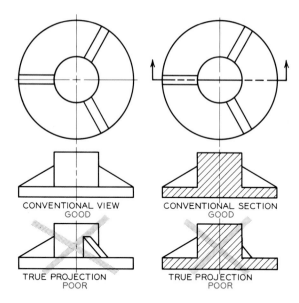

FIG. 9.44 Symmetrically located ribs are shown revolved in both orthographic and sectional views as a conventional practice.

can be drawn in either position. The revolved cutting plane can be used to show the path of the cutting plane in more complex parts.

Symmetrically spaced spokes are rotated and are not section-lined, in the same manner as ribs in section. Figure 9.47 illustrates the preferred and poor-

FIG. 9.45 A part with symmetrically located ribs and holes is shown in section with both ribs and holes rotated to the cutting plane.

FIG. 9.43 Symmetrically spaced holes in a circular plate should be revolved in their sectional views to show them at their true radial distance from the center. At (a), no hole is shown at the center, but a hole is shown at the center in (b), since the hole is through the center of the plate. At (c), one of the holes is rotated to the cutting plane so that the sectional view will be symmetrical and more descriptive.

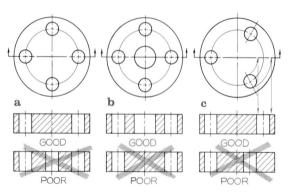

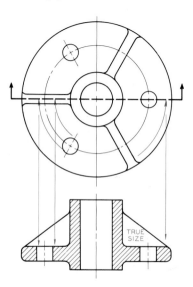

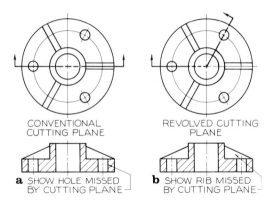

FIG. 9.46 Symmetrically located ribs are shown in section in revolved positions to show the ribs true size. Foreshortened ribs are omitted.

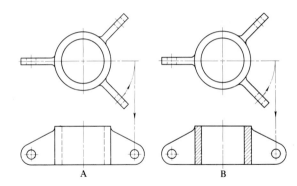

FIG. 9.49 Symmetrically spaced lugs (flanges) are revolved to show the front view and the sectional view as symmetrical.

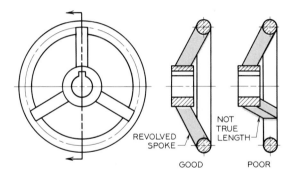

FIG. 9.47 Symmetrically positioned spokes are revolved to show their true size in section. Spokes are not section-lined in section.

FIG. 9.48 Solid webs in sections of the type at A are section-lined. Spokes are not section-lined when the cutting plane passes through them, as shown at B.

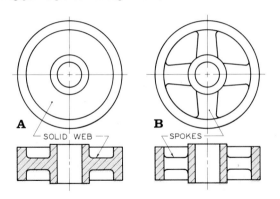

practice methods of representing spokes. Only the revolved, true-size spokes are drawn; the intermediate spokes are omitted.

The reason for not section-lining spokes can be seen in Fig. 9.48. If the spokes at B had been section-lined, the cross section of the part would be confused with the part at A, where there are no spokes but a continuous web.

Lugs that are symmetrically positioned about the central hub of the object in Fig. 9.49 are revolved to show them true size in both views and sections. A more complex object that involves the same principle of rotation can be seen in Fig. 9.50. The oblique arm

FIG. 9.50 A part with an oblique feature attached to the circular hub is revolved so it will appear true shape in the front view, which is a sectional view.

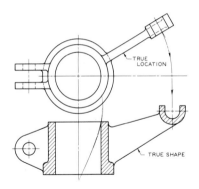

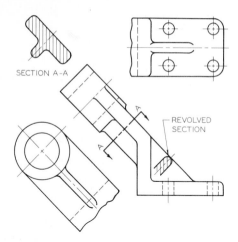

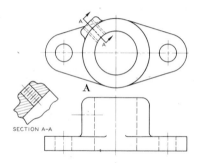

FIG. 9.52 A partial auxiliary section projected from the top view to show a threaded boss.

FIG. 9.51 Sectional views can be shown as auxiliary views for added clarity.

is drawn in the section as if it had been revolved to the center line in the top view and then projected to the sectional view.

9.16
Auxiliary sections

Auxiliary sections can be used to supplement the principal views used in orthographic projections, as shown in Fig. 9.51. Auxiliary cutting plane A-A is passed through the front view, and the auxiliary view is projected from the cutting plane as indicated by the sight arrows. Sectional view A-A gives the cross-sectional description of the part.

Another auxiliary section is given in Fig. 9.52 to show a section through a threaded boss. The location of this threaded hole is located by a center line in the front view. In Fig. 9.53, section A-A is drawn to clarify the front view where the cutting plane is positioned. The right-side view is drawn as a partial view.

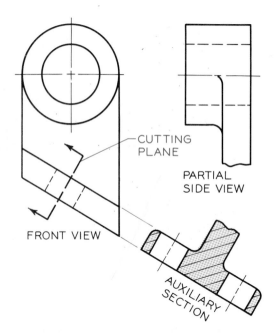

FIG. 9.53 An auxiliary section projected from the front view. The right-side view is a partial view since it is supplemented by the auxiliary section.

Problems

These problems can be solved on Size A or Size B sheets.

1–7. (Fig. 9.54) Full sections: Draw two of these problems per Size A sheet. Each grid is equal to ¼ inch or 5 mm. Complete the front views as full sections.

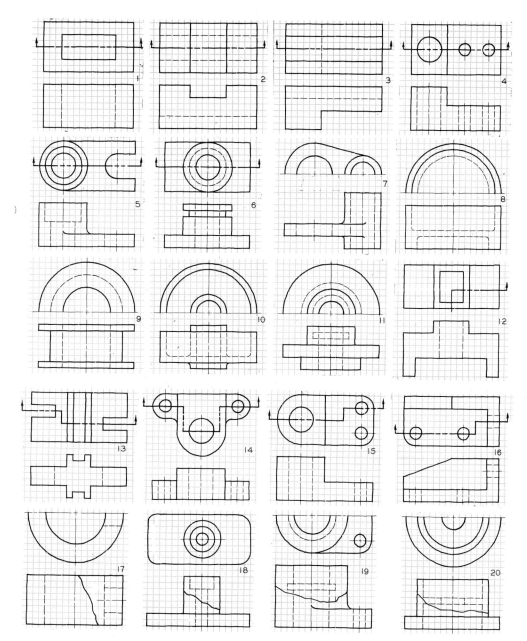

FIG. 9.54 Problems 1–20: Introductory sections.

8–12. (Fig. 9.54) Half-sections: Draw two of these problems per Size A sheet. Each grid is equal to ¼ inch or 5 mm. Complete the front views as half-sections.

13–16. (Fig. 9.54) Offset sections: Draw two of these problems per Size A sheet. Each grid is equal to ¼ inch or 5 mm. Complete the front views as off-set sections.

17–20. (Fig. 9.54) Broken-out sections: Draw two of these problems per Size A sheet. Each grid is equal to ¼ inch or 5 mm. Complete the broken-out sections.

21–30. (Figs. 9.55–9.57) Half-sections: The views given are rectangular views of cylindrical parts.

Draw one problem per Size A sheet. Show the circular view and draw the rectangular view as a half-section. Omit dimensions and show the cutting plane. Select an appropriate scale.

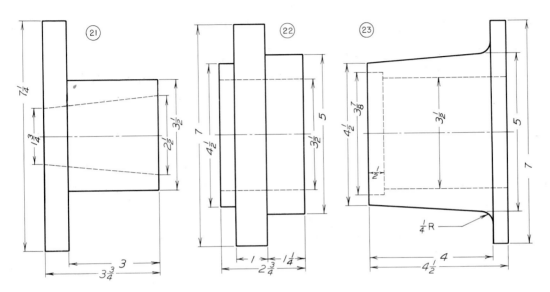

FIG. 9.55 Problems 21–23: Half-sections.

FIG. 9.56 Problems 24–26: Half-sections.

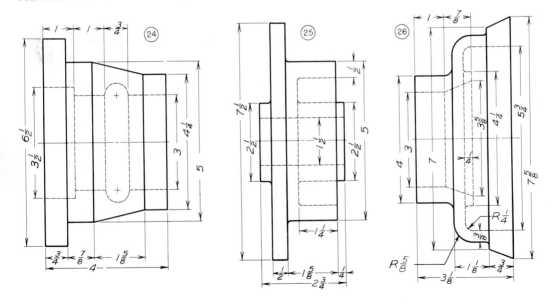

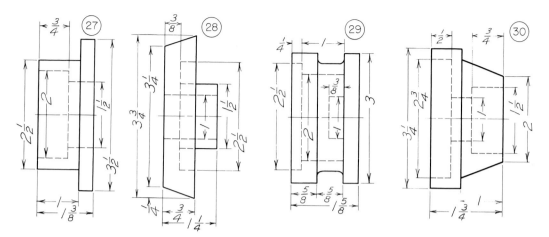

FIG. 9.57 Problems 27–30: Half-sections.

31–36. (Figs. 9.58–9.60) Full and offset sections: Draw the given views on Size A sheets, two per sheet. Complete the sections as indicated by the cutting planes. Problem 36 will require an offset section.

FIG. 9.58 Problems 31–32: Full sections.

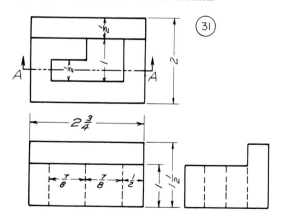

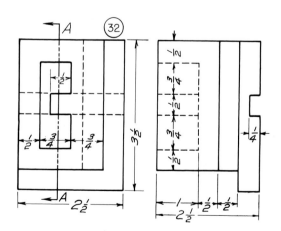

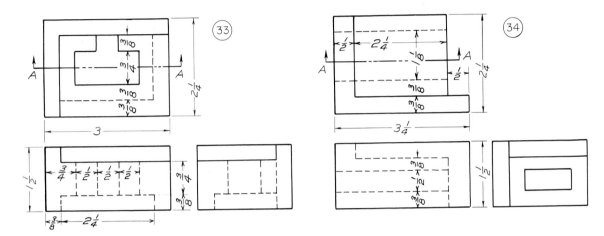

FIG. 9.59 Problems 33–34: Full sections.

FIG. 9.60 Problems 35–36: Offset sections.

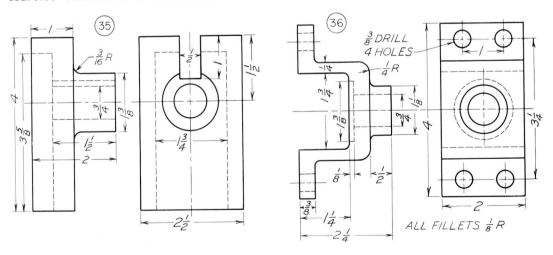

37–43. (Figs. 9.61–9.65) Sections and conventions: Draw each problem on a Size B sheet. Complete the sectional views as indicated by the cutting planes or as good practice demands. Omit the dimensions.

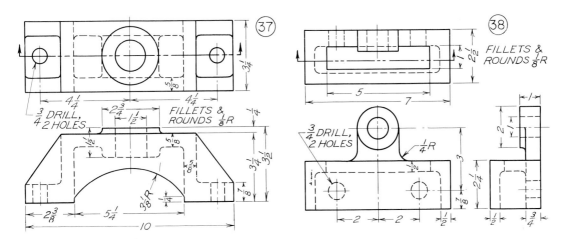

FIG. 9.61 Problems 37–38: Full sections.

FIG. 9.62 Problems 39–40: Full sections.

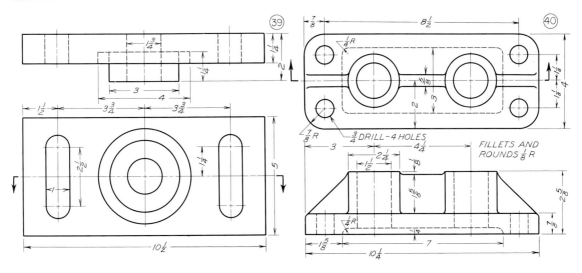

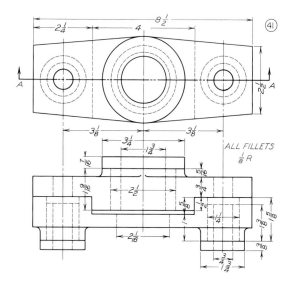

FIG. 9.63 Problem 41: Full section.

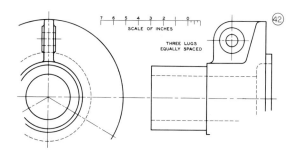

FIG. 9.64 Problem 42: Sections and conventions. (Show the lugs in all three positions and show the entire view.)

FIG. 9.65 Problem 43: Section and conventions. (Show a full section through this part and show the entire view.)

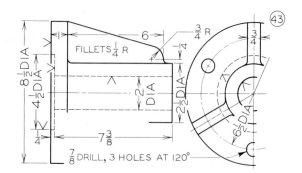

44–46. (Figs. 9.66–9.68) Sections: Draw the necessary views to describe the parts by using sections and conventional practices. Draw one problem per Size B sheet. Omit dimensions.

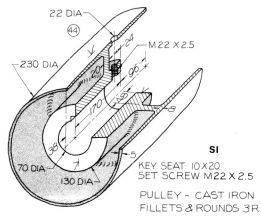

FIG. 9.66 Problem 44: Section.

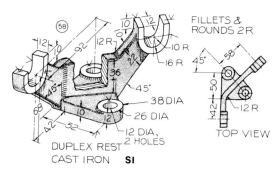

FIG. 9.67 Problem 45: Section.

FIG. 9.68 Problem 46: Section.

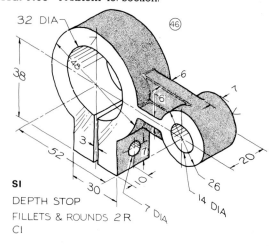

10

Screws, Fasteners, and Springs

10.1
Threaded fasteners

Screw threads provide a fast and easy method of fastening two parts together and of exerting a force that can be used to adjust movable parts. For a screw thread to function, there must be two parts: an internal thread and an external thread. The internal threads may be tapped inside a part, such as a motor block, or more commonly they may be tapped inside a nut. Whenever possible, the nuts and bolts used in industrial projects should be stock parts that can be obtained from many sources. This reduces manufac-

turing expenses and improves the interchangeability of parts, which is very important for repair or replacement of damaged fasteners.

Threaded fasteners made in different countries or by different manufacturers may have threads of different specifications that will not match. Progress has been made toward establishing standards that will unify threads both in this country and abroad by the introduction of metric standards. Other efforts have led to the adoption of the *Unified Screw Thread* by the United States, Britain, and Canada (ABC Standards), which is a modification of both the American Standard thread and the Whitworth thread.

FIG. 10.1 Thread terminology.

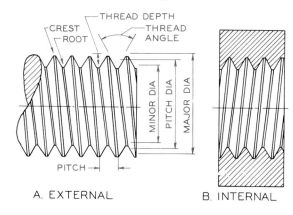

A. EXTERNAL   B. INTERNAL

10.2
Definitions of thread terminology

Succeeding sections will discuss the uses and methods of representing screw threads. The terms used and defined below are illustrated in Fig. 10.1.

EXTERNAL THREAD is a thread on the outside of a cylinder such as a bolt (Fig. 10.2).

INTERNAL THREAD is a thread cut on the inside of a part such as a nut (Fig. 10.2).

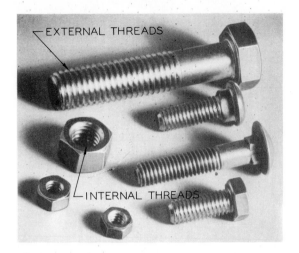

FIG. 10.2 Examples of external threads (bolts) and internal threads (nuts). (Courtesy of Russell, Burdsall & Ward Bolt and Nut Company.)

MAJOR DIAMETER is the largest diameter on an internal or external thread.

MINOR DIAMETER is the smallest diameter that can be measured on a screw thread.

PITCH DIAMETER is the diameter of an imaginary cylinder passing through the threads at the points where the thread width is equal to the space between the threads.

LEAD is the distance that a screw will advance when turned 360°.

PITCH is the distance between crests of threads. Pitch is found mathematically by dividing one inch by the number of threads per inch of a particular thread.

CREST is the peak edge of a screw thread.

THREAD ANGLE is the angle between threads cut by the cutting tool.

ROOT is the bottom of the thread cut into a cylinder.

THREAD FORM is the shape of the thread cut into a threaded part.

THREAD SERIES is the number of threads per inch for a particular diameter, grouped into three series:

coarse, fine, and extra fine. Coarse series provides for rapid assembly, and extra-fine series provides for fine adjustment.

THREAD CLASS is a closeness of fit between two mating threaded parts. Class 1 represents a loose fit and Class 3 a tight fit.

RIGHT-HAND THREAD is a thread that will assemble when turned clockwise. A right-hand thread will slope downward to the right on an external thread when the axis is horizontal, and in the opposite direction on an internal thread.

LEFT-HAND THREAD is a thread that will assemble when turned counterclockwise. A left-hand thread slopes downward to the left on an external thread when the axis is horizontal, and in the opposite direction on an internal thread.

10.3
Thread specifications (English system)

Form

Thread form is the shape of the thread cut into a part, as illustrated in Fig. 10.3. The Unified form, a combination of the American National and the British Whitworth, is the most widely used, because it is a standard in several countries. It is referred to as UN in abbreviations and thread notes. The American National is signified by the letter *N*.

A new thread form, the UNR, was introduced into the 1974 ANSI standards. This designation is specified only for external threads—there is no UNR designation for internal threads. Figure 10.4 compares the profiles of the external UN and UNR threads. The UN form has a flat root (rounded root is optional) in part A, whereas the UNR thread *must* have a rounded root, as shown in part B. The rounded root of the UNR thread is designed to reduce the wear of the threading tool and to improve the fatigue strength of the thread. UNR threads are usually made by rolling.

The transmission of power is achieved by the use of the *Acme*, *square*, *buttress*, and *worm* threads. These are commonly used in gearing and other pieces of machinery. The *sharp V* is used for set screws and in ap-

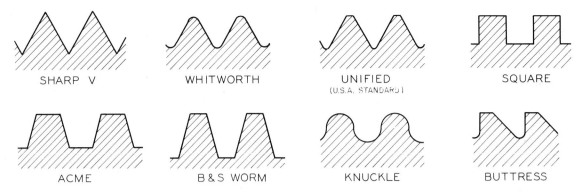

FIG. 10.3 Standard thread forms.

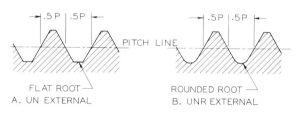

FIG. 10.4 The UN external thread (A) has a flat root (rounded root is optional); the UNR has a rounded root formed by rolling (B). The UNR form does not apply to internal threads.

plications where friction in assembly is desired. The *knuckle* form is a fast-assembling thread used for light assemblies, such as light bulbs and bottle caps.

Series

Thread series is closely related to thread form. It designates the type of thread specified for a given application and is abbreviated as C, F, and EF.

There are eleven standard series of threads listed under the American National form and the Unified National (UN/UNR) form. There are three series, with abbreviations coarse (C), fine (F), and extra-fine (EF), and eight with constant pitches (4, 6, 8, 12, 16, 20, 28, and 32 threads per inch).

A Unified National form for a coarse-series thread is specified as UNC or UNRC, which is a combination of form and series in a single note. Similarly, an American National form for a coarse thread is written NC. The *coarse-thread* series (UNC/UNRC or NC) is suitable for bolts, screws, nuts, and general use with

cast iron, soft metals, or plastics when rapid assembly is desired. The *fine-thread* series (NF or UNF/UNRF) is suitable for bolts, nuts, or screws when a high degree of tightening is required. The *extra-fine* series (UNEF/UNREF or NEF) is used for applications that will have to withstand high stresses. This series is suitable for sheet metal, thin nuts, ferrules, or couplings when length of engagement is limited.

The 8-thread series (8 UN), 12-thread series, (12 N or 12 UN/UNR), and 16-thread series (16 N or 16 UN/UNR) are threads with a uniform pitch for large diameters. The 8 UN is used as a substitute for the coarse-thread series on diameters larger than 1" when a medium-pitch thread is required. The 12 UN is used on diameters larger than 1½", with a thread of a medium-fine pitch as a continuation of the fine-thread series. The 20 UN series is used on diameters larger than 1¹¹/₁₆", with threads of an extra-fine pitch as a continuation of the extra-fine series.

Class of fit

Thread classes are used to indicate the tightness of fit between a nut and a bolt or any two mating threaded parts. This fit is determined by the tolerances and allowances applied to threads. Classes of fit are indicated by the numbers 1, 2, or 3 followed by the letters *A* or *B*. For UN forms, the letter *A* represents an external thread, while the letter *B* represents an internal thread. These letters are omitted when the American National form is used.

CLASS 1A AND 1B threads are used on parts that require assembly with a minimum of binding.

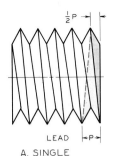
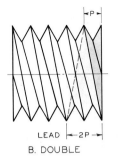
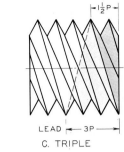

FIG. 10.5 Single and multiple threads.

LEAD |←P→|
A. SINGLE

LEAD |←2P→|
B. DOUBLE

LEAD |←3P→|
C. TRIPLE

CLASS 2A AND 2B threads are general-purpose threads for bolts, nuts, screws, and nominal applications in the mechanical field and are widely used in the mass production industries.

CLASS 3A AND 3B threads are used in precision assemblies where a close fit is desired to withstand stresses and vibration.

Single and multiple threads

A *single thread* (Fig. 10.5A) is a thread that will advance the distance of its pitch in one full revolution of 360°. In other words, its pitch is equal to its lead. In the drawing of a single thread, the crest line of the thread will slope ½P, since only 180° of the revolution is visible in a single view. A double thread is composed of two threads, resulting in a lead equal to 2P, meaning that the threaded part will advance a distance of 2P in a single revolution of 360° (Fig. 10.5B). The crest line of a double thread will slope a distance equal to P in the view in which 180° can be seen. Similarly, a triple thread will advance 3P in 360° with a crest line slope of 1½P in the view in which 180° of the cylinder is visible (Fig. 10.5C). The lead of a double thread is 2P; that of a triple thread, 3P. Although power on multiple threads is limited, they are used wherever quick motion is required.

Thread notes

Drawings of threads are only symbolic representations that are inadequate unless thread notes are applied to give the thread specifications (Fig. 10.6). The major diameter is given first, followed by the number of threads per inch, the form and series, the class of fit, and a letter denoting whether the thread is external or internal. This completes the note for a single right-

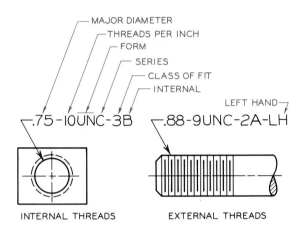

FIG. 10.6 Parts of a thread note for an external thread.

hand thread. However, if the thread is left-hand or double, the word *DOUBLE* or *TRIPLE* is included in the note along with the letters *LH* for left-hand thread.

The UNR thread note is shown for the external thread in Fig. 10.7A (UNR does not apply to internal threads). When inches are used as the unit of measurement, fractions can be written as decimals or as common fractions. Decimal fractions are preferred.

FIG. 10.7 The UNR thread notes apply to external threads only. Notes can be given as decimal fractions or as common fractions, as shown in part B.

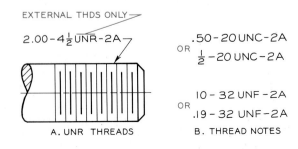

A. UNR THREADS

B. THREAD NOTES

10.4 Using thread tables

The UN/UNR thread table is given in Appendix 15. A portion of this table is shown in Table 10.1. If an external thread (bolt) with a 1.500-inch diameter is to have a "fine" thread, it will have 12 threads per inch. Therefore the thread note can be written as follows:

1½–12 UNF–2A or 1.500–12 UNF–2A.

If the thread had been an internal one (nut), the thread note would have been the same, but the letter *B* would have been used instead of the letter *A*.

A constant-pitch thread series can be selected for the larger diameters. The constant-pitch thread notes are written with the letters *C, F,* or *EF* omitted. For example, a 1¾-inch diameter bolt with a fine thread could be noted in constant-pitch series as

1¾–12 UN–2A or 1.750–12 UN–2A.

This table can also be used for the UNR thread form (for external threads only) by substituting UNR in place of UN; for example, UNEF can be written as UNREF.

10.5 Metric thread specifications (ISO)

Metric thread specifications are recommended by the ISO (International Organization for Standardization). Thread specifications can be given with a *basic desig-*

TABLE 10.1
AMERICAN NATIONAL STANDARD UNIFIED INCH SCREW THREADS (UN AND UNR THREAD FORM)*

Sizes			Threads per Inch											
			Series with Graded Pitches			Series with Constant Pitches								
Primary	Secondary	Basic Major Diameter	Coarse UNC	Fine UNF	Extra-Fine UNEF	UN	6 UN	8 UN	12 UN	16 UN	20 UN	28 UN	32 UN	Sizes
1		1.0000	8	12	20	—	—	UNC	UNF	16	UNEF	28	32	1
	1 1/16	1.0625	—	—	18	—	—	8	12	16	20	28	—	1 1/16
1 1/8		1.1250	7	12	18	—	—	8	UNF	16	20	28	—	1 1/8
	1 3/16	1.1875	—	—	18	—	—	8	12	16	20	28	—	1 3/16
1 1/4		1.2500	7	12	18	—	—	8	UNF	16	20	28	—	1 1/4
	1 5/16	1.3125	—	—	18	—	—	8	12	16	20	28	—	1 5/16
1 3/8		1.3750	6	12	18	—	UNC	8	UNF	16	20	28	—	1 3/8
	1 7/16	1.4375	—	—	18	—	6	8	12	16	20	28	—	1 7/16
1 1/2		1.5000	6	12	18	—	UNC	8	UNF	16	20	28	—	1 1/2
	1 9/16	1.5625	—	—	18	—	6	8	12	16	20	—	—	1 9/16
1 5/8		1.6250	—	—	18	—	6	8	12	16	20	—	—	1 5/8
	1 11/16	1.6875	—	—	18	—	6	8	12	16	20	—	—	1 11/16
1 3/4		1.7500	5	—	—	—	6	8	12	16	20	—	—	1 3/4
	1 13/16	1.8125	—	—	—	—	6	8	12	16	20	—	—	1 13/16
1 7/8		1.8750	—	—	—	—	6	8	12	16	20	—	—	1 7/8
	1 15/16	1.9375	—	—	—	—	6	8	12	16	20	—	—	1 15/16

*By using this table, a diameter of 1½ inches that is to be threaded with a fine thread would have the following thread note: 1½–12 UNF–2A. (Courtesy of ANSI B1.1-1974.)

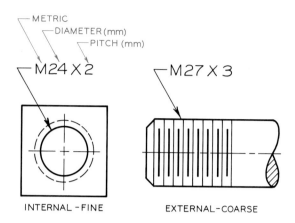

FIG. 10.8 Basic designations for metric threads.

separated by the "×" sign. The pitch can be omitted in notes for coarse threads, but U.S. standards prefer that it be shown. Table 10.2 shows the commercially available ISO threads recommended for general use. Additional ISO specifications are given in Appendix 18.

Complete designation

For some applications it is necessary to show a complete thread designation, as shown in Fig. 10.9. The first part of this note is the same as the basic designa-

nation that is suitable for general applications, or the *complete designation* can be used where detailed specifications are needed.

Basic designation

Examples of metric screw thread notes are shown in Fig. 10.8. Each note begins with the letter *M*, which designates the note as a metric note, followed by the diameter in millimeters and the pitch in millimeters

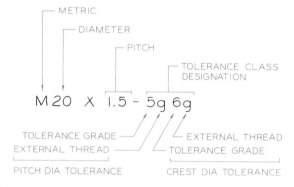

FIG. 10.9 A complete designation note for metric threads.

TABLE 10.2

BASIC THREAD DESIGNATIONS FOR COMMERCIAL SERIES OF ISO METRIC THREADS

Nominal Size (mm)	Pitch, P (mm)	Basic Thread Designation*	Nominal Size (mm)	Pitch, P (mm)	Basic Thread Designation*	Nominal Size (mm)	Pitch, P (mm)	Basic Thread Designation*
1.6	0.35	M1.6	8 <	1.25	M8	22 <	2.5	M22
1.8	0.35	M1.8		1	M8 × 1		1.5	M22 × 1.5
2	0.4	M2	10 <	1.5	M10	24 <	3	M24
2.2	0.45	M2.2		1.25	M10 × 1.25		2	M24 × 2
2.5	0.45	M2.5	12 <	1.75	M12	27 <	3	M27
3	0.5	M3		1.25	M12 × 1.25		2	M27 × 2
3.5	0.6	M3.5	14 <	2	M14	30 <	3.5	M30
4	0.7	M4		1.5	M14 × 1.5		2	M30 × 2
4.5	0.75	M4.5	16 <	2	M16	33 <	3.5	M33
5	0.8	M5		1.5	M16 × 1.5		2	M33 × 2
6	1	M6	18 <	2.5	M18	36 <	4	M36
7	1	M7		1.5	M18 × 1.5		3	M36 × 3
			20 <	2.5	M20	39 <	4	M39
				1.5	M20 × 1.5		3	M39 × 3

*U.S. practice is to include the pitch symbol even for the coarse pitch series. Basic descriptions shown are as specified in ISO Recommendations.

Source: Courtesy of Greenfield Tap and Die Corporation.

TABLE 10.3
TOLERANCE GRADES, ISO THREADS

External Thread		Internal Thread	
Major Diameter (d_1)	Pitch Diameter (d_2)	Minor Diameter (D_1)	Pitch Diameter (D_2)
—	3	—	—
4	4	4	4
—	5	5	5
6	6	6	6
—	7	7	7
8	8	8	8
—	9	—	—

Grade 6 is medium; smaller numbers are finer, and larger numbers are coarser. (Courtesy of ANSI B1-1972.)

tion, but in addition it has a tolerance class designation separated by a dash. The 5g represents the pitch diameter tolerance, and 6g represents the crest diameter tolerance.

The numbers 5 and 6 are *tolerance grades*. Grade 6 is commonly used for a medium general-purpose thread that is nearly equal to the 2A and 2B classes of fit specified under the Unified system. Grades with numbers smaller than 6 are used for "fine"-quality fits and short lengths of engagement. Grades represented by numbers greater than 6 are recommended for "coarse"-quality fits and long lengths of engagement. Table 10.3 gives the tolerance grades for internal and external threads for the pitch diameter and the major and minor diameters.

The letters following the grade numbers designate *tolerance positions*. Lowercase letters represent external threads (bolts), as shown in Fig. 10.10. The lowercase letters *e*, *g*, and *h* represent large allowance, small allowance, and no allowance, respectively. Uppercase letters are used to designate internal threads (nuts); *G*

designates small allowance, and *H* designates no allowance. The letters are placed after the tolerance grade number. For example, 5g designates a medium tolerance with small allowance for the pitch diameter of an external thread, and 6H designates a medium tolerance with no allowance for the minor diameter of an internal thread.

Tolerance classes are fine, medium, and coarse, as listed in Table 10.4. These classes of fit are combinations of tolerance grades, tolerance positions, and lengths of engagement—short (S), normal (N), and long (L). The length of engagement can be determined by referring to Appendix 17. Once it has been decided to use either a fine, medium, or coarse class of fit for a particular application, the specific designation should be selected first from the classes shown in large print in Table 10.4, second from the classes shown in medium-size print, and third from the classes shown in small print. Classes shown in boxes are for commercial threads.

Variations in the complete designation thread notes are shown in Fig. 10.11. The tolerance class symbol is written as *6H* if the crest and pitch diameters have identical grades (part A). Since an uppercase *H* is used, this is an internal thread. Where considered necessary, the length-of-engagement symbol may be added to the tolerance class designation, as shown in part B.

PITCH AND CREST DIA
TOLERANCE SYMBOL
(TOLERANCE EQUAL)

LENGTH OF ENGAGEMENT
GROUP SYMBOL

M22 X 1.5-6H
A

M24 X 3-7g6gL
B

FIG. 10.11 When both pitch and crest diameter tolerance grades are the same, the tolerance class symbol is shown only once (A). Letters S, N, and L are used to indicate the length of the thread engagement (B).

Designations for the desired fit between mating threads can be specified as shown in Fig. 10.12. A slash is used to separate the tolerance class designations of the internal and external threads.

Additional information pertaining to ISO threads may be obtained from *ISO Metric Screw Threads*, a booklet of standards published by ANSI in 1972. These standards were used as the basis for most of this section.

FIG. 10.10 Symbols used to represent tolerance grade, position, and class.

TOLERANCE POSITIONS

EXTERNAL THREADS
LOWER-CASE LETTERS

e = LARGE ALLOWANCE
g = SMALL ALLOWANCE
h = NO ALLOWANCE

INTERNAL THREADS
UPPER-CASE LETTERS

G = SMALL ALLOWANCE
H = NO ALLOWANCE

LENGTH OF ENGAGEMENT

S = SHORT   N = NORMAL   L = LONG

TABLE 10.4
PREFERRED TOLERANCE CLASSES, ISO THREADS*

Quality	External Threads (Bolts)									Internal Threads (Nuts)					
	Tolerance position e (large allowance)			Tolerance position g (small allowance)			Tolerance position h (no allowance)			Tolerance position G (small allowance)			Tolerance position H (no allowance)		
	Length of engagement			Length of engagement			Length of engagement			Length of engagement			Length of engagement		
	Group S	Group N	Group L	Group S	Group N	Group L	Group S	Group N	Group L	Group S	Group N	Group L	Group S	Group N	Group L
Fine							3h4h	4h	5h4h				4H	5H	6H
Medium	**6e**	7e6e	5g6g	**6g**	7g6g	5h6h	6h	7h6h	5G	6G	7G	**5H**	**6H**	**7H**
Coarse					8g	9g8g					7G	8G		7H	8H

*In selecting tolerance class, select first from the large bold print, second from the medium-size print, and third from the small print. Classes shown in boxes are for commercial threads.

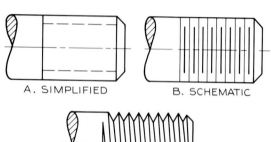

FIG. 10.12 A slash mark is used to separate the tolerance class designations of mating internal and external threads.

FIG. 10.13 Three major types of thread representations.

A. SIMPLIFIED   B. SCHEMATIC

C. DETAILED

10.6
Thread representation

The three major types of thread representations are (A) simplified, (B) schematic, and (C) detailed (Fig. 10.13). The detailed representation is the most realistic approximation of the true appearance of a thread, while the simplified representation is the most symbolic.

10.7
Detailed UN/UNR threads

Examples of detailed representations of internal and external threads are shown in Figs. 10.14A and

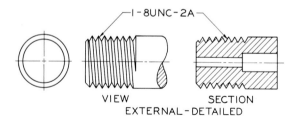

FIG. 10.14A Detailed thread representations of external threads.

VIEW   SECTION
EXTERNAL - DETAILED

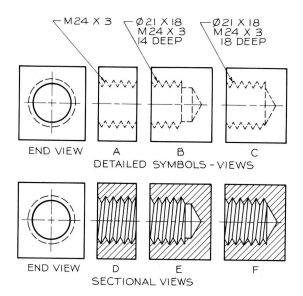

FIG. 24 X 3   Ø21 X 18 M24 X 3 14 DEEP   Ø21 X 18 M24 X 3 18 DEEP

END VIEW   A   B   C
DETAILED SYMBOLS – VIEWS

END VIEW   D   E   F
SECTIONAL VIEWS

FIG. 10.14B Detailed thread representations of internal threads.

10.14B. Instead of helical curves, straight lines are used to indicate crest and root lines.

The construction of a detailed thread representation is shown in Fig. 10.15. The pitch is found by dividing 1″ by the number of threads per inch. This can be done graphically, as shown in Step 1. However, in most cases this construction is unnecessary, since the pitch can be approximated by using an existing scale or by using dividers for spacing. Where threads are close, they should purposely be drawn at a larger spacing to facilitate the drawing process. Note in Step 4 that a 45° chamfer is used to indicate a bevel of the threaded end to improve ease of assembly of the threaded parts.

Metric threads would be drawn in the same manner; however, the pitch would not need to be computed. It is given in millimeters in the metric thread table.

10.8
Detailed square threads

The method of drawing a detailed representation of a square thread is shown in four steps in Fig. 10.16. This method gives an approximation of the true projection of a square thread.

In Step 1, the *major* diameter is laid off. The number of threads per inch is taken from the table in Appendix 19 for this size of thread. The pitch (P) is found by dividing 1″ by the number of threads per inch. Distances of $P/2$ are marked off with dividers.

FIG. 10.15 Detailed thread representation

Step 1 To draw a detailed representation of a 1.75–5 UNC–2A thread, the pitch is determined by dividing 1″ by the number of threads per inch, 5 in this case. The pitch is laid off the length of the thread.

Step 2 Since this is a right-hand thread, the crest lines slope downward to the right equal to ½P. The crest lines will be final lines drawn with an H or F pencil.

Step 3 The root lines are found by constructing 60° vees between the crest lines. The root lines are drawn from the bottom of the vees. Root lines are parallel to each other, but not to crest lines.

Step 4 A 45° chamfer is constructed at the end of the thread from the minor diameter. Strengthen all lines and add a thread note.

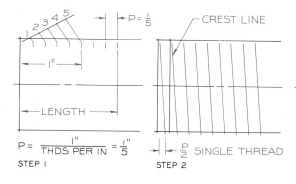

$$P = \frac{1''}{\text{THDS PER IN}} = \frac{1''}{5}$$

STEP 1   STEP 2

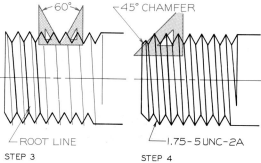

STEP 3   STEP 4

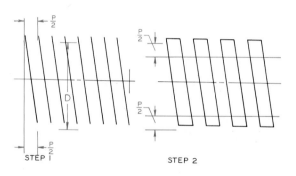
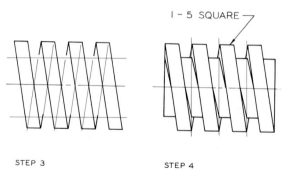

FIG. 10.16 Drawing the square thread

Step 1 Lay out the major diameter. Space the crest lines ½P apart. Slope them downward to the right for right-hand threads.

Step 2 Connect every other pair of crest lines. Find the minor diameter by measuring ½P inward from the major diameter.

Step 3 Connect the opposite crest lines with light construction lines. This will establish the profile of the thread form.

Step 4 Connect the inside crest lines with light construction lines to locate the points on the minor diameter where the thread wraps around the minor diameter.

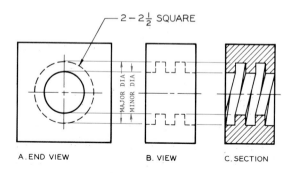

FIG. 10.17 Internal square threads.

FIG. 10.18 Conventional method showing square threads without drawing each thread.

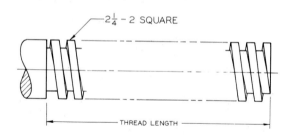

Steps 2, 3, and 4 are completed, and a thread note is added to complete the thread representation.

Square internal threads are drawn in the same manner, as shown in Fig. 10.17. Note that the threads in the section view are drawn in a slightly different way. The thread note for an internal thread is placed in the circular view whenever possible, with the leader pointing toward the center.

When a square thread is rather long, it need not be drawn continuously but can be represented using the symbol shown in Fig. 10.18.

10.9
Detailed Acme threads

The method involved in preparing detailed drawings of Acme threads is shown in Fig. 10.19 in four steps.

The length and the major diameter are laid off with light construction lines. From the table in Appendix 19, the pitch is found by dividing the number of threads per inch into 1″ to begin Step 1.

Steps 2, 3, and 4 complete the thread representation. The thread note is added in the last step.

Internal Acme threads are shown in Fig. 10.20. Note that in the section view, left-hand internal

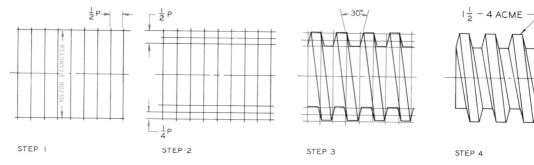

STEP 1   STEP 2   STEP 3   STEP 4

FIG. 10.19 Drawing the Acme thread

Step 1 Lay out the major diameter and the thread length, and divide the shaft into equal divisions ½P apart.

Step 2 Locate the minor diameter a distance ½P inside the major diameter. Locate the pitch diameter between the major and minor diameters.

Step 3 Draw construction lines at 15° with the vertical along the pitch diameter as shown to make a total angle of 30°. Draw the crest lines and the thread profile.

Step 4 Darken the lines, draw the root lines, and add the thread note to complete the drawing.

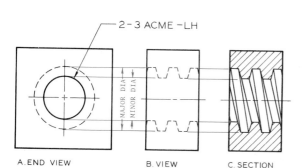

A. END VIEW   B. VIEW   C. SECTION

FIG. 10.20 Internal Acme threads.

FIG. 10.21 Cutting an Acme thread on a lathe. (Courtesy of Clausing Corporation.)

threads are sloped so that they look the same as right-hand external threads.

Figure 10.21 shows a shaft that is being threaded on a lathe. These Acme threads are being cut as the tool travels the length of the shaft.

10.10
Schematic representation

Examples of schematic representations of internal and external threads are shown in Figs. 10.22A and 10.22B. Note that the threads are indicated by parallel nonsloping lines that do not show whether the threads are right-hand or left-hand. This information is given in the thread note. Since this representation is easy to construct and gives a good symbolic representation of threads, it is the most generally used thread symbol.

The method of constructing schematic threads is illustrated in Fig. 10.23 in four steps.

The method of drawing a schematic representation of metric threads is shown in Fig. 10.24. The pitch (in millimeters) can be taken directly from the metric thread table as the distance used to separate the crest lines.

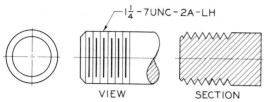

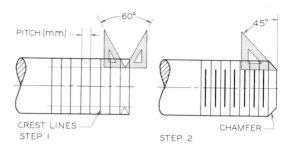

A External—Schematic

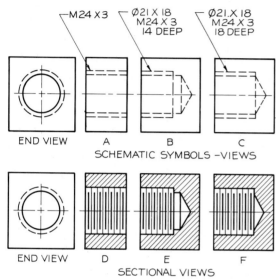

END VIEW A B C

SCHEMATIC SYMBOLS –VIEWS

END VIEW D E F

SECTIONAL VIEWS

B Internal—Schematic

FIG. 10.22 Schematic representations of external threads are shown in part A. Schematic representations of internal threads are shown in part B.

FIG. 10.24 Schematic metric threads

Step 1 The pitch of metric threads can be taken directly from the metric tables, which can be used to find the minor diameter.

Step 2 The root lines are drawn in heavy between the crest lines. The end of the thread is chamfered.

10.11
Simplified threads

Figures 10.25A and 10.25B illustrate the use of simplified representations with notes to specify thread details. Of the three types of thread representations covered, this is the easiest to draw. Hidden lines can be positioned by eye to approximate the minor diameter.

The steps involved in constructing a simplified thread drawing are shown in Fig. 10.26.

FIG. 10.23 Drawing schematic threads

Step 1 Lay out the major diameter and divide the shaft into divisions of a distance P apart. These crest lines should be drawn as thin lines.

Step 2 Find the minor diameter by drawing a 60° angle between two crest lines on each side.

Step 3 Draw heavy root lines between the crest lines.

Step 4 Chamfer the end of the thread from the minor diameter, and give a thread note.

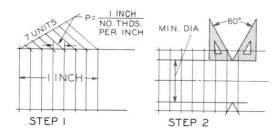

STEP 1   STEP 2

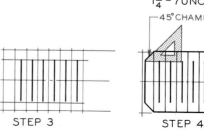

STEP 3

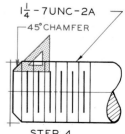

STEP 4

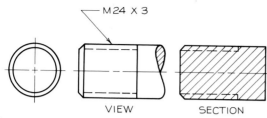

M24 X 3

VIEW   SECTION

A External—Simplified

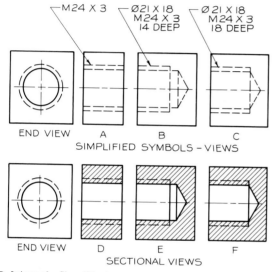

M24 X 3   Ø21 X 18   Ø21 X 18
           M24 X 3   M24 X 3
           14 DEEP   18 DEEP

END VIEW   A   B   C
SIMPLIFIED SYMBOLS – VIEWS

END VIEW   D   E   F
SECTIONAL VIEWS

B Internal—Simplified

FIG. 10.25 Simplified thread representations of external threads (A). Simplified thread representations of internal threads (B).

FIG. 10.26 Drawing simplified threads

Step 1 Lay out the major diameter. Find the pitch (*P*) and lay out two lines a distance *P* apart.

Step 2 Find the minor diameter by constructing a 60° angle between the two lines on both sides.

10.12
Drawing small threads

Instead of using exact measurements to draw small threads, minor diameters can be drawn smaller by approximation in order to separate the root and crest lines, as illustrated in Fig. 10.27. This procedure makes the drawing easier to draw and more readable. Exactness is unnecessary, since the drawing is only a symbolic representation of a thread. Schematic threads are drawn with crest lines farther apart to prevent crowding of lines.

For both internal and external threads, a thread note is added to the symbolic drawing to give the necessary specifications and to complete the description of the threaded part.

10.13
Nuts and bolts

Nuts and bolts come in many forms and sizes for different applications (Fig. 10.28). Drawings of the more common types of threaded fasteners are shown in Fig. 10.29. A *bolt* is a threaded cylinder with a head and a nut for holding two parts together (Fig. 10.29A). A *stud* does not have a head but is screwed into one part with a nut attached to the other end (Fig. 10.29B). A *cap screw* is similar to a bolt, but it does not have a nut; instead it is screwed into a member with internal threads for greater strength (Fig. 10.29C). A *machine*

Step 3 Draw a 45° chamfer from the minor diameter to the major diameter.

Step 4 Show the minor diameter as a dashed line. Add a thread note.

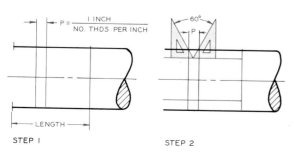

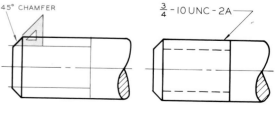

$P = \dfrac{1 \text{ INCH}}{\text{NO. THDS PER INCH}}$

LENGTH

45° CHAMFER

$\frac{3}{4}$ - 10 UNC - 2A

STEP 1   STEP 2   STEP 3   STEP 4

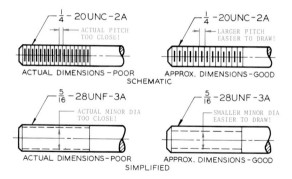

FIG. 10.27 Simplified and schematic threads should be drawn by using larger dimensions if the actual dimensions would result in lines drawn too close together.

FIG. 10.28 Examples of nuts and bolts. (Courtesy of Russell, Burdsall & Ward Bolt and Nut Company.)

screw is similar to a cap screw, but it is smaller (Fig. 10.29D). A *set screw* is used to adjust one member with respect to another, usually to prevent a rotational movement (Fig. 10.29E).

The types of heads used on standard bolts and nuts are illustrated in Fig. 10.30. These heads are used on both types of bolts: *regular* and *heavy*. The thickness of the head is the primary difference between the two series. Heavy-series bolts have the thicker heads and are used at points where bearing loads are heaviest. Bolts and nuts are classified as *finished* and *unfinished*. Figure 10.30 shows an unfinished head; that is, none of the surfaces of the head are machined. The finished head has a washer face that is 1/64″ thick to provide a circular boss on the bearing surface of the bolt head or the nut.

Other standard forms of bolt and screw heads are shown in Fig. 10.31. These heads are used primarily on cap screws and machine screws. Standard types of nuts are illustrated in Fig. 10.30. Finished nuts have washer faces for more accurate assembly. A hexagon *jam nut* does not have a washer face, but it is chamfered on both sides.

Although ANSI tables are provided in Appendixes 21–25 to indicate the standard bolt lengths and their corresponding thread lengths, the following can be used as a general guide for square- and hexagon-head bolts:

Hexagon bolt lengths are available in increments of ¼″ up to 8″ long, in ½″ increments from 8″ to 20″ long, and in 1″ increments from 20″ to 30″ long.

Square-head bolt lengths are available in increments of ⅛″ from lengths of ½″ to ¾″ long, ¼″ increments from ¾″ to 5″ long, ½″ increments from 5″ to 12″ long, and 1″ increments from 12″ to 30″ long.

The lengths of the threads on both hexagon- and square-head bolts up to 6″ long can be found by the formula: Thread length = $2D + ¼″$, where D is the diameter of the bolt. The threaded length for bolts

FIG. 10.29 Types of threaded bolts and screws.

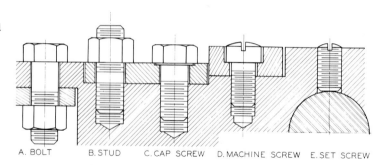

A. BOLT   B. STUD   C. CAP SCREW   D. MACHINE SCREW   E. SET SCREW

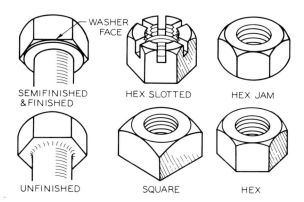

FIG. 10.30 Types of finishes for bolt heads and types of nuts.

over 6″ long can be found by the formula: Thread length = $2D + \frac{1}{2}''$.

The threads for bolts can be coarse, fine, or 8-pitch threads. It is understood that the class of fit for bolts and nuts will be 2A and 2B if no class of fit is specified.

Standard square- and hexagon-head bolts are designated by notes in the following form:

$\frac{3}{8}$–16 × $1\frac{1}{2}$ SQUARE BOLT—STEEL, or
$\frac{1}{2}$–13 × 3 HEX CAP SCREW, SAE GRADE 8—STEEL, or
0.75 × 5.00 HEX LAG SCREW—STEEL

The numbers represent bolt diameter, threads per inch (omit for lag screws), length, name of screw, and material. It is understood that these will have a class 2 fit.

FIG. 10.31 Common types of bolt and screw heads.

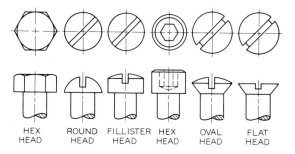

| HEX HEAD | ROUND HEAD | FILLISTER HEAD | HEX HEAD | OVAL HEAD | FLAT HEAD |

Nuts are designated by notes in the following form:

$\frac{1}{2}$–13 SQUARE NUT—STEEL, or
$\frac{3}{4}$–16 HEAVY HEX NUT, SAE GRADE 5—STEEL, or
1.00–8 HEX THICK SLOTTED NUT—CORROSION RESISTANT STEEL

When not noted as "HEAVY," nuts are assumed to be regular nuts. The class of fit is assumed to be 2B for nuts when not noted.

10.14
Drawing the square bolt head

Detailed tables are available in Appendix 20 and in published standards for various types of threaded parts. In most cases it is sufficient to draw nuts and bolts with only general proportions.

The first step in drawing a bolt head or a nut is to determine whether it is to be *across corners* or *across flats*. In other words, are the outlines at either side of the view going to represent corners, or are they going to be edge views of flat surfaces of the part? The square head in Fig. 10.32 is drawn across corners. Nuts and bolts should be drawn across corners whenever possible.

10.15
Drawing the hexagon bolt head

It is desirable that nuts and bolts be drawn across corners since this gives a better impression of the parts. An example of constructing the head of a hexagon-head bolt is shown in Fig. 10.33

Note that the diameter of the bolt is D. The thickness of the head is drawn equal to $\frac{2}{3}D$. The top view of the head is drawn as a circle with a radius of $\frac{3}{4}D$. For most applications, this proportionality based on D is sufficient for drawing bolt heads.

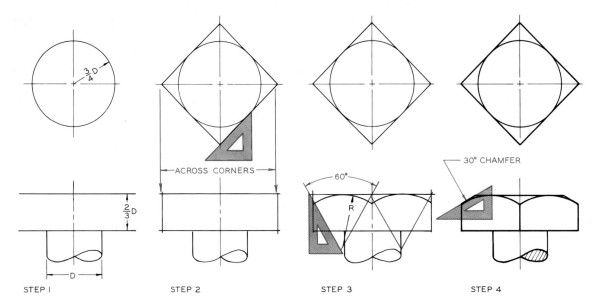

STEP I STEP 2 STEP 3 STEP 4

FIG. 10.32 Drawing the square head

Step 1 Draw the diameter of the bolt. Use the diameter to establish the head diameter and thickness.

Step 2 Draw the top view of the square head with a 45° triangle to give an across-corners view.

Step 3 Show the chamfer in the front view by using a 30°–60° triangle to find the centers for the radii.

Step 4 Show a 30° chamfer tangent to the arcs in the front view. Strengthen the lines.

FIG. 10.33 Drawing the hexagon head

Step 1 Draw the diameter of the bolt. Use the diameter to establish the head diameter and thickness.

Step 2 Construct a hexagon with a 30°–60° triangle to give an across-corners view.

Step 3 Find arcs in the front view to show the chamfer of the head.

Step 4 Draw a 30° chamfer that is tangent to the arcs in the front view. Strengthen the lines.

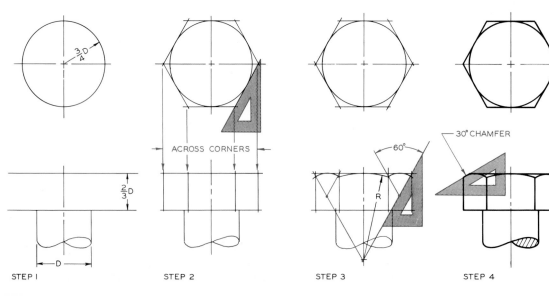

STEP I STEP 2 STEP 3 STEP 4

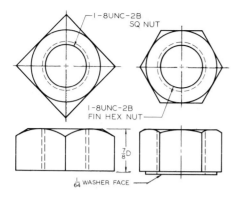

FIG. 10.34 Drawings of hexagon and square nuts are constructed in the same manner as drawings of bolt heads. Standard notes are added to give nut specifications.

10.16
Drawing nuts

The construction of a drawing of a square nut or a hexagon nut across corners is exactly the same as the construction of a drawing of a bolt head across corners. The only variation is the thickness of the nut. The regular nut thickness is $\frac{7}{8}D$, and for the heavy

FIG. 10.35 Examples of hexagon and square nuts drawn across flats. Notes are added to give nut specifications.

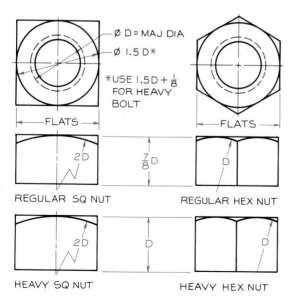

nut, the thickness is equal to the diameter (D).

Examples of square and hexagon nuts drawn across corners are shown in Fig. 10.34. Hidden lines are shown in the front view to indicate threads. Since it is understood that nuts are threaded, these hidden lines may be omitted in general applications.

Note that a $\frac{1}{64}''$ washer face is shown on the hexagon nut. This is usually drawn thicker than $\frac{1}{64}''$ so that the face will be more noticeable in the drawing. Thread notes are placed in the circular views where possible. In the case of the square nut, the note tells us that the major diameter of the thread is 1″, that the nut has 8 threads per inch, that the thread is of the Unified National form and coarse series, with a fit of 2, and that it is a regular square nut since it is not labeled as "HEAVY." The hexagon nut is similar except that it is a finished hexagon nut.

The leader from the note is directed toward the center of the circular view, but the arrow stops at the first visible circle it makes contact with.

Nuts can be drawn across flats in situations where doing so improves the drawing. Examples of nuts drawn across flats are shown in Fig. 10.35.

For regular nuts, the distance across flats is $1\frac{1}{2} \times D$ (D is the major diameter of the thread). For heavy nuts this distance is increased by $\frac{1}{8}''$. The top views are drawn in the same manner as in across-corners drawings, except that they are positioned to give different front views.

In the case of the square nut (Fig. 10.35), the front view is a simple rectangle, with only the arc formed by the chamfer giving a hint that it is a nut.

The hexagon nut drawn across flats looks more like a nut in the front view than does the square nut. Still, the hexagon nut drawn across corners is a better representation (Fig. 10.35). Notes are added with leaders to complete the representation of the nuts. A washer face should be added to a nut if it is finished, except in the case of the square nut. Square nuts are always unfinished.

10.17
Drawing nuts and bolts in combination

The same rules followed in drawing nuts and bolts separately apply when drawing nuts and bolts in assembly. Examples are shown in Fig. 10.36.

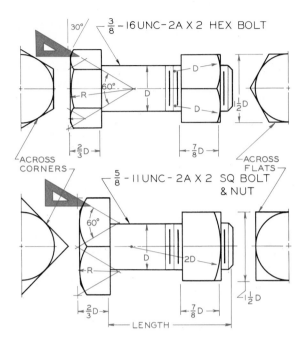

$\frac{3}{8}$–16UNC–2A X 2 HEX BOLT

$\frac{5}{8}$–11UNC–2A X 2 SQ BOLT & NUT

FIG. 10.36 Construction of nuts and bolts in assembly.

The diameter of the bolt is used as the basis for other dimensions. The note is added to give the specifications of the nut and bolt. In the figure, the bolt heads are drawn across corners and the nuts across flats. The half-end views have been included to show how the front views were found by projection.

10.18
Cap screws

Cap screws are used to hold two parts together without the use of a nut. One of these two parts has a threaded cylindrical hole and thus serves the same function as the nut. The other part is drilled with an oversize hole so that the cap screw will pass through it freely. When the cap screw is tightened, the two parts are held securely together.

The standard types of cap screws are illustrated in Fig. 10.37. Tables are available in Appendixes 22 and 23 to give the dimensions of several of these types of cap screws.

The cap screws in Fig. 10.37 are drawn on a grid in order to show the proportions of each type. The

proportions shown here can be used for drawing cap screws of all sizes. These types of cap screws range in diameter from No. 0 (0.060″) to 1½″.

10.19
Machine screws

Machine screws are smaller than most cap screws, usually less than 1″ in diameter. The machine screw is used to attach parts together; it is screwed either into another part or into a nut. Machine screws are threaded their full length when their length is 2″ or shorter.

Drawings of common machine screws and their notes are given in Fig. 10.38. Many other types of machine screws are available. The dimensions of round-head machine screws are given in Appendix 24.

The four types of machine screws in Fig. 10.38 are drawn on a grid to give the proportions of the head in relation to the major diameter of the screw. Machine screws range in diameter from No. 0 (0.060″) to ¾″.

When slotted-head screws are drawn, it is conventional practice to show the slots positioned at a 45° angle in the circular view, as illustrated in Fig. 10.31.

FIG. 10.37 The proportions of the standard types of cap screws are shown here for drawing cap screws of all sizes. Notes are given to provide typical specifications.

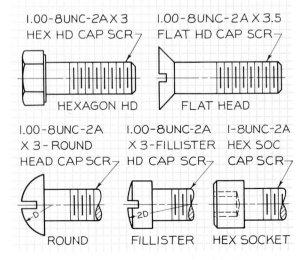

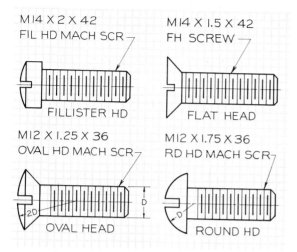

M14 X 2 X 42
FIL HD MACH SCR

FILLISTER HD

M14 X 1.5 X 42
FH SCREW

FLAT HEAD

M12 X 1.25 X 36
OVAL HD MACH SCR

OVAL HEAD

M12 X 1.75 X 36
RD HD MACH SCR

ROUND HD

FIG. 10.38 Standard types of machine screws. The proportions shown here can be used for drawing machine screws of all sizes.

Even though the slot is turned at 45° in the top view, the front view of the slot is drawn to show the width and depth of the slot.

10.20
Set screws

Parts such as wheels or pulleys are commonly attached to shafts. Set screws or keys are used to afix these parts on a shaft. Examples of various types of set screws are shown in Fig. 10.39.

Table 10.5 shows the dimensions of the various features of the set screws shown in Fig. 10.39. Drawings of set screws need not employ these dimensions precisely; like the other fasteners discussed in this chapter, set screw threads can be drawn as approximations.

Note that the points and the heads of these set screws are of different types. Set screws are available in any desired combination of point and head. The shaft against which the set screw is tightened may have a flat surface machined to give a good bearing surface for the set screw point. In this case, a dog point

FIG. 10.39 Types of set screws. Set screws are available with various combinations of heads and points. Notes give their specifications. Dimensions are given in Table 10.5.

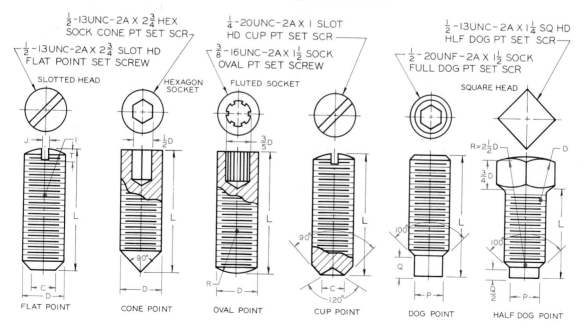

$\frac{1}{2}$-13UNC-2A X 2$\frac{3}{4}$ HEX SOCK CONE PT SET SCR

$\frac{1}{2}$-13UNC-2A X 2$\frac{3}{4}$ SLOT HD FLAT POINT SET SCREW

$\frac{1}{4}$-20UNC-2A X 1 SLOT HD CUP PT SET SCR

$\frac{3}{8}$-16UNC-2A X 1$\frac{1}{2}$ SOCK OVAL PT SET SCREW

$\frac{1}{2}$-13UNC-2A X 1$\frac{1}{4}$ SQ HD HLF DOG PT SET SCR

$\frac{1}{2}$-20UNF-2A X 1$\frac{1}{2}$ SOCK FULL DOG PT SET SCR

SLOTTED HEAD HEXAGON SOCKET FLUTED SOCKET SQUARE HEAD

FLAT POINT CONE POINT OVAL POINT CUP POINT DOG POINT HALF DOG POINT

TABLE 10.5
DIMENSIONS FOR THE SET SCREWS SHOWN IN FIG. 10.39 (ALL DIMENSIONS GIVEN IN INCHES)

D	I	J	T	R	C		P		Q	q
					Diameter of Cup and Flat Points		Diameter of Dog Point		Length of Dog Point	
Nominal Size	Radius of Headless Crown	Width of Slot	Depth of Slot	Oval Point Radius	Max	Min	Max	Min	Full	Half
5 0.125	0.125	0.023	0.031	0.094	0.067	0.057	0.083	0.078	0.060	0.030
6 0.138	0.138	0.025	0.035	0.109	0.074	0.064	0.092	0.087	0.070	0.035
8 0.164	0.164	0.029	0.041	0.125	0.087	0.076	0.109	0.103	0.080	0.040
10 0.190	0.190	0.032	0.048	0.141	0.102	0.088	0.127	0.120	0.090	0.045
12 0.216	0.216	0.036	0.054	0.156	0.115	0.101	0.144	0.137	0.110	0.055
¼ 0.250	0.250	0.045	0.063	0.188	0.132	0.118	0.156	0.149	0.125	0.063
⁵⁄₁₆ 0.3125	0.313	0.051	0.076	0.234	0.172	0.156	0.203	0.195	0.156	0.078
⅜ 0.375	0.375	0.064	0.094	0.281	0.212	0.194	0.250	0.241	0.188	0.094
⁷⁄₁₆ 0.4375	0.438	0.072	0.109	0.328	0.252	0.232	0.297	0.287	0.219	0.109
½ 0.500	0.500	0.081	0.125	0.375	0.291	0.270	0.344	0.344	0.250	0.125
⁹⁄₁₆ 0.5625	0.563	0.091	0.141	0.422	0.332	0.309	0.391	0.379	0.281	0.140
⅝ 0.625	0.625	0.102	0.156	0.469	0.371	0.347	0.469	0.456	0.313	0.156
¾ 0.750	0.750	0.129	0.188	0.563	0.450	0.425	0.563	0.549	0.375	0.188

Source: Courtesy of ANSI B18.6.2-1956.

FIG. 10.40   Miscellaneous types of bolts.

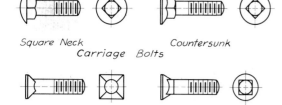

Square Neck
Carriage Bolts

Countersunk

Square Head
Countersunk Plow Bolts

Round Head, Square Neck

Step Bolt

Button Head
Machine Bolt

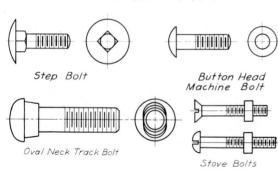

Oval Neck Track Bolt

Stove Bolts

or a flat point would be most effective to press against the flat surface. The cup point gives good friction when applied to a round shaft.

Specifications for set screws are given in Appendixes 27, 28, and 29.

10.21
Miscellaneous screws

A few of the many types of specialty screws are shown in Fig. 10.40, each having its own special application. Others shown in Fig. 10.41 are: lag bolt, hanger bolt, Phillips head screw, and drive screw.

FIG. 10.41   Examples of specialty bolts and screws.

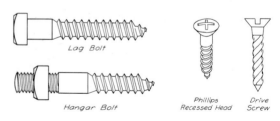

Lag Bolt

Hangar Bolt

Phillips
Recessed Head

Drive
Screw

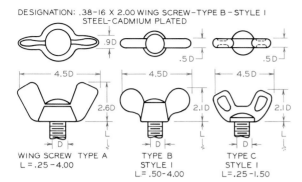

DESIGNATION: .38-16 X 2.00 WING SCREW–TYPE B – STYLE I
STEEL-CADMIUM PLATED

WING SCREW TYPE A
L = .25 – 4.00

TYPE B
STYLE I
L= .50-4.00

TYPE C
STYLE I
L=.25 –1.50

FIG. 10.42 Wing screw proportions are shown for screw diameters of about 5/16″. These proportions can be used to draw wing screws of any size diameter. Type A is available in screw diameters of 4, 6, 8, 10, 12, 0.25, 0.313, 0.375, 0.438, 0.50, and 0.625. Type B is available in diameters of 10 to 0.625; and Type C in diameters of 6 to 0.375.

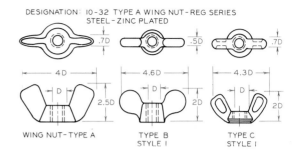

DESIGNATION: 10-32 TYPE A WING NUT–REG SERIES
STEEL – ZINC PLATED

WING NUT– TYPE A

TYPE B
STYLE I

TYPE C
STYLE I

FIG. 10.44 Wing nut proportions are given for screw diameters of 3/8″. These proportions can be used to draw thumb screws of any size. Type A wing nuts are available in screw diameters (in inches) of 3, 4, 5, 6, 8, 10, 12, 0.25, 0.313, 0.375, 0.438, 0.50, 0.583, 0.625, and 0.75. Type B nuts are available in sizes from 5 to 0.75, and Type C nuts in sizes from 4 to 0.50.

Wing screws of three types are shown in Fig. 10.42. They are available in incremental lengths of 1/8″. Wing screws are used to join parts that are assembled and disassembled by hand.

Thumb screws of two types are shown in Fig. 10.43. These serve the same purpose as wing screws.

Wing nuts of three types are shown in Fig. 10.44. They can be used with screws of types other than wing screws or thumb screws.

The manufacture and design of specialized threaded fasteners is a career field within itself, with an ever-growing need for additional fasteners of various types.

10.22
Wood screws

A wood screw is a pointed screw having a sharp thread of coarse pitch for insertion in wood. The three most common types of wood screws are shown in Fig. 10.45. These examples are drawn on a grid to show the proportions of the various heads in relation to the

FIG. 10.45 Standard types of wood screws. The proportions shown here can be used for drawing wood screws of all sizes.

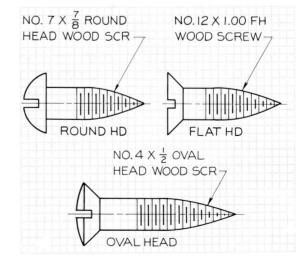

NO. 7 X 7/8 ROUND HEAD WOOD SCR — ROUND HD

NO.12 X 1.00 FH WOOD SCREW — FLAT HD

NO. 4 X 1/2 OVAL HEAD WOOD SCR — OVAL HEAD

FIG. 10.43 Thumb screw proportions are given for screw diameters of about 1/4″. These proportions can be used to draw thumb screws of any screw diameter. Type A is available in diameters of 6, 8, 10, 12, 0.25, 0.313, and 0.375. Type B thumb screws are available in diameters of 6 to 0.50.

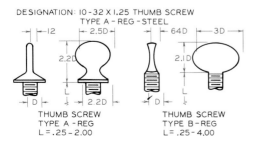

DESIGNATION: 10-32 X 1.25 THUMB SCREW
TYPE A – REG -STEEL

THUMB SCREW
TYPE A - REG
L =.25 – 2.00

THUMB SCREW
TYPE B – REG
L= .25 – 4.00

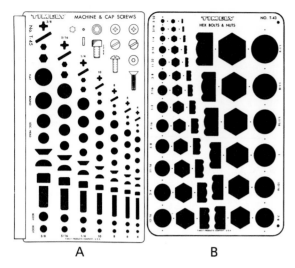

FIG. 10.46 Examples of templates that can be used for drawing threaded fasteners. (Courtesy of Timely Products Company.)

major diameter of the screw. Detailed dimensions for wood screws are given in tables published by the American National Standards Institute.

Sizes of wood screws are specified by single numbers, such as 0, 6, or 16. From 0 to 10 each digit represents a different size. Beginning at 10, only even-numbered sizes are standard; that is, 10, 12, 14, 16, 18, 20, 22, and 24. The following formula can be used

FIG. 10.47 Three types of taps for threading internal holes.

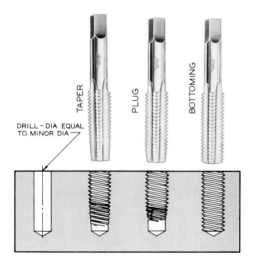

to relate these numbered sizes to the actual diameter of the screws:

Actual DIA = 0.060 + screw number × 0.013.

For example, the diameter of a No. 5 screw is

0.060 + 5(0.013) = 0.125.

10.23
Use of templates

Templates are available for drawing threads, nuts, and threaded fasteners. They are available for a range of sizes and are satisfactory for most applications, since thread representations are approximations at best.

Two typical templates are shown in Fig. 10.46. The black areas represent the holes cut into the thin plastic templates. The template is laid on the drawing, and the threaded features are drawn by using the template as a guide. Templates are also available for drawing nuts and bolts in pictorials. These are used primarily by the technical illustrator.

10.24
Tapping a hole

A threaded hole is called a *tapped hole*, since the tool used to cut the threads is called a tap. The types of taps available for threading small holes by hand are shown in Fig. 10.47.

The taper, plug, and bottoming hand taps are identical in size, length, and measurements, their only difference being the chamfered portion of their ends. The taper tap has a long chamfer (8 to 10 threads), the plug tap has a chamfer of 3 to 5 threads, and the bottoming tap has a short chamfer of only 1 to 1 ½ threads.

When tapping by hand in open or "through" holes, the taper should be used for coarse threads, since it ensures straighter starting. The taper tap is also recommended for the harder metals. The plug tap can be used in soft metals or for fine-pitch threads. When it is desirable to tap a hole to the very bottom, all three taps—taper, plug, and bottoming—should be used in this order.

Notes are added to specify the depth of the drilled hole and the depth of the threads. For example, a note reading ⅞ DIA, 3 DEEP, 1–8 UNC–2A, 2 DEEP means that the hole will be drilled deeper than it is threaded, and the last usable thread will be 2″ deep in the hole. Note that the drill point has an angle of 120°.

10.25
Washers, lock washers, and pins

Washers, called *plain washers*, are used with nuts and bolts to improve the assembly and the strength of the fastening. Plain washers are noted on a working drawing in the following manner:

0.938 × 1.750 × 0.134 TYPE A PLAIN WASHER

These numbers represent the inside diameter, the outside diameter, and the thickness, in that order. These dimensions can be found in Appendix 36.

A *lock washer* is a washer that prevents a nut or cap screw from loosening as a result of vibration or movement. Two of the more common types, shown in Fig. 10.48, are the *external-tooth lock washer* and the *helical-spring lock washer*. Other, more varied types

FIG. 10.48 Two types of lock washers for preventing a bolt from unscrewing.

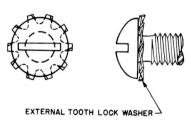

EXTERNAL TOOTH LOCK WASHER

SPRING LOCK WASHER

of locking washers and devices are shown in Fig. 10.49.

Tables for spring lock washers are given in Appendix 37 for regular and extra-heavy-duty helical-spring lock washers. These are designated on drawings with notes in the following form:

HELICAL-SPRING LOCK WASHER–¼ REGULAR–
PHOSPHOR BRONZE.

(The ¼ is the washer's inside diameter).

Tooth lock washers are designated with notes in the following form:

INTERNAL-TOOTH LOCK WASHER–¼-TYPE A–
STEEL, or

EXTERNAL-TOOTH LOCK WASHER–562-TYPE B–
STEEL.

Straight pins and taper pins are used to fix parts together in a specified alignment. Dimensions for these are given in Appendix 34.

Other locking devices are cotter pins, split taper pins, and straight pins (Fig. 10.50). Tables of specifications for cotter pins are given in Appendix 38.

10.26
Pipe threads

Pipe threads are used in connecting pipes, tubing, lubrication fittings, and other applications. The most commonly used pipe thread is tapered at a ratio of 1 to 16, but straight pipe threads are available (without a taper). Since pipe threads are usually tapered, the threads will engage only for an effective length determined by the formula below:

$$L = (0.80D + 6.8)P,$$

where D is the outside diameter of the pipe and P is the pitch of the thread.

Methods of representing tapered threads are shown in Fig. 10.51. Notice that a taper of 1 to 16 is shown to call attention to the fact that the threads are tapered.

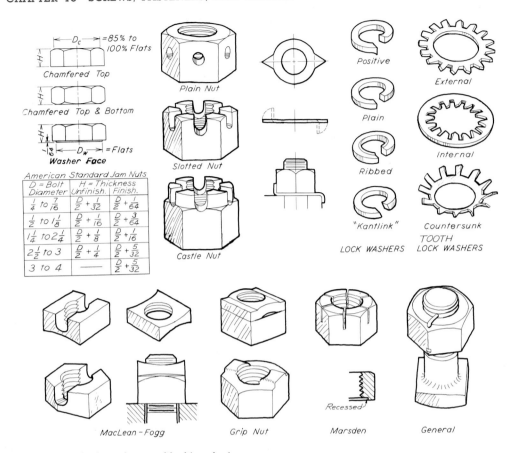

American Standard Jam Nuts		
D = Bolt Diameter	H = Thickness	
	Unfinish.	Finish.
$\frac{1}{4}$ to $\frac{7}{16}$	$\frac{D}{2}+\frac{1}{32}$	$\frac{D}{2}+\frac{1}{64}$
$\frac{1}{2}$ to $1\frac{1}{8}$	$\frac{D}{2}+\frac{1}{16}$	$\frac{D}{2}+\frac{3}{64}$
$1\frac{1}{4}$ to $2\frac{1}{4}$	$\frac{D}{2}+\frac{1}{8}$	$\frac{D}{2}+\frac{1}{16}$
$2\frac{1}{2}$ to 3	$\frac{D}{2}+\frac{1}{4}$	$\frac{D}{2}+\frac{5}{32}$
3 to 4	—	$\frac{D}{2}+\frac{5}{32}$

FIG. 10.49 Types of lock washers and locking devices.

FIG. 10.50 Types of pins used to fix parts together.

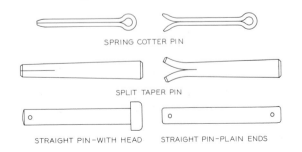

FIG. 10.51 Schematic and simplified techniques of representing external and internal pipe threads.

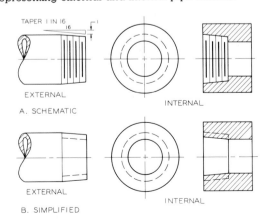

The following abbreviations are associated with pipe threads:

N = National	G = Grease
P = Pipe	I = Internal
T = Taper	M = Mechanical
C = Coupling	L = Locknut
S = Straight	H = Hose coupling
F = Fuel and oil	R = Railing fittings

These abbreviations are used in combination for the following ANSI symbols:

NPT = National pipe taper
NPTF = National pipe thread (dryseal—for pressure-tight joints)
NPS = Straight pipe thread
NPSC = Straight pipe thread in couplings
NPSI = National pipe straight internal thread
NPSF = Straight pipe thread (dryseal)
NPSM = Straight pipe thread for mechanical joints
NPSL = Straight pipe thread for locknuts and locknut pipe threads
NPSH = Straight pipe thread for hose couplings and nipples
NPTR = Taper pipe thread for railing fittings

To specify a pipe thread in note form, the nominal pipe diameter (the internal diameter), the number of threads per inch, and the symbol that denotes the type of thread are given. For example:

$$1\frac{1}{4}-11\frac{1}{2} \text{ NPT} \quad \text{or} \quad 3-8 \text{ NPTR.}$$

FIG. 10.52 Typical pipe-thread notes.

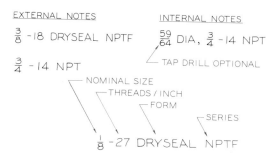

These specifications can be taken from Appendix 10. Examples of external and internal thread notes are shown in Fig. 10.52. The dryseal thread is used in applications where a pressure-tight joint is required without the use of a lubricant or sealer. Dryseal threads may be straight or tapered. No clearance between the mating parts of the joint is permitted, giving the highest-quality fit.

The tap drill is sometimes given in the internal pipe thread note (Fig. 10.52), but this is optional.

10.27
Keys

Keys are used to attach parts to shafts in order to transmit power to pulleys, gears, or cranks. Several types of keys are shown pictorially and orthographically in Fig. 10.53. The four types illustrated here are the most commonly used keys. To specify a key, notes must be given for the keyway, the key, and the keyseat, as shown in Fig. 10.53A, C, E, and G. The notes given are typical of the notes used to give key specifications. These dimensions may be found in Appendixes 31 and 32 for various types of keys.

10.28
Rivets

Rivets are fasteners used to join thin materials in a permanent joint. Rivets are designed to fit into holes that are slightly larger than the diameter of the rivet. The rivet is inserted in the hole, and the headless end is formed into the specified shape by applying pressure to the projecting end. This forming operation is done when the rivets are either hot or cold, depending on the application.

Typical shapes and proportions of small rivets are shown in Fig. 10.54. These rivets vary in diameter from 1/16" to 1¾". Rivets are used extensively in pressure-vessel fabrication, in heavy structures such as bridges and buildings, and in construction with sheet metal.

The proportions for large rivets, shown in Fig. 10.55, are based on the diameters of the rivet bodies. Many ANSI tables of standard dimensions are available for sizing rivets.

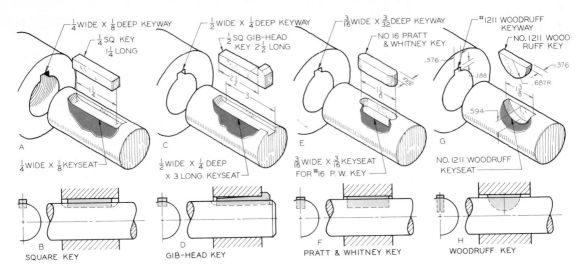

FIG. 10.53 Standard keys used to hold parts on a shaft.

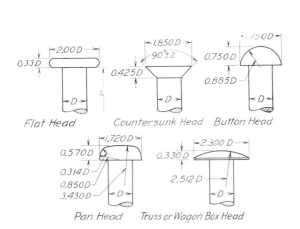

FIG. 10.54 Types and proportions of small rivets. Small rivets have shank diameters up to ½".

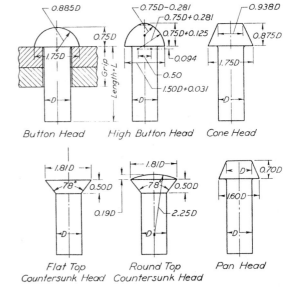

FIG. 10.55 Types and proportions of large rivets. Large rivets have shank diameters of ½" DIA to 1¾" DIA.

Three types of lap joints are shown in Fig. 10.56. The joints are held secure by one, two, or three rivets, as shown in the sectional view. Note that the bodies of the rivets are drawn as hidden circles in the top views.

Single and double riveted butt joints are illustrated in Fig. 10.57.

The standard symbols recommended by ANSI for representing rivets are shown in Fig. 10.58. Rivets that

are driven in the shop are called shop rivets, and those assembled on the job at the site are called field rivets.

10.29
Springs

Of the many types of springs that are available, some of the more commonly used ones are (1) compression, (2) torsion, (3) extension, (4) flat, and (5) constant

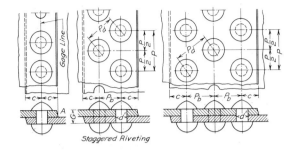

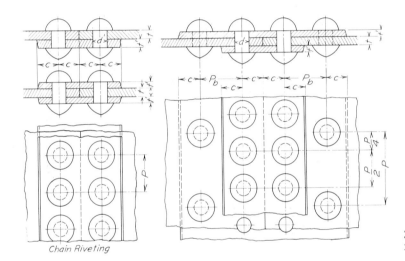

FIG. 10.56 Examples of lap joints using single rivets, double rivets, and triple rivets. The fourth example shows three plates that are fastened by double rivets.

force. Figure 10.59 shows the single-line conventional representation of the first three of these. Also shown are the types of ends that can be used on compression springs and also the simplified single-line representation of coil springs.

A typical working drawing of a compression spring is shown in Fig. 10.60. The ends of the spring are drawn by using the double-line representation, and conventional lines are used to indicate the undrawn portion of the spring. Only the diameter and the free length of the spring are given on the drawing itself. The remaining specifications are given in tabular form near the drawing.

FIG. 10.57 Single and double riveted butt joints.

FIG. 10.58 The symbols used to represent rivets in a drawing. (Courtesy of ANSI 14.14.)

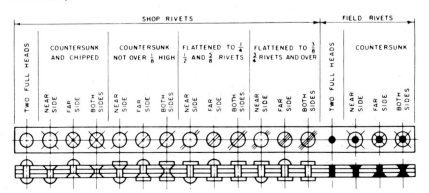

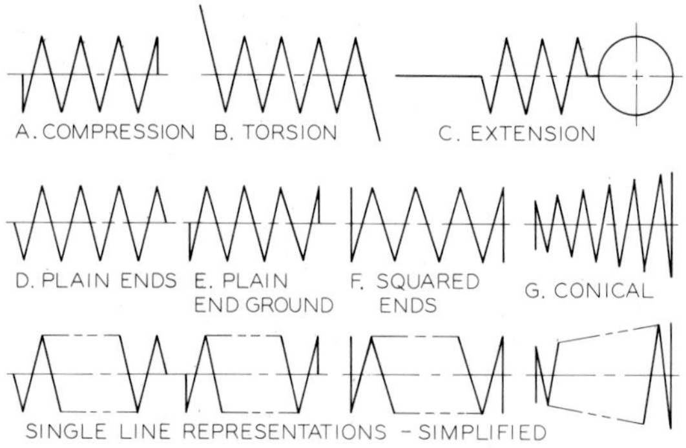

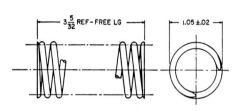

FIG. 10.59 Single-line representations of various types of springs.

A. COMPRESSION B. TORSION C. EXTENSION

D. PLAIN ENDS E. PLAIN END GROUND F. SQUARED ENDS G. CONICAL

SINGLE LINE REPRESENTATIONS – SIMPLIFIED

$3\frac{5}{32}$ REF – FREE LG   1.05 ±.02

FIG. 10.60 A conventional double-line drawing of a compressions spring and its specifications. (Courtesy of the U.S. Department of Defense.)

WIRE DIA   .120
DIRECTION OF HELIX   OPTIONAL
TOTAL COILS $12\frac{1}{2}$ REF
LOAD AT COMPRESSED LG OF 2.05 IN.
= 39 LB ± 3.9 LB
LOAD AT COMPRESSED LG OF 1.69 IN.
= 51.5 LB ± 5.2 LB

FIG. 10.61 A conventional double-line drawing of an extension spring and its specifications. (Courtesy of the U.S. Department of Defense.)

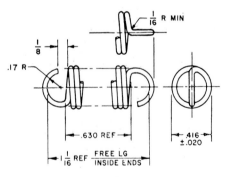

$\frac{1}{16}$ R MIN

$\frac{1}{8}$

.17 R

.630 REF   .416 ±.020

$\frac{1}{16}$ REF   FREE LG INSIDE ENDS

WIRE DIA   .042
DIRECTION OF HELIX   OPTIONAL
TOTAL COILS   14 REF
RELATIVE POSITION OF ENDS 180° ± 20°
EXTENDED LG INSIDE ENDS
WITHOUT PERMANENT SET   2.45 IN. (MAX)
INITIAL TENSION   1.0 LB ± .10 LB
LOAD   4 LB ± .4 LB AT 1.56 IN.
EXTENDED LG INSIDE ENDS
LOAD   6.3 LB ± .63 LB AT 1.95 IN.
EXTENDED LG INSIDE ENDS

A working drawing of an extension spring (Fig. 10.61) is very similar to that of a compression spring. In a drawing of a helical torsion spring (Fig. 10.62), angular dimensions must be shown to specify the initial and final positions of the spring as torsion is applied to it. All types of springs require a table of specifications to describe their details.

10.30
Drawing springs

Springs may be drawn as schematic representations using single lines to represent the springs. Examples of single-line drawings are shown in Fig. 10.63. Each is drawn by laying out the diameter of the coils and the lengths of the springs, and then the number of active coils are drawn by using the diagonal-line method.

In part B, the two end coils are "dead" coils, and only four are active. An extension spring is drawn at C.

For applications where more realism is desired, a double-line drawing of a thread can be made as shown in Fig. 10.64. The end result is a good approximation of the spring.

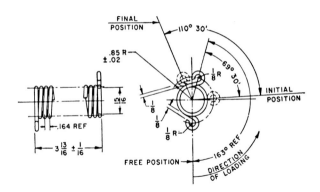

WIRE DIA .148
DIRECTION OF HELIX LEFT HAND
TOTAL COILS 20.55 REF
TORQUE 15 LB IN. ± 1.5 LB IN. AT INITIAL POSITION
TORQUE 33 LB IN. ± 3.3 LB IN. AT FINAL POSITION
MAXIMUM DEFLECTION WITHOUT SET BEYOND FINAL POSITION 56°
SPRING RATE .16 LB IN. / DEG REF

FIG. 10.62 A conventional double-line drawing of a helical torsion spring and its specifications. (Courtesy of the U.S. Department of Defense.)

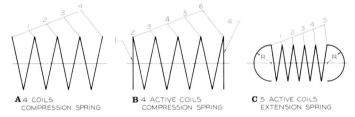

A 4 COILS
COMPRESSION SPRING

B 4 ACTIVE COILS
COMPRESSION SPRING

C 5 ACTIVE COILS
EXTENSION SPRING

FIG. 10.63 In part A, a schematic drawing of a spring with 4 active coils as shown. The diagonal-line method is used to divide it into 4 equally spaced coils. The spring at B has 6 coils, but only 4 of them are active cells. An extension spring with 5 active coils is shown at C.

FIG. 10.64 Detailed drawing of a spring

Step 1 Lay out the diameter and the length of the spring, and locate the coils by the diagonal-line technique.

Step 2 Locate the coils on the lower side along the bisectors of the spaces between the coils on the upper side.

Step 3 Connect the coils on each side. This is a right-hand coil; a left-hand spring would slope in the opposite direction.

Step 4 Construct the back side of the spring and the end coils to complete the detailed drawing of a compression spring.

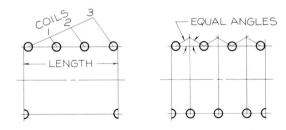

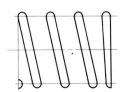

Problems

These problems are to be completed on Size A sheets. Problems 1 through 16 are to be laid out one problem per sheet. Problems 17 through 30 are to be laid out two problems per sheet. The boxes drawn around the figures representing the problems are approximately 6″ wide × 5″ high, which equals about half a Size A sheet.

Fasteners

1. The layout in Fig. 10.65 is to be used for constructing a detailed representation of an Acme thread with a major diameter of 3″. The thread note specifications are 3–1½ ACME. Show both external and internal thread representations. Show the thread note. Use inches or millimeters as instructed.

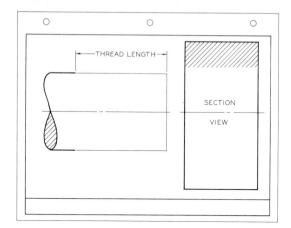

FIG. 10.65   Construction of thread symbols (Problems 1, 2, and 3).

2. Repeat Problem 1, but draw internal and external detailed representations of a square thread that is 3″ in diameter. The note specifications are 3–1½ SQUARE. Apply notes to both parts.

3. Repeat Problem 1, but draw internal and external detailed thread representations of an American National thread form. The major diameter of each part is 3″. The note specifications are 3–4 NC–2. Apply notes to both parts.

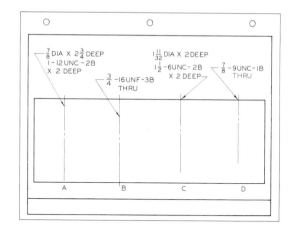

FIG. 10.66   Internal threads (Problems 4, 5, and 6).

4. Notes are given in Fig. 10.66 to specify the depth of the holes that are to be drilled and the threads that are to be tapped in the holes. Following these notes, draw detailed representations of the threads as views according to specifications.

5. Repeat Problem 4, but use schematic representations.

6. Repeat Problem 4, but use simplified thread representations.

7. Figure 10.67 shows a layout of two external threaded parts and their end views. Also shown is

FIG. 10.67   Internal and external threads (Problems 7, 8, and 9).

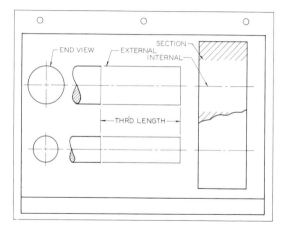

a piece into which the external threads will be screwed. Complete all three views of each of the parts. Use detailed threads and apply notes to the internal and external threads. Use the table in Appendix 15 for thread specifications. Use UNC threads with a 2A fit.

8. Repeat Problem 7, but use schematic thread representations.

9. Repeat Problem 7, but use simplified thread representations.

10. Referring to Fig. 10.68, complete the drawing with instruments as a finished hexagon bolt and nut. The bolt head is a heavy nut drawn across corners. Use detailed thread representations. Show notes to specify the parts of the assembly. Thread specifications are 1½–6 UNC–3 or M36 × 4.

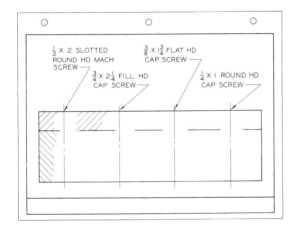

FIG. 10.68 Nuts and bolts in assembly (Problems 10, 11, and 12).

11. Referring to Fig. 10.68, complete the drawing with instruments as an unfinished squarehead bolt and nut. The bolt head is drawn across corners. The regular nut is to be drawn across corners. Use schematic thread representations. Show notes to specify the parts. Use the table in Appendix 15 or 18 for thread specifications (English or SI).

12. Referring to Fig. 10.68, complete the drawing with instruments as a finished hexagon nut and bolt. The regular bolt and nut are to be drawn across flats. Use simplified thread representations. Show notes to specify the parts. Use the table in Appendix 15 or 18 for specifications.

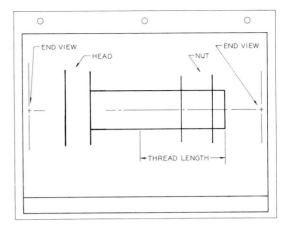

FIG. 10.69 Cap screws and machine screws (Problems 13, 14, and 15).

13. The notes in Fig. 10.69 apply to machine and cap screws that are to be drawn in the section view of the two parts. The holes in which the screws are to be drawn are through holes. Complete the drawings and show notes as given. Show remaining section lines. Use detailed thread symbols.

14. Repeat Problem 13, but use schematic thread symbols.

15. Repeat Problem 13, but use simplified thread symbols.

16. *Design.* The pencil pointer shown in Fig. 10.70 has a shaft of ¼″ that fits into a bracket designed to

FIG. 10.70 Design involving threaded parts (Problem 16).

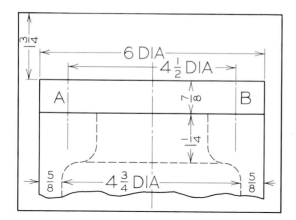

FIG. 10.71 Problem 17.

19. (Fig. 10.73). Draw a ⅞″ (22 mm) DIA special stud with a 1¼″ DIA collar on axis AB. Show a plain washer and a regular nut at end B. On axes CD,

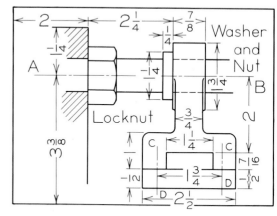

FIG. 10.73 Problem 19.

clamp onto a desk top. A set screw holds the shaft in position. Make a drawing of the bracket, estimating its dimensions. Show the details and the method of using the set screw to hold the shaft. Give the specifications for the set screw.

17. (Fig. 10.71). On axes A and B, construct ⅝″ (16 mm) hexagon-head bolts across flats with a coarse thread, UNC. Convert the view to a half-section to show how the parts are assembled together.

18. (Fig. 10.72). On axes E and F, draw a stud with a hexagon-head nut shown across corners. The stud is to have a diameter of ⅞″ (22 mm) and should be a UNC form and series.

draw a machine screw of your selection to hold the two parts together.

20. (Fig. 10.74). Draw a cap screw on axis ab to fasten the parts together. The following dimensions are bolt DIA 1¼″ or 32 mm; C, 1.25″; E, 1.12″; H, 0.25″R; hexagon head across corners, G, 4.50″; and D, 2.00″.

FIG. 10.72 Problem 18.

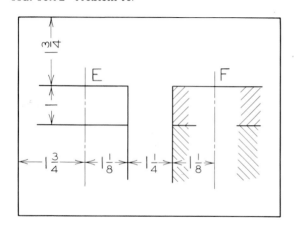

FIG. 10.74 Problem 20.

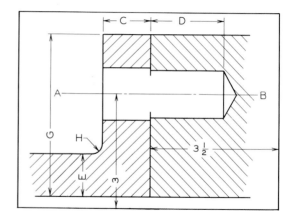

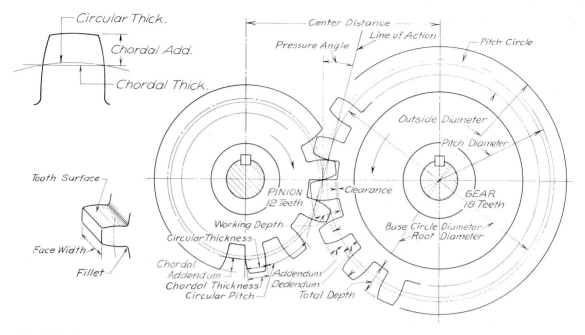

FIG. 11.2 Gear terminology for spur gears.

ameter will have a diametral pitch of 5, which means that there are 5 teeth per inch of diameter: $DP = N / PD$ (where N = number of teeth).

CIRCULAR PITCH (CP) is the circular measurement from one point on a tooth to the corresponding point on the next tooth measured along the pitch circle: $CP = 3.14 / DP$.

FIG. 11.3 Gear terminology for spur gears.

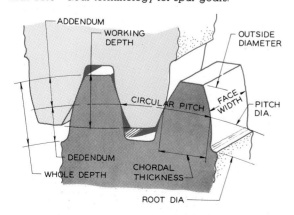

CENTER DISTANCE (CD) is the distance from the center of a gear to its mating gear's center: $CD = (N_p + N_s) / (2DP)$ (where N_p and N_s are the number of teeth in the pinion and spur, respectively).

ADDENDUM (A) is the height of a gear above its pitch circle: $A = 1 / DP$.

DEDENDUM (D) is the depth of a gear below the pitch circle: $D = 1.157 / DP$.

WHOLE DEPTH (WD) is the total depth of a gear tooth: $WD = A + D$.

WORKING DEPTH (WKD) is the depth to which a tooth fits into a meshing gear: $WKD = 2 / DP$; or $WKD = 2A$.

CIRCULAR THICKNESS (CRT) is the circular distance across a tooth measured along the pitch circle: $CRT = 1.57 / DP$.

CHORDAL THICKNESS (CT) is the straight-line distance across a tooth at the pitch circle: $CT = PD (\sin 90° / N)$ (where N = number of teeth).

CHORDAL ADDENDUM (CA) is the radial distance from the chordal thickness at the pitch circle to the top of the tooth. $CA = A + (AT)^2/4(PD)$ where AT = arc thickness of a tooth along the pitch circle.

FACE WIDTH (FW) is the width across a gear tooth parallel to its axis. This is a variable dimension, but it is usually 3 to 4 times the circular pitch: $FW = 3$ to $4(CP)$.

OUTSIDE DIAMETER (OD) is the maximum diameter of a gear across its teeth: $OD = PD + 2A$.

ROOT DIAMETER (RD) is the diameter of a gear measured from the bottom of its gear teeth: $RD = PD - (2D)$.

PRESSURE ANGLE (PA) is the angle between the line of action and a line perpendicular to the center line of two meshing gears. Angles of 14.5° and 20° are standard angles for involute gears.

BASE CIRCLE (BC) is the circle from which an involute tooth curve is generated or developed: $BC = PD (\cos PA)$.

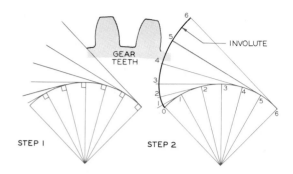

FIG. 11.4 Construction of an involute

Step 1 The base arc is divided into equal divisions with radial lines from the center. Tangents are drawn perpendicular to the radial lines on the arc.

Step 2 The chordal distance from 1 to 0 is used as the radius and 1 as the center to find point 1 on the involute curve. The distance from 2 to newly found 1 is revolved to the tangent line through 2 to locate a second point, and the process is continued.

10.3
Tooth forms

The most common gear tooth is an *involute tooth* with a 14.5° pressure angle. The 14.5° angle is the angle of contact between two gears when the tangents of both gears pass through the point of contact. Gears with pressure angles of 20° and 25° are also used. Gear teeth with larger pressure angles are wider at the base and thus are stronger than the standard 14.5° teeth.

The standard gear face is an involute that keeps the meshing gears in contact as the gear teeth are revolved past one another. The principle of constructing an involute is illustrated in Fig. 11.4.

An involute curve can be thought of as the path of a string that is kept taut as it is unwound from the base arc. It is unnecessary to use this procedure in drawing gear teeth since most detail drawings employ only approximations of gear teeth, if teeth are shown at all.

11.4
Gear ratios

The diameters of two meshing spur gears establish ratios that are important to the function of the gears. Examples of these ratios are given in Fig. 11.5.

If the radius of a gear is twice that of its pinion (the small gear), then the diameter is twice that of the pinion, and the gear has twice as many teeth as the pinion. In this case, the pinion must make twice as many turns as the larger gear. In other words, the rev-

FIG. 11.5 Ratios between meshing spur gears.

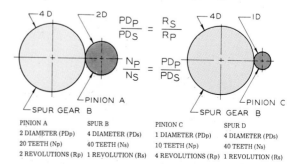

$$\frac{PD_P}{PD_S} = \frac{R_S}{R_P}$$

$$\frac{N_P}{N_S} = \frac{PD_P}{PD_S}$$

PINION A	SPUR B
2 DIAMETER (PDp)	4 DIAMETER (PDs)
20 TEETH (Np)	40 TEETH (Ns)
2 REVOLUTIONS (Rp)	1 REVOLUTION (Rs)

PINION C	SPUR D
1 DIAMETER (PDp)	4 DIAMETER (PDs)
10 TEETH (Np)	40 TEETH (Ns)
4 REVOLUTIONS (Rp)	1 REVOLUTION (Rs)

olutions per minute (RPM) of the pinion is twice that of the larger gear.

When the diameter of the gear is four times the diameter of the pinion, there must be four times as many teeth on the gear as on the pinion, and the number of revolutions of the pinion will be four times that of the larger gear.

The relationship between two meshing spur gears can be developed in formula form by finding the velocity of a point on the small gear that is equal to $\pi PD \times$ RPM of the pinion. The velocity of a point on the large gear is equal to $\pi PD \times$ RPM of the spur. Since the velocity of points on each gear must be equal, the equation may be written as

$$\pi PD_p(\text{RPM}) = \pi PD_s(\text{RPM});$$

therefore,

$$\frac{PD_p}{PD_s} = \frac{\text{RPM}_s}{\text{RPM}_p}.$$

If the radius of the pinion is 1 inch, the radius of the spur is 4 inches, and the RPM of the pinion is 20 RPM, then the RPM of the spur can be found as follows:

$$\frac{2(1)}{2(4)} = \frac{\text{RPM}_s}{20 \ \text{RPM}_p}; \ \text{RPM}_s = \frac{2(20)}{2(4)} = 5 \ \text{RPM}$$

(one-fourth of the revolutions per minute of the pinion).

The number of teeth on each gear is also proportional to the radii and diameters of a pair of meshing gears. This relationship can be written:

$$\frac{N_p}{N_s} = \frac{PD_p}{PD_s}$$

(where N_p and N_s are the numbers of teeth on the pinion and spur, respectively, and PD_p and PD_s are their pitch diameters).

11.5
Gear calculations

Before a working drawing of a gear can be started, the drafter must perform a series of calculations to determine the dimensions of the gear using the definitions and formulas previously introduced. The following problem is given as an example.

PROBLEM 1 Calculate the dimensions for a spur gear that has a pitch diameter of 5 inches, a diametral pitch of 4, and a pressure angle of 14.5°. (The diametral pitch is the same for meshing gears.)

SOLUTION

No. of teeth = $PD \times DP$ = 5×4 = 20
Addendum = $1 / 4$ = 0.25
Dedendum = $1.157 / 4$ = 0.2893
Circular thickness = $1.5708 / 4$ = 0.3927
Outside diameter = $5 + 2(0.25)$ = 5.50
Root diameter = $5 - 2(0.2893)$ = 4.421
Chordal thickness = $5(\sin 90° / 20)$ = $5 \times (0.079)$ = 0.392
Chordal addendum = $0.25 + (0.3927^2/(4 \times 5))$ = 0.2577
Face width = $3.5(0.79)$ = 2.75
Circular pitch = $3.14 / 4$ = 0.785
Working depth = $2 / 4$ = 0.500"
Whole depth = $0.250 + 0.289$ = 0.539

These dimensions can be used to draw the spur gear and to provide specifications necessary for its manufacture.

A second problem is given as an example of the method of determining the design information for two meshing gears when their working ratios are known.

PROBLEM 2 Find the number of teeth and other specifications for a pair of meshing gears with a driving gear that turns at 100 RPM and a driven gear that turns at 60 RPM. The diametral pitch for each is 10. The center-to-center distance between the gears is 6 inches.

SOLUTION

STEP 1 Find the sum of the teeth on both gears:

Total teeth = $2 \times$ (C-to-C dist.) $\times$ DP
$= 2 \times 6 \times 10 = 120$ teeth

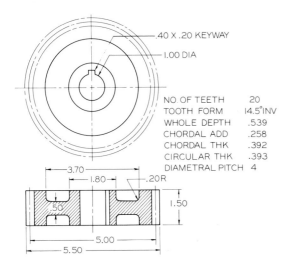

NO. OF TEETH 20
TOOTH FORM 14.5°INV
WHOLE DEPTH .539
CHORDAL ADD .258
CHORDAL THK .392
CIRCULAR THK .393
DIAMETRAL PITCH 4

FIG. 11.6 A detail drawing of a spur gear with a table of cutting data to supplement the dimensions shown on the drawing.

STEP 2 Find the number of teeth for the driving gear by two steps:

A. $$\frac{\text{Driver RPM}}{\text{Driven RPM}} + 1 = \frac{100}{60} + 1 = 2.667$$

B. $$\frac{\text{Total teeth}}{\frac{100}{60} + 1} = \frac{120}{2.667} = 45 \; teeth^*$$

STEP 3 Find the number of teeth for the driven gear:

Total teeth minus teeth on driver = teeth on driver
120 − 45 = 75 teeth

STEP 4 The other specifications for the gears can be computed as shown in Problem 1 by using the formulas in Section 11.2.

11.6
Drawing spur gears

A conventional drawing of a spur gear is shown in Fig. 11.6. The teeth need not be drawn, since this is time-consuming and unnecessary. It is possible to

*The number of teeth must be a whole number since there cannot be fractional teeth on a gear. It may be necessary to adjust the center distance to yield a whole number of teeth.

omit the circular view and to show only a sectional view of the gear with a table of dimensions that have been calculated. Dimensions of this type are called *cutting data.*

When the circular view is drawn, circular center lines are drawn to represent the root circle, pitch circle, and outside circle of the gear.

A table of dimensions is a necessary part of a gear drawing, as shown in Fig. 11.7. These data can be calculated by formula or taken from tables of standards in gear handbooks, such as *Machinery's Handbook.*

11.7
Bevel gear terminology

Bevel gears are gears whose axes intersect at angles. Although the angle of intersection is usually 90°, other angles are also used. The smaller of the two bevel gears is called the *pinion,* as in spur gearing.

The terminology of bevel gearing is illustrated in Fig. 11.8. The corresponding formulas for each feature are given here:

PITCH ANGLE OF PINOIN (SMALL GEAR) (PA$_p$):

$$\tan PA_p = \frac{N_p}{N_g}.$$

(N_g and N_p are the number of teeth on the gear and the pinion, respectively.)

PITCH ANGLE OF GEAR (PA$_g$):

$$\tan PA_g = \frac{N_g}{N_p}.$$

PITCH DIAMETER (PD) is the number of teeth *(N)* divided by the diametral pitch *(DP)*:

$$PD = N / P.$$

ADDENDUM (A) is measured at the large end of the tooth: $A = 1 / DP.$

DEDENDUM (D) is measured at the large end of the tooth: $D = 1.157 / DP.$

WHOLE TOOTH DEPTH (WD): $WD = 2.157 / DP.$

THICKNESS OF TOOTH (TT) at pitch circle:
$TT = 1.571 / DP.$

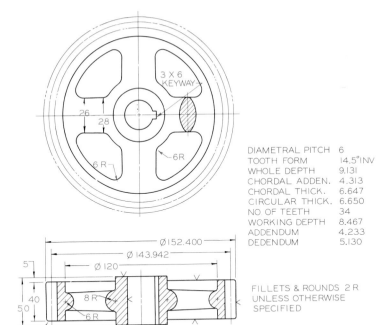

DIAMETRAL PITCH	6
TOOTH FORM	14.5°INV
WHOLE DEPTH	9.131
CHORDAL ADDEN.	4.313
CHORDAL THICK.	6.647
CIRCULAR THICK.	6.650
NO OF TEETH	34
WORKING DEPTH	8.467
ADDENDUM	4.233
DEDENDUM	5.130

FILLETS & ROUNDS 2 R
UNLESS OTHERWISE
SPECIFIED

FIG. 11.7 A detail drawing of a spur gear that has been converted to metric units by multiplying dimensions in inches by 25.4 to yield millimeters.

FIG. 11.8 The terminology and definitions of bevel gears. (Courtesy of Philadelphia Gear Corporation.)

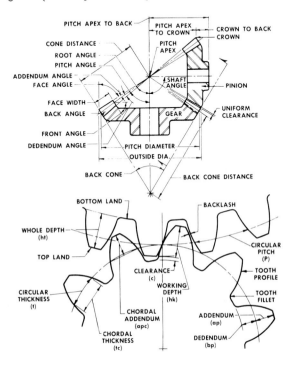

DIAMETRAL PITCH: $DP = N / PD$; (N = number of teeth).

ADDENDUM ANGLE (AA) is the angle formed by the addendum and the pitch cone distance: *(PCD)*:
$$\tan AA = \frac{A}{PCD}.$$

PITCH CONE DISTANCE: $PCD = PD / (2 \times \sin PA)$.

DEDENDUM ANGLE (DA) is the angle formed by the dedendum and the pitch cone distance:
$$\tan DA = \frac{D}{PCD}.$$

FACE ANGLE (FA) is the angle between the gear's center line and the top of its teeth:
$$FA = 90° - (PCD + AA).$$

CUTTING ANGLE (OR ROOT ANGLE) (CA) is the angle between the gear's axis and the roots of the teeth: $CA = PCD - D$.

OUTSIDE DIAMETER (OD) is the greatest diameter of a gear across its teeth: $OD = PD + 2A$.

APEX-TO-CROWN DISTANCE (AC) is the distance from the crown of the gear to the apex of the cone measured parallel to the axis of the gear: $AC = OD / (2 \tan FA)$.

CHORDAL ADDENDUM (CA):

$$CA = A + \frac{TT^2 \cos PA}{4(PD)}$$

CHORDAL THICKNESS (CT) at the large end of the tooth:

$$CT = PD \times \sin \frac{90°}{N}.$$

FACE WIDTH (FW) can vary, but it is recommended that it be approximately equal to the pitch cone distance divided by 3: $FW = PCD / 3$.

Gear handbooks can be used for finding many of these dimensions from tables rather than using the formulas given above.

11.8 _____
Bevel gear calculations

The following example of calculating the specifications for two bevel gears is given to demonstrate how the formulas in the previous section are used. You will note that some of the formulas result in the same specifications that apply to both the gear and the pinion.

PROBLEM 3 Two bevel gears intersect at right angles. They have a diametral pitch of 3, 60 teeth on the gear, 45 teeth on the pinion, and a face width of 4 inches. Find the dimensions of the gear.

SOLUTION

Pitch cone angle of gear: tan $PCA = 60 / 45 = 1.33$; $PCA = 53° 7'$.
Pitch cone angle of pinion: tan $PCA = 45 / 60$; $PCA = 36° 52'$.
Pitch diameter of gear: $60 / 3 = 20.00''$.
Pitch diameter of pinion: $45 / 3 = 15.00''$.
The following formulas are the same for both the gear and the pinion:
Addendum: $1 / 3 = 0.333''$.

Dedendum: $1.157 / 3 = 0.3857''$.
Whole depth: $2.157 / 3 = 0.719''$.
Tooth thickness on pitch circle: $1.571 / 3 = 0.5237''$.
Pitch cone distance: $20 / (2 \sin 53° 7') = 12.5015''$.
Addendum angle: tan $AA = 0.333 / 12.5015 = 1° 32'$.
Dedendum angle: $DA = 0.3857 / 12.5015 = 0.0308 = 1° 46'$.
Face width: $PCD / 3 = 4.00''$.

The remainder of the formulas must be applied separately to the gear and the pinion.

Chordal addendum of gear:

$$0.333 + \frac{0.5237^2 \times \cos 53°7'}{4 \times 20} = 0.336''.$$

Chordal addendum of pinion:

$$0.333 + \frac{0.5237^2 \times \cos 36°52'}{4 \times 15} = 0.338''.$$

Chordal thickness of gear:

$$\sin \frac{90°}{60} \times 20 = 0.524''.$$

Chordal thickness of pinion:

$$\sin \frac{90°}{45} \times 15 = 0.523''.$$

Face angle of gear:

$$90° - (53° 7' + 1° 32') = 35° 21'.$$

Face angle of pinion:

$$90° - (36° 52' + 1° 32') = 51° 36'.$$

Cutting angle of gear:

$$53° 7' - 1° 46' = 51° 21'.$$

Cutting angle of pinion:

$$36° 52' - 1° 46' = 35° 6'.$$

Angular addendum of gear:

$$0.333 \times \cos 53° 7' = 0.1999''.$$

Angular addendum of pinion:

$$0.333 \times \cos 36° 52' = 0.2667''.$$

Outside diameter of gear:

$$20 + 2 (0.1999) = 20.4000''.$$

Outside diameter of pinion:

$$15 + 2 (0.2667) = 15.533''.$$

Apex-to-crown distance of gear:

$$\frac{20.400}{2} \times \tan 35° 7' = 7.173''.$$

Apex-to-crown distance of pinion:

$$\frac{15.533}{2} \times \tan 51° 36' = 9.800''.$$

11.9
Drawing bevel gears

The dimensions calculated above are used to lay out the bevel gears in a detail drawing. Many of the calculated dimensions would be difficult to measure on a drawing within a high degree of accuracy; therefore, it is important to provide a table of "cutting data" for each gear.

The steps of drawing the bevel gears are shown in Fig. 11.9. The finished drawings are shown with a combination of dimensions and a table of dimensions, which is not shown.

FIG. 11.9 Construction of bevel gears

Step 1 Lay out the pitch diameters and axes of the two bevel gears.

Step 2 Draw construction lines to establish the limits of the teeth by using the addendum and dedendum dimensions.

Step 3 Draw the pinion and the gear using the specified dimensions or those that were calculated by formula.

Step 4 Complete the detail drawings of both gears and provide a table of cutting data.

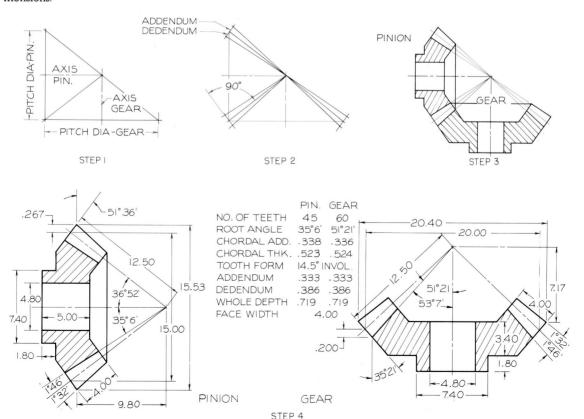

	PIN.	GEAR
NO. OF TEETH	45	60
ROOT ANGLE	35°6'	51°21'
CHORDAL ADD.	.338	.336
CHORDAL THK.	.523	.524
TOOTH FORM	14.5° INVOL.	
ADDENDUM	.333	.333
DEDENDUM	.386	.386
WHOLE DEPTH	.719	.719
FACE WIDTH	4.00	

STEP 4

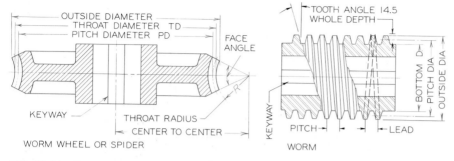

FIG. 11.10 The terminology and definitions of worm gears.

11.10

Worm gears

A worm gear is composed of a thread shaft called a *worm* and a circular gear called a *spider* (Fig. 11.10). The worm is revolved in a continuous motion, which causes the spider to revolve about its axis.

The following terminology is illustrated in Figs. 11.10 and 11.11. The following formulas can be used to calculate the dimensions associated with these terms.

Worm specifications and formulas

LINEAR PITCH (P) is the distance from one thread to the next measured parallel to the worm's axis: $P = L/N$ (where N is number of threads: 1 if a single thread, 2 if a double thread, etc.).

LEAD (L) is the distance that a thread advances in a turn of 360°.

ADDENDUM OF TOOTH: $AW = 0.3183\ P$.

FIG. 11.11 A detail drawing of a worm gear (spider) and the table of cutting data.

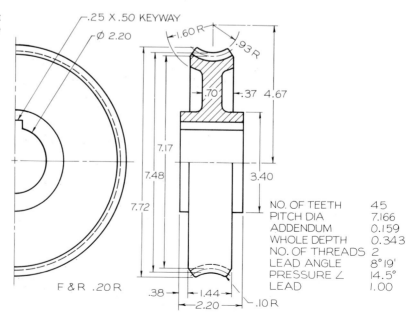

NO. OF TEETH	45
PITCH DIA	7.166
ADDENDUM	0.159
WHOLE DEPTH	0.343
NO. OF THREADS	2
LEAD ANGLE	8°19'
PRESSURE ∠	14.5°
LEAD	1.00

PITCH DIAMETER: $PDW = OD - 2AW$. (OD is the outside diameter.)

WHOLE DEPTH OF TOOTH: $WDT = 0.6866 \times P$.

BOTTOM DIAMETER OF WORM:
$$BD = OD - 2\,WDT.$$

WIDTH OF THREAD AT ROOT: $WT = 0.31P$.

MINIMUM LENGTH OF WORM:
$$MLW = \sqrt{8\,PDS \times AW}.$$
(PDS is the pitch diameter of the spider.)

HELIX ANGLE OF WORM:
$$\cot \beta = \frac{3.14\,PDW}{L}.$$

OUTSIDE DIAMETER: $OD = PD + 2A$.

Spider specifications and formulas

PITCH DIAMETER OF SPIDER: $PDS = \dfrac{N(P)}{3.14}$. (N is the number of teeth on the spider.)

THROAT DIAMETER OF SPIDER: $TD = PDS + 2A$.

RADIUS OF SPIDER THROAT:
$$RST = \frac{OD \text{ of worm}}{2} - 2A.$$

FACE ANGLE (FA) may be selected to be between 60° and 80° for the average application.

CENTER-TO-CENTER DISTANCE (between the worm and spider): $CD = \dfrac{PDW + PDS}{2}$.

OUTSIDE DIAMETER OF SPIDER:
$$ODS = TD + 0.4775\,P.$$

Face width of gear: $FW = 2.38\,(P) + 0.25$.

11.11
Worm gear calculations

The following is an example problem that has been solved for a worm gear by using the formulas given above.

PROBLEM 4 Calculate the specifications for a worm and worm gear (spider). The gear has 45 teeth, and the worm has an outside diameter of 2.50″. The worm has a double thread and a pitch of 0.5″.

SOLUTION

Lead: $L = 0.5″ \times 2 = 1″$.

Worm addendum: $AW = 0.3183P = 0.1592″$.

Pitch diameter of worm:
$$PDW = 2.50″ - 2\,(0.1592″) = 2.1818″.$$

Pitch diameter of gear:
$$PDS = (45 \times 0.5)\,/\,3.14 = 7.166″.$$

Center distance between worm and gear:
$$CD = \frac{(2.182 + 7.166)}{2} = 4.674″.$$

Whole depth of worm tooth:
$$WDT = 0.687 \times 0.5 = 0.3433″.$$

Bottom diameter of worm:
$$BD = 2.50 - 2(0.3433) = 1.813″.$$

Helix angle of worm:
$$\cot \beta = \frac{3.14\,(2.1816)}{1} = 8°\,19′.$$

Width of thread at root: $WT = 0.31\,(1) = 0.155″$.

Minimum length of worm:
$$MLW = \sqrt{8(0.1592)\,(7.1656)} = 3.02″.$$

Throat diameter of gear:
$$TD = 7.1656 + 2(0.1592) = 7.484″.$$

Radius of gear throat:
$$RST = (2.5\,/\,2) - (2 \times 0.1592) = 0.9318″.$$

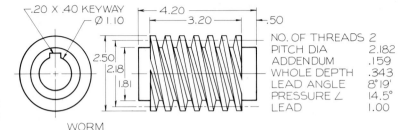

FIG. 11.12 A detail drawing of a worm using the dimensions that were calculated.

NO. OF THREADS	2
PITCH DIA	2.182
ADDENDUM	.159
WHOLE DEPTH	.343
LEAD ANGLE	8°19'
PRESSURE ∠	14.5°
LEAD	1.00

WORM

Face width:

$$FW = 2.38\ (0.5) + 0.25 = 1.44''.$$

Outside diameter of gear:

$$ODS = 7.484 + 0.4775\ (0.5) = 7.723''.$$

11.12
Drawing worm gears

The worm and worm wheel (spider) are drawn and dimensioned as shown in Figs. 11.11 and 11.12. Each gear must be dimensioned and supplemented with cutting data.

It is advantageous to use full or partial sections to show the details of the gears. In some instances, the circular views are omitted.

The specifications derived by the formulas in the previous section must be used for scaling and laying out the drawings.

11.13
Cams

Cams are irregularly shaped machine elements that produce motion in a single plane, usually up and down (Fig. 11.13). As the cam revolves about its center, the variation in the cam's shape produces a rise or fall in the follower that is in contact with it. The shape of the cam is determined graphically prior to the preparation of manufacturing specifications.

Only plate cams are covered in the brief review of this type of mechanism. Cams utilize the principle of the inclined wedge, with the surface of the cam causing a change in the slope of the plane, thereby producing the desired motion.

FIG. 11.13 Examples of machined cams. (Courtesy of Ferguson Machine Company.)

11.14
Cam motion

Cams are designed primarily to produce (1) uniform or linear motion, (2) harmonic motion, (3) gravity motion, or (4) combinations of these. Some cams are designed to serve special needs that do not fit these patterns but are instead based on particular design requirements. Displacement diagrams are used to represent the travel of the follower relative to the rotation of the cam.

Uniform motion

Uniform motion is shown in the displacement diagram in Fig. 11.14A. Displacement diagrams represent the motion of the cam follower as the cam rotates through 360°. The uniform-motion curve has sharp corners, indicating abrupt changes of velocity at two points, causing the follower to bounce. Hence this motion is usually modified with arcs that smooth this change of velocity. The radius of the modifying arc is varied up to a radius of one-half the total displacement of the

FIG. 11.14 Displacement diagrams

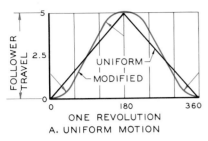

A. UNIFORM MOTION

Uniform-motion diagrams are modified with arcs of one-fourth to one-third of total displacement to smooth out the velocity of the follower at these points of abrupt change.

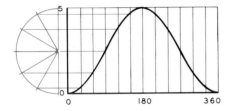

B. HARMONIC MOTION

Harmonic motion is plotted by projecting from a semicircle whose diameter is equal to the rise of the follower. The semicircle must be divided into the same number of sectors as the divisions on the x-axis of the graph to the point of maximum rise.

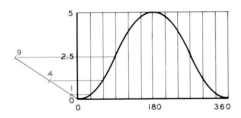

C. GRAVITY MOTION

Gravity-motion diagrams are constructed so that the rise of the follower is relative to the square of the units on the x-axis: 1^2, 2^2, 3^2, etc.

follower, depending on the speed of operation. Usually a radius of about one-third to one-fourth total displacement is best.

Harmonic motion

Harmonic motion, plotted in Fig. 11.14B, is a smooth, continuous motion based on the change of position of the points on the circumference of a circle. At moder-

ate speeds, this displacement gives a smooth operation. A semicircle is drawn with its diameter equal to the total motion of the follower to locate the points on the curve.

Gravity motion

Gravity motion (uniform acceleration), plotted in Fig. 11.14C, is used for high-speed operation. The variation of displacement is analogous to the force of gravity exerted on a falling body, with the difference in displacement being 1, 3, 5, 5, 3, 1, based on the square of the number. For instance, $1^2 = 1$; $2^2 = 4$; $3^2 = 9$. This same motion is repeated in reverse order for the remaining half of the motion of the follower. Intermediate points can be found by squaring fractional increments, such as $(2.5)^2$.

11.15
Construction of a cam

Cam followers

Three basic types of cam followers are (A) the flat surface, (B) the roller, and (C) the knife edge, as shown in Fig. 11.15. The flat-surface and knife-edge followers are limited to use with slow-moving cams where minor force will be exerted during rotation. The roller is the most-often used form of follower, since it can withstand higher speeds.

Plate cam—harmonic motion

The steps of constructing a plate cam with harmonic motion are shown in Fig. 11.16. The drafter must know the following before designing a cam: the mo-

FIG. 11.15 Three basic types of cam followers—the flat surface, the roller, and the knife-edge.

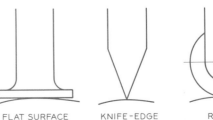

FLAT SURFACE   KNIFE-EDGE   ROLLER

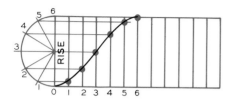

Step 1 Construct a semicircle whose diameter is equal to the rise of the follower. Divide the semicircle into the same number of divisions as there are between 0° and 180° on the horizontal axis of the displacement diagram. Plot half of the displacement curve.

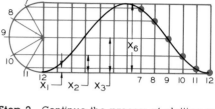

Step 2 Continue the process of plotting points by projecting from the semicircle, starting from the top of the semicircle and proceeding to the bottom. Complete the curve symmetrical to the left half.

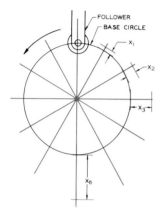

Step 3 Draw the base circle and draw the follower. Divide the base circle into the same number of sectors as there are divisions in the displacement diagram. Transfer distances from the displacement diagram to the irrespective radial lines and measure outward from the base circle.

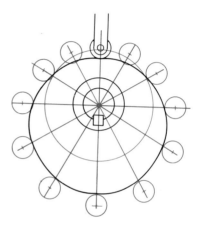

Step 4 Draw circles to represent the positions of the roller as the cam revolves in a counterclockwise direction. Draw the cam profile tangent to all the rollers to complete the drawing.

FIG. 11.16 Construction of a plate cam with harmonic motion

tion of the follower, the rise of the follower, the size and the type of the follower, the position of the follower, the diameter of the base circle, and the direction of rotation.

The specifications for the cam in Fig. 11.16 are given graphically. The four steps of construction are followed as illustrated to construct the cam profile shown in Step 4.

Plate cam—uniform acceleration

The steps of constructing a cam with uniform acceleration is done in the same manner as in the previous example except for a different displacement diagram and a knife-edge follower.

The graphical layout of the problem is shown in Fig. 11.17. The profile of the cam is found by following the four steps of construction.

Plate cam—combination

In Fig. 11.18, a knife-edge follower is used with a plate cam to produce a 4″ rise with harmonic motion from 0° to 180°, a 4″ fall with a uniform acceleration from 180° to 300°, and dwell (no follower motion) from 300° to 360°. You are to draw the cam that will give this motion from the base circle.

The displacement diagram is drawn with a harmonic curve with a full rise of 4″. The curve is then drawn with a uniform acceleration drop of 4″. The

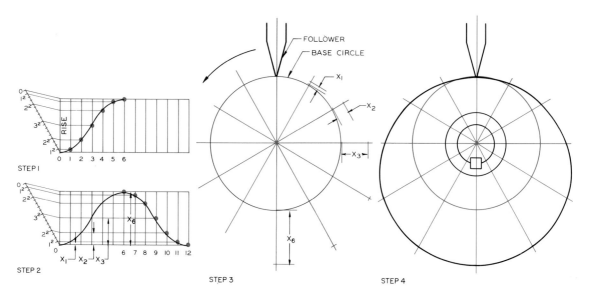

FIG. 11.17 Construction of a plate cam with uniform acceleration

Step 1 Construct a displacement diagram to represent the rise of the follower. Divide the horizontal axis into angular increments of 30°. Draw a construction line through point 0; locate the 1^2, 2^2, and 3^2 divisions and project them to the vertical axis to represent half of the rise. The other half of the rise is found by laying off distances along the construction line with descending values.

Step 2 Use the same construction to find the right half of the symmetrical curve.

Step 3 Construct the base circle and draw the knife-edge follower. Divide the circle into the same number of sectors as there are divisions in the displacement diagram. Transfer distances from the displacement diagram to the respective radial lines of the base circle, measuring outward from the base circle.

Step 4 Connect the points found in Step 3 with a smooth curve to complete the cam profile. Show also the cam hub and keyway.

FIG. 11.18 Construction of a plate cam with combination motions

Step 1 The cam is to rise 4" in 180° with harmonic motion, fall 4" in 120° with uniform acceleration, and dwell for 60°. These motions are plotted on the displacement diagram.

Step 2 Construct the base circle and draw the knife-edge follower. Transfer distances from the displacement diagram to the respective radial lines of the base circle, measuring outward from the base circle.

Step 3 Draw a smooth curve through the points found in Step 2 to complete the profile of the cam.

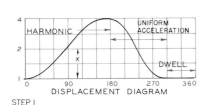

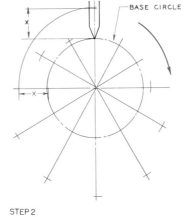

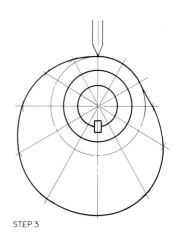

values of rise must be known before beginning the problem. The dwell is a horizontal line to complete the diagram of the 360° rotation of the cam.

The profile of the plate cam is found by completing Steps 2 and 3 in the same manner as in the previous two examples.

11.16
Construction of a cam with an offset follower

The cam in Fig. 11.19 is required to produce harmonic motion through 360°. This motion is plotted directly from the follower rather than by using a displacement diagram, since no combinations of motion are involved.

A semicircle is drawn with its diameter equal to the total motion of the follower. The base circle is drawn to pass through the center or roller of the follower. The center line of the follower is extended downward, and a circle is drawn tangent to the extension with its center at the center of the base circle. The small circle is divided into 30° intervals to establish points through which construction lines will be drawn tangent to the circle.

The distances from the tangent points to the position points along the path of the follower are laid out along the tangent lines that were drawn at 30° inter-

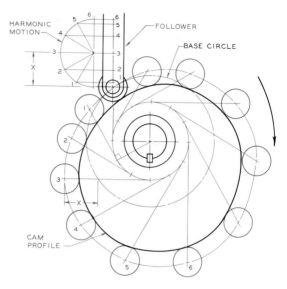

FIG. 11.19 Construction of a plate cam with an offset roller follower.

vals. Alternatively, these points can be located by measuring from the base circle as shown in the example, where point 3 was located distance X from the base circle.

Draw the circular roller in all views and construct the profile of the cam tangent to the rollers at all positions.

Problems

Gears

Use Size A (8½″ × 11″) sheets for the following gear problems. Select the most appropriate scale so that the drawings will utilize the available space.

    1–5. Calculate the following dimensions for the following spur gears and make a detail drawing of each. Give the dimensions and cutting data for each gear.

Problem	Gear Teeth	Diametral Pitch	14.5° Involute
1	20	5	″
2	30	3	″
3	40	4	″
4	60	6	″
5	80	4	″

Provide any other dimensions that are needed, using your judgment.

6–10. Calculate the gear sizes and number of teeth (similar to Problem 2 in Section 11.5), using the ratios and data below:

Problem	RPM Pinion	RPM Gear	Center to Center	Diametral Pitch
6	100 (driver)	60	6.0″	10
7	100 (driver)	50	8.0″	9
8	100 (driver)	40	10.0″	8
9	100 (driver)	35	12.0″	7
10	100 (driver)	25	14.0″	6

11–20. Make detail drawings of each of the gears for which calculations were made in Problems 6 through 10. Provide a table of cutting data and other dimensions that are needed to complete the specifications.

21–25. Calculate the specifications for the bevel gears that intersect at 90°, and make detail drawings of each with the necessary dimensions and cutting data.

Problem	Diametral Pitch	No. of Teeth on Pinion	No. of Teeth on Gear
21	3	60	15
22	4	100	40
23	5	100	60
24	6	100	50
25	7	100	30

26–30. Calculate the specifications for worm gears and make detail drawings of each, providing the necessary dimensions and cutting data.

Problem	No. of Teeth in Spider Gear	Outside DIA of Worm	Pitch of Worm	Thread of Worm
26	45	2.50	0.5	double
27	30	2.00	0.80	single
28	60	3.00	0.80	double
29	30	2.00	0.25	double
30	80	4.00	1.00	single

Cams

Draw the following on Size B sheets (11″ × 17″) with the following standard dimensions: base circle, 3.50″; roller follower, 0.60″ diameter; shaft, 0.75″ diameter; hub, 1.25″ diameter; direction of rotation, clockwise. The follower is positioned vertically over the center of the base circle except in Problems 40 and 41. Lay out the problems and displacement diagrams as shown in Fig. 11.20.

31. Draw a plate cam with a knife-edge follower for uniform motion and a rise of 1.00″.

32. Draw a displacement diagram and a cam that will give a modified uniform motion to a knife-edge follower with a rise of 1.7″. Modify the uniform motion with an arc of one-quarter of the rise in the displacement diagram.

33. Draw a displacement diagram and a cam that will give a harmonic motion to a roller follower with a rise of 1.60″.

34. Draw a displacement diagram and a cam that will give a harmonic motion to a knife-edge follower with a rise of 1.00″.

35. Draw a displacement diagram and a cam that will give uniform acceleration to a knife-edge follower with a rise of 1.70″.

36. Draw a displacement diagram and a cam that will give a uniform acceleration to a roller follower with a rise of 1.40″.

37. Draw a displacement diagram and a cam that will give the following motion to a knife-edge follower: rise 1.25″ with harmonic motion in 120°, dwell for 120°; and fall 1.25″ with uniform acceleration.

38. Draw a displacement diagram and a cam that will give the following motion to a knife-edge follower: dwell for 70°; rise 1″ with a modified uniform motion in 100°; fall 1″ with a harmonic motion in 100°; and dwell for 90°.

39. Draw a displacement diagram and a cam that will give the following motion to a roller follower: rise 1.25″ with a harmonic motion in 120°; dwell for 120°; and fall 1.25″ with a uniform acceleration in 120°.

40. Repeat Problem 32, but offset the follower 0.60″ to the right of the vertical center line.

41. Repeat Problem 33, but offset the follower 0.60″ to the left of the vertical line.

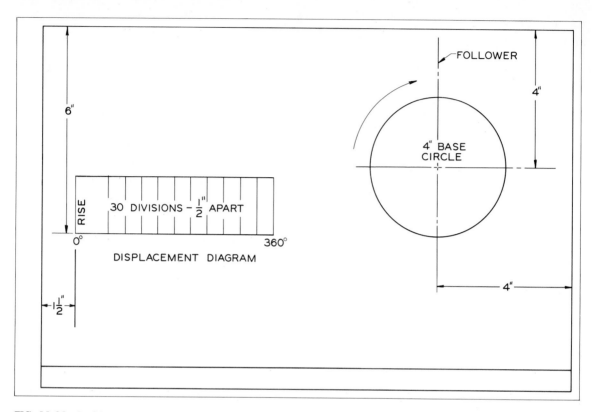

FIG. 11.20 Problem layout for the cam problems on Size B sheets.

12 Materials and Processes

12.1
Introduction

This chapter presents an overview of the materials and processes used in manufacturing. The major emphasis is placed on the forming and fabrication of metals since these processes are fundamental to manufacturing (Fig. 12.1).

The study of metals, called *metallurgy*, is a highly complex area that is constantly changing as new processes and alloys are being developed. The guidelines for designations of various types of metals have been standardized by three associations: the American Iron and Steel Institute (AISI), the Society of Automotive Engineers (SAE), and the American Society for Testing Materials (ASTM).

12.2
Iron*

Metals that contain iron, even in small quantities, are called *ferrous metals*. The three types of iron are *gray iron*, *white iron*, and *ductile iron*. Iron is used in the production of machine parts by the casting process;

*This section on iron was developed by Dr. Tom Pollock, a metallurgist at Texas A&M University.

consequently, iron is often called *cast iron*. Iron is cheaper than steel and it is easy to machine, but it does not have the ability to withstand the shock and forces that steel can withstand.

GRAY IRON contains flakes of graphite, which results in low strength and ductility, but makes the ma-

FIG. 12.1 This furnace operator is pouring an aluminum alloy of manganese into ingots (shown at the right) that will be remelted and cast. (Courtesy of the Aluminum Company of America.)

TABLE 12.1

ATSM GRADE (1000 psi)	SAE Grade	Typical Uses
ASTM 25 CI	G 2500 CI	Small engine blocks, pump bodies, transmission cases, clutch plates
ASTM 30 CI	G 3000 CI	Auto engine blocks, flywheels, heavy castings
ASTM 35 CI	G 3500 CI	Deisel engine blocks, tractor transmission cases, heavy and high-strength parts
ASTM 40 CI	G 4000 CI	Deisel cylinders, pistons, cam shafts

terial easy to machine. Gray iron will resist vibrations better than will other types of iron. Types of gray iron with two designations and their typical applications are given in Table 12.1.

WHITE IRON contains carbon in the form of hard carbide particles that are very hard and brittle, which enables it to withstand wear and abrasion. There are no designated grades of white iron, but there are differences in composition from one supplier to another. White iron is used for parts on grinding and crushing machines, digging teeth on earthmovers and mining equipment, and wear plates on reciprocating machinery used in textile mills.

DUCTILE IRON, also called *nodular* or *spheroidized* iron, contains carbon in the form of tiny graphite spheres. It is usually stronger and tougher than *gray iron* of a similar composition, but it is more expensive to produce. The numbering system for ductile iron is given by three sets of numbers, as shown below:

$$\text{60-40-18}$$

Tensile strength in 1000 psi } / Yield strength in 1000 psi } { Elongation in % (a measure of ductility)

Commonly used alloys of ductile iron and their applications are shown in Table 12.2.

MALLEABLE IRON is made from *white iron* by a heat-treatment process that converts carbides into carbon nodules (similar to ductile iron). Malleable iron is used for many of the same purposes as ductile iron. The numbering system for designating the grades of

TABLE 12.2

Grade	Typical Uses
60-40-18 CI	Valves, steam fittings, chemical plant equipment, pump bodies
65-45-12 CI	Machine components that are shock loaded, disc brake calipers
80-55-6 CI	Auto crankshafts, gears, rollers
100-70-3 CI	High-strength gears and machine parts
120-90-2 CI	Very high strength gears, rollers, and slides

malleable iron is shown below:

A. 325 10

Tensile strength in MPa } Elongation in %}

B. M 32 10

Stands for "martensite," which is not essential } { Tensile strength in MPa × 0.1 { Elongation in %

Some of the commonly used grades of malleable iron and their applications are given in Table 12.3.

12.3
Steel

Steel is an alloy of iron with the addition of other materials, but carbon is the ingredient that has the greatest effect on the grade of the steel. The three broad

TABLE 12.3

ASTM Grade	Typical Uses
35018 CI	Marine and railroad valves and fittings, "black-iron" pipe fittings (similar to 60-40-18 ductile CI)
45006 CI	Machine parts (similar to 80-55-6 ductile CI)
M3210 CI	Low-stress components, brackets
M4504 CI	Crankshafts, hubs
M7002 CI	High-strength parts, connecting rods, universal joints
M8501 CI	Wear-resistant gears and sliding parts

types of steel are *plain carbon steels, free-cutting carbon steels,* and *alloy steels.* The types of steels and their designations by four-digit numbers are shown in Table 12.4.

The number designations begin with a digit that indicates the type of steel: 1 is carbon steel, 2 is nickel steel, and so on. The second digit gives the percentage content of the material represented by the first digit. The last two digits give the percentage of carbon in the alloy, where 100 is equal to 1% and 50 is equal to 0.50%.

TABLE 12.4
NUMBERING AND APPLICATIONS
OF TYPES OF STEEL

Type of Steel	Number	Application
Carbon steels		
Plain carbon	10XX	Tubing, wire, nails
Resulphurized	11XX	Nuts, bolts, screws
Manganese steel	13XX	Gears, shafts
Nickel steel	23XX	Keys, levers, bolts
	25XX	Carburized parts
Nickel-chromium	31XX	Axles, gears, pins
	32XX	Forgings
	33XX	Axles, gears
Molybdenum steel	40XX	Gears, springs
Chromium-molybdenum	41XX	Shafts, tubing
Nickel-chromium	43XX	Gears, pinions
Nickel-molybdenum	46XX	Cams, shafts
	48XX	Roller bearings, pins
Chromium steel	51XX	Springs, gears
	52XX	Ball bearings
Chromium vanadium	61XX	Springs, forgings
Silicon manganese	92XX	Leaf springs

Source: The Society of Automotive Engineers.

Some of the more of often used SAE (Society of Automotive Engineers) steels are: 1010, 1015, 1020, 1030, 1040, 1070, 1080, 1111, 1118, 1145, 1320, 2330, 2345, 2515, 3135, 3130, 3240, 3310, 4023, 4042, 4063, 4140, and 4320.

12.4
Copper

Copper was one of the first metals discovered. It is a soft metal that can be easily formed and bent without breakage. Since it has a high resistance to corrosion and because of its high level of conductance, it is used in the manufacture of pipes, tubing, and electrical wiring. It is an excellent roofing and screening material since it withstands the weather well.

Copper has a number of alloys, including brasses, tin bronzes, nickel silvers, and copper-nickel alloys. A few of the numbered designations of wrought copper are: C11000, C11100, C11300, C11400, C11500, C11600, C10200, C12000, and C12200.

Brass is an alloy of copper and zinc, and bronze is an alloy of copper and tin.

Copper and copper alloys can be easily finished by buffing or plating. All of these can be joined by soldering, brazing, and welding and can be easily machined and used for casting.

Applications of copper are bushings, tubes, weatherstripping, marine hardware, tanks, electrical sockets, rivets, screws, and similar uses where weather resistance is required.

12.5
Aluminum

Aluminum is a corrosion-resistant, lightweight metal that has applications for many industrial products. Most of the materials that are called aluminum are actually alloys of aluminum. Alloys of aluminum possess greater strength than the pure metal, and more applications are possible.

The types of wrought aluminum alloys are designated by four digits, as shown in Table 12.5. (Wrought aluminum has properties that permit it to be formed by hammering.) The first digit, from 2 through 9, indicates the alloying element that is combined with aluminum. The last two digits identify the other alloy-

TABLE 12.5
NUMBERING DESIGNATIONS FOR WROUGHT
ALUMINUM AND ALUMINUM ALLOYS

Composition	Alloy Number	Applications
Aluminum (99% pure)	1XXX	Tubing, tank cars
Aluminum alloys		
Copper	2XXX	Aircraft parts, screws, rivets
Manganese	3XXX	Tanks, siding, gutters
Silicon	4XXX	Forging, wire
Magnesium	5XXX	Tubes, welded vessels
Magnesium and silicon	6XXX	Auto body, pipe
Zinc	7XXX	Aircraft structures
Other elements	8XXX	

ing materials or indicate the aluminum purity. The second digit indicates modifications of the original alloy or impurity limits.

A four-digit numbering system, with the last digit to the right of the decimal point, is used to designate types of cast aluminum and aluminum alloys. (When used for castings, the aluminum is melted and poured into a mold to form it.) The first digit indicates the alloy group as shown in Table 12.6. The next two digits identify the aluminum alloy or the aluminum purity. The numeral 1 to the right of the decimal point represents ingot aluminum, and 0 represents aluminum for casting. Ingots are blocks of cast metal that are to be remelted. Billets are castings of aluminum that are to be formed by forging.

TABLE 12.6
ALUMINUM CASTING AND
INGOT DESIGNATIONS

Composition	Alloy Number
Aluminum, (99 % pure)	1XX.X
Aluminum alloys	
Copper	2XX.X
Silicon with copper and/or magnesium	3XX.X
Silicon	4XX.X
Magnesium	5XX.X
Zinc	7XX.X
Tin	8XX.X
Other elements	9XX.X

Source: Society of Automotive Engineers.

12.6
Magnesium

Magnesium is a light metal that is considered to be available in an inexhaustible supply since it is extracted from seawater and natural brines. It is approximately half the weight of aluminum, and therefore it is an excellent material for aircraft parts, clutch housing, crankcases for air-cooled engines, and other applications where lightness of weight is desirable.

Magnesium and its alloys can be joined by bolting, riveting, and welding. Some numbered designations of magnesium alloys are: M10100, M11630, M11810, M11910, M11912, M12390, M13320, M16410, and M16620.

Magnesium is used for die casting, sand castings, extruded tubing, sheet metal, and forging.

12.7
Properties of materials

All materials have different properties that the designer must use to his or her best advantage. Consequently, it is important to be familiar with the terms used to describe these properties.

DUCTILITY is a softness present in some materials, such as copper and aluminum, that permits them to be formed by stretching (drawing) or hammering without breaking. Wire is made of ductile materials that can be drawn through a die.

BRITTLENESS is a characteristic of metals that will not stretch without breaking, such as cast irons and hardened steels.

MALLEABILITY is the ability of a metal to be rolled or hammered without breaking.

HARDNESS is a metal's ability to resist being dented when it receives a blow.

TOUGHNESS is the property of being shock-resistant while remaining malleable and resisting cracking or breaking.

ELASTICITY is the characteristic of a metal to return to its original shape after being bent or stretched.

12.8
Heat treatment of metals

The properties of different metals can be changed by various forms of heat treating. Steels are more affected by heat treating than are other materials.

HARDENING of steel is performed by heating the material to a prescribed temperature depending on its content and then quenching the hot steel in oil or water.

QUENCHING is the process of rapidly cooling heated metal by immersing it in liquids, gases, or solids (such as sand, limestone, or asbestos).

TEMPERING is the process of reheating previously hardened steel and then cooling it, usually by air. This increases the steel's toughness.

ANNEALING is the process of heating and cooling metals to soften them, to release their internal stresses, and to make them easier to machine and work with.

NORMALIZING is achieved by heating metals and letting them cool in air to room temperature, relieving their internal stresses.

CASE HARDENING is the process of hardening a thin outside layer of a metal. In this process, the outer layer is placed in contact with carbon or nitrogen compounds that become absorbed by the metal as it is heated. Afterward, it is quenched to complete the case hardening.

FLAME HARDENING is the method of heating a metal to a prescribed range with a flame and then quenching the metal.

12.9
Castings

Two major methods of forming shapes are *casting* and *pressure forming*. Casting involves the preparation of a mold into which is poured molten metal that cools and forms the part. The types of casting are *sand casting*, *permanent-mold casting*, *die casting*, and *investment casting*.

FIG. 12.2 A large casting of a landing-gear mechanism of an aircraft is being removed from its mold. (Courtesy of Cameron Iron Works.)

Sand casting

Sand casting is the most commonly used casting method. In the first step, a form or pattern is made that is representative of the final part to be cast. It is made of wood or metal. The pattern is placed in a metal box called a flask, and molding sand is packed around the pattern. When the pattern is withdrawn from the sand, it leaves a void forming the mold. Molten metal is poured into the mold through sprues, or gates. After cooling, the casting is removed and cleaned (Fig. 12.2).

Cores are parts that are formed in sand and placed within a mold to leave holes or hollow portions within the finished casting (Fig. 12.3). Once the casting has been formed and set, the sand cores can be broken apart and removed, leaving behind the desired void within the casting and thereby reducing its weight. Cores add considerably to the cost of a casting; they should not be used unless they are adequately offset by savings in materials.

Since the patterns must be placed in sand and then withdrawn before the metal is poured, it is necessary to taper the sides of the patterns for ease of withdrawal from the sand (Fig. 12.4). This taper is called *draft*. The amount of draft depends on the depth of the pattern in the sand; it usually varies from 2° to 8° for most applications. Also, patterns are made oversize to compensate for shrinkage that will occur when the metal cools.

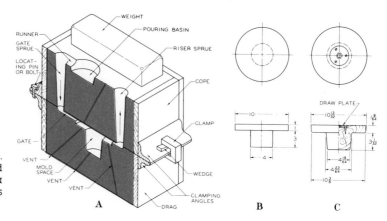

FIG. 12.3 A typical sand mold. The opening that will be filled with molten metal is formed by a wood or metal pattern that is pressed into the sand.

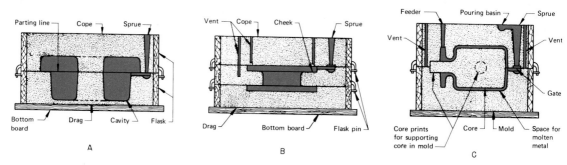

FIG. 12.4 Three types of sand molds: (A) two-section mold, (B) three-section mold, (C) two-section mold with a core. (Courtesy of General Motors Corporation.)

The sand casting has a rough surface that is not desirable for contacting other moving parts or surfaces. Consequently, it is common practice to machine portions of a casting by drilling, grinding, shaping, or other machining operations (Fig. 12.5). When this is to be done, the pattern should be made larger in these areas to compensate for removal of the metal caused by the machining operation.

Fillets and rounds at all intersections will increase the strength of a casting. Also, it is necessary to use fillets and rounds since it is difficult to form square corners by the sand-casting process.

Permanent-mold castings

Permanent molds are made for the mass production of parts. The molds are generally made of cast iron and are coated to prevent fusing with the molten metal that is poured into them. Permanent molds are often used with the manufacture of aluminum and magnesium parts. An example of a permanent mold is shown in Fig. 12.6.

Die castings

Die castings are generally used for the mass production of parts made of aluminum, magnesium, zinc alloys, and copper; however, other materials are used also. Die castings are made by forcing molten metal into dies (or molds) under pressure. The dies are permanent molds that are used over and over again. Die casting can be produced at a low cost and a high rate of production, at close tolerances, and with good surface qualities.

The same general principles recommended for sand castings—using fillets and rounds, allowing for shrinkage, and specifying draft angles—apply to die castings also. An example of a die casting for a simple part is shown in Fig. 12.7.

FIG. 12.5 This casting of the outer cylinder of an aircraft's landing gear is being bored on a horizontal boring mill. (Courtesy of Cameron Iron Works.)

Investment casting

Investment casting is a process used to produce complicated parts that would be difficult to form with uniformity by any other method. This technique is even used to form the complicated shapes of artistic sculptures.

A mold or die is made to cast a master pattern that is made of wax since a new pattern must be used

FIG. 12.6 Permanent molds are made of metal for repetitive usage when parts are mass produced. In this example, a sand core is made from another mold and is placed in the permanent mold to give a hollow void within the casting. (Courtesy of General Motors Corporation.)

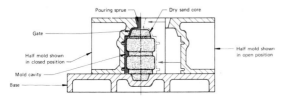

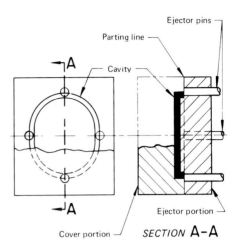

FIG. 12.7 A die for casting a simple part. Unlike the sand casting, the metal is forced into the die to form the die casting. (Courtesy of General Motors Corporation.)

for each investment casting. The wax pattern will be identical to the finished casting. The wax pattern is placed inside a container, and a plaster mixture or sand is poured (or invested) around the wax pattern.

The wax pattern is melted with heat once the investment has cured. This step removes the wax and leaves a hollow cavity that will serve as the mold for the molten metal. When filled and set, the plaster or sand is broken away from the finished investment casting.

A working drawing of a casting is shown in Fig. 12.8.

12.10
Forgings

Forging is the process of shaping or forming metal by hammering or squeezing the heated metal into a die. The resulting forging possesses high strength and a resistance to loads, vibrations, and impacts.

When preparing forging drawings, the following must be considered: (1) draft angles and parting lines, (2) fillets and rounds, (3) forging tolerances, (4) allowance of extra material for machining, and (5) heat treatment of the finished forging (Fig. 12.9). Some of the standard steels used for forging are designated by the following SAE numbers: 1015, 1020, 1025, 1045, 1137, 1151, 1335, 1340, 4620, 5120, and 5140. Other

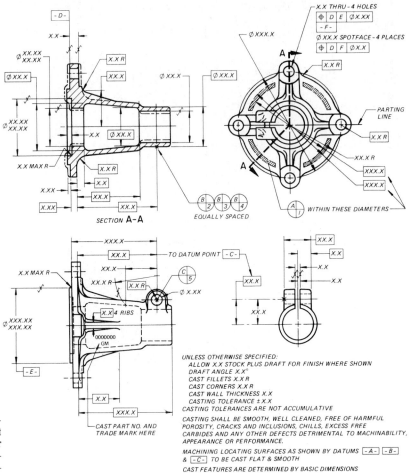

FIG. 12.8 A dimensioned drawing of a casting. Note that the part has been cast oversize to allow for finished surfaces that must be machined. (Courtesy of General Motors Corporation.)

FIG. 12.9 Three stages of manufacturing a turbine fan are shown here. The blank is first formed by forging; it is then machined; and the fan blades are then attached in their machined slots. (Courtesy of Avco Lycoming.)

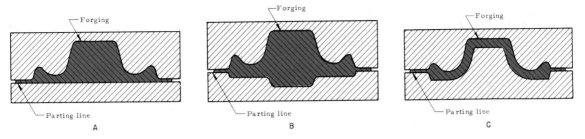

FIG. 12.10 Three types of forging dies: (A) single-impression die, (B) double-impression die, (C) interlocking die. (Courtesy of General Motors Corporation.)

materials that can be forged are iron, copper, and aluminum.

Drop forges and press forges are used to hammer the metal (called billets) into the forging dies by multiple blows or forces. These repetitive blows sequentially form the metal into the desired shape.

Examples of dies are shown in Fig. 12.10. A single-impression die gives an impression on one side of the parting line between the mating dies; a double-impression die gives an impression on both sides of the parting line. Interlocking dies result in a forging whose impression may cross the parting line on either side.

An example of an object that is forged with auxiliary rams to hollow the forging is shown in Fig. 12.11. A drawing of a forged part is illustrated in Fig. 12.12.

FIG. 12.11 Auxiliary rams can be used to form internal features on a part. (Courtesy of General Motors Corporation.)

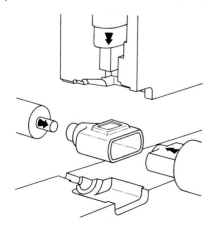

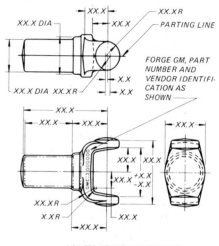

UNLESS OTHERWISE SPECIFIED:
DRAFT ANGLES $X°$.
ALL FILLETS X.XR, CORNERS X.XR.
+X.X– X.X TOLERANCES ON
FORGING DIM.

SNAG AND REMOVE SCALE.

SAMPLE FORGINGS ARE TO BE
APPROVED BY METALLURGICAL
AND ENGRG DEPTS FOR GRAIN
FLOW STRUCTURE.

FORGING DRAWING

FIG. 12.12 A drawing of a forging. The blank is forged oversize to allow for machining operations that will remove metal from it. (Courtesy of General Motors Corporation.)

Rolling

Rolling is a type of forging in which the stock is rolled between two rollers to give it a desired shape. Rolling can be done at right angles to the axis of the part or

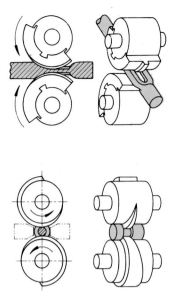

FIG. 12.13 Features on parts may be formed by rolling. In these examples, parts are being rolled parallel and perpendicular to their axes. (Courtesy of General Motors Corporation.)

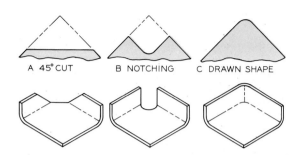

FIG. 12.14 Box-shaped parts formed by stamping. (A) A corner cut of 45° permits folding flanges and may require no further trim. (B) Notching has the same effect as the 45° cut, but it is often more attractive. (C) A continuous corner flange requires that the blank be developed so that it can be drawn into shape. (Courtesy of General Motors Corporation.)

parallel to its axis (Fig. 12.13). The stock is usually heated before rolling if a high degree of shaping is required. If the forming requires only a slight change in configuration, the rolling can be performed when the metal is cold. This process is called cold rolling.

12.11
Stamping

Stamping is a method of forming flat metal stock into three-dimensional shapes. In many applications, these can serve a design requirement more economically than a casting can. The first step of stamping is to cut out the shapes, called *blanks*, that are to be bent.

Blanks are formed into shape by bending and pressing them against forms. Examples of box-shaped parts are shown in Fig. 12.14. An example of a flange stamping is illustrated in Fig. 12.15. Holes in stampings are made by punching, extruding, or piercing, as shown in Fig. 12.16.

12.12
Machining operations

After the metal has been formed into the shape of the final product, machining operations usually have to be performed to complete the part. The machining operations involve several basic types of equipment: the *lathe, drill press, milling machine, shaper,* and *planer.*

FIG. 12.15 A sheet metal flange design, with notes calling attention to design details. (Courtesy of General Motors Corporation.)

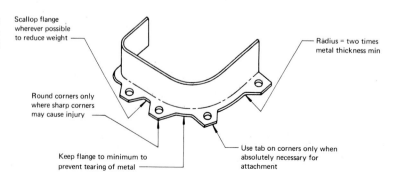

Scallop flange wherever possible to reduce weight

Radius = two times metal thickness min

Round corners only where sharp corners may cause injury

Keep flange to minimum to prevent tearing of metal

Use tab on corners only when absolutely necessary for attachment

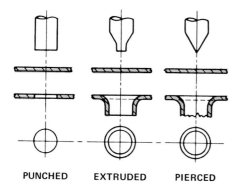

PUNCHED   EXTRUDED   PIERCED

FIG. 12.16 Three methods of forming holes in sheet metal. (Courtesy of General Motors Corporation.)

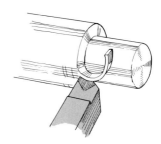

FIG. 12.19 Turning is the most basic of all operations performed on the lathe. A continuous chip is removed by a cutting tool as the part is rotated.

FIG. 12.17 A typical small-size metal lathe that holds the work pieces between centers as it is rotated about its axis. (Courtesy of the Clausing Corporation.)

FIG. 12.18 The fundamental operations performed on a lathe are illustrated on the two parts shown here.

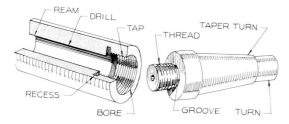

The lathe

The lathe is a machine that shapes cylindrical parts or holes by rotating the work piece between the centers of the lathe (Fig. 12.17). The more fundamental operations performed on the lathe are *turning, facing, drilling, boring, reaming, threading,* and *undercutting* (Fig. 12.18).

TURNING is the process of forming a cylinder by a tool that advances against the cylinder being turned and moves parallel to its axis (Fig. 12.19).

FACING is the process of forming flat surfaces that are perpendicular to the axis of rotation of the part being rotated by the lathe. This is often done on the end of a cylinder.

DRILLING is performed by mounting a drill in the tail stock of the lathe and rotating the work while the bit is advanced into the part (Fig. 12.20).

BORING is the process of making large holes that are too big to be drilled. Large holes are bored by enlarging smaller holes that have been drilled, or cored holes that have been formed by casting (Fig. 12.21).

REAMING is the removal of only a few thousandths of an inch of material inside a drilled hole to bring it to its required level of tolerance. Conical as well as cylindrical reaming can be performed on the lathe (Fig. 12.22).

THREADING of external and internal holes can be done on the lathe. The die used for cutting internal holes is called a *tap* (Fig. 12.23).

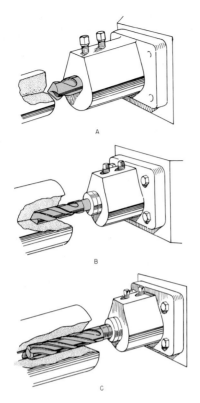

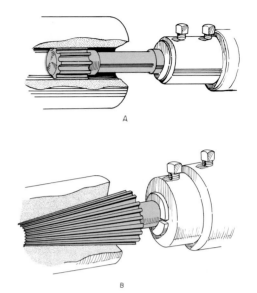

FIG. 12.22 Fluted reamers can be used to finish inside cylindrical and conical holes within a few thousandths of an inch.

FIG. 12.20 Three steps of drilling a hole in the end of a cylinder involve (A) start drilling, (B) twist drilling, and (C) core drilling. These steps should be performed in sequence. Core drilling enlarges the previously drilled hole to the required size.

UNDERCUTTING is a method of cutting a recess inside a cylindrical hole with a tool mounted on a boring bar. The groove is cut as the tool advances from the center of the axis of revolution into the part (Fig. 12.24).

The turret lathe is a programmable lathe that can perform sequential operations on the same part, such as drilling a series of holes, boring them, and then reaming them in succession. The turret is mounted in such a way as to rotate each tool into position for its particular operation (Fig. 12.25).

FIG. 12.21 Boring is the method of enlarging holes that are usually larger than available drill bits. The cutting tool is attached to the boring bar on the lathe.

FIG. 12.23 External and internal threads (shown here) can be cut on a lathe. The die being used to cut the threads is called a tap. Note that a recess has been formed at the end of the threaded hole prior to threading.

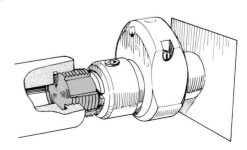

FIG. 12.24 A recess (undercut) can be formed by using the boring bar with the cutting tools attached as shown. As the boring bar is moved off center of the axis, the tool will form the recess.

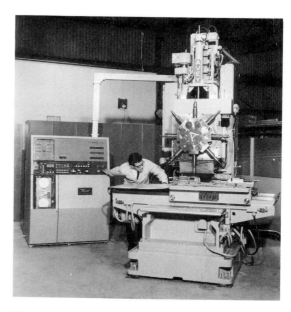

FIG. 12.26 A multiple-head drill press that can be programmed to perform a series of drill press operations in a desired sequence.

FIG. 12.25 A turret lathe that performs a sequence of operations as the turret head rotates in succession.

The drill press

The drill press is used to drill small- and medium-sized holes into stock that is held on the bed of the press. The work is held in position by a fixture or a clamp while the drilling is performed (Fig. 12.26).

In addition to drilling holes, the drill press can be used for *counterdrilling, countersinking, counterboring, spotfacing,* and *threading.* Examples of these operations are shown in Fig. 12.27.

BROACHING Cylindrical holes can be converted into square holes or hexagonal holes by using a tool called a *broach.* The broach is similar to a drill bit and is mounted in an overhead press.

The broach has a series of teeth along its axis, beginning with teeth that are nearly the size of the hole to be broached and tapering to the size of the finished hole that is to be broached. The broach is forced through the hole, with each tooth cutting more from the hole as it passes through.

This mass production process enables shops to form holes, other than cylindrical holes, in rapid succession.

FIG. 12.27 The basic operations that can be performed on the drill press are (left to right): drilling, reaming, boring, counterboring, spotfacing, countersinking, and tapping (threading).

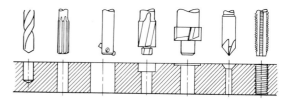

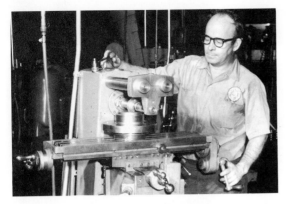

FIG. 12.28 This small milling machine is being used to cut a slot in the work piece. (Courtesy of the Clausing Corporation.)

The milling machine

The milling machine uses a series of cutting tools that are rotated about a shaft (Fig. 12.28). The work piece is passed under the cutters and in contact with them to remove the metal.

The milling machine has a wide variety of cutters to form different grooved slots, threads, and gear teeth. Irregular grooves in cams can be cut by the milling machine. It can be used to finish a surface on a part within a high degree of tolerance.

FIG. 12.29 The shaper moves back and forth across the part, removing metal as it advances. It can be used to finish surfaces, cut slots, and for many other operations.

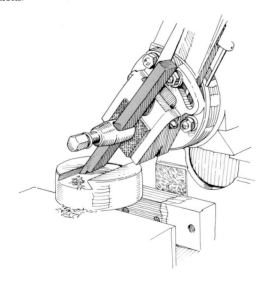

The shaper

The shaper is a machine that holds the work stationary while the cutter on the machine passes back and forth across the work to finish the surface or to cut a groove, one stroke at a time (Fig. 12.29). With each stroke of the cutting tool, the material is shifted slightly so as to align the part for the next overlapping stroke. The shaper can be used to cut a variety of slots and grooves in a part.

The planer

The planer is similar to the shaper except that the work is passed under the cutters by the planer rather than the work remaining stationary as in the case of the shaper (Fig. 12.30). Like the shaper, the planer can cut grooves or slots and finish surfaces that must meet tolerance specifications.

FIG. 12.30 The planer has stationary cutters, and the work is fed past them to finish larger surfaces. This planer has a 30-foot bed. (Courtesy of Simmons Machine Tool Corporation.)

12.13
Surface finishing

The process of finishing a surface to the desired uniformity is called surface finishing. It may be accomplished by several methods, including *grinding*,

FIG. 12.31 The upper surface of this part is being ground to a smooth finish by a grinding wheel. (Courtesy of the Clausing Corporation.)

polishing, lapping, buffing, and *honing.* Each method removes a portion of the metal from the surface to produce a smoother finish than before the finishing.

GRINDING is the finishing of a flat surface by holding it against a rotating abrasive wheel (Fig. 12.31). Grinding is used to smooth surfaces and to sharpen edges that are used for cutting, such as drill bits.

POLISHING is performed in the same manner as grinding except the polishing wheel is flexible since it is made of felt, leather, canvas, or fabric.

LAPPING is used to produce very smooth surfaces by holding the surfaces to be lapped against another large, flat surface, called a *lap,* which has been coated with a fine abrasive powder. As the lap rotates in contact with the work piece, the surface is finished to a high degree of smoothness. Lapping is done only after the surface has been previously finished by a less accurate technique, such as grinding or polishing. Cylindrical parts can be lapped by using a lathe in conjunction with the lapping surface.

BUFFING is a method of removing scratches from a surface with a rotating buffer wheel that is made of wool, cotton, or other fabric. Sometimes the buffer is a cloth or felt belt that is applied to the surface being buffed. An abrasive mixture is applied to the buffed surface from time to time to enhance the buffing. Polishing machines can be used for buffing by changing the wheels for each operation.

HONING is the process of finishing the outside or the inside of holes within a high degree of tolerance. The honing tool is rotated as it is passed through the holes to give the sort of finishes found in gun barrels, engine cylinders, and similar products where a high degree of smoothness is required.

13

Dimensioning

13.1
Introduction

Working drawings are dimensioned drawings used to describe the details of a part or a project so that construction can be performed in accordance with desired specifications.

Dimensions are needed to provide specifications and sizes of the various features. When properly ap-

plied, dimensions and notes will supplement the drawings so they can be used as legal contracts for construction.

The techniques of dimensioning presented in this chapter are based primarily on the standards of the American National Standards Institute (ANSI) and especially their standards, Y14.5M, *Dimensioning and Tolerancing for Engineering Drawings*. Various industrial standards from major corporations, such as the General Motors Corporation, have been utilized also.

Imperial (English) and metric (SI) units and the methods of using both are presented in keeping with current standards.

FIG. 13.1 This typical working drawing is dimensioned in millimeters.

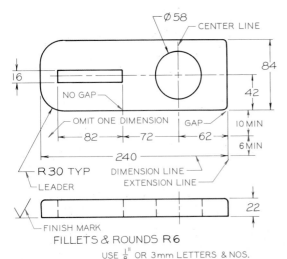

USE $\frac{1}{8}$" OR 3mm LETTERS & NOS.

13.2
Dimensioning terminology

The Guide Slide in Fig. 13.1 has been properly dimensioned and is used as an example to identify some of the terms of dimensioning.

DIMENSION LINES are thin lines (2H–4H pencil) with arrows at each end. Numbers placed near their midpoints specify a part's size.

EXTENSION LINES are thin lines (2H–4H pencil) that extend from a view of an object for dimensioning the part. The arrowheads of dimension lines end at extension lines.

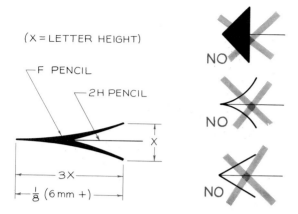

(X = LETTER HEIGHT)

F PENCIL

2H PENCIL

NO

NO

NO

FIG. 13.2 Arrowheads are drawn as long as the height of the letters used on a drawing. They are one-third as wide as they are long.

CENTER LINES are thin lines (2H–4H pencil) used to locate the centers of cylindrical parts, such as cylindrical holes.

LEADERS are thin lines (2H–4H pencil) drawn from a note to a feature to which the note applies.

ARROWHEADS are placed at the ends of dimension lines and leaders to indicate the endpoints of these lines. The arrowheads are drawn the same length as the height of the letters or numerals, ⅛″ in most cases. The form of the arrowhead is shown in Fig. 13.2.

DIMENSION NUMBERS are placed near the middle of the dimension line and are usually ⅛″ in height; units of measurement (″, IN, or mm) are omitted.

PLACEMENT TECHNIQUES of applying dimension and extension lines to parts are shown in Fig. 13.3. Dimension lines should not be closer than ⅜″ (10 mm) from a part.

13.3
Units of measurement

The two most commonly used units of measurement are the decimal inch, in the English (imperial) system, and the millimeter, in the metric (SI) system.

The inch in its common fraction form can be used, but it is preferable to give fractions in decimal form. Common fractions make addition, division, and general arithmetic very hard to perform. A comparison of dimensions in millimeters with those in inches is shown in Fig. 13.3.

Examples are shown in Fig. 13.4 where the units are given in millimeters, decimal inches, and fractional inches. Dimensions in millimeters are usually rounded off to whole numbers without decimal fractions. When a metric dimension is less than a millimeter, a zero precedes the decimal point.

Regardless of the units of measurement, the units are omitted from the dimension numbers, and the units are normally understood to be in millimeters or inches (not feet and inches) as specified by the scale on the drawing. For example: 112, not 112 mm; and 67, not 67″ or 5′–7″.

FIG. 13.3 Dimension lines should be placed at least ⅜″ (10 mm) from an object. Other rows of dimensions should be located at least ¼″ (6 mm) apart.

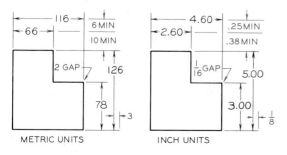

METRIC UNITS   INCH UNITS

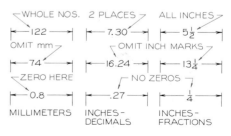

FIG. 13.4 When using SI units, dimensions are usually given to the nearest whole millimeter. When decimal inches are used, the fractions are carried to two decimal places. If fractions are given as common fractions, the fractions are twice as tall as whole numbers.

Architects still use a combination of feet and inches, but the inch units are omitted, such as 7'–2 to represent seven feet and two inches. Feet and decimal fractions of feet are used by engineers to dimension large-scale projects such as road designs. These dimensions will be expressed as 252.7', for example.

When using decimal inches, it is common practice to show all dimensions with two-place decimal fractions even though the last numbers are zeros. For dimensions of less than an inch, no zero precedes the decimal point.

13.4
English/metric conversions

Dimensions in inches can be converted to millimeters by multiplying by 25.4. Similarly, dimensions in millimeters can be converted to inches by dividing by 25.4.

For most applications, the millimeter does not need more than a one-place decimal when it is found by conversion from inches. The last digit retained in a conversion of millimeters is unchanged if it is followed by a number less than 5. For example, 34.43 is rounded off to 34.4.

The last digit to be retained is increased by one if it is followed by a number greater than 5. The number 34.46 is rounded off to 34.5.

The last digit to be retained is increased by one if it is odd and is followed by exactly 5. The number 34.75 is rounded off to 34.8.

These same rules apply to decimal units that are taken to more than two decimal places, since decimal inches are customarily carried to two decimal places.

13.5
Dual dimensioning

Some drawings require that both metric and English units be shown on each dimension. This method of dimensioning is called *dual dimensioning.*

Dual dimensioning can be accomplished by placing the inch equivalent of millimeters either under or over the other units, as shown in Fig. 13.5. If the drawing was originally dimensioned in inches, then the inch dimensions are placed on top of the millimeters. When converted to millimeters, inches are mul-

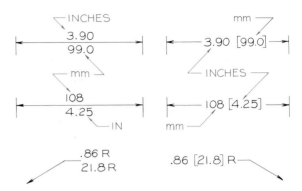

FIG. 13.5 Some dimensions are dual-dimensioned where both inches and millimeters are given on a single dimension line. If the drawing was originally made in inches, then the equivalent measurement in millimeters is placed under the inches or to the right in parentheses. If the drawing was originally made in millimeters, then the inch equivalents would be placed under or to the right. When inches are converted to millimeters, the millimeters may need to be written as decimal fractions.

tiplied by 25.4, and the equivalent millimeters are given under the inch measurements.

If the drawing was originally dimensioned in millimeters and then converted to inches, the millimeters would be placed over the equivalent in inches. When drawings are originally made in millimeters, whole numbers with no fractions are used.

The other method of dual dimensioning involves the use of brackets placed around the converted dimensions (Fig. 13.5).

It is important that you be consistent with whichever method you decide to use. Do not mix these two methods on the same drawing.

13.6
Metric designation

A comparison of the metric system with the English system is shown in Fig. 13.6. The metric system is the Système Internationale d'Unités and is denoted by the letters SI. This system utilizes the European system of the first-angle of projection, which locates the front view over the top view and the right side view to the left of the front view.

When drawings are made for international circulation, it is customary to use one of the symbols at C or D in Fig. 13.6 to designate the angle of projection

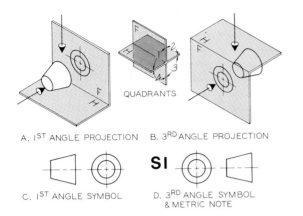

A. 1ST ANGLE PROJECTION B. 3RD ANGLE PROJECTION

C. 1ST ANGLE SYMBOL D. 3RD ANGLE SYMBOL & METRIC NOTE

FIG. 13.6 The European system of orthographic projection places the top view under the front view, the opposite of the American system. The system used on a drawing should be indicated by one of the symbols shown in parts C and D. If metric units are used, the large letters *SI* should be placed near the title block, or else the word *METRIC* should be used on the drawing.

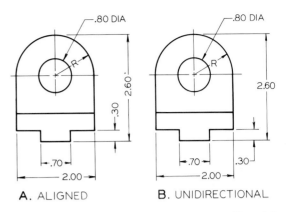

A. ALIGNED B. UNIDIRECTIONAL

FIG. 13.7 (A) Aligned dimensions are positioned to read from the bottom and right side of the sheet. (B) When the dimensions are positioned so all of them read from the bottom, they are unidirectional.

used. The large letters *SI* are used to indicate that the measurements are metric, or the word *METRIC* can be written prominently on the drawing or in the title block.

13.7
Aligned and unidirectional numbers

The two methods of positioning dimension numbers on a dimension line are the *aligned* and *unidirectional* methods.

The unidirectional system is more widely accepted since it is easier to apply numerals that all read from the bottom of the sheet, as shown in Fig. 13.7B.

The aligned system places the numerals in an aligned direction with the dimension lines (Fig. 13.7A). The numbers must be readable from the bottom or the right side of the page.

Examples of aligned dimensions on angular dimension lines are shown in Fig. 13.8. Avoid placing aligned dimensions in the "trouble zone," since these numerals would read from the left instead of the right side or bottom of the sheet.

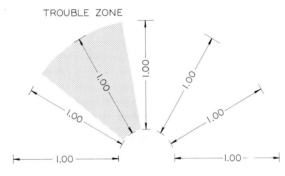

FIG. 13.8 Numbers on angular dimension lines should not be placed in the area called the trouble zone. This causes them to be read from the left side of the sheet rather than the right and bottom.

13.8
Placement of dimensions

It is good practice to dimension the views that are most descriptive.

You can see that the front view of Fig. 13.9 is more descriptive than the top view; therefore, this is the view that should be dimensioned.

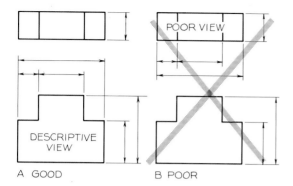

A GOOD   B POOR

FIG. 13.9 It is a rule of dimensioning to place the dimensions on the most descriptive views where the true contour of the object can be seen.

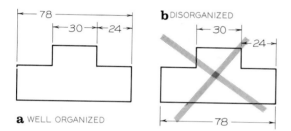

a WELL ORGANIZED

FIG. 13.10 Dimensions should be placed on the views in a well-organized manner to make them as readable as possible.

The dimensions should be applied to the views in an organized manner (Fig. 13.10). Locate the dimension lines by beginning with the smaller ones to avoid crossing dimension and extension lines.

FIG. 13.11 The first row of dimensions should be placed at least 0.40 inches (⅜″) from the view, and successive rows should be at least 0.25 inches (¼″) from the first row. If greater spaces are used, these general proportions should be used.

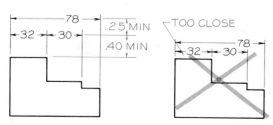

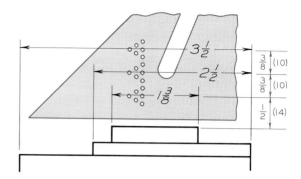

FIG. 13.12 When common fractions are used, the center holes of the triangular arrangements on the Braddock-Rowe triangle are aligned with the dimension lines to automatically space the lines as well as to draw the guidelines.

> Leave at least 0.4 inches between the object and the first row of dimensions (Fig. 13.11). The successive rows of dimensions should be at least 0.25 inches apart. If greater spaces are used, these same general proportions should be applied.

The Braddock-Rowe triangle can be used to space the dimension lines as well as used as a lettering guide (Fig. 13.12).

When a row of dimensions is placed on a drawing, one of the dimensions is omitted, as in Fig. 13.13A, since the overall dimension supplements the omitted dimension. If it is felt that the omitted dimension needs to be given as a reference dimension, it is

FIG. 13.13 (A) One intermediate dimension is customarily omitted since the overall dimension provides this measurement. (B) If all the intermediate dimensions are given, one should be placed in parentheses to indicate that it has been given as a reference dimension.

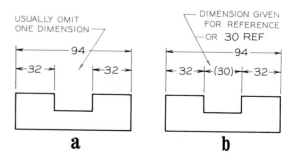

a   **b**

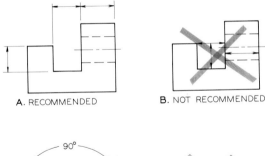

A. RECOMMENDED   B. NOT RECOMMENDED

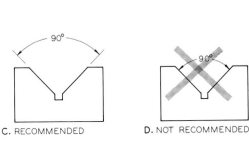

C. RECOMMENDED   D. NOT RECOMMENDED

FIG. 13.14 Dimensioning rules

(A) It is preferred practice to place the dimensions outside a part. Dimension lines should not be used as extension lines as at B.

(C and D) Angles are dimensioned with arcs and extension lines rather than showing the angle inside the angular cut as shown at D.

placed in parentheses to indicate that it is a reference dimension. In some cases the abbreviation REF is placed after a dimension to indicate reference.

The recommended techniques of dimensioning features are shown in Fig. 13.14.

FIG. 13.15 Extension lines extend from the edge of an object, leaving a small gap. They do not have gaps where they cross object lines or other extension lines.

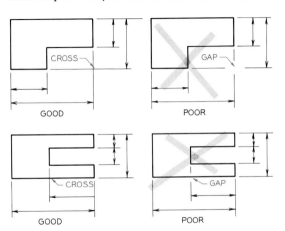

GOOD   POOR

GOOD   POOR

Examples of the placement of extension lines are shown in Fig. 13.15. Extension lines may cross other extension lines or object lines with no gaps. Extension lines are also used to locate theoretical points outside of curved surfaces (Fig. 13.16).

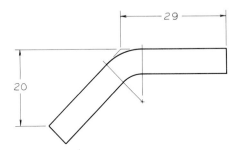

FIG. 13.16 A curved surface is dimensioned by locating the theoretical point of intersection with extension lines.

13.9
Dimensioning in limited spaces

Several examples of dimensioning in limited spaces are shown in Fig. 13.17. Regardless of space limitations, the numerals should not be drawn smaller than they appear elsewhere on the drawing.

When numbers are crowded where dimension lines are closely grouped, they should be staggered to make them more readable (Fig. 13.18).

FIG. 13.17 Where room permits, the numerals and arrows should be placed inside the extension lines. Other placements are shown as the spacing becomes smaller.

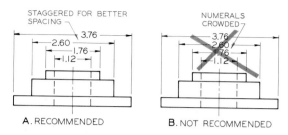

FIG. 13.18 When close spacing tends to crowd dimensioning numerals, they should be staggered for better spacing.

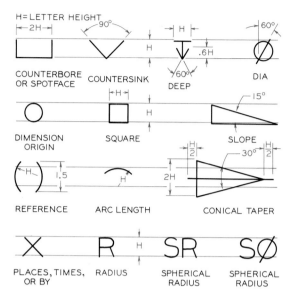

FIG. 13.19 These symbols can be used to dimension parts. The proportions of the symbols are based on the letter height, H, which is usually ⅛ in.

13.10
Dimensioning symbology

A number of symbols used in dimensioning are shown in Fig. 13.19. The sizes of the symbols are based on the height of the lettering that is used in dimensioning an object, usually one-eighth inch. Symbols are used to reduce time in preparing notes while adding to the clarity of a dimension.

13.11
Dimensioning prisms

The most basic element to be dimensioned is the prism, which is no more than a block when drawn in its simplest form. The following rules for dimensioning prisms are illustrated in Fig. 13.20.

1. Dimensions should extend from the most descriptive view (Fig. 13.20A).

2. Dimensions that apply to two views should be placed between these two views (Fig. 13.20A).

3. The first row of dimension lines should be placed a minimum of 0.40″ (10 mm) from the view. Successive rows are placed at least 0.25″ (6 mm) apart.

4. Extension lines may cross, but dimension lines should not cross another line unless absolutely necessary.

5. In order to dimension each measurement in its most descriptive view, you may have to place dimensions in more than one view (Fig. 13.20B).

6. In the case of notches, clarity may be achieved more effectively by placing the dimension line inside the notch (Fig. 13.20C).

7. Whenever possible, dimensions should be applied to visible lines rather than hidden lines (Fig. 13.20D).

8. Dimensions should not be repeated, nor should unnecessary information be given.

13.12
Dimensioning angles

Angles can be dimensioned by using either coordinates to locate the ends of angular lines or planes or angular measurements in degrees (Fig. 13.21). The two methods should not be mixed when dimensioning the same angle, since the measurements may not agree.

Units for angular measurements are degrees, minutes, and seconds, as shown in Fig. 13.21C. There are 60 minutes in a degree and 60 seconds in a minute. Seldom will angular measurements need to be measured to the nearest second.

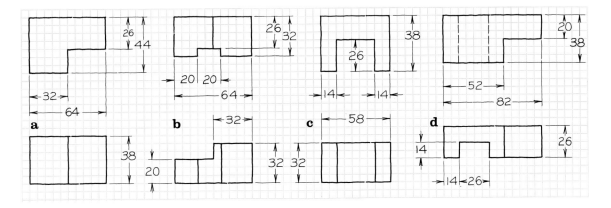

FIG. 13.20 Dimensioning prisms (metric)

A. Dimensions should extend from the most descriptive view and be placed between the views to which they apply.

B. One intermediate dimension is not given. Extension lines may cross object lines.

C. It is permissible to dimension a notch inside the object if this improves clarity.

D. Whenever possible, dimensions should be placed on visible lines, not hidden lines.

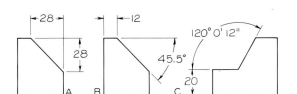

FIG. 13.21 Angular planes can be dimensioned by using coordinates (A), or an angle measured in degrees from the located vertex (B). Angles can be measured in decimal fractions (B) or in degrees, minutes, and seconds (C).

13.13
Dimensioning cylinders

Cylinders that are dimensioned may be either solid cylinders or cylindrical holes. All diametrical dimensions should be preceded by the symbol ∅ (Fig. 13.22).

Solid cylinders are dimensioned in their rectangular views by using diameters (not radii).

FIG. 13.22 (A) it is preferred that cylinders be dimensioned in their rectangular views using a diameter rather than a radius. (B) Dimensions should be placed between the views when possible. (C) The circular view can be omitted if DIA or ∅ is written before the dimensions of the diameters.

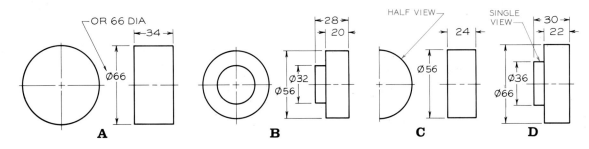

Parts composed of several concentric cylinders are dimensioned with diameters, beginning with the smallest cylinder (Fig. 13.22B). Parts of this type are assumed to be concentric unless otherwise noted.

A cylindrical part may be sufficiently dimensioned with only one view, the circular view being omitted.

Likewise, an external micrometer caliper can be used for measuring the outside diameters of a part (Fig. 13.24). The choice of diameter rather than radius makes it possible to measure diameters during machining when the part is held between centers on a lathe.

13.14
Measuring cylindrical parts

Cylindrical parts are dimensioned with diameters rather than radii because diameters are easier to measure.

An internal cylindrical hole is measured with an internal micrometer caliper that has a built-in gage permitting greater accuracy of measurement (Fig. 13.23).

13.15
Cylindrical holes

Cylindrical holes may be dimensioned by one of the methods shown in Fig. 13.25.

The preferred method of dimensioning cylindrical holes is to draw a leader from the circular view, and then add the dimension, preceded by $\varnothing$, to indicate that the dimension is a diameter (Fig. 13.26). The English system often uses DIA after the diametrical dimension.

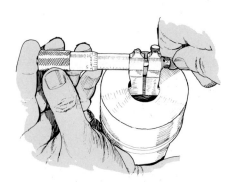

FIG. 13.23 A micrometer caliper is used to measure internal cylindrical diameters. This is why holes are dimensioned with diameters instead of radii.

FIG. 13.24 An external micrometer caliper with a built-in gage for measuring the diameter of a cylinder.

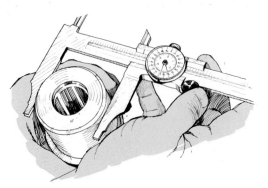

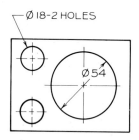
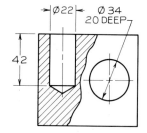

FIG. 13.25 Several acceptable methods of dimensioning cylindrical holes and shapes. The symbol $\varnothing$ is always placed in front of the diametrical dimension.

FIG. 13.26 The preferable method of dimensioning cylindrical holes is with a leader, a dimension, and the symbol $\varnothing$ to indicate that the dimension is a diameter. Previous standards recommended the abbreviation DIA after the dimension.

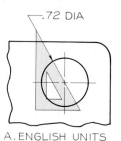
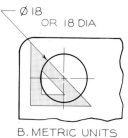

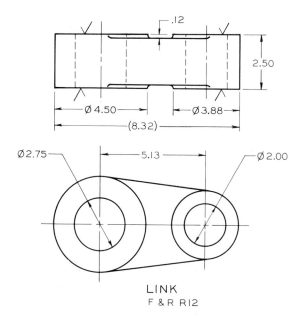

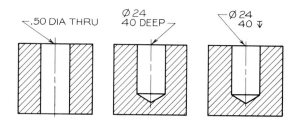

FIG. 13.28 Three methods of dimensioning holes, with combinations of notes and symbols.

FIG. 13.27 This part composed of cylindrical features has been dimensioned using several approved methods. (F&R R.12 means that fillets and rounds have a 0.12 radius.)

LINK
F & R R12

Sometimes the note DRILL or BORE is added to specify the shop operation, but current standards prefer the use of ⌀ instead.

A part containing cylindrical features is dimensioned in Fig. 13.27 in both the circular and the rectangular views to illustrate various methods of dimensioning.

Three methods of dimensioning holes are shown in Fig. 13.28. The depth of a hole is its usable depth, not the depth to the point left by the drill bit.

13.16
Pyramids, cones, and spheres

Three methods of dimensioning pyramids are shown in Fig. 13.29A, B, and C. The pyramids at B and C are truncated (portions removed). Note that in all three examples, the apex of the pyramid is located in the rectangular view.

Two acceptable methods of dimensioning cones are shown in Fig. 13.29D and E.

A sphere is dimensioned by using its diameter if it is a complete sphere (Fig. 13.29F); a radius is used if it is less than a full sphere (Fig. 13.29G). Only one view is necessary to describe a sphere.

13.17
Leaders

Leaders are used to apply notes and dimensions to a feature that they describe. Leaders are drawn at a standard angle of a triangle, as shown in Fig. 13.26.

FIG. 13.29 Methods of dimensioning pyramids, cones, and spheres are shown here.

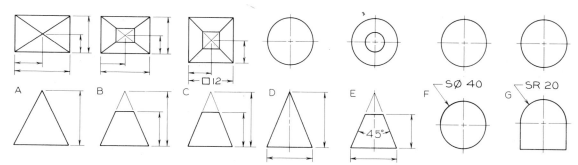

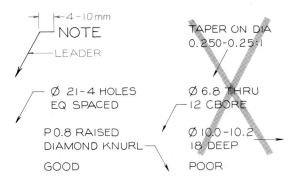

FIG. 13.30 Leaders from notes should extend with a horizontal bar from the first or last word of the note, not from the middle of the note.

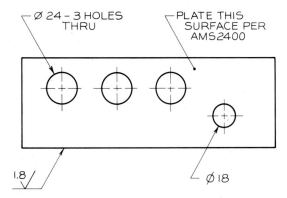

FIG. 13.31 Examples of notes applied to a part with leaders.

FIG. 13.32 When space permits, the dimension and arrow should be placed between the center and the arc. If room is not available for the number, the arrow is placed between the center and the arc with the number on the outside. If there is no room for the arrow, then both the dimension and the arrow are placed outside the arc with a leader.

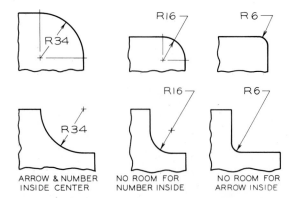

Examples of notes using leaders are shown in Fig. 13.30. The leader should be drawn from either the first word of the note or the last word of the note. Each leader should begin with a short horizontal line near the note.

Applications of leaders on a part are shown in Fig. 13.31. Notice that a dot is used instead of an arrowhead when the note is applied to a surface that does not appear as an edge.

13.18
Dimensioning arcs

Cylindrical parts that are less than a full circle are dimensioned with radii, as shown in Fig. 13.32. Current standards recommend that radii be dimensioned with

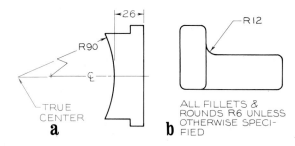

FIG. 13.33 When a radius is very long, it may be shown with a false radius with a "zigzag" to indicate that it is not true length. It should end on the center line of the true center. Fillets and rounds may be noted to reduce repetitive dimensions of small arcs.

FIG. 13.34 When several arcs are dimensioned with the same radii, it is preferable that separate leaders be used rather than combining the leaders as shown at B.

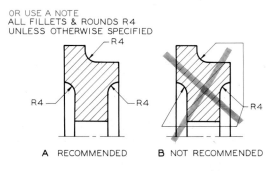

an *R* preceding the dimension, such as R 10. Previous standards recommended that the *R* follow the dimension, such as 10 R. Consequently, both methods will be seen in practice.

When the arc being dimensioned is a long arc, it may be dimensioned with a false radius, as shown in Fig. 13.33A. A "zigzag" is placed in the radius to indicate that it is not a true radius. The center of the arc should lie on the center line of the extension of the line on which the true center lies.

13.19
Fillets & rounds, and TYP

Fillets and *rounds* are rounded corners conventionally used on castings. A fillet is an internal rounding, and a round is an external rounding.

When fillets and rounds are equal in radii, a note may be placed on the drawing to eliminate the need for repetitive dimensioning. The note may read as follows: ALL FILLETS AND ROUNDS R 6. If most, but not all, of the fillets and rounds have equal radii, the following note may be used: ALL FILLETS AND ROUNDS R 6 UNLESS OTHERWISE SPECIFIED. In this case, only the fillets and rounds of different radii are dimensioned (Fig. 13.33B). The notes may be abbreviated, such as F & R 10.

Fillets and rounds should be dimensioned with separate leaders, as shown in Fig. 13.34, rather than using long, confusing leaders, as shown at B.

Repetitive features on a drawing may be noted as shown in Fig. 13.35. The note TYPICAL or TYP means that although only one of these features is dimensioned, the dimensions are typical of those that are undimensioned. The term *PLACES* is sometimes used to specify the number of places in which a similar feature appears. Similarly, the number of holes sharing the same dimension may be indicated by including the number of holes in the note.

13.20
Curved surfaces

An irregular shape composed of a number of tangent arcs of varying sizes (Fig. 13.36) can be dimensioned by using a series of radii.

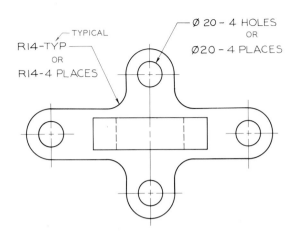

FIG. 13.35 Notes can be used to indicate that similar features and dimensions are repeated on drawings without having to show the dimensions repetitively.

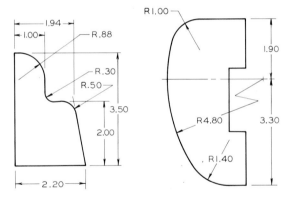

FIG. 13.36 Examples of dimensioned parts that are composed of a series of tangent arcs.

When the curve is irregular rather than composed of arcs (Fig. 13.37), the coordinate method can be used to locate a series of points along the curve from two datum lines. The drafter must use judgment to determine the proper spacing for the points. Note that extension lines may be placed at angles to provide additional space for dimensions.

A special case of an irregular curve is the symmetrical curve shown in Fig. 13.38. Note that dimension lines are used as extension lines in violation of a previously established rule.

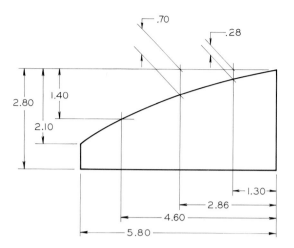

FIG. 13.37 This object with an irregular curve is dimensioned by using coordinates.

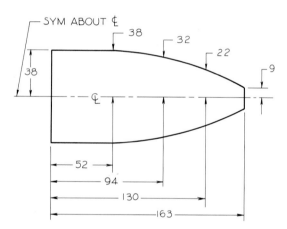

FIG. 13.38 This symmetrical part is dimensioned about its center line.

FIG. 13.39 Symmetrical parts may be dimensioned about their center lines as shown at A. It is better to dimension symmetrical parts of this type as shown at B.

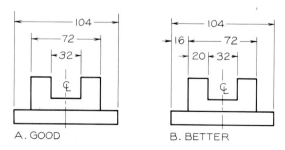

A. GOOD   B. BETTER

13.21
Symmetrical objects

Symmetrical objects may be dimensioned as shown in Fig. 13.39A, where a center line is used with the initials *CL* to identify it. In this case, it is assumed that the dimensions are each centered about the center line.

The better method is the one shown in Fig. 13.39B, where the need for an assumption is eliminated.

13.22
Finished surfaces

Many parts are formed as castings in a mold that gives their exterior surfaces a rough finish. If the part is designed to come in contact with another surface, the rough finish must be machined by grinding, shaping, lapping, or a similar process.

To indicate that a surface is to be finished, finish marks are drawn on the surface where it appears as an edge (Fig. 13.40).

Finish marks should be repeated in every view where the surface appears as an edge, even if it is a hidden line.

Three methods of drawing finish marks are shown in Fig. 13.40. The simple V mark is preferred

FIG. 13.40 Finish marks of these types can be used to indicate that a surface has been machined to a smooth surface. The traditional V can be used for general applications, but the new finish mark at B is suggested as the new symbol. The f shown at C is also acceptable to indicate finished surfaces.

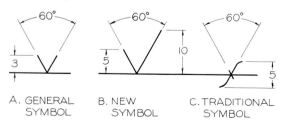

A. GENERAL
SYMBOL

B. NEW
SYMBOL

C. TRADITIONAL
SYMBOL

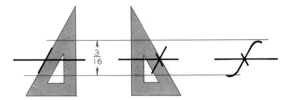

FIG. 13.41 The steps of drawing the f finish mark.

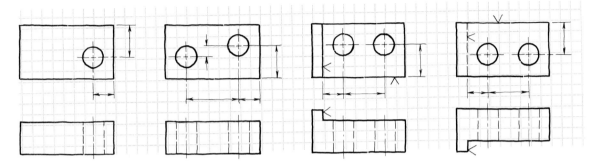

FIG. 13.42 Location dimensions

A. Cylindrical holes should be located in the circular view from two surfaces of the object.

B. When more than one hole is to be located, the other holes should be located in relation to the first from center to center.

C. A more accurate location is possible if holes are located from finished surfaces.

D. Holes should be located in the circular view and from finished surfaces even if the finished surfaces are hidden.

in general cases. The uneven V (Fig. 13.40B) is a newly recommended symbol that is related to surface texture and will be discussed in the next chapter. The steps of constructing the traditional f that is used as a finish mark are shown in Fig. 13.41.

When an object is finished on all surfaces, the note FINISHED ALL OVER is placed on the drawing. This can be abbreviated as FAO without periods after the letters. Cylindrical holes, although finished, are not specified with finish marks.

The centers of the holes are located with coordinates in the circular view when possible.

It is desirable that the holes be located from finished surfaces since holes can be located more accurately from a smooth, machined surface than from an unfinished surface. This rule is followed even if the finished surface is a hidden line, as in Fig. 13.43.

FIG. 13.43 These cylindrical holes are located from center to center in their circular view and from a finished surface, if one is available.

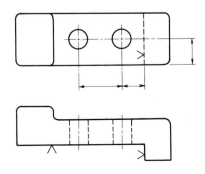

13.23
Location dimensions

Location dimensions are used to locate the positions, not the sizes, of geometric elements, such as cylindrical holes. Figure 13.42 illustrates the basic rules for locating cylindrical parts. Since location dimensions do not involve the sizes of various features, the dimensions of the holes are omitted for clarity in Fig. 13.42.

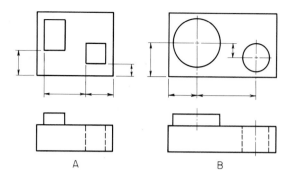

FIG. 13.44 Prisms and cylinders are located with co-ordinates in the view where both coordinates can be seen.

Prisms are located with respect to each other, as shown in Fig. 13.44. Only a single corner of a prism need be located with respect to another when the sides of the prisms are parallel.

Location dimensions should be placed on views where both dimensions can be shown, as in the top view in Fig. 13.44. Cylinders are located in their circular views (Figs. 13.43B and 13.45).

13.24
Location of holes

When holes must be located accurately, the dimensions should originate from a common reference plane on the part to reduce the accumulation of errors in measurement. Examples of holes that are located from reference planes with coordinates are shown in Fig. 13.46A and B.

When several holes in a series are to be equally spaced, as in Fig. 13.46C, a note specifying that they are equally spaced can be used to locate the holes. The first and last holes of the series are determined by the usual location dimensions.

Holes through circular plates (Fig. 13.47) may be located by coordinates or by a note. When a note is used, the diameter of the circle passing through the centers of the holes must be given. This circle is referred to as the "bolt circle," or circle of centers.

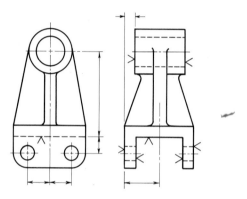

FIG. 13.45 An example of location dimensions applied to a part to locate its geometric features.

FIG. 13.46 Location of holes

A. Holes can be more accurately located if common datum planes are used, from which all measurements are made.

B. A diagonal dimension can be used to locate a hole of this type from another hole's center.

C. A note can be used to specify the spacing between the centers of holes in an equally spaced series.

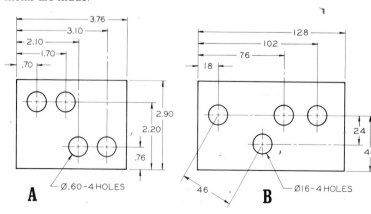

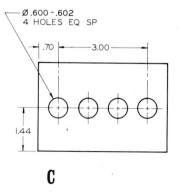

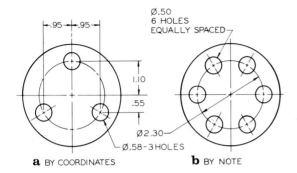

a BY COORDINATES   **b** BY NOTE

FIG. 13.47 Holes may be located in circular plates by coordinates (A) or by note (B).

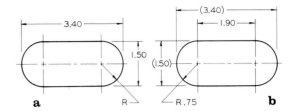

a   R   R.75   **b**

FIG. 13.49 The preferred method of dimensioning objects with rounded ends is shown at (a). A less desirable method is shown at (b).

A similar method of locating holes is the polar system illustrated in Fig. 13.48. The radial distances from the point of concurrency and their angular measurements (in degrees) between the holes locate the centers.

13.25
Objects with rounded ends

Objects with rounded ends should be dimensioned from end to end, as shown in Fig. 13.49. This dimension provides the overall length without using arithmetic. The radius is shown with the letter *R* without a dimension to specify that the end is formed by an arc. Since the height is given, the radius is understood to be half the height (1.50 in this case).

FIG. 13.48 Methods of locating cylindrical holes on concentric arcs.

If the object is dimensioned from center to center of the rounded ends, the overall dimension should be given as a reference dimension (3.40) to eliminate the calculations required to determine the overall dimension. In this case the radius must be given.

A part with partially rounded ends is dimensioned as shown in Fig. 13.50A. The overall dimension is given, and the dimensions of the radii are given in order that their centers may be located by measuring from the ends.

When an object has a rounded end that is less than a semicircle (Fig. 13.50B), location dimensions must be used to locate the center of the arc.

Slots with rounded ends are dimensioned in Fig. 13.51. Only one slot is dimensioned in part A, with a note indicating that there are two slots. The slot at B is dimensioned giving the overall dimension and the

FIG. 13.50 Examples of dimensioned parts consisting of rounded ends and cylindrical features.

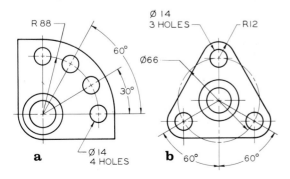

a   **b** 60°   60°

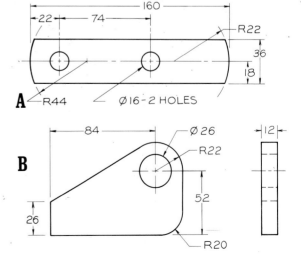

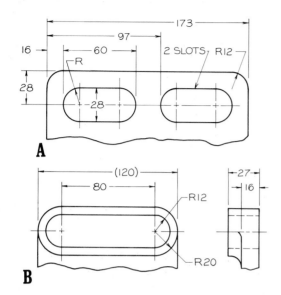

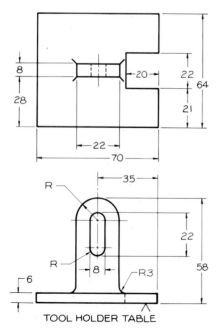

FIG. 13.51 Methods of dimensioning slots.

two arcs, which are understood to apply to both ends. The distance between the centers is given as a reference dimension.

The dimensioned views of the tool holder table in Fig. 13.52 show examples of arcs and slots. To prevent

FIG. 13.52 This part is dimensioned by using the previously presented principles.

TOOL HOLDER TABLE

crossing of dimension lines, it is often necessary to place dimensions in a view that is not as descriptive as might be desired.

13.26
Machined holes

Machined holes are holes that are made or refined by a machine operation, such as drilling or boring. These operations are specified by notes and leaders applied to the holes (Fig. 13.53).

It is preferred to give the diameter of the hole with the symbol ∅ in front of the dimension, with no reference to the machining method used. However, the note 32 DRILL may be used in some cases instead of ∅ 32. In practice, you will see diameters dimensioned as XX DIA, since this was recommended by the previous standards.

FIG. 13.53 Cylindrical holes may be dimensioned by either of the methods shown at A and B. Note that when one view is given at C, it is necessary to indicate in the note that the holes are "thru" holes.

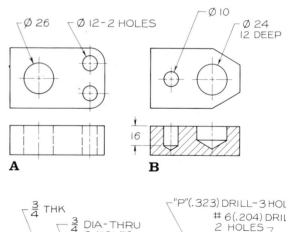

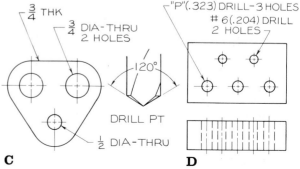

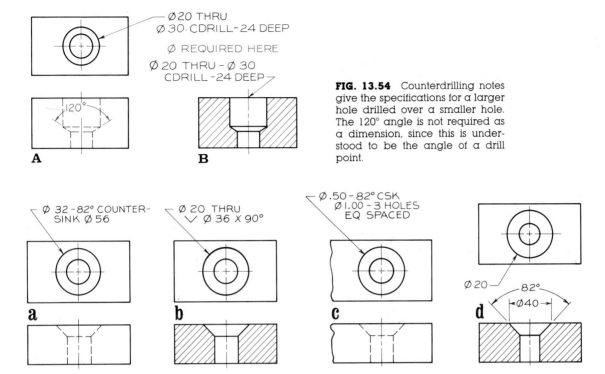

FIG. 13.54 Counterdrilling notes give the specifications for a larger hole drilled over a smaller hole. The 120° angle is not required as a dimension, since this is understood to be the angle of a drill point.

FIG. 13.55 Examples of notes and symbols specifying countersunk holes are shown here in various acceptable forms.

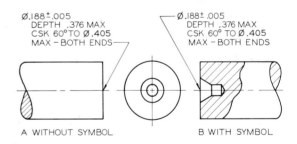

FIG. 13.56 Methods of specifying countersinks in the ends of cylinders for mounting them on a lathe between centers.

DRILLED HOLES Several methods of specifying drilled holes are shown in Fig. 13.53. The depth of drilled holes can be specified in the note or it may be dimensioned in the rectangular view. The depth of a drilled hole is dimensioned as the usable part of the hole; the conical point is disregarded, as shown in part B. The standard angle for the point of the drill is 120°.

COUNTERDRILLED HOLES A large hole drilled inside a smaller drilled hole to enlarge it is a counterdrilled hole, as illustrated in Fig. 13.54. (The 120° dimension is given to indicate the angle of the drill point; it need not be shown on the drawing.)

COUNTERSUNK HOLES Countersinking is the process of forming a conical hole at the end of a cylindrical hole for flat head screws (Fig. 13.55). The diameter of the countersunk hole (the maximum diameter on the surface) and the angle of the countersink are given in the notes.

Countersinking is also used to provide center holes in shafts, spindles, and other cylindrical parts that must be held between the centers of a lathe (Fig. 13.56).

SPOTFACING Spotfacing is a machining process used to finish the surface around the top of a hole in order to provide a level seat for a washer or a fastener head (Fig. 13.57a and b).

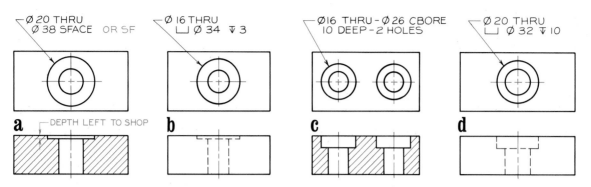

FIG. 13.57 Spotfaces can be specified as shown at (a) and (b). The depth of the spotface can be specified as shown if needed. Counterbores are dimensioned by giving the diameters of both holes and the depth of the larger hole (C and D).

A spotfacing tool is shown in Fig. 13.58, where it has spotfaced a *boss* (a raised cylindrical element).

COUNTERBORING   Counterbored holes are made by enlarging the drilled hole to a larger diameter (Fig. 13.57). Note that the bottoms of the counterbored holes are flat with no taper as in counterdrilled and countersunk holes.

BORED HOLES   Boring is a machine operation that is usually performed on a lathe with a boring bar (Fig. 13.59) for making large holes.

REAMED HOLES   Reamed holes are holes that have been finished or slightly enlarged after having been drilled or bored. This operation uses a ream, which is similar to a drill bit.

FIG. 13.58 A spotfacing tool being used to spotface the cylindrical boss to provide a smooth bearing for a bolt.

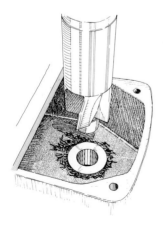

FIG. 13.59 Boring a large hole on a lathe with a boring bar. (Courtesy of Clausing Corporation.)

13.27
Chamfers

Chamfers are beveled edges that are used on cylindrical parts, such as shafts and threaded fasteners. They eliminate sharp edges and facilitate the assembly of parts.

When a chamfer angle is 45°, a note can be used in either of the forms shown in Fig. 13.60A. When the chamfer is at an angle other than 45°, the angle and the length are given as shown in part B.

Chamfers can also be specified at the openings of holes, as shown in Fig. 13.61.

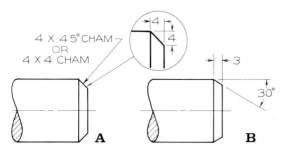

FIG. 13.60 Chamfers can be dimensioned by using either type of note shown at A when the angle is 45°. If the angle is other than 45°, it should be dimensioned as shown at B.

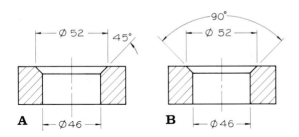

FIG. 13.61 Inside chamfers are dimensioned by using one of these methods.

FIG. 13.62 Methods of dimensioning slots in a shaft and a slot for a Woodruff key that will hold the part on the shaft are shown here. The dimensions for these features are given in Appendix 32.

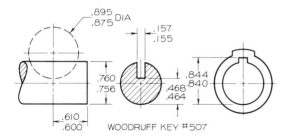

13.28
Keyseats

A keyseat is a slot cut into a shaft for the purpose of aligning the shaft with a part mounted on it. This part may be a pulley or a collar. The method of dimensioning a keyway and keyseat is shown in Fig. 13.62. The double dimensions are tolerances, which will be discussed in the next chapter.

13.29
Knurling

Knurling is the operation of cutting diamond-shaped or parallel patterns on cylindrical surfaces for gripping, for decoration, or for a press fit between two parts that will be permanently assembled.

A *diamond knurl* and a *straight knurl* are drawn and dimensioned in Fig. 13.63. Knurls should be dimensioned with specifications that give type, pitch, and diameter.

The "96 DP" means diametrical pitch, which is the ratio of the number of grooves on the circumference (N) to the diameter (D), which is found by the equation $DP = N/D$. The preferred diametrical pitches for knurling are 64DP, 96DP, 128DP, and 160DP. For diameters of 1 inch, knurling of 64DP, 96DP, 128DP, and 160DP will have 64, 96, 128, and 160 teeth, respectively on the circumference. The note P 0.8 means that the knurling grooves are 0.8 mm apart. (Note:

FIG. 13.63 A diamond knurl with a diametrical pitch of 96 is shown at the left. The diametrical pitch is the number of teeth about the circumference divided by the diameter. A straight knurl is shown at the right where the pitch (P) is 0.8 mm. Pitch is the distance between the grooves on the circumference.

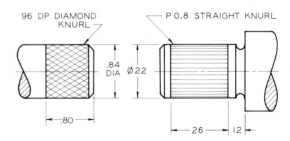

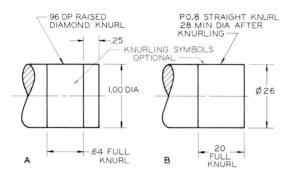

FIG. 13.64 Knurls need not be drawn if they are dimensioned as shown here.

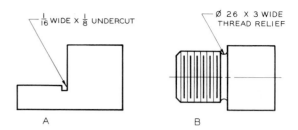

FIG. 13.66 An undercut can be dimensioned as shown at A. A thread relief, which is a type of neck, is dimensioned at B.

Calculations must be made using inches. Conversion to millimeters can be made afterward.)

Knurls for press fits are specified with diameters before knurling, and the minimum diameter after knurling. A simplified method of representing knurls is shown in Fig. 13.64, where notes are used and the knurls are not drawn.

13.30
Necks and undercuts

A *neck* is a recess cut into a cylindrical part. A recess is used where cylinders of different diameter join (Fig. 13.65) to ensure that a part assembled on the smaller shaft will fit flush against the shoulder of the larger cylinder.

Undercuts are similar to necks and serve the same purpose (Fig. 13.66A). An undercut ensures that a part fitting in the corner of the surface will fit flush against both surfaces, and it permits space for trash to drop out of the way when entrapped in the corner. An undercut could also be a recessed neck on the inside of a cylindrical hole. A thread relief (Fig. 13.66B) is used to square the threads where they intersect a larger cylinder.

13.31
Tapers

Tapers can be either conical surfaces or flat planes. Examples of conical and flat tapers are dimensioned in Fig. 13.67. A conical taper can be dimensioned by (1)

FIG. 13.67 Examples of flat and conical tapers. Taper is the ratio of the diameters (or heights) at each end of a sloping surface to the length of the taper.

FIG. 13.65 Necks are recesses in cylinders that are used where cylinders of different sizes join together.

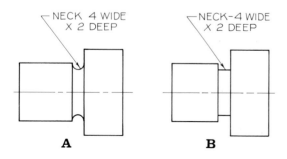

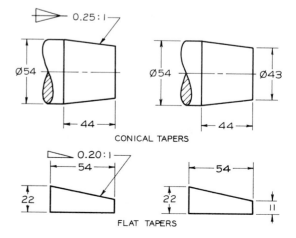

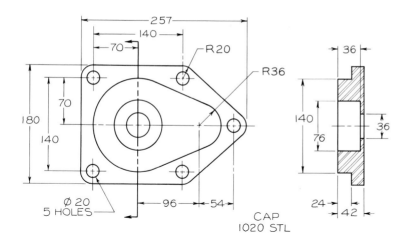

CAP
1020 STL

FIG. 13.68 An example of a part dimensioned with a variety of dimensioning principles.

FIG. 13.69 (A and B) Threaded holes are sometimes dimensioned by giving the tap drill size in addition to the thread specifications. The tap drill size is not required, but it is permissible. The part at B is dimensioned to indicate a neck, a taper, and a break corner, which is a slight round to remove the sharpness from a corner. The collar at C has a knurl note, a chamfer note, and a note indicating the insertion of a #2 taper pin. The part at D gives a note for dimensioning a keyway.

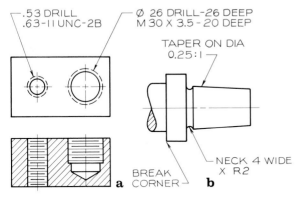

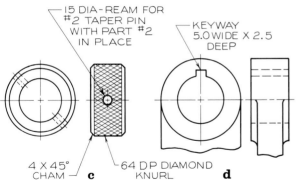

a diameter, taper, and length of taper or (2) a diameter at each end of the cone and the length of the cone

Taper is the ratio of the difference in the diameters of a cone to the distance between two diameters (Fig. 13.67). Taper can be expressed as inches per inch (.25" per inch), inches per foot (3.00 per foot), or millimeters per millimeter (0.25:1).

Flat taper is the ratio of the difference in the heights at each end of a feature to the distance between the heights. Flat tapers can be expressed as inches per inch (.20 per inch), inches per foot (2.40 per foot), or millimeters per millimeter (0.20:1), as shown in Fig. 13.67.

13.32
Dimensioning sections

Sections are dimensioned in the same manner as regular views. A dimensioned section is shown in Fig. 13.68. Most of the principles of dimensioning covered in this chapter have been applied to this part.

13.33
Miscellaneous notes

A variety of notes are used on detail drawings to provide information and specifications that would be difficult to represent by other means (Figs. 13.69 and 13.70).

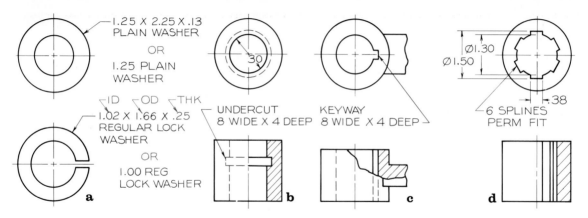

FIG. 13.70

(a and b) Methods of dimensioning washers and lock washers are shown. These dimensions can be found in the tables in Appendixes 36 and 37. An undercut is dimensioned at (b). A keyway is dimensioned at (c), and a spline inside a hole is dimensioned at (d).

Notes are placed horizontally on the sheet with short dashes between the lines in the notes if they lie on the same horizontal line. The common abbreviations seen in these notes can be used to save space. Refer to Appendix 1 for the standard approved abbreviations.

Note in Figs. 13.69 and 13.70 that the leaders for the notes originate at the first word of the note or the last word of the note.

13.34
Tabular drawings

When parts have similar features except for variations in size, a table of values can be given to provide the variations in dimensions, as in Fig. 13.71. The part dimensions are given with their respective dimensions A through Z. Tabular dimensions shorten drafting time while providing the essential information.

FIG. 13.71 This object is dimensioned by using tabular values that apply to the same drawing. The letters in balloons on each dimension line are given in the table below the drawing.

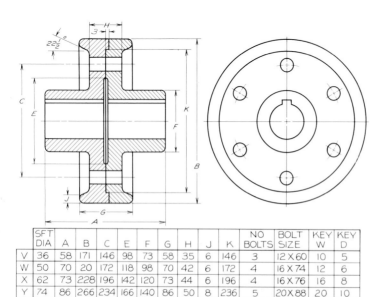

	SFT DIA	A	B	C	E	F	G	H	J	K	NO BOLTS	BOLT SIZE	KEY W	KEY D
V	36	58	171	146	98	73	58	35	6	146	3	12 X 60	10	5
W	50	70	20	172	118	98	70	42	6	172	4	16 X 74	12	6
X	62	73	228	196	142	120	73	44	6	196	4	16 X 76	16	8
Y	74	86	266	234	166	140	86	50	8	236	5	20 X 88	20	10
Z	88	92	286	254	184	158	92	58	8	254	6	20 X 96	22	11

Problems

1–50. These problems, shown in Figs. 13.72 and 13.73, are to be solved on Size A paper if they are drawn half-size. If drawn full size, Size B paper should be used. Note that the dimensions of the views are found by using your dividers to transfer drawing sizes to the scales beneath the problems. This will provide you with either metric or imperial units of measurement.

You will need to vary the spacing between the views to provide adequate room for the dimensions. It would be desirable to sketch the views and the dimensions in order to determine the spacing required before laying out the problems with instruments.

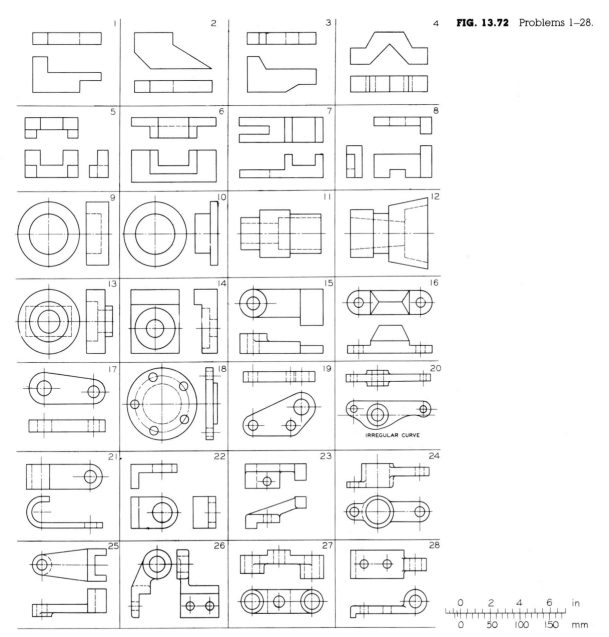

FIG. 13.72 Problems 1–28.

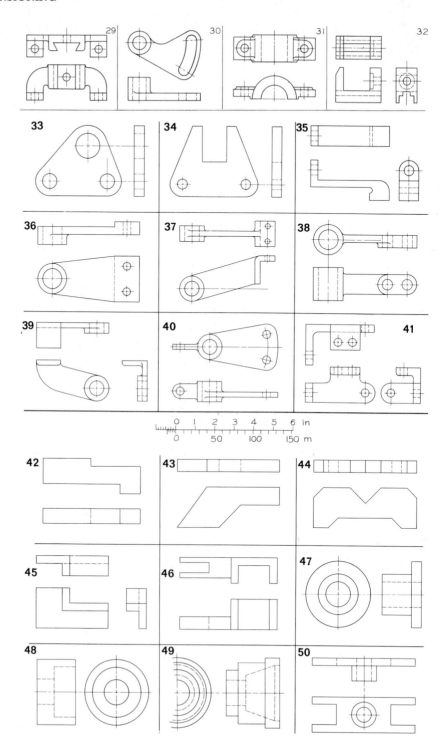

FIG. 13.73 Problems 29–50.

14
Tolerances

14.1
Introduction

Parts produced today require more accurate dimensions than did those produced in the past because many of today's parts are made by different companies in different locations. Therefore, these parts must be specified so they will be interchangeable.

The techniques of dimensioning parts to ensure interchangeability is called **tolerancing.** Each dimension is allowed a certain degree of variation within a specified zone called a tolerance. For example, a part's dimension might be expressed as 100 ± 0.50, which yields a tolerance of 1.00 mm.

It is desirable that all dimensions be given as large a tolerance as possible without interfering with the function of a part. This reduces production costs, since manufacturing to close tolerances is expensive.

14.2
Tolerance dimensions

Several acceptable methods of specifying tolerances are shown in Fig. 14.1. When "plus-and-minus" tolerancing is used, tolerances are applied to a basic di-

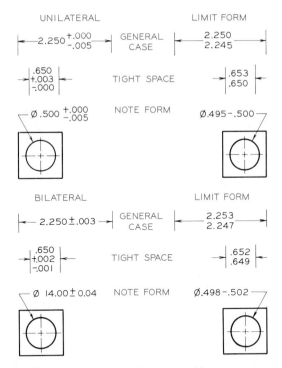

FIG. 14.1 Methods of positioning and indicating tolerances in unilateral, bilateral, and limit forms.

mension. When plus-and-minus dimensions allow variation in only one direction, the tolerancing is *unilateral.* Tolerancing that permits variation in either direction from the basic dimension is *bilateral.*

The customary methods of indicating toleranced dimensions are shown in Fig. 14.2. The positioning

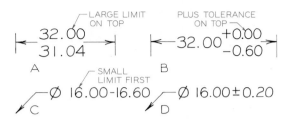

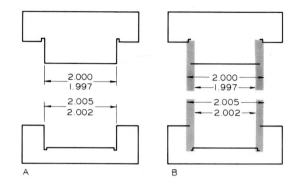

FIG. 14.2 When limit dimensions are given, the large limits are placed either above the small limits or to their right. The plus limits are placed above the minus limits in plus-and-minus tolerancing.

FIG. 14.4 Each of these mating parts has a tolerance of 0.003″ (variation in size). The allowance between the assembled parts (tightest fit) is 0.002″.

FIG. 14.3 Positioning and spacing of numerals used to specify tolerances.

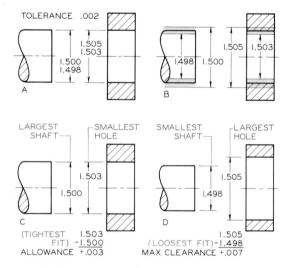

FIG. 14.5 The allowance (tightest fit) between these assembled parts is +0.003″. The maximum clearance is 0.007″.

and spacing of numerals of toleranced dimensions are shown in Fig. 14.3.

Tolerances may be given in the form of limits; that is, two dimensions are given that represent the largest and smallest sizes permitted for a feature of the part. When limits for dimensions are compared with their counterparts in plus-and-minus form, we see that both methods result in the same tolerance.

14.3
Mating parts

Mating parts are parts that fit together within a prescribed degree of accuracy (Fig. 14.4). The upper piece is dimensioned with two measurements that indicate the upper and lower limits of the size. The notch is slightly larger, allowing the parts to be assembled with a clearance fit.

An example of mating cylindrical parts is shown in Fig. 14.5A. Part B of the figure illustrates the meaning of the tolerance dimensions. The size of the shaft can vary in diameter from 1.500″ (its maximum size) to 1.498″ (its minimum size). The difference between these limits on a single part is a tolerance, 0.002″ in this case. The dimensions of the hole in part A are given with limits of 1.503 and 1.505, for a tolerance

of 0.002 (the difference between the limits as illustrated in part B).

14.4
Terminology of tolerancing

The meaning of most of the terms used in tolerancing can be seen by referring to Fig. 14.5.

TOLERANCE is the difference between the limits prescribed for a single part. The tolerance of the shaft in Fig. 14.5 is 0.002".

LIMITS OF TOLERANCE are the extreme measurements permitted by the maximum and minimum sizes of a part. The limits of tolerance of the shaft in Fig. 14.5 are 1.500 and 1.498.

ALLOWANCE is the tightest fit between two mating parts. The allowance between the largest shaft and the smallest hole in Fig. 14.5 is 0.003 (negative for an interference fit).

NOMINAL SIZE is an approximate size that is usually expressed with common fractions. The nominal sizes of the shaft and hole in Fig. 14.5 are 1.50" or 1½".

BASIC SIZE is the exact theoretical size from which limits are derived by the application of plus-and-minus tolerances. There is no basic diameter if this is expressed with limits.

ACTUAL SIZE is the measured size of the finished part.

FIT signifies the type of fit between two mating parts when assembled. The types of fit are clearance, interference, transition, and line fits.

CLEARANCE FIT is a fit that gives a clearance between two assembled mating parts. The fit between the shaft and the hole in Fig. 14.5 is a clearance fit that permits a minimum clearance of 0.003" and a maximum clearance of 0.007".

INTERFERENCE FIT is a fit that results in an interference between the two assembled parts. The shaft in Fig. 14.6A is larger than the hole, which requires a force or press fit that has an effect similar to that of a weld between the two parts.

TRANSITION FIT can result in either an interference or a clearance. The shaft in Fig. 14.6B can be either smaller or larger than the hole and still be within the prescribed tolerances.

LINE FIT can result in a contact of surfaces or a clearance between them. The shaft in Fig. 14.6C can have contact or clearance when the limits are approached.

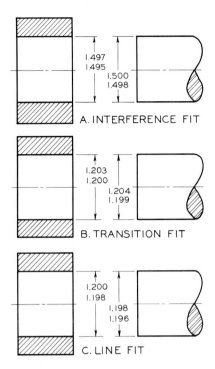

FIG. 14.6 Types of fits between mating parts. The clearance fit is not shown.

SELECTIVE ASSEMBLY is a method of selecting and assembling parts by trial and error. Using this method, parts can be assembled that have greater tolerances and consequently can be produced at a reduced cost. This process of hand assembly represents a compromise between a high degree of manufacturing accuracy and ease of assembly of interchangeable parts.

SINGLE LIMITS are dimensions that are designated by MIN or MAX (minimum or maximum) instead of being labeled by both (Fig. 14.7). Depths of holes, lengths, threads, corner radii, chamfers, and so on, are dimensioned in this manner in some cases. Caution should be exercised to prevent substantial deviations from the single limit.

FIG. 14.7 Single tolerances can be given in some applications in MAX or MIN form.

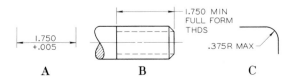

14.5
Basic hole system

The basic hole system is a widely used system of dimensioning holes and shafts to give the required allowance between the two assembled parts. The smallest hole is taken as the basic diameter from which the limits of tolerance and allowance are applied.

The hole is used because many of the standard drills, reamers, and machine tools are designed to give standard hole sizes. Therefore, it is advantageous to use this diameter as the basic dimension.

If the smallest diameter of a hole is 1.500", the allowance (0.003 in this example) can be subtracted from this diameter to find the diameter of the largest shaft (1.497"). The smallest limit for the shaft can then be found by subtracting the tolerance from 1.497".

14.6
Basic shaft system

Some industries use the basic shaft system of applying tolerances to dimensions of shafts, since many shafts come in standard sizes. In this system, the largest diameter of the shaft is used as the basic diameter from which the tolerances and allowances are applied.

For example, if the largest permissible shaft is 1.500", the allowance can be added to this dimension to yield the smallest possible diameter of the hole into which the shaft must fit. Therefore, if the parts are to have an allowance of 0.004", the smallest hole would have a diameter of 1.504".

14.7
Metric limits and fits

This section will cover the metric system as recommended by the International Standards Organization (ISO), which has been presented in ANSI B4.2-1978. These fits usually apply to cylinders—holes and shafts. However, these tables can also be used to determine the fits between any parallel surfaces, such as a key in a slot.

Metric definitions of limits and fits

Most of the definitions given below are illustrated in Fig. 14.8.

BASIC SIZE is the size from which the limits or deviations are assigned. Basic sizes, usually diameters, should be selected from Table 14.1 under the heading of "first choice."

DEVIATION is the difference between the hole or shaft size and the basic size.

UPPER DEVIATION is the difference between the maximum permissible size of a part and its basic size.

LOWER DEVIATION is the difference between the minimum permissible size of a part and its basic size.

FUNDAMENTAL DEVIATION is the deviation that is closest to the basic size. In the note 40H7, the letter *H* (an uppercase letter) represents the fundamental deviation for a hole. In the note 40g6, the letter *g* (a lowercase letter) represents the fundamental deviation for a shaft.

TOLERANCE is the difference between the maximum and minimum allowable sizes of a single part.

FIG. 14.8 Definition of terms related to metric fits.

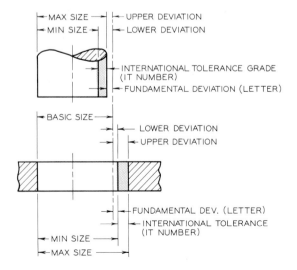

TABLE 14.1
PREFERRED SIZES

Basic Size, mm		Basic Size, mm		Basic Size, mm	
First Choice	Second Choice	First Choice	Second Choice	First Choice	Second Choice
1		10		100	
	1.1		11		110
1.2		12		120	
	1.4		14		140
1.6		16		160	
	1.8		18		180
2		20		200	
	2.2		22			220
2.5		25		250	
	2.8		28		280
3		30		300	
	3.5		35			350
4		40		400	
	4.5		45		450
5		50		500	
	5.5		55		550
6		60		600	
	7		70		700
8		80		800	
	9		90		900
				1000	

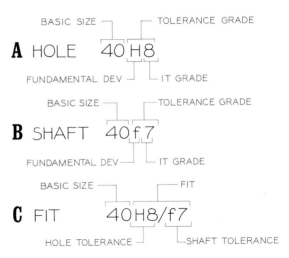

FIG. 14.9 Symbols and their definitions as applied to holes and shafts.

INTERNATIONAL TOLERANCE (IT) GRADE is a group of tolerances that vary in accordance with the basic size and provide a uniform level of accuracy within a given grade. In the note 40H7, the number 7 represents the IT grade. There are 18 IT grades: IT01, IT0, IT1, . . . , IT16.

TOLERANCE ZONE is the zone that represents the tolerance grade and its position in relation to the basic size. This is a combination of the fundamental deviation (represented by a letter) and the International Tolerance grade (IT number). For example, in note 40H8, the H8 portion indicates the tolerance zone.

HOLE BASIS system is a system of fits based on the minimum hole size as the basic diameter. The fundamental deviation for a hole basis system is an uppercase letter, *H*, for example (Fig. 14.9).

SHAFT BASIS system is a system of fits based on the maximum shaft size as the basic diameter. The funda-

mental deviation for a shaft basis system is a lowercase letter, *f*, for example (Fig. 14.9).

CLEARANCE FIT is a fit that results in a clearance between the two assembled parts under all tolerance conditions.

INTERFERENCE FIT is a fit between two parts which requires that they be forced together when assembled because of their interference with each other.

TRANSITION FIT is a fit that results in a clearance or an interference fit between two assembled parts.

TOLERANCE SYMBOLS are notes that are used to communicate the specifications of tolerance and fit. Examples of these are given in Fig. 14.9. The basic size is the primary dimension from which the tolerances are determined; therefore, it is the first part of the symbol. It is followed by the fundamental position letter and the IT number to give the tolerance zone. Uppercase letters are used to indicate the fundamental deviation for holes, and lowercase letters for shafts.

Three methods of specifying the tolerance information are shown in Fig. 14.10. The information in

FIG. 14.10 Three methods of giving tolerance symbols. The numbers in parentheses are for reference.

40H8   40H8 (40.039 / 40.000)   40.039 / 40.000 (40H8)

A   **B**   **C**

TABLE 14.2
DESCRIPTION OF PREFERRED FITS

	ISO symbol		Description
	Hole Basis	Shaft Basis	
Clearance fits	H11/c11	C11/h11	*Loose running* fit for wide commercial tolerances or allowances on external members.
	H9/d9	D9/h9	*Free running* fit not for use where accuracy is essential, but good for large temperature variations, high running speeds, or heavy journal pressures.
	H8/f7	F8/h7	*Close running* fit for running on accurate machines and for accurate location at moderate speeds and journal pressures.
	H7/g6	G7/h6	*Sliding* fit not intended to run freely, but to move and turn freely and locate accurately.
Transition fits	H7/h6	H7/h6	*Locational clearance* fit provides snug fit for locating stationary parts, but can be freely assembled and disassembled.
	H7/k6	K7/h6	*Locational transition* fit for accurate location, a compromise between clearance and interference.
	H7/n6	N7/h6	*Locational transition* fit for more accurate location where greater interference is permissible.
Interference fits	H7/p6[1]	P7/h6	*Locational interference* fit for parts requiring rigidity and alignment with prime accuracy of location but without special bore pressure requirements.
	H7/s6	S7/h6	*Medium drive* fit for ordinary steel parts or shrink fits on light sections, the tightest fit usable with cast iron.
	H7/u6	U7/h6	*Force* fit suitable for parts that can be highly stressed or for shrink fits where the heavy pressing forces required are impractical.

↑ More clearance — More interference ↓

[1]Transition fit for basic sizes in range from 0 through 3 mm.

FIG. 14.11 Types of fits. (A) A clearance fit; (B) a transition fit where there can be an interference or a clearance; (C) an interference fit where the parts must be forced together.

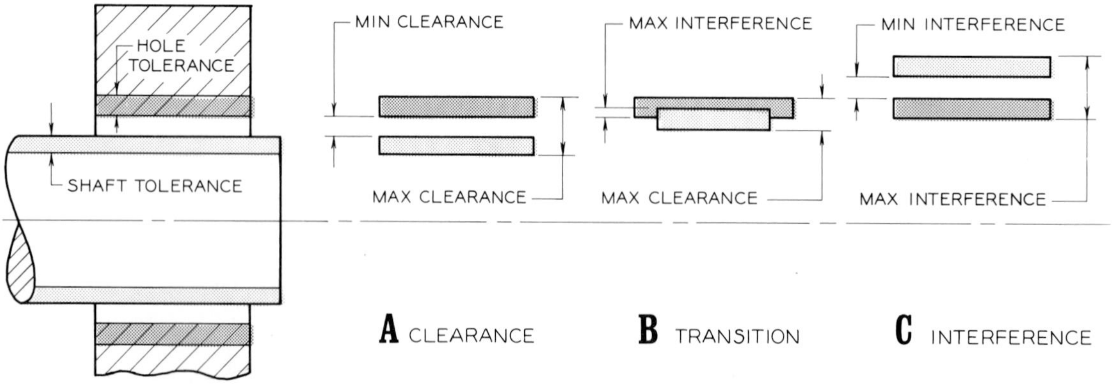

parentheses indicates that it is for reference only. The upper and lower limits are found in the tables in Appendix 44. This method will be covered in the next section.

14.8
Preferred sizes and fits

The preferred basic sizes for computing tolerances are shown in Table 14.1. Under the "First Choice" heading, the numbers increase by about 25% of the preceding numbers. Those in the "Second Choice" column increase at approximately 12% increments.

Where possible, you should select basic diameters from the first column since these correspond to standard sizes for round, square, and hexagonal metal products. This makes it possible to use stock sizes and reduce expenses.

Preferred fits are shown in Table 14.2 for clearance, transition, and interference fits. This table gives the tolerance symbols for hole basis and shaft basis fits. Where possible, fits should be taken from this table for mating parts. The tables in Appendixes 45–48 correspond to these fits.

PREFERRED FITS—HOLE BASIS SYSTEM Figure 14.11 illustrates the symbols used to show the combinations of fits that are possible when using the hole basis system. There is a clearance between the two parts at A, a transition fit at B, and an interference fit at C. This technique of representing fits is used in Fig. 14.12 to show a series of fits for a hole basis system. Note that the lower deviation of the hole is zero. In other words, the smallest size of the hole is the basic size.

The sizes of the shafts are varied to give a variety of fits from c11 to u6, where there is a maximum of interference. These fits correspond to those given in Table 14.2.

FIG. 14.12 The preferred fits for a hole basis system. These fits correspond to those given in Table 14.2.

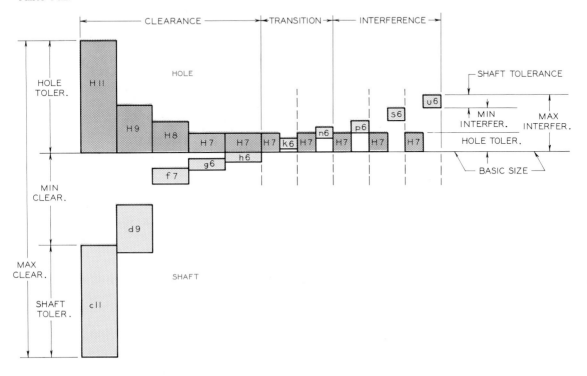

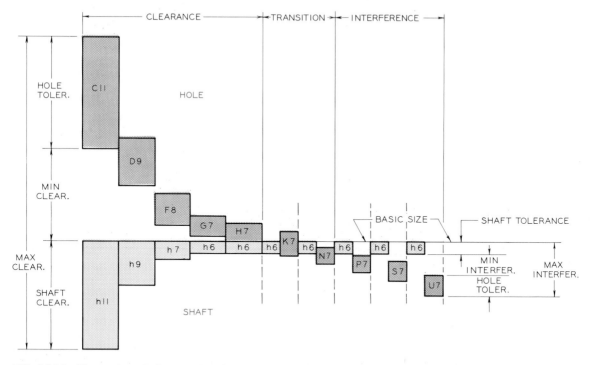

FIG. 14.13 The preferred fits for a shaft basis system. These fits correspond to those given in Table 14.2.

PREFERRED FITS—SHAFT BASIS SYSTEM Figure 14.13 illustrates the preferred fits based on the shaft basis system, where the largest shaft size is the basic diameter. The variation in the fit between the parts is caused by varying the size of the holes. This results in a range of fits from a clearance fit of C11/h11 to an interference fit of U7/h6.

14.9
Example problems— metric system

The following problems are given and solved as examples of determining the sizes and limits and the application of the proper symbols to mating parts. The solution of these problems requires the use of the tables in Appendix 45, the table of preferred sizes (Table 14.1), and the table of preferred fits (Table 14.2).

EXAMPLE 1 (Fig. 14.14) Given: Hole basis system, close running fit, basic diameter = 39 mm.

Solution: Use a basic diameter of 40 mm (Table 14.1) and fit of H8/f7 (Table 14.2).

Hole: Find the upper and lower limits of the hole in Appendix 45 under H8 and across from 40 mm. These limits are 40.000 and 40.039 mm.

Shaft: The upper and lower limits of the shaft are found under f7 and across from 40 mm in Appendix 45. These limits are 39.950 and 39.975 mm.

Symbols: The methods of noting the drawings are shown in Fig. 14.14. Any of these methods is appropriate.

EXAMPLE 2 (Fig. 14.15) Given: Hole basis system, locational transition fit, basic diameter = 57 mm.

Solution: Use a basic diameter of 60 mm (Table 14.1) and a fit of H7/k6 (Table 14.2).

Hole: Find the upper and lower limits of the hole in Appendix 46 under H7 and across from 60 mm. These limits are 60.000 and 60.030 mm.

Shaft: The upper and lower limits of the shaft are found under k6 and across from 60 mm in Appendix 46. These limits are 60.021 and 60.002 mm.

Symbols: Two methods of applying the tolerance symbols to a drawing are shown in Fig. 14.15.

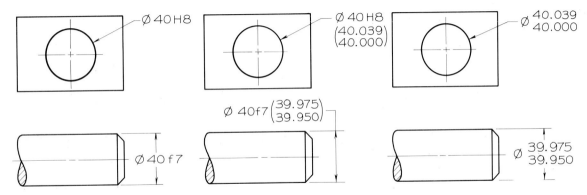

FIG. 14.14 Any of these three methods can be used to apply symbols to a detail drawing of two mating parts.

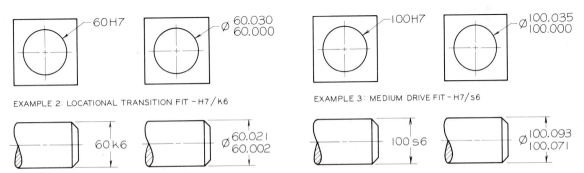

EXAMPLE 2: LOCATIONAL TRANSITION FIT – H7/k6

EXAMPLE 3: MEDIUM DRIVE FIT – H7/s6

FIG. 14.15 Examples of tolerance symbols applied to a transition fit. Both methods are acceptable.

FIG. 14.16 Examples of tolerance symbols applied to an interference fit. Both methods are acceptable.

EXAMPLE 3 (Fig. 14.16) Given: Hole basis system, medium drive fit, basic diameter = 96 mm.

Solution: Use a basic diameter of 100 mm (Table 14.1) and a fit of H7/s6 (Table 14.2).

Hole: Find the upper and lower limits of the hole in Appendix 46 under H7 and across from 100 mm. These limits are 100.035 and 100.000 mm.

Shaft: The upper and lower limits of the shaft are found under s6 and across from 100 mm in Appendix 46. These limits are found to be 100.093 and 100.071 mm. From the table, you can see that the tightest fit is an interference of 0.093 mm and the loosest fit is an interference of 0.036 mm. An interference is indicated by a minus sign in front of the numbers.

EXAMPLE 4 (Fig. 14.17) Given: Shaft basis system, loose running fit, basic diameter = 116 mm.

Solution: Use a basic diameter 120 mm (Table 14.1) and a fit of C11/h11 (Table 14.2).

Hole: Find the upper and lower limits of the hole in Appendix 47 under C11 and across from 120 mm. These limits are 120.400 and 120.180.

Shaft: The upper and lower limits of the shaft are found under h11 and across from 120 mm in Appendix 47. These limits are 119.780 and 120.000 mm.

FIG. 14.17 Examples of tolerance symbols applied to a clearance fit. Both methods are acceptable.

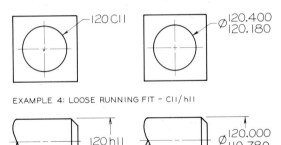

EXAMPLE 4: LOOSE RUNNING FIT – C11/h11

```
FIT: H8/f7      Ø 45 BASIC    FROM APPENDIX
                                H8        f7
                              0.046    -0.025
                              0.000    -0.050
              45.046
HOLE LIMITS
              45.000
                               HOLE     SHAFT
              44.975
SHAFT LIMITS
              44.950
```

FIG. 14.18 The limits of a nonstandard diameter, 45 mm, and an H8/f7 fit are calculated by using values from Appendixes 49 and 50.

14.10
Preferred metric fits— nonpreferred sizes

Limits of tolerances for preferred fits, shown in Table 14.2, can be computed for nonstandard sizes. Limits of tolerances for nonstandard diameters appear in Appendix 49 for nonstandard hole sizes and in Appendix 50 for nonstandard shaft sizes.

The hole and shaft limits for an H8/f7 fit and a 45-mm DIA are computed in Fig. 14.18. The tolerance limits of 0.000 and 0.046 for an H8 hole are taken from Appendix 49 across from the size range of 40–50 mm. The tolerance limits 0.000 and 0.050 mm are taken from Appendix 50. The limits of sizes for the hole and shaft are computed by applying these limits of tolerance to the 45-mm basic diameter.

14.11
Standard fits (English units)

The ANSI B4.1-1955 standard specifies a series of fits between cylindrical parts that are based on the basic hole system in inches. The tolerances placed on the holes are unilateral and positive.

The types of fit covered in this standard are:

RC   Running and sliding fits

LC   Clearance locational fits

LT   Transition locational fits

LN   Interference locational fits

FN   Force and shrink fits

Each of these types of fit is listed in tables in Appendixes 39 to 43. There are several classes of fit under each of the types given above. The definition of each type of fit given in the standards is listed below.

RUNNING AND SLIDING FITS (RC) are fits for which limits of clearance are specified to provide a similar running performance, with suitable lubrication allowance throughout the range of sizes. The clearance for the first two classes (RC 1 and RC 2), used chiefly as slide fits, increases more slowly with diameter size than that of the other classes, so that accurate location is maintained even at the expense of free relative motion.

LOCATIONAL FITS (LC) are fits intended to determine only the location of the mating parts; they may provide rigid or accurate location (interference fits) or some freedom of location (clearance fits). They are divided into three groups: clearance fits (*LC*), transition fits (*LT*), and interference fits (*LN*).

FORCE FITS (FN) are special types of interference fits, typically characterized by maintenance of constant bore pressures throughout the range of sizes. The interference therefore varies almost directly with diameter, and the difference between its minimum and maximum values is small, to maintain the resulting pressures within reasonable limits.

Figure 14.19 illustrates how one uses the values from these tables in Appendix 39 for an RC 9 fit. The basic diameter for the hole and shaft is 2.5000", which is between the range of 1.97 and 3.15 given in the last column of the table. Since all limits are given in thousandths, the values can be converted by moving the decimal point three places to the left. For example, $+0.7$ is $+0.0007''$.

The upper and lower limits of the shaft (2.4910" and 2.4865") are found by subtracting the two limits ($-0.0090''$ and $-0.0135''$) from the basic diameter. The upper and lower limits of the hole (2.5007" and 2.5000") are found by adding the two limits ($+0.007''$ and $0.000''$) to the basic diameter.

When the two parts are assembled, the tightest fit ($+0.0205''$) and the loosest fit ($+0.0090''$) are found by subtracting the maximum and minimum sizes of the holes and shafts. Note that these values are provided in the second column of the table as a check on the limits.

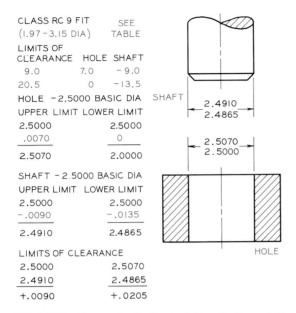

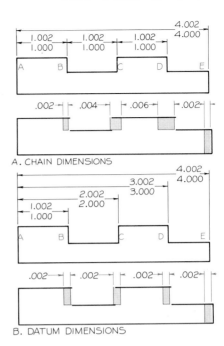

FIG. 14.19 The method of calculating limits and allowances for an RC9 fit between a shaft and a hole. The basic diameter is 2.5000 inches.

FIG. 14.20 When dimensions are given as chain dimensions, the tolerances can accumulate to give a variation of 0.006" at D instead of 0.002". When dimensioned from a single datum plane, the variations of X and Y cannot deviate more than the specified 0.002" from the datum.

The same method (but different tables) is used for calculating the limits for all types of fit. Plus values of clearance indicate that there is clearance and minus values that there will be interference between the assembled parts.

These values should be converted to millimeters when using the metric system by multiplying inches by 25.4 or by using corresponding metric tables.

14.12
Chain dimensions

When parts are dimensioned to locate surfaces or geometric features by a chain of dimensions (Fig. 14.20A), variations may occur that exceed the tolerances specified. As successive measurements are made, with each based on the preceding one, the tolerances may accumulate, as shown in Fig. 14.20A. For example, the tolerance between surfaces A and B is 0.002; between A and C, 0.004; and between A and D, 0.006.

This accumulation of tolerances can be eliminated by measuring from a single plane called a datum

plane. A datum plane is usually a plane on the object, but it could be a plane on the machine used to make the part.

Note that the tolerances between the intermediate planes in Fig. 14.20B are uniform since each of the planes was located with respect to a single datum. In our example, this is 0.002, which represents the maximum tolerance.

14.13
Origin selection

In some cases, there will be a need to specify a surface as the origin for locating another surface. An example is illustrated in Fig. 14.21, where the origin surface is the shorter one at the base of an object. The result is that the angular variation permitted is less for the longer surface that it would be if the origin plane had been the longer surface.

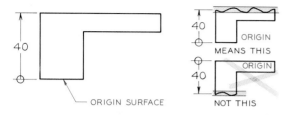

FIG. 14.21 This method is used to indicated the origin surface for locating one feature of a part with respect to another.

14.14
Conical tapers

A method of specifying a conical taper is shown in Fig. 14.22 by giving a basic diameter and a basic taper. The basic diameter of 20 mm is located midway in the length of the cone with a toleranced dimension. Taper is a ratio of the difference in the diameters of two circular sections of a cone to the distance between the sections. The radial tolerance zone can be found as shown in Fig. 14.22.

14.15
Tolerance notes

All dimensions on a drawing are toleranced either by the previously covered rules or by a general note on a drawing placed in or near the title block. For example,

FIG. 14.22 Taper is indicated with a combination of tolerances and taper symbols. The variation in diameter at any point is 0.06 mm or 0.03 in radius.

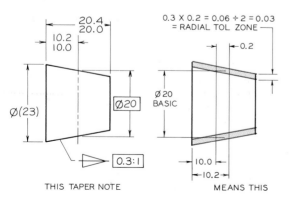

THIS TAPER NOTE   MEANS THIS

the note TOLERANCE ± 1/64 (or its decimal equivalent, 0.40 mm) might be given on a drawing for the less critical dimensions. These tolerances are for surfaces that do not come into contact with other surfaces.

Some industries may give dimensions in inches where decimals are carried out to two, three, and four places. A note might be given on the drawing in this manner: TOLERANCES XX.XX ± 0.10; XX.XXX ± 0.005. Tolerances of four places would be given directly on the dimension lines, but those dimensions with two and three decimal places would have the tolerances indicated in the note.

The most common method of noting tolerances is to use as large a tolerance as feasible expressed in a note such as TOLERANCES ± 0.05 (± 1 mm when using metrics), and give the tolerances for the mating dimensions that require smaller tolerances on the dimension lines.

Angular tolerances should be given in a general note in or near the title block, such as ANGULAR TOLERANCES ± 0.5° or ± 30'. When angular tolerances less than this are specified, they should be given on the drawing where these angles are dimensioned. Techniques of tolerancing angles are shown in Fig. 14.23.

FIG. 14.23 Angles can be toleranced by any of these methods using limits or the plus-and-minus method.

14.16
General tolerances—metric

All dimensions on a drawing are understood to have tolerances, and the amount of tolerance must be noted. This section is based on the metric system, with the millimeter as the unit of measurement as outlined in ANSI B4.3-1978.

Tolerances may be specified by (1) applying tolerances directly to the dimensions, (2) giving tolerances in specification documents that are associated with the drawings, or (3) applying a general note on the drawing.

TABLE 14.3
IT GRADES AND THEIR RELATIONSHIP TO MACHINING PROCESSES

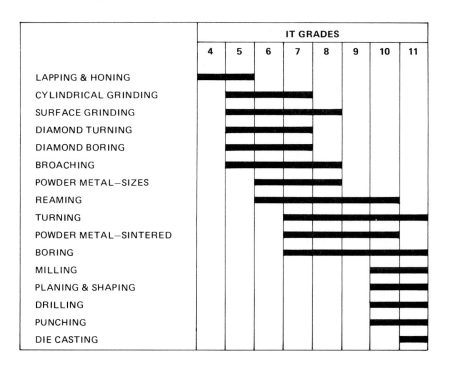

LINEAR DIMENSIONS may be toleranced by indicating ± one-half of an International Tolerance (IT) grade as given in Appendix 44. The appropriate IT grade can be selected from the graph in Fig. 14.24 when the nature of the application is known.

You can see that IT grades for mass produced items range from IT12 through IT16. When the machining process is known, the IT grades can be selected from Table 14.3.

General tolerances using IT grades may be expressed in a note as follows:

UNLESS OTHERWISE SPECIFIED ALL

$$\text{UNTOLERANCED DIMENSIONS ARE} \pm \frac{\text{IT14}}{2}.$$

FIG. 14.24 The International Tolerance grades and their applications.

This means that tolerance of ± 0.700 mm is allowed for a dimension between 315 and 400 mm. The value of the tolerance is listed in Appendix 44 as 1.400.

Table 14.4 shows recommended tolerances for fine, medium, and coarse series for dimensions of graduated sizes. A medium tolerance, for example, can be specified by the following note:

GENERAL TOLERANCE SPECIFIED IN
ANSI B4.3 MEDIUM SERIES APPLY.

This same information can be given on the drawing in table form by selecting the grade—medium in this example—from Table 14.4 and presenting it on the drawing as shown in Fig. 14.25.

General tolerances can be expressed in a table of values that gives the tolerances for dimensions with one or no decimal places. An example of this table is shown in Fig. 14.26.

Another method of giving general tolerances is a note in this form:

UNLESS OTHERWISE SPECIFIED ALL UNTOLER-
ANCED DIMENSIONS ARE ± 0.8 mm.

TABLE 14.4
FINE, MEDIUM, AND COARSE SERIES: GENERAL TOLERANCE—LINEAR DIMENSIONS

Variations in mm								
Basic Dimensions, mm		0.5 to 3	Over 3 to 6	Over 6 to 30	Over 30 to 120	Over 120 to 315	Over 315 to 1,000	Over 1,000 to 2,000
Permissible variations	Fine series	±0.05	±0.05	±0.1	±0.15	±0.2	±0.3	±0.5
	Medium series	±0.1	±0.1	±0.2	±0.3	±0.5	±0.8	±1.2
	Coarse series		±0.2	±0.5	±0.8	±1.2	±2	±3

DIMENSIONS IN mm							
GENERAL TOLERANCE UNLESS OTHERWISE SPECIFIED THE FOLLOWING TOLERANCES ARE APPLICABLE							
LINEAR	OVER TO	0.5 6	6 30	30 120	120 315	315 1000	1000 2000
TOL	±	0.1	0.2	0.3	0.5	0.8	1.2

FIG. 14.25 This table is a medium series of values taken from Table 14.4. It is placed on the working drawing to provide the tolerances for various ranges of sizes.

This method should be used only where the dimensions on a drawing have slight differences in magnitude.

ANGULAR TOLERANCES are expressed (1) as an angle in decimal degrees or in degrees and minutes, (2) as a taper expressed in percentage (number of millimeters per 100 mm), or (3) as milliradians. A milliradian is found by multiplying the angle in degrees by 17.45. The suggested tolerances for each of these units are shown in Table 14.5. Note that the angular tolerances are based on the length of the shorter leg of the angle.

General angular tolerances may be given on the drawing with a note in the following form:

UNLESS OTHERWISE SPECIFIED THE GENERAL TOLERANCES IN ANSI B4.3 APPLY.

A second method shows the portion of Table 14.5 on the drawing using the units desired, as shown in Fig. 14.27.

ANGULAR TOLERANCE				
LENGTH OF SHORTER LEG – mm	UP TO 10	OVER 10 TO 50	OVER 50 TO 120	OVER 120 TO 400
TOL	±1°	±0° 30'	±0° 20'	±0° 10'

FIG. 14.27 This table can be placed on a drawing to indicate the general tolerances for angles. These values were extracted from Table 14.5.

A third method is a note with a single tolerance such as:

UNLESS OTHERWISE SPECIFIED THE GENERAL ANGULAR TOLERANCES ARE ±0° 30' (or ±0.5°).

FIG. 14.26 This table of tolerances can be placed on a drawing to indicate the tolerances for dimensions with one decimal and those with no decimals.

DIMENSIONS IN mm					
GENERAL TOLERANCE UNLESS OTHERWISE SPECIFIED THE FOLLOWING TOLERANCES ARE APPLICABLE					
LINEAR	OVER TO	– 120	120 315	315 1000	1000 –
TOL	ONE DECIMAL ±	0.3	0.5	0.8	1.2
	NO DECIMALS ±	0.8	1.2	2	3

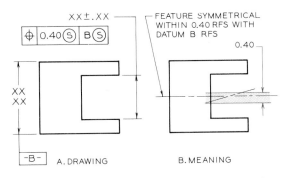

A. DRAWING B. MEANING

FIG. 14.33 Tolerances of position should include a note of M, S, or L to indicate maximum material condition, regardless of feature size, or least material condition.

RULE 3 (ALL OTHER GEOMETRIC TOLERANCES) RFS applies for all other geometric tolerances for individual tolerances and datum references, if no modifying symbol is given in the feature control symbol. If a feature is to be at maximum material condition, MMC must be specified.

14.21

The three-datum plane concept

A datum plane is used as the origin of a part's dimensioned features that have been toleranced. Datum planes are usually associated with the manufacturing

FIG. 14.34 When a part is referenced to a primary datum plane, the object contacts the plane with its three highest points. The vertical surface contacts the secondary vertical datum plane with two points. The third datum plane is contacted by one point on the object. The datum planes are listed in this order in the feature control symbol.

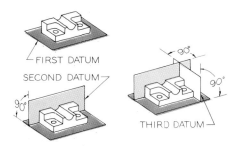

FIRST DATUM
SECOND DATUM
THIRD DATUM

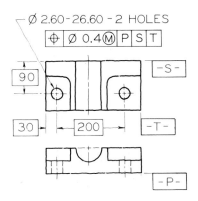

FIG. 14.35 The sequence of the three-plane reference system (shown in Fig. 14.34) is labeled where the planes appear as edges. Note that the primary datum plane, P, is listed first in the feature control symbol; the secondary plane, S, next; and the third plane, T, last.

equipment, such as machine tables, or with the locating pins.

Three mutually perpendicular datum planes are used to dimension a part accurately. For example, the part in Fig. 14.34 is placed in contact with the primary datum plane at its base where three points must make contact with the datum. The part is further related to the secondary plane with two contacting points. The third (tertiary) datum is contacted by a single point on the object.

The priority of these datum planes is noted on the drawing of the part by feature control symbols as shown in Fig. 14.35. The primary datum is surface P, the secondary is surface S, and the tertiary is surface T. Examples of feature control symbols are given in Fig. 14.36, where the primary, secondary, and tertiary datum planes are listed in order of priority in the symbol.

FIG. 14.36 Feature control symbols may have from one to three datum planes indicated. These are listed in order of priority.

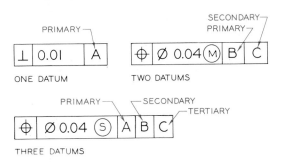

PRIMARY

ONE DATUM

SECONDARY
PRIMARY

TWO DATUMS

PRIMARY SECONDARY
TERTIARY

THREE DATUMS

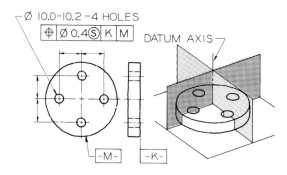

FIG. 14.37 These true-position holes are located with respect to primary datum *K* and datum *M*. Since datum *M* is a circle, this implies that the holes are located about two intersecting datum planes formed by the crossing center lines in the circular view. This satisfies the three-plane concept.

14.22
Cylindrical datum features

Figure 14.37 illustrates a part with a cylindrical datum feature that is the axis of a true cylinder. Primary datum *K* establishes the first datum. Datum *M* is associated with two theoretical planes—the second and third in a three-plane relationship.

The two theoretical planes are represented on the drawing by perpendicular center lines. The intersection of the center-line planes coincides with the datum axis. All dimensions originate from the datum axis, which is perpendicular to datum *K*; the two intersecting datum planes are used to indicate the direction of measurements in an *x* and *y* direction.

The sequence of the datum reference in the feature control symbol is significant to the manufacturing and inspection processes. The part in Fig. 14.38 is dimensioned with an incomplete feature control symbol; it does not specify the primary and secondary datum planes. The schematic drawing at B illustrates the effect of specifying diameter *A* as the primary and surface *B* as the secondary datum plane. This means that the part is centered about cylinder *A* by mounting the part in a chuck, mandrel, or centering device on the processing equipment, which centers the part at *RFS* (regardless of feature size). Surface *B* is assembled to contact at least one point of the third datum plane.

If surface *B* were specified as the primary datum feature, it would be assembled to contact datum plane *B* in at least three points. The axis of datum feature *A* will be gaged by the smallest true cylinder that is perpendicular to the first datum that will contact surface *A* at *RFS*.

In Fig. 14.38D, plane *B* is specified as the primary datum feature, and cylinder *A* is specified as the secondary datum feature at *MMC* (maximum material condition). The part is mounted on the processing equipment where at least three points on feature *B* are in contact with datum *B*. The second and third planes intersect at the datum axis to complete the three-plane relationship. This utilization of the modifier to specify MMC gives a more liberal tolerance zone than would be otherwise acceptable when RFS was specified.

FIG. 14.38 Three examples are shown to illustrate the effects of selection of the datum planes in order of priority and the effect of RFS and MMC.

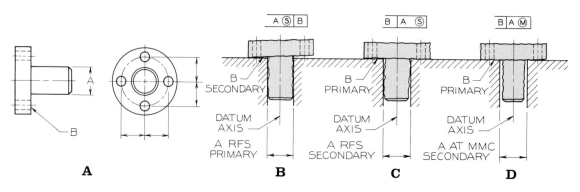

14.23
Datum features at RFS

When dimensions of size are applied to a part at RFS, the datum is established by contact between surfaces on the processing equipment with surfaces of the part. Variable machine elements, such as chucks or center devices, are adjusted to fit the external or internal features of a part and thereby establish datums.

PRIMARY DIAMETER DATUMS For an external cylinder (shaft), the datum axis is the axis of the smallest circumscribed cylinder that contacts the cylindrical feature of the part. In other words, the largest diameter of the part will make contact with the smallest contacting cylinder of the machine element that holds the part (Fig. 14.39).

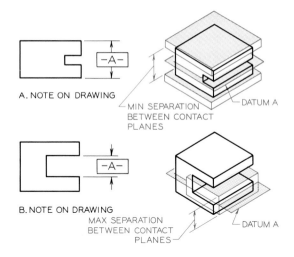

FIG. 14.40 The datum plane for external parallel surfaces is the center plane between two contacting parallel planes at their minimum separation. The datum plane for internal parallel surfaces is the center plane between two contacting parallel surfaces at maximum separation.

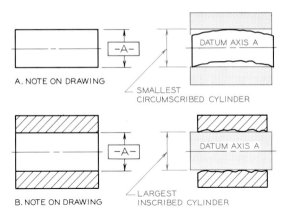

FIG. 14.39 The datum axis of a shaft is the smallest circumscribed cylinder that contacts the shaft. The datum axis of a hole is the center line of the largest inscribed cylinder that contacts the hole.

For an internal cylinder (hole), the datum axis is the axis of the largest inscribed cylinder that contacts the inside of the hole. In other words, the smallest diameter of the hole will make contact with the largest cylinder of the machine element inserted in the hole (Fig. 14.39).

PRIMARY EXTERNAL PARALLEL DATUMS The datum for external features is the center plane between two parallel planes at their minimum separa-

tion that contact the planes of the object (Fig. 14.40A). These planes are planes of a viselike device that holds the part; therefore, the planes of the part are at maximum separation, while the planes of the device are at minimum separation.

PRIMARY INTERNAL PARALLEL DATUMS The datum for internal features is the center plane between two parallel planes at their maximum separation (Fig. 14.40B) that contact the planes of the object. This is the condition in which the slot is at its smallest opening size.

SECONDARY DATUMS The secondary datum (axis or center plane) for both external and internal diameters or distances between parallel planes is found as covered in the previous two paragraphs, but with an additional requirement. This requirement is that the contacting cylinder or the contacting parallel planes must be oriented perpendicular to the primary datum. Figure 14.41 illustrates how datum *B* is the axis of a cylinder. This principle also can be applied for parallel planes in addition to cylinders.

TERTIARY DATUMS The third datum (axis or center plane) for both external and internal features is found in the same manner as covered in the previous three

ON DRAWING

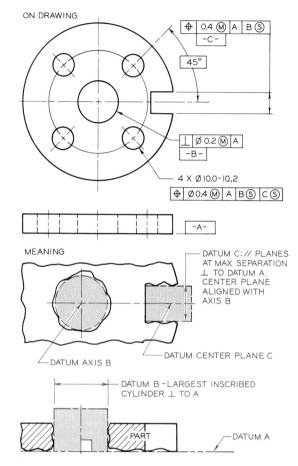

MEANING

DATUM C: // PLANES AT MAX SEPARATION ⊥ TO DATUM A. CENTER PLANE ALIGNED WITH AXIS B

DATUM CENTER PLANE C

DATUM AXIS B

DATUM B – LARGEST INSCRIBED CYLINDER ⊥ TO A

PART

DATUM A

FIG. 14.41 The illustration of a part located with respect to primary, secondary, and tertiary datum planes.

paragraphs, but with an additional requirement. This requirement is that the contacting cylinder or parallel planes must be angularly oriented with respect to the secondary datum. Datum C in Fig. 14.41 is the tertiary datum plane.

14.24
Datum targets

Instead of using a plane surface as a datum, specified datum targets are indicated on the surface of a part where the part is supported by spherical or pointed locating pins. The symbol X is used to indicate target points that are supported by locating pins at specified

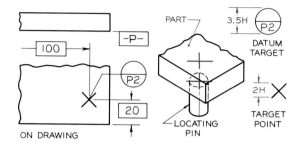

ON DRAWING

FIG. 14.42 Target points from which a datum point is established, are located with an "X" and a target symbol.

points (Fig. 14.42). Datum target symbols are placed outside the outline of the part with a leader that is directed toward the target. When the target is on the near (visible) surface, the leaders are solid lines; when the target is on the far (invisible) surface, as shown in Fig. 14.44, the leaders are hidden.

Three target points are required to establish the primary datum plane, two for the secondary, and one for the tertiary. Notice that the target symbol in Fig. 14.42 is labeled as $P2$ to match the designation of the primary datum, P. The other two points (not shown) would be labeled $P1$ and $P2$ to establish the primary datum.

A datum target line is specified in Fig. 14.43 for a part that is supported on a datum line instead of a datum point. An X and a phantom line are used to locate the line of support.

Target areas are specified for cases where spherical or pointed locating pins are inadequate to support a part. The diameters of the targets are specified with crosshatched circles surrounded by phantom lines as shown in Fig. 14.44. The target symbols give both the diameter of the targets and their number designations.

FIG. 14.43 An "X" and a phantom line are used to locate target lines on a drawing.

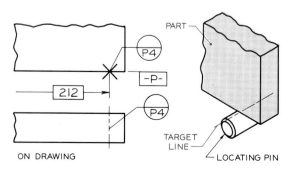

ON DRAWING

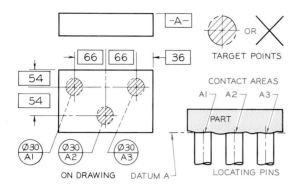

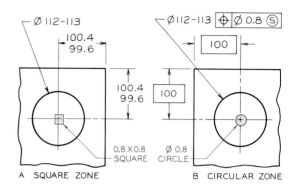

FIG. 14.44 Target points with areas are located with basic dimensions and target symbols that give the diameters of the targets. Hidden leaders indicate that the targets are on the hidden side of the plane.

FIG. 14.45 The dimensions at A give a square tolerance zone for the axis of the hole. At B, basic dimensions locate the true center of the circle about which a circular tolerance zone of 0.8 mm is specified.

The X symbol could be used as an alternative method of indicating targets with areas.

The part is located on its datum plane by placing it on the three locating pins with 30-mm diameters, as shown in Fig. 14.44. Notice that the leaders from the target areas to the target symbols are hidden to indicate that the targets are on the hidden side of the object.

14.25
Tolerances of location

Tolerances of location deal with *position, concentricity,* and *symmetry.*

True-position tolerancing

Whereas toleranced location dimensions give a square tolerance zone for locating the center of a hole, *true-position dimensions* locate the exact position of a hole's center about which a circular tolerance zone is given (Fig. 14.45). *Basic dimensions* are used to locate true positions. Basic dimensions are exact untoleranced dimensions which are indicated by boxes drawn around them (Part B).

In both methods, the diameters of the holes are toleranced by notes. The true-position method (Part B) uses a feature control symbol to specify the diameter of the circular tolerance zone inside of which the center of the hole must lie. A circular zone gives a more uniform tolerance of the hole's true position than a square zone.

In Fig. 14.46, you can see an enlargement of the square tolerance zone. The diagonal across the square zone is greater than the specified tolerance by a factor of 1.4. The true-position method shown in Fig. 14.47 can have a larger circular tolerance zone by a factor of 1.4 and still have the same degree of accuracy of position as the specified 0.1 square zone.

If the coordinate method could accept a variation of 0.014 across the diagonal of the square tolerance zone, then the true-position tolerance should be acceptable with a circular zone of 0.014, which is a greater tolerance than the square zone permitted (Fig. 14.46). True-position tolerances can be applied by symbol, as in Fig. 14.47.

The circular tolerance zone specified in the circular view of a hole is assumed to extend the full depth of the hole. Therefore, the tolerance zone for the cen-

FIG. 14.46 The coordinate method of tolerancing gives a square tolerance zone. The diagonal of the square exceeds the specified tolerance by a factor of 1.4.

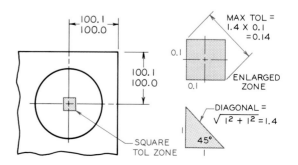

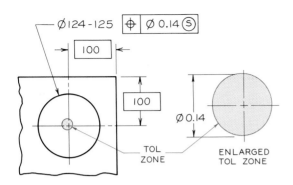

FIG. 14.47 The true-position method of tolerancing gives a circular tolerance zone with equal variations in all directions from the true axis of the hole.

ter line of the hole is a cylindrical zone inside which the axis must lie. The size of the hole and its position are both toleranced; consequently, these two tolerances are used to establish the diameter of a gage cylinder that can be used to check for the conformance of the dimensions to the specifications (Fig. 14.48).

The circle is found by subtracting the true position tolerance from the hole at maximum material condition (the smallest permissible hole). This zone represents the least favorable condition when the part is gaged or assembled with a mating part. When the hole is not at MMC, it is larger and permits a greater tolerance and easier assembly.

Gaging a two-hole pattern

Gaging is a technique of checking dimensions to determine whether they have met the specifications of tolerance. The effects of the size and positional toler-

ances can be seen when two holes are dimensioned as shown in Fig. 14.49.

The two holes are positioned 26.00 mm apart with a basic dimension. The holes have limits of 12.70 and 12.84 for a tolerance of 0.14. They are located at true position within a diameter of 0.18. The gage pin diameter is calculated to be 12.52 mm (the smallest hole's size minus the true-position tolerance), as illustrated in part B. This means that two pins with diameters of 12.52 mm that are spaced exactly 26.00 mm apart could be used to check the diameters and positions of the holes at MMC, the most critical size. If the pins can be inserted in the holes, then the holes are properly sized and located.

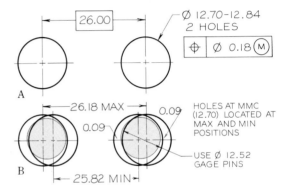

FIG. 14.49 When two holes are located at true position at MMC, they may be gaged with pins 12.52 mm in diameter that are located 26.00 mm apart.

FIG 14.50 When two holes are at their maximum size, the centers of the holes can be spaced as far as 26.32 mm apart and still be acceptable. (A) The holes can be placed as close as 25.68 apart when the holes are at maximum size (B).

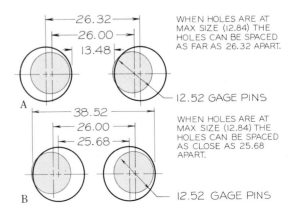

FIG. 14.48 When a hole is located at true position at MMC, no element of the hole shall be inside the imaginary cylinder, cylinder A.

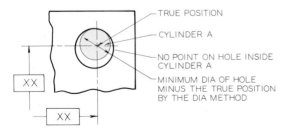

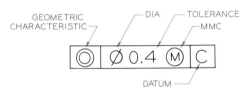

FIG. 14.51 A typical feature control symbol. This one indicates that a surface is concentric to datum C within a diameter of 0.4 mm at maximum material condition.

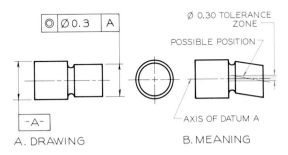

A. DRAWING   B. MEANING

FIG. 14.52 Concentricity is a tolerance of location. The feature control symbol specifies that the smaller cylinder should be concentric to cylinder A within 0.3 mm about the axis of A.

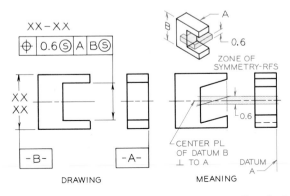

DRAWING   MEANING

FIG. 14.53 Symmetry is a tolerance of position that specifies that a part's feature be symmetrical about the center plane between parallel surfaces on the part.

FIG. 14.54 Flatness is a tolerance of form that specifies two parallel planes inside of which the object's surface must lie.

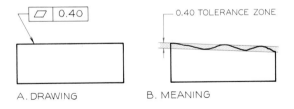

A. DRAWING   B. MEANING

When the holes are not at MMC (when they are larger than their minimum size), these gage pins will permit a greater range of variation (Fig. 14.50). When the holes are at their maximum size of 12.84 mm, they can be located as close as 25.68 from center to center or as far apart as 26.32 center to center. When not specified, true-position tolerances are assumed to apply at MMC.

Concentricity

Feature control symbols will be used to specify concentricity and other geometric characteristics throughout the remainder of this chapter (Fig. 14.51).

Concentricity is a feature of location because it specifies the relationship of one cylinder with another; that is, both share the same axis. In Fig. 14.52, the large cylinder is "flagged" as datum *A*. This means that the large diameter is used as the datum for measuring the variation of the smaller cylinder's axis.

Symmetry

Symmetry is a feature of location. A part or a feature is symmetrical when it has the same contour and size on opposite sides of a central plane. A symmetry tolerance locates features with respect to a datum plane.

The method of noting symmetry is shown in Fig. 14.53. Datum plane *B* is used to establish the position of the symmetry of the notch. The feature control symbol notes that the notch is symmetrical about datum *B* within a zone of 0.6 mm.

14.26
Tolerances of form

Flatness

A surface is flat when all its elements are in one plane. A feature control symbol is used to specify flatness within a 0.4 mm zone RFS in Fig 14.54. No point on the surface may vary more than 0.40 from the highest to the lowest point on the surface.

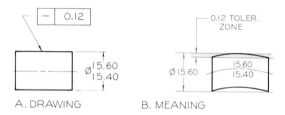

FIG. 14.55 Straightness is a tolerance of form that indicates that the elements of a surface are straight lines. The symbol must be applied where the elements appear as straight lines.

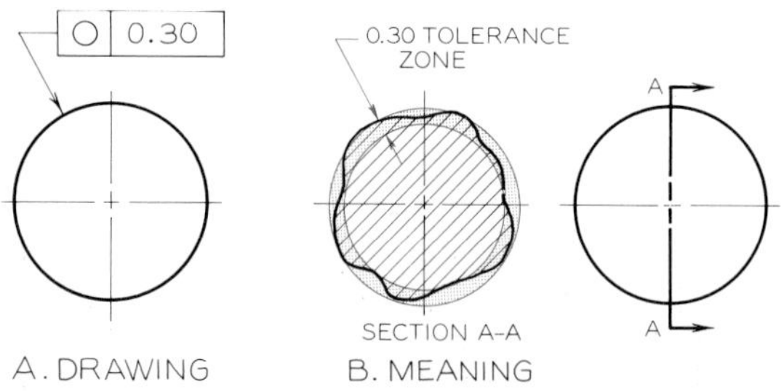

Straightness

A surface is straight if all its elements are straight lines. A feature control symbol is used to specify straightness of a cylinder in Fig. 14.55. A total of 0.12 mm RFS is permitted as the elements are gaged in a vertical plane parallel to the axis of the cylinder.

Roundness

A surface of revolution (a cylinder, cone, or sphere) is round when all points on the surface intersected by a plane are equidistant from the axis. A feature control symbol is used to specify roundness of a cone and cylinder in Fig. 14.56. This symbol permits a tolerance of 0.34 mm RFS on the radius of each part.

The roundness of a sphere is specified in Fig. 14.57.

FIG. 14.56 Roundness is a tolerance of form that indicates that a cross section through a surface of revolution is round and lies within two concentric circles.

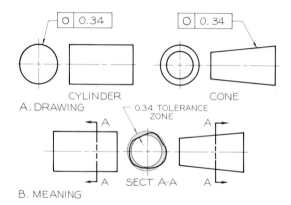

FIG. 14.57 Roundness of a sphere is indicated in this manner, which means that any cross section through it is round within the specified tolerance in the symbol.

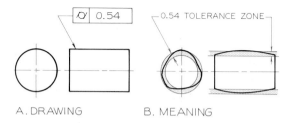

FIG. 14.58 Cylindricity is a tolerance of form that indicates that the surface of a cylinder lies within an envelope formed by two concentric cylinders.

Cylindricity

A surface of revolution is cylindrical when all its elements form a cylinder. A cylindricity tolerance zone is specified in Fig. 14.58, where a tolerance of 0.54 mm RFS is permitted on the radius of the cylinder. Cylindricity is a combination of tolerances of roundness and straightness applied to a cylindrical object.

14.27
Tolerances of profile

Profile tolerancing is used to specify tolerances about a contoured shape formed by arcs or by irregular curves. Profile can apply to a single line or to a surface.

The surface in Fig. 14.59 is given a profile tolerance that is unilateral. (It can only be smaller than the points located.) Examples of specifying bilateral and unilateral tolerance zones are shown at B.

A profile tolerance for a single line can be specified as shown in Fig. 14.60. In this example, the curve

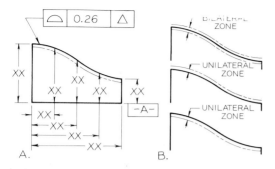

FIG. 14.59 Profile is a tolerance of form that is used to tolerance irregular curves of planes. The curving plane is located by coordinates, and the tolerance is located by any of the methods shown at B.

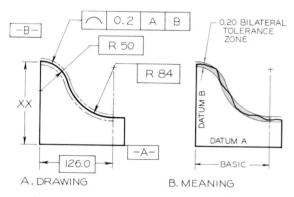

FIG. 14.60 Profile of a line is a tolerance of form that specifies the variation allowed from the path of a line. In this case, the line is formed by tangent arcs. The tolerance zone may be either bilateral or unilateral, as shown in Fig. 14.59B.

FIG. 14.61 Parallelism is a tolerance of form that specifies that a plane is parallel to another within specified limits. Plane *B* is the datum plane in this case.

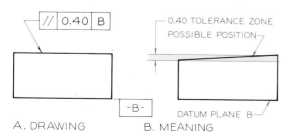

is formed by tangent arcs whose radii are given as basic dimensions. The radii are permitted to vary by plus or minus 0.10 mm about the basic radii.

14.28
Tolerances of orientation

Tolerances of orientation include *parallelism, perpendicularity,* and *angularity*.

Parallelism

A surface or a line is parallel when all its points are equidistant from a datum plane or axis. Two types of parallelism are:

1. A tolerance zone between planes parallel to a datum plane within which the axis or surface of the feature must lie (Fig. 14.61). This tolerance also controls flatness.

2. A cylindrical tolerance zone parallel to a datum feature within which the axis of a feature must lie (Fig. 14.62).

The effect of specifying parallelism at MMC can be seen in Fig. 14.63, where the modifier *M* is given in the feature control symbol. Tolerances of form apply regardless of feature size when not specified. By specifying parallelism at MMC, this means that the axis of the cylindrical hole must vary no more than

FIG. 14.62 Parallelism of one center line to another can be specified by using the diameter of one of the holes as the datum.

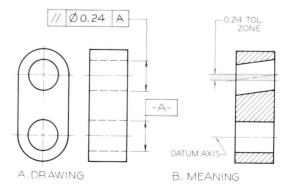

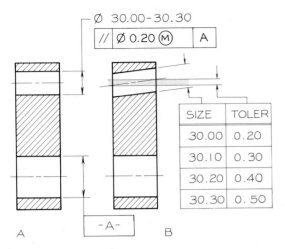

A   -A-   B

SIZE	TOLER
30.00	0.20
30.10	0.30
30.20	0.40
30.30	0.50

FIG. 14.63 The most critical tolerance will exist when features are at MMC. In this example, the upper hole must be parallel to the hole used as datum A within 0.20 DIA. As the hole approaches its maximum size of 30.30 mm, the tolerance zone approaches 0.50 mm.

0.20 mm when the holes are at the smallest permissible size.

As the hole approaches its upper limit of 30.30, the tolerance zone increases until it reaches 0.50 DIA. Therefore, a greater variation is given at MMC than at RFS.

Perpendicularity

Surfaces, axes, or lines that are at right angles to each other are perpendicular to each other. The perpendicularity of two planes is specified in Fig. 14.64. Note that datum plane C is "flagged," and the feature control symbol is applied to the perpendicular surface.

FIG. 14.64 Perpendicularity is a tolerance of form that gives a tolerance zone for a plane that is perpendicular to a specified datum plane.

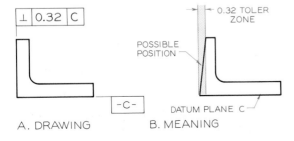

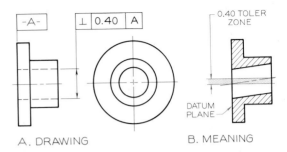

FIG. 14.65 Perpendicularity can apply to the axis of a feature, such as the center line of a cylinder.

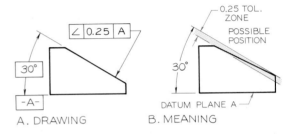

FIG. 14.66 Angularity is a tolerance of form that specifies the tolerance for an angular surface with respect to a datum plane. The 30° angle is a true angle, a basic angle. The tolerance of 0.25 mm is from this basic angle.

A hole is specified as perpendicular to a surface in Fig. 14.65, where surface A is indicated as the datum plane.

Angularity

A surface or line is angular when it is at a specified angle (other than 90°) from a datum or an axis. The angularity of a surface is specified in Fig. 14.66, where the angle is given a basic dimension of 30°. The angle is permitted to vary within a tolerance zone of 0.25 mm about the angle.

14.29
Tolerances of runout

Runout tolerance is a means of controlling the functional relationship between one or more parts to a common datum axis. The features controlled by runout are surfaces of revolution about an axis and surfaces that are perpendicular to the axis.

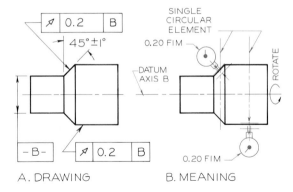

A. DRAWING   B. MEANING

FIG. 14.67 Runout tolerance of a surface is a tolerance of form that is a composite of several form characteristics. It is used to specify concentric cylindrical parts. The part is mounted on one of the axes, the datum axis, and the part is gaged as it is rotated about the datum axis.

The datum axis is established by using a functional cylindrical feature that rotates about the axis, such as diameter *B* in Fig. 14.67. When the part is rotated about this axis, the features of rotation must fall within the prescribed tolerance at *full indicator movement (FIM)*.

The two types of runout are circular runout and total runout. One arrow in the feature control symbol indicates circular runout, and two arrows indicate total runout.

FIG. 14.68 The runout tolerance in this example is measured by mounting the object on the primary datum plane, surface *C*, and the secondary datum plane, cylinder *D*. The cylinder and conical surface is gaged to determine if it conforms to a tolerance zone of 0.03 mm. The end of the cone could have been noted to specify its runout (perpendicularity to the axis).

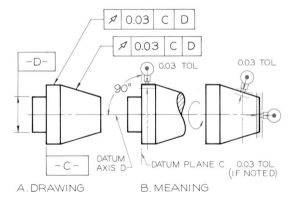

A. DRAWING   B. MEANING

CIRCULAR RUNOUT (one arrow) of an object is measured by rotating it about its axis for 360° to determine whether a circular cross section at any point exceeds the permissible runout tolerance. This same technique is used to measure the amount of wobble that exists in surfaces that are perpendicular to the axis of rotation.

TOTAL RUNOUT (two arrows) is used to specify cumulative variations of circularity, straightness, coaxiality, angularity, taper, and profile of a surface (Fig. 14.68). Total runout is applied to all circular and profile positions as the part is rotated 360°. When applied to surfaces that are perpendicular to the axis, total runout controls variations in perpendicularity and flatness of the surface.

The part shown in Fig. 14.69 is dimensioned by using a composite of several techniques of geometric tolerancing. Symbols and dimensions are used to specify the geometric characteristics of the part.

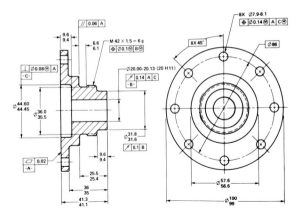

FIG. 14.69 A part dimensioned with a combination of notes and symbols to describe its geometric features. (Courtesy of ANSI Y14.5M—1982.)

14.30
Surface texture

The surface texture of a part will affect its function; consequently, this must be more precisely specified than by the general V that does not elaborate on the finish desired. Most of the terms of surface texture defined below are shown in Fig. 14.70.

SURFACE TEXTURE is the variation in the surface, including roughness, waviness, lay, and flaws.

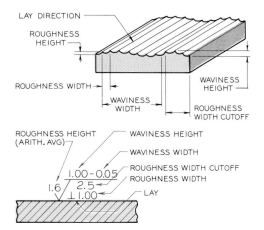

FIG. 14.70 Characteristics of surface texture.

ROUGHNESS describes the finest of the irregularities in the surface. These are usually caused by the manufacturing process used to smooth the surface.

ROUGHNESS HEIGHT is the average deviation from the mean plane of the surface. It is measured in microinches (μin) or micrometers (μm), which are millionths of an inch or of a meter.

ROUGHNESS WIDTH is the width between successive peaks and valleys that form the roughness measured in microinches or micrometers.

ROUGHNESS WIDTH CUTOFF is the largest spacing of repetitive irregularities that includes average roughness height (measured in inches or millimeters). When not specified, a value of 0.8 mm (0.030 in) is assumed.

WAVINESS is a widely spaced variation that exceeds the roughness width cutoff. Roughness may be considered as superimposed on a wavy surface. Waviness is measured in inches or millimeters.

WAVINESS HEIGHT is the peak-to-valley distance between waves. It is measured in inches or millimeters.

WAVINESS WIDTH is the spacing between peaks or wave valleys measured in inches or millimeters.

LAY is the direction of the surface pattern, which is determined by the production method used.

FLAWS are irregularities or defects that occur infrequently or at widely varying intervals on a surface. These include cracks, blow holes, checks, ridges, scratches, and the like. Unless otherwise specified, the effect of flaws are not included in roughness height measurements.

CONTACT AREA is the surface that will make contact with its mating surface.

The symbols used to specify surface texture are given in Fig. 14.71. The point of the V must contact the edge view of the surface being specified, an extension line from the surface, or a leader that points to the surface.

In Fig. 14.72, values of surface texture that can be applied to surface texture symbols, individually or in combination, are given. The roughness height values are related to manufacturing processes used to finish the surface (Fig. 14.73).

Lay symbols that indicate the direction of texture of a surface are given in Fig. 14.74. These symbols can

FIG. 14.71 Surface control symbols for specifying surface finish. General notes (I, J, and K) can be used to specify the recommended machining operation for finishing a surface.

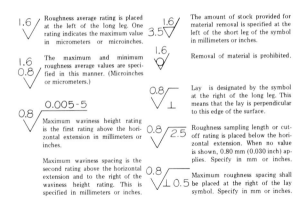

1.6 / Roughness average rating is placed at the left of the long leg. One rating indicates the maximum value in micrometers or microinches.

1.6 / 0.8 / The maximum and minimum roughness average values are specified in this manner. (Microinches or micrometers.)

0.8 / 0.005-5 Maximum waviness height rating is the first rating above the horizontal extension in millimeters or inches.

Maximum waviness spacing is the second rating above the horizontal extension and to the right of the waviness height rating. This is specified in millimeters or inches.

3.5 1.6 / The amount of stock provided for material removal is specified at the left of the short leg of the symbol in millimeters or inches.

1.6 / Removal of material is prohibited.

0.8 / ⊥ Lay is designated by the symbol at the right of the long leg. This means that the lay is perpendicular to this edge of the surface.

0.8 / 2.5 Roughness sampling length or cut-off rating is placed below the horizontal extension. When no value is shown, 0.80 mm (0.030 inch) applies. Specify in mm or inches.

0.8 / ⊥ 0.5 Maximum roughness spacing shall be placed at the right of the lay symbol. Specify in mm or inches.

FIG. 14.72 Values can be added to surface control symbols for more precise specifications. These may be in combinations other than those shown in these examples.

FIG. 14.73 The surface roughness heights produced by various types of production methods are shown here in micrometers (microinches). (Courtesy of the General Motors Corporation.)

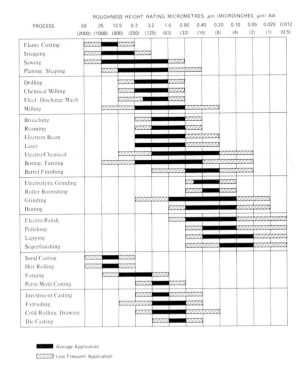

The ranges shown above are typical of the processes listed.

Higher or lower values may be obtained under special conditions.

FIG. 14.74 These symbols are used to indicate the direction of lay with respect to the surface where the control symbol is placed. (Courtesy of ANSI B46.1.)

Lay Symbol	Meaning	Example Showing Direction of Tool Marks
—	Lay approximately parallel to the line representing the surface to which the symbol is applied.	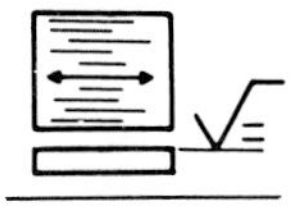
⊥	Lay approximately perpendicular to the line representing the surface to which the symbol is applied.	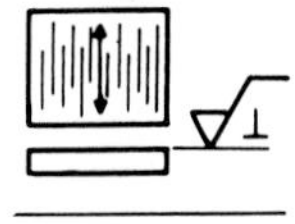
X	Lay angular in both directions to line representing the surface to which the symbol is applied.	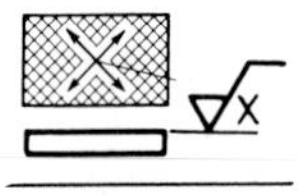
M	Lay multidirectional.	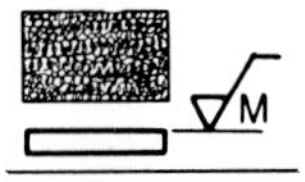
C	Lay approximately circular relative to the center of the surface to which the symbol is applied.	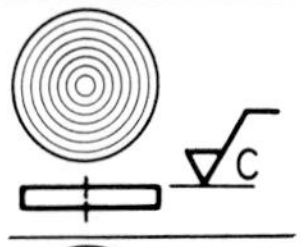
R	Lay approximately radial relative to the center of the surface to which the symbol is applied.	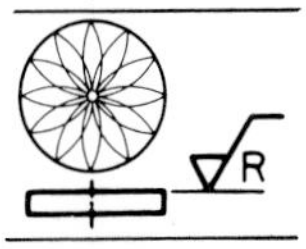
P	Lay particulate, non-directional, or protuberant.	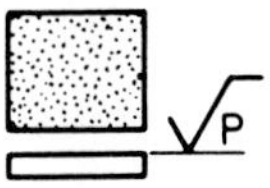

be incorporated into surface texture symbols as shown in Fig. 14.75. An example of a part with a variety of surface texture symbols applied to it is shown in Fig. 14.76.

FIG. 14.75   Examples of fully specified surface control symbols.

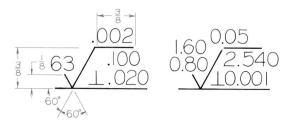

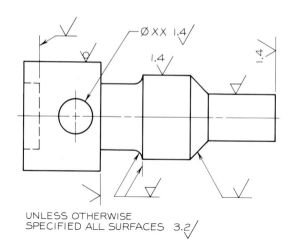

UNLESS OTHERWISE
SPECIFIED ALL SURFACES   3.2/

FIG. 14.76   This drawing illustrates the techniques of applying surface texture symbols to a part.

Problems

These problems can be solved on Size A sheets. The problems are laid out on a grid of 0.20 inches (5 mm).

Cylindrical fits

1. Construct the drawing of a shaft and hole as shown in Fig. 14.77 (it need not be drawn to scale), give the limits for each diameter, and complete the table of values. Use a basic diameter of 1.00 inches (25 mm) and a class RC 1 fit, or a corresponding metric fit.

2. Same as Problem 1, but use a basic diameter of 1.75 inches (45 mm) and a class RC 9 fit, or a corresponding metric fit.

3. Same as Problem 1, but use a basic diameter of 2.00 inches (51 mm) and a class RC 5 fit, or a corresponding metric fit.

4. Same as Problem 1, but use a basic diameter of 12.00 inches (305 mm) and a class LC 11 fit, or a corresponding metric fit.

5. Same as Problem 1, but use a basic diameter of 3.00 inches (76 mm) and a class LC 1 fit, or a corresponding metric fit.

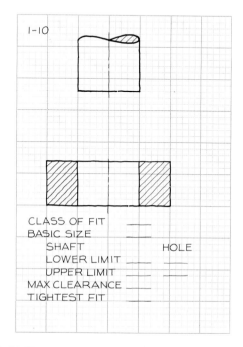

FIG. 14.77   Problems 1–10.

6. Same as Problem 1, but use a basic diameter of 8.00 inches (203 mm) and a class LC 1 fit, or a corresponding metric fit.

7. Same as Problem 1, but use a basic diameter of 102 inches (2 591 mm) and a class LN 3 fit, or a corresponding metric fit.

8. Same as Problem 1, but use a basic diameter of 11.00 inches (279 mm) and a class LN 2 fit, or a corresponding metric fit.

9. Same as Problem 1, but use a basic diameter of 6.00 inches (152 mm) and a class FN 5 fit, or a corresponding metric fit.

10. Same as Problem 1, but use a basic diameter of 2.60 inches (66 mm) and a class FN 1 fit, or a corresponding metric fit.

Tolerances of position

11. Make an instrument drawing on Size A paper of the part shown in Fig. 14.78. Locate the two holes with a size tolerance of 1.00 mm and a true-position tolerance of 0.50 DIA. Show the

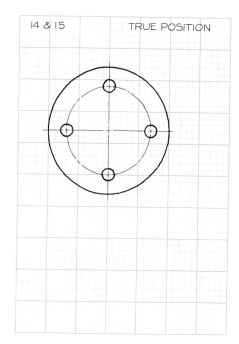

FIG. 14.79 Problems 14 and 15.

FIG. 14.78 Problems 11–13.

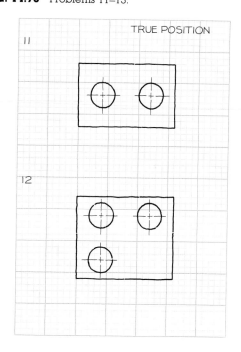

proper symbols and dimensions for this arrangement.

12. Same as Problem 11, except locate three holes using the same tolerances for size and position.

13. Give the specifications for a two-pin gage that can be used to gage the correctness of the two holes specified in Problem 11. Make a sketch of the gage and show the proper dimensions on it.

14. Using true positioning, locate the holes and properly note them to provide a size tolerance of 1.50 mm and a locational tolerance of 0.60 DIA (Fig. 14.79).

15. Same as Problem 14, except locate six equally spaced holes of the same size using the same tolerances of position.

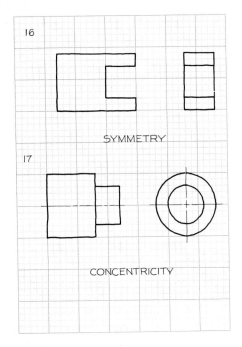

FIG. 14.80 Problems 16 and 17.

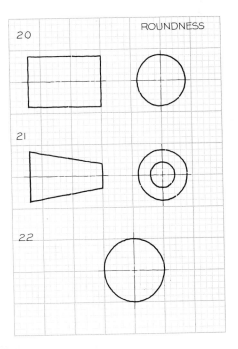

FIG. 14.82 Problems 20–22.

FIG. 14.81 Problems 18 and 19.

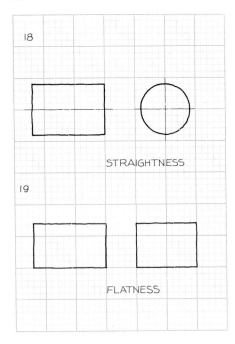

16. Using a feature control symbol and the necessary dimensions, indicate that the notch is symmetrical to the left-hand end of the part in Fig. 14.80 within 0.60 mm.

17. Using a feature control symbol and the necessary dimensions, indicate that the small cylinder is concentric with the large one (the datum cylinder) within a tolerance of 0.80 (Fig. 14.80).

18. Using a feature control symbol and the necessary dimensions, indicate that the elements of the cylinder are straight within a tolerance of 0.20 mm (Fig. 14.81).

19. Using a feature control symbol and the necessary dimensions, indicate that the upper surface of the object is flat within a tolerance of 0.08 mm (Fig. 14.81).

20–22. Using feature control symbols and the necessary dimensions, indicate that the cross sections of the cylinder, cone, and sphere are round within a tolerance of 0.40 mm (Fig. 14.82).

23. Using a feature control symbol and the necessary dimensions, indicate that the profile of the irreg-

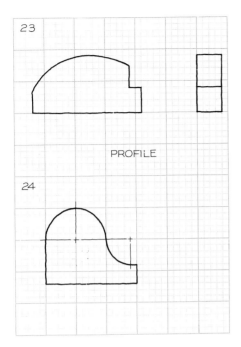

FIG. 14.83 Problems 23 and 24.

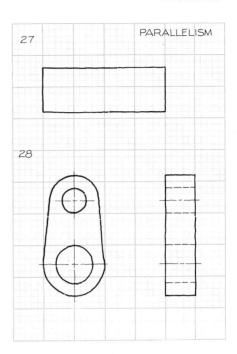

FIG. 14.85 Problems 27 and 28.

FIG. 14.84 Problems 25 and 26.

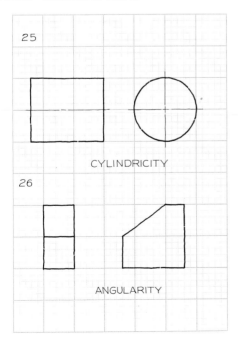

ular surface of the object lies within a tolerance zone of 0.40 mm, either bilateral or unilateral (Fig. 14.83).

24. Using a feature control symbol and the necessary dimensions, indicate that the profile of the line formed by tangent arcs has a profile that lies within a tolerance zone of 0.40 mm, either bilateral or unilateral (Fig. 14.83).

25. Using a feature control symbol and the necessary dimensions, indicate that the cylindricity of the cylinder is 0.90 mm (Fig. 14.84).

26. Using a feature control symbol and the necessary dimensions, indicate that the angularity tolerance of the inclined plane is 0.7 mm from the bottom of the object, the datum plane (Fig. 14.84).

27. Using a feature control symbol and the necessary dimensions, indicate that the upper surface of the object is parallel to the lower surface, the datum, within 0.30 mm (Fig. 14.85).

28. Using a feature control symbol and the necessary dimensions, indicate that the small hole is parallel to the large hole, the datum, within a tolerance of 0.80 mm (Fig. 14.85).

29. Using a feature control symbol and the necessary dimensions, indicate that the vertical surface is perpendicular to the bottom of the object, the datum, within a tolerance of 0.20 mm (Fig. 14.86).

30. Using a feature control symbol and the necessary dimensions, indicate that the hole is perpendicular to datum *A* within a tolerance of 0.08 mm (Fig. 14.86).

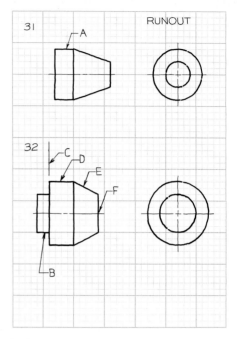

FIG. 14.87 Problems 31 and 32.

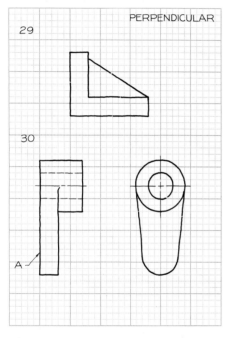

FIG. 14.86 Problems 29 and 30.

31. Using a feature control symbol and cylinder *A* as the datum, indicate that the conical feature has a runout of 0.80 mm (Fig. 14.87).

32. Using a feature control symbol with cylinder *B* as the primary datum and surface *C* as the secondary datum, indicate that surfaces *D*, *E*, and *F* have a runout of 0.60 mm (Fig. 14.87).

15

Welding

15.1

Introduction

Welding is the process of joining metal by heating a joint to a suitable temperature with or without the application of pressure, and with or without the use of filler material. Welding is used to join assemblies permanently when it will be unnecessary to disassemble them for maintenance or other purposes.

The welding practices presented in this chapter are in compliance with the standards developed by the American Welding Society and the American National Standards Institute (ANSI). Reference is also made to drafting standards used by General Motors Corporation.

Some of the advantages of welding over other methods of fastening are: (1) simplified fabrication, (2) economy, (3) increased strength and rigidity, (4) ease of repair, (5) creation of gas- and liquid-tight joints, and (6) reduction in weight and/or size.

The various welding processes are given in Fig. 15.1. The three main types of welding are *gas welding*, *arc welding*, and *resistance welding*.

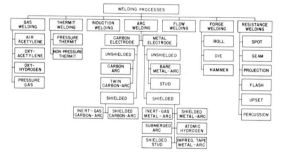

FIG. 15.1 The types of welding processes available to the designer. The three major types are gas welding, arc welding, and resistance welding. (Courtesy of General Motors Corporation.)

GAS WELDING is a process in which gas flames are used to melt and fuse metal joints. Gases such as acetylene or hydrogen are mixed within a torch and are burned with air or oxygen (Fig. 15.2). The oxyacetylene method is the best-known process of gas welding, and it is widely used for repair work and field construction.

Most oxyacetylene welding is done manually with a minimum of equipment. Filler material in the form of welding rods is used to deposit metal at the joint as it is heated. Most metals, except for low- and medium-carbon steels, will require fluxes that aid in the process of melting and fusing the metals together.

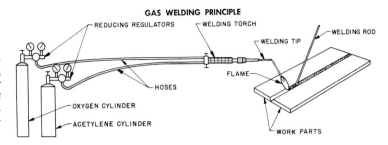

FIG. 15.2 The gas welding process burns gases, such as oxygen and acetylene, in a torch to apply heat to a joint while the welding rod supplies the filler material. (Courtesy of General Motors Corporation.)

ARC WELDING is a process that uses an electric arc to heat and fuse the joints (Fig. 15.3). In some cases, pressure may be applied in addition to the heat; in others, pressure will not be required. The filler material is supplied by a consumable electrode through which the electric arc is transmitted or through a nonconsumable electrode to fuse the metals. Metals that are well suited to arc welding are wrought iron, low- and medium-carbon steels, stainless steel, copper, brass, bronze, aluminum, and some nickel alloys.

FLASH WELDING is a form of arc welding that is similar to resistance welding since both pressure and an electric current are used to join two pieces (Fig. 15.4). The two parts are brought together, causing a heat buildup between them. As the metal burns, the current is turned off and the pressure between the parts is increased to fuse the parts together.

RESISTANCE WELDING is a group of processes where metals are fused together by heat produced from the resistance of the parts to the flow of electric current and by the application of pressure. Fluxes and filler materials are normally not used. All resistance welds are either lap- or butt-type welds.

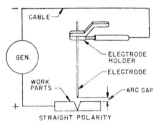

FIG. 15.3 The arc welding process can use either DC or AC current that is passed through an electrode to heat the joint to be welded. (Courtesy of General Motors Corporation.)

FIG. 15.4 Flash welding is a type of arc welding that uses a combination of electric current and pressure to fuse two parts together. (Courtesy of General Motors Corporation.)

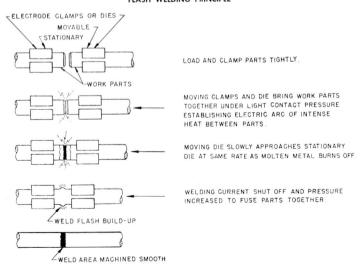

SPOT WELDING PRINCIPLE

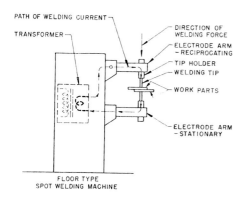

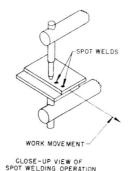

FIG. 15.5 Resistance welding can be used to join lap and butt joints to fuse metals by heat generated by passing an electrical current through them. (Courtesy of General Motors Corporation.)

TABLE 15.1
RECOMMENDED RESISTANCE WELDING PROCESSES

Material	Spot Welding	Flash Welding
Low-carbon mild steel:		
SAE 1010	R	R
SAE 1020	R	R
Medium-carbon steel:		
SAE 1030	R	R
SAE 1050	R	R
Wrought-alloy steel:		
SAE 4130	R	R
SAE 4340	R	R
High-alloy austenitic Stainless steel:		
SAE 30301–30302	R	R
SAE 30309–30316	R	R
Ferritic and martensitic Stainless steel:		
SAE 51410–51430	S	S
Wrought heat resisting alloys:*		
10–9–DL	S	S
16–25–6	S	S
Cast iron	NA	NR
Gray iron	NA	NR
Aluminum & aluminum alloys	R	S
Nickel & nickel alloys	R	S

S—Satisfactory   NA—Not applicable
R—Recommended   NR—Not recommended
*For composition, see **American Society of Metals Handbook**.
Source: Courtesy of General Motors Corporation.

Figure 15.5 illustrates how resistance spot welding is performed on a lap joint. The two parts are lapped, and pressed together, and an electric current fuses the parts together where they join. A series of spots spaced at intervals, called *spot welds,* are used to secure the parts. Other welds besides spot welds can be produced by resistance welding. Table 15.1 suggests the processes of welding that can be used for various types of materials.

15.2
Weld joints

The five standard weld joints are illustrated in Fig. 15.6. The butt joint can be joined with the following types of welds: *square groove, V-groove, bevel groove, U-groove,* and *J-groove.* The corner joint can be joined

FIG. 15.6 The standard types of joints encountered in the welding process.

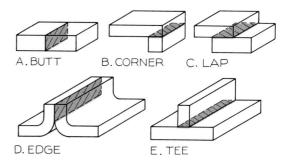

A. BUTT   B. CORNER   C. LAP

D. EDGE   E. TEE

with the same welds as the butt joint, but with the addition of the fillet weld as well.

The tee joint can be joined with the following welds: bevel groove, J-groove, and fillet welds. Welds used to join lap joints are fillet, bevel groove, J-groove, slot, plug, spot, projection, and seam welds. The edge joint uses the same welds that are used for lap joints with the addition of the square groove, V-groove, U-groove, and seam welds.

15.3
Welding symbols

The specification of welds on a working drawing is done by the application of symbols. If a drawing has a general note such as ALL JOINTS WELDED or WELDED THROUGHOUT, the designer has transferred the design responsibility to the welder. Welding is too important to be left to chance; it must be completely specified.

The method of providing specifications on a drawing is by the use of a welding symbol as shown in Fig. 15.7. This example gives the symbol in its entirety, which is seldom needed in its complete form. Instead, the symbol is usually modified to a simpler form when not all the specifications are necessary for a particular application.

The scale of the welding symbol is shown in Fig. 15.8, where it is drawn on a 3 mm (⅛″) grid. Its size can be scaled down using these same proportions

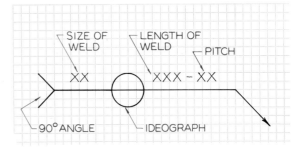

FIG. 15.8 When the grid is drawn as a full-size ⅛″ (3 mm) grid, the size of the welding symbol can be determined. It can be drawn smaller or larger at these same proportions when necessary.

where space is limited. The lettering used is the standard height of ⅛″ or 3 mm.

> The *ideograph* is the symbol used to denote the type of weld desired. In general, the ideograph depicts the cross section of the type of weld used.

The more often used ideographs are drawn to scale on the ⅛″ (3 mm) grid to represent their full size when added to the welding symbol (Fig. 15.9).

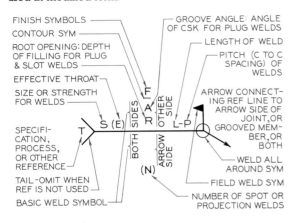

FIG. 15.7 The welding symbol. It is not necessary to show the entire symbol in all applications. It may be used in modified form.

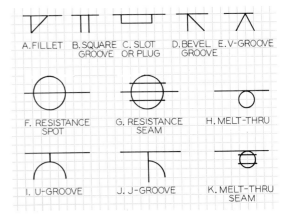

FIG. 15.9 The sizes of the ideographs are shown on the ⅛″ (3 mm) grid. These sizes should be used in conjunction with the symbol shown in the previous figure.

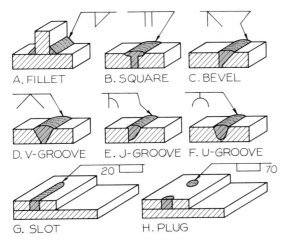

FIG. 15.10 The standard types of welds and their corresponding ideographs.

15.4
Types of welds

The more commonly used welds are shown in Fig. 15.10, along with their corresponding ideographs. The fillet weld is a built-up weld at the angular intersection between two surfaces. The square, bevel, V-groove, J-groove, and U-groove welds involve grooves inside of which the weld is placed. Slot and plug welds have intermittent holes or openings where the parts are welded. Holes are unnecessary when resistance

welding is used. Note that the symbols (ideographs) are symbolic of the cross sections of the welds and grooves.

15.5
Application of symbols

Fillet welds are applied to two parts in Fig. 15.11 to indicate three methods of welding. At A, the fillet ideograph is placed on the lower side of the horizontal line of the symbol, which indicates that the weld is to be at the joint on the *arrow side*, the side of the arrow.

> The vertical leg of the ideograph is *always* on the left side.

By placing the ideograph on the upper side of the horizontal, the weld is specified to be on the *other side*, the joint on the other side of the part away from the arrow. Again, the vertical leg is on the left side.

When the part is to be welded on both sides of the vertical part, the ideograph at C is used. Note that the tail has been omitted from the symbol along with other written specifications. This is permitted when detailed specifications are given in another form to specify the details of the welds used.

A single arrow can be used to specify a weld that is to be all around two joining parts (Fig. 15.12). A circle of 6 mm diameter placed at the bend in the leader of the symbol gives this specification. If this process is to be done *in the field*, a black circle of 3 mm

FIG. 15.11 Fillet welds are indicated by abbreviated symbols. When the ideograph is on the lower side, it refers to the arrow side; when it is placed above the horizontal line, it refers to the other side.

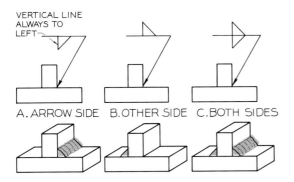

FIG. 15.12 Symbols for indicating fillet welds all around two types of parts.

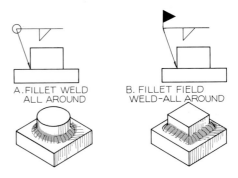

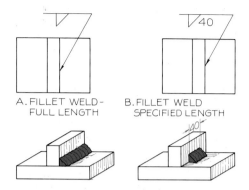

A. FILLET WELD—
FULL LENGTH

B. FILLET WELD
SPECIFIED LENGTH

FIG. 15.13 Symbol for indicating full-length fillet welds and fillet welds less than full length.

diameter can be used to denote this joint. This means that the parts will be assembled on the site rather than in a shop. Both symbols can be used separately as well as together.

When a fillet weld is to be the full length of the two parts, it may be specified as shown in Fig. 15.13. Since the ideograph is on the lower side, the weld will be on the arrow side. If the weld is to be less than full length, it can be specified as shown at B where the number, 40, represents its length in millimeters, and it is centered about the approximate location of the arrow.

When fillet welds are of different lengths and are positioned on both sides, they may be specified as shown in Fig. 15.14A. The dimensions on the lower side of the horizontal give the length of the weld on

FIG. 15.14 Symbols for specifying varying and intermittent welds.

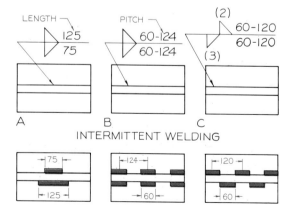

INTERMITTENT WELDING

the arrow side, and the number on the upper side gives the length on the other side.

Intermittent welds are welds of a given length that are spaced uniformly apart from center to center by a distance called the *pitch*. Since the welds are on both sides, are 60 mm long, and have pitches of 124 mm, this can be indicated by a symbol as shown in Fig. 15.14B. If the intermittent welds are to be staggered to alternate positions on both sides, this can be specified by the symbols shown in part C.

15.6
Groove welds

The more standard groove welds are illustrated in Fig. 15.15 with their respective symbols. When the depth of the grooves, the angle of the chamfer, and the root openings are not given on a symbol, this information needs to be specified elsewhere on the drawings or in supporting documents. At B and E, the depth of the chamfer of the prepared joint is given in parentheses above the size dimension of the weld, which takes into account the penetration of the weld beyond this chamfer. The size of the joint is equal to the depth of the prepared joint when only one number is given.

When the chamfer is different on each side of the joint it can be noted with a symbol as shown in Fig. 15.15C. If the spacing between the two parts, the *root opening,* is to be specified, this is done by placing its dimensions in millimeters between the groove angle number and the weld ideograph, part E.

A bevel weld is groove beveled from one of the parts being joined; consequently, the symbol must indicate which is to be beveled, as shown in Fig. 15.15E. To call attention to this operation, the leader from the symbol is bent and aimed toward the piece to be beveled. This practice also applies to J-welds, where one side is grooved and the other is not (Fig. 15.16).

15.7
Example welds

A series of joints and welds is shown in Fig. 15.17 that incorporates the previously covered principles. Letters are used instead of numerals to represent specified dimensions.

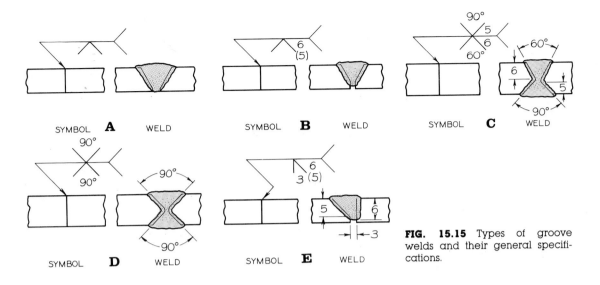

SYMBOL **A** WELD

SYMBOL **B** WELD

SYMBOL **C** WELD

SYMBOL **D** WELD

SYMBOL **E** WELD

FIG. 15.15 Types of groove welds and their general specifications.

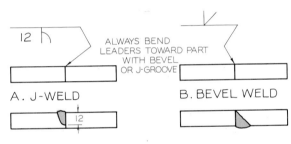

A. J-WELD

B. BEVEL WELD

ALWAYS BEND LEADERS TOWARD PART WITH BEVEL OR J-GROOVE

FIG. 15.16 J-welds and bevel welds are specified by bent arrows pointing to the side of the joint that is to be grooved.

FIG. 15.17 An assortment of welds and grooves with their accompanying symbols. Letters are used instead of numbers to give the weld specification.

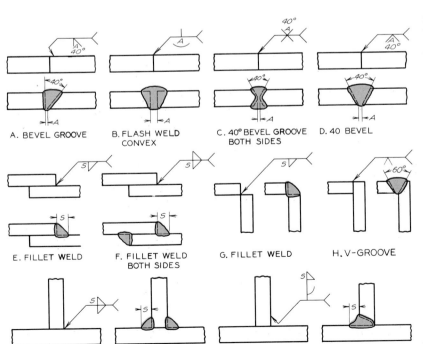

A. BEVEL GROOVE

B. FLASH WELD CONVEX

C. 40° BEVEL GROOVE BOTH SIDES

D. 40 BEVEL

E. FILLET WELD

F. FILLET WELD BOTH SIDES

G. FILLET WELD

H. V-GROOVE

I. FILLET WELD-BOTH SIDES

J. J-GROOVE & FILLET WELD OTHER SIDE

15.8
Surface-contoured welds

Contour symbols are used to indicate which of the three types of contours is desired on the surface of the weld: flush, concave, or convex. Flush welds are those that are smooth with the surface or flat across the hypotenuse of a fillet weld. A concave contour is a weld that bulges inward with a curve, and a convex contour is one that bulges outward with a curve (Fig. 15.18).

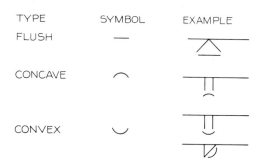

FIG. 15.18 The contour symbols that are used to specifiy the surface finish of a weld.

In many cases, it is necessary to finish the weld by a supplementary process to bring it to the desired contour. These processes may be added to the contour symbols to specify their operations in finishing the welds. These processes and their letter specifications are: *chipping* (C), *grinding* (G), *hammering* (H), *machining* (M), *rolling* (R), and *peening* (P). Examples of these contour symbols are given in Fig. 15.19.

FIG. 15.19 Examples of contoured surface symbols and letters of finishing applied to them.

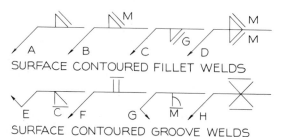

15.9
Seam welds

A seam weld is a weld that joins two lapping parts with a continuous weld or a series of closely spaced spot welds. The process used for seam welds must be given by abbreviations placed in the tail of the weld symbol. The ideograph for a resistance weld is about 12 mm in diameter, and it is placed with the horizontal line of the symbol through its center. The weld's width, length, and pitch are indicated by numbers, as shown in Fig. 15.20.

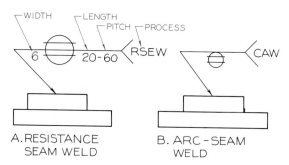

FIG. 15.20 Resistance and arc seam welds with the processes are indicated in the tail of the symbol for each. The arc weld must specify arrow side or other side in the symbol.

When the seam weld is made by arc welding, the diameter of the ideograph is about 6 mm, and it is placed on the upper or lower side of the horizontal bar of the symbol to indicate whether the seam will be on the arrow or the other side when applied (part B). When the numeral that represents the length of the weld is omitted from the symbol, it is understood that the seam weld extends between abrupt changes in the seam or as it is dimensioned.

Spot welds are specified in a similar manner with ideographs. Specifications, as shown in Fig. 15.21, are given by diameter, number of welds, and pitch between the welds. The process, resistance spot welding (RSW), is noted in the tail of the symbol. For arc welding, the arrow side or other side must be indicated by a symbol, as shown in part B.

(The abbreviations of various welding processes can be seen in Table 15.2.)

TABLE 15.2
WELDING PROCESS ABBREVIATIONS

CAW	Carbon arc welding	FRW	Friction welding	PGW	Pressure gas welding
CW	Cold welding	FW	Flash welding	RB	Resistance brazing
DB	Dip brazing	GMAW	Gas metal arc welding	RPW	Projection welding
DFW	Diffusion welding	GTAW	Gas tungsten welding	RSEW	Resistance seam welding
EBW	Electron beam welding	IB	Induction brazing	RSW	Resistance spot welding
ESW	Electroslag welding	IRB	Infrared brazing	RW	Resistance welding
EXW	Explosion welding	OAW	Oxyacetylene welding	TB	Torch brazing
FB	Furnace brazing	OHW	Oxyhydrogen welding	UW	Upset welding
FOW	Forge welding				

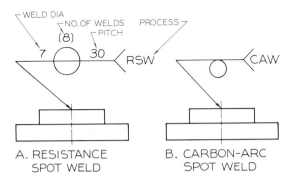

A. RESISTANCE SPOT WELD

B. CARBON-ARC SPOT WELD

FIG. 15.21 Resistance and arc spot welds with the process indicated in tail of the symbol. The arc weld must specify arrow- or other-side location in the symbol.

15.10
Built-up welds

When the surface of a part is to be enlarged by welding, called *building up*, this can be indicated by the symbol shown in Fig. 15.22. The width of the built-

FIG. 15.22 The method of applying a symbol to a built-up weld on a surface.

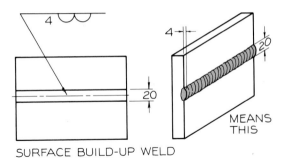

SURFACE BUILD-UP WELD

MEANS THIS

up weld is dimensioned in the view, and the height of the weld above the surface is specified in the symbol to the left of the ideograph. The radius of the circular segment is 6 mm.

15.11
Welding standards

Figure 15.23 (pages 322 and 323) gives an overview of the welding symbols and specifications that have been covered in the previous sections. The chart was prepared by the American Welding Society of Miami, Florida. It can be used as a single reference for most general types of welding and their associated symbols.

15.12
Brazing

Brazing is a method of joining pieces of metal that is similar to welding. It is a process in which the joints are heated above 800 degrees Fahrenheit and a non-ferrous filler material with a melting point below the base materials is distributed by capillary action between the closely fit parts.

Prior to brazing, the parts to be brazed must be cleansed, and flux is added to the joints. The brazing filler is also added prior to or just as the joints are heated beyond the melting point of the filler. There are two basic joints for brazing: lap and butt joints, as shown in Fig. 15.24. The filler material is allowed to flow between the parts to form the joint after it has melted.

Brazing is used to hold parts together, to provide gas- and liquid-tight joints, to assure electrical conduc-

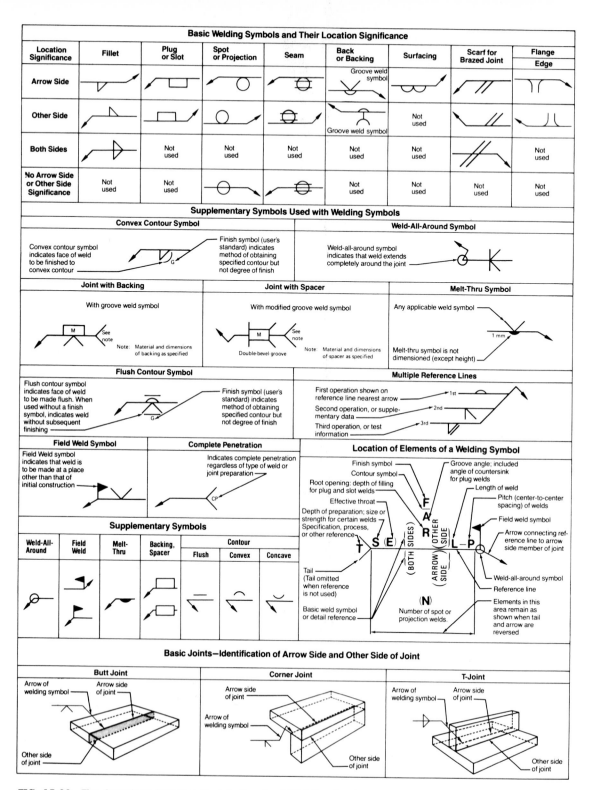

FIG. 15.23 The American Welding Society Standard Welding Symbols. This chart serves as a review of the principles covered in this chapter.

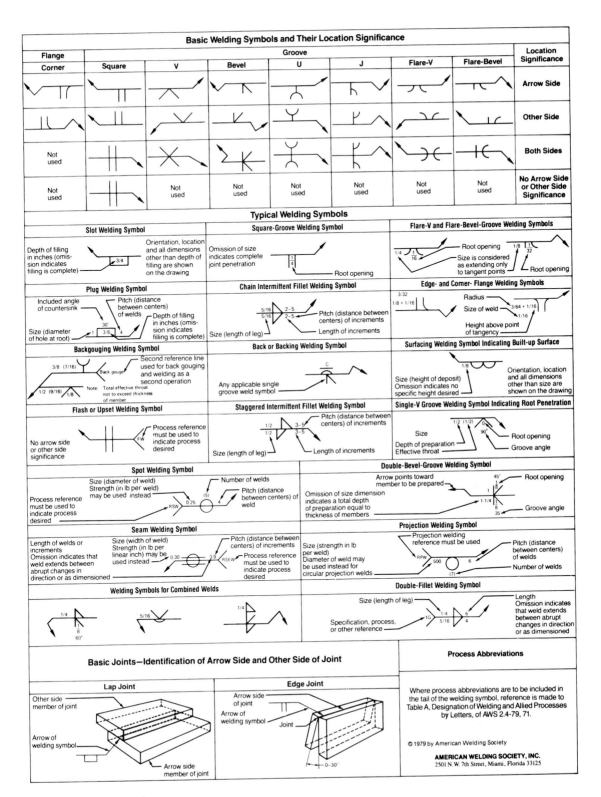

FIG. 15.23 continued

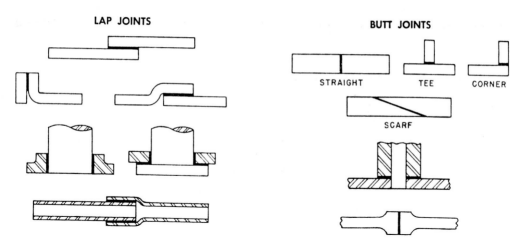

FIG. 15.24 Examples of the two basic types of brazing joints: lap joints and butt joints. (Courtesy of General Motors Corporation.)

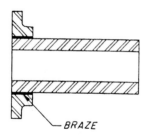

FIG. 15.25 The method of indicating a brazed joint in a drawing. (Courtesy of General Motors Corporation.)

joining is widely used in the automotive and electrical industries.

Soldering is one of the basic techniques of welding and is often done by hand with a soldering iron of the type shown in Fig. 15.26. The iron is placed on the joint to heat it and to melt the solder that fuses the joint. The method of indicating a soldered joint is shown in Fig. 15.26 where a heavy, dark line is used with a note.

FIG. 15.26 A typical hand-held soldering iron used to soft-solder two parts together, and the method of indicating a soldered joint on a drawing. (Courtesy of General Motors Corporation.)

tivity, and to aid in repair and salvage. Brazed joints will withstand more stress, higher temperature, and more vibration than will soft-soldered joints.

The method of noting a brazed joint in a drawing is illustrated in Fig. 15.25. A heavy, dark line is used to indicate the brazed joint.

15.13
Soft-soldering

Soldering is the process of joining two metal parts with another metal that melts below the temperature of the metals being joined. Solders are alloys of non-ferrous metals that melt below 800° F. This method of

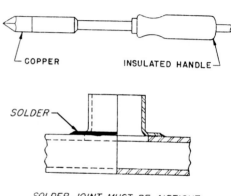

Problems

Make working drawings of the following problems given in Chapter 16. Wherever feasible, change the joints of the features of the parts to be welded instead of joined by one-piece casting. These drawings should be made on Size B sheets.

1. Use Fig. 16.36.
2. Use Fig. 16.40.
3. Use Fig. 16.45.
4. Use Fig. 16.47.
5. Use Fig. 16.51.
6. Use Fig. 16.53.
7. Use Fig. 16.65.
8. Use Fig. 16.67.
9. Use Fig. 16.72, part 1.
10. Use Fig. 16.78, part 1.

Working Drawings

16

16.1
Introduction

Working drawings are drawings from which a design is constructed. The set may contain any number of sheets from one to over one hundred, depending on the complexity of the project. The written instructions that accompany working drawings are called *specifications*. When the project can be represented on several drawing sheets, the written specifications are often written on the drawings to consolidate the information into a single format. Although much of the work in preparing working drawings is done by the drafter, the designer, who is most often an engineer, is responsible for their correctness.

> A *working drawing* is often called a *detail drawing* because it describes and gives the dimensions of the details of the parts being presented.

All the principles of orthographic projection, sections, conventions, dimensioning, tolerancing, pictorials, and practically all other types of graphical techniques are utilized to communicate the details in a working drawing.

16.2
Working drawings–inch system

An example of a working drawing is shown in Fig. 16.1, where the base-plate mount is detailed in three orthographic views. Dimensions and notes are used to give the necessary information to construct the piece without misinterpretation. This particular drawing is dimensioned with decimal inches.

Decimal inches are preferable to common fractions, although both systems are still in widespread use. The English system, which is based on the inch, is giving way to the metric system, which is based on the millimeter. Decimal inches make it possible to handle arithmetic with much greater ease than is possible with fractions. Inch marks such as X.XX" are omitted from dimensions on a working drawing since it is understood that these dimensions are in inches.

An example of a part represented in a set of working drawings dimensioned with common fractions using the inch as the unit of measurement is the revolving clamp assembly shown in Fig. 16.2. The detail drawings of the parts of the assembly are shown in Figs. 16.3 through 16.5. Several dimensioned parts are shown on each sheet as orthographic views. The arrangement of these parts on the sheet has no rela-

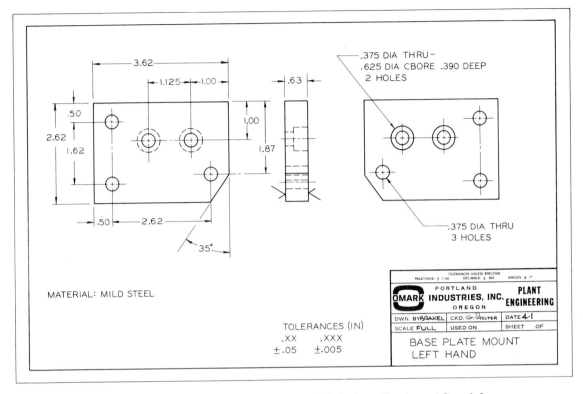

FIG. 16.1 A working drawing of a single part dimensioned in inches. (Courtesy of Omark Industries, Inc.)

tionship to how they fit together. They are simply positioned to take advantage of the available space. Each part is given a number and a part name for identification purposes. The material that each part is made of is indicated along with the notes to explain fully any manufacturing procedures that are necessary.

Each sheet is numbered in the title block, and the other title block information is completed.

An orthographic assembly is given on sheet 3 (Fig. 16.5), which explains how the parts fit together. The parts are numbered to correspond to the part numbers in the parts list, which serves as a bill of materials.

FIG. 16.2 A revolving clamp assembly manufactured to hold parts stationary while they are being machined. (Courtesy of Jergens, Inc.)

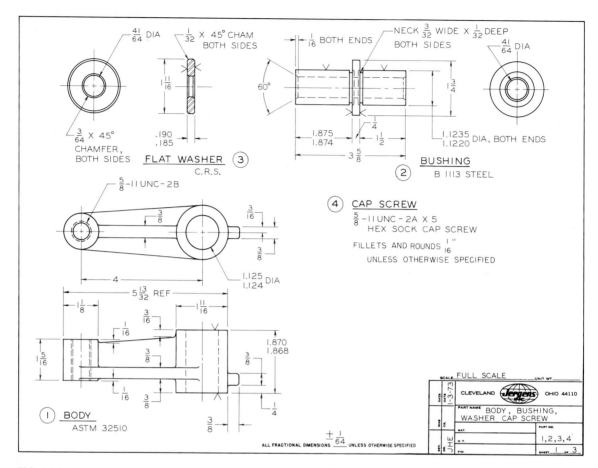

FIG. 16.3 Detail drawings showing parts of the clamp assembly. (Courtesy of Jergens, Inc.)

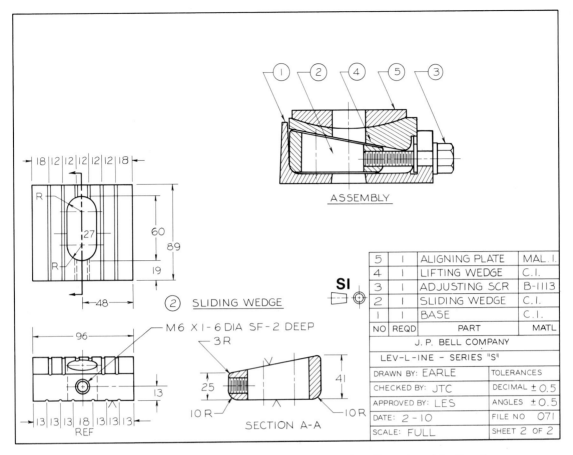

FIG. 16.10 A working drawing and assembly of the lifting device dimensioned in SI units. (Courtesy of Unisorb Machinery Installation Systems.)

FIG. 16.11 A dual-dimensioned drawing. The dimensions are given in inches, with their equivalents in millimeters given in parentheses. (Courtesy of General Motors Corporation.)

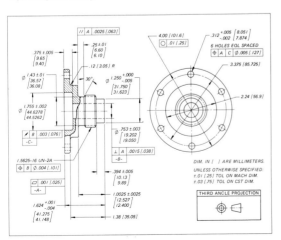

16.4
Working drawings–dual dimensions

Some working drawings are dimensioned in both inches and millimeters for those who may work in both systems. A typical example is shown in Fig. 16.11, where the dimensions in parentheses are millimeters and the dimensions above the parentheses are inches. Converting from one unit to the other will result in fractional units that must be rounded off.

The units may first be made in millimeters and then converted to inches equally as well. Usually, the converted units are shown in parentheses. An explanation of the system used should be noted in the title block.

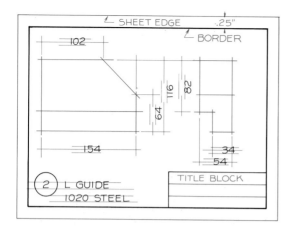

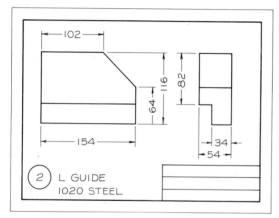

FIG. 16.12 Laying out a detail drawing

Step 1 The border and title strip are drawn on a preliminary sheet that will be traced. The views are positioned to allow adequate room for their dimensions. Guidelines are constructed for all dimensions and notes.

Step 2 The layout is overlaid with vellum or film. Then the lines are drawn to their proper weights, the dimensions and notes are lettered, and the title block is completed to finish the detail drawing.

16.5
Laying out a working drawing

The working drawing is laid out by beginning with the border, if a printed border is not provided. At least 0.25″ (7 mm) should be allowed for the border at the edge of the sheet (Fig. 16.12, Step 1). The title block

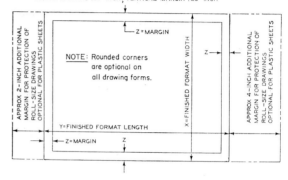

SIZE DES LTR	FLAT SIZES			SIZE DES LTR	ROLL SIZES			
	X WIDTH	Y LENGTH	Z MARGIN		X WIDTH	Y MIN LENGTH	Y MAX LENGTH	Z MARGIN
A(HORIZ)	8.50	11	.25 & .38*	G	11	42	144	.38
A(VERT)	11	8.50	.25 & .38*	H	28	48	144	.50
B	11	17	.38	J	34	48	144	.50
C	17	22	.50	K	40	48	144	.50
D	22	34	.50					
E	34	44	.50					
F	28	40	.50					

*HORIZONTAL MARGINS .38-INCH; VERTICAL MARGIN .25-INCH

FIG. 16.13 The standard sheet sizes for working drawings dimensioned in inches. (Courtesy of the U.S. Department of Defense.)

FIG. 16.14 A typical parts list and title strip suitable for most student assignments.

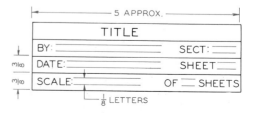

is drawn to size at the lower right corner of the sheet, and views and dimensions are drawn lightly to take advantage of the available space. When using tracing paper or film, it is more efficient to lay out the views and dimensions on a different sheet of paper and then overlay this drawing with vellum or film for tracing the final working drawing. Guidelines for lettering are drawn for each dimension line.

The lines are darkened to their proper weight, and the drawing is completed as shown in Step 2. Properly positioning the views to provide adequate space is one of the major concerns of the drafter in laying out a drawing.

The standard sheet sizes of working drawings are shown in Fig. 16.13. Papers, films, cloths, and reproduction materials are available in these modular sizes. Most of your student assignments will be completed on Size A and Size B sheets

16.6
Title blocks and parts lists

A typical parts list and title block for student assignments are shown in Fig. 16.14. These blocks are usually located in the lower right-hand corner of the drawing sheet against the borders. You will notice from observation that practically all title blocks contain the following information: title or part name, drafter, date, scale, company, and sheet number. Other information such as tolerances, checkers, and materials may be shown in more elaborate title blocks.

FIG. 16.15 A title block typical of those used in industry. (Courtesy of General Motors Corporation.)

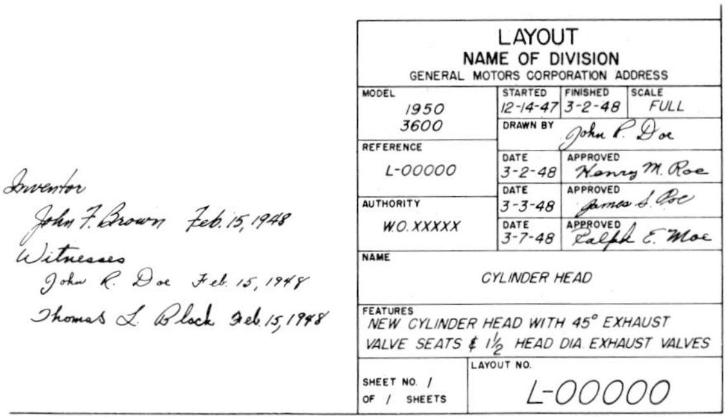

The standard parts list is shown in Fig. 16.14 with the usual elements.

The *parts list* should be placed directly over and in contact with the title block in the lower right corner of the drawing.

A title block used by General Motors Corporation is shown in Fig. 16.15. A note to the left of the title block lists John F. Brown as an inventor. One or two associates are asked to date and sign the drawings as witnesses of the work of the inventor. This procedure establishes the ownership of the ideas and dates the time of their development, in case this becomes an issue in obtaining a patent at a later date.

An example of a title block and revision block is shown in Fig. 16.16. This block is typical of those that are printed on sheets by various industries. Revision blocks are used to indicate modifications of parts at later dates to improve the design.

It is important to give the number of sheets in the set on each sheet. For example: sheet 2 of 6, sheet 3 of 6, and so on.

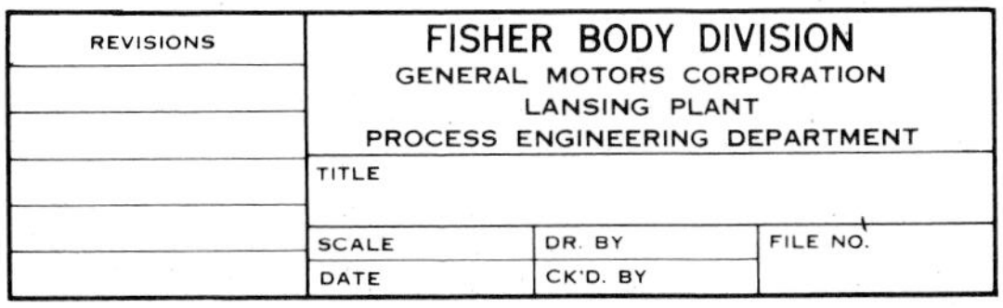

FIG. 16.16 An example of a title block with a revision block used in industry. (Courtesy of General Motors Corporation.)

16.7
Scale specification

The scale of a working drawing should be indicated in or near the title block when all drawings are the same scale. If several drawings are made at different scales, the scales used should be indicated in the drawings.

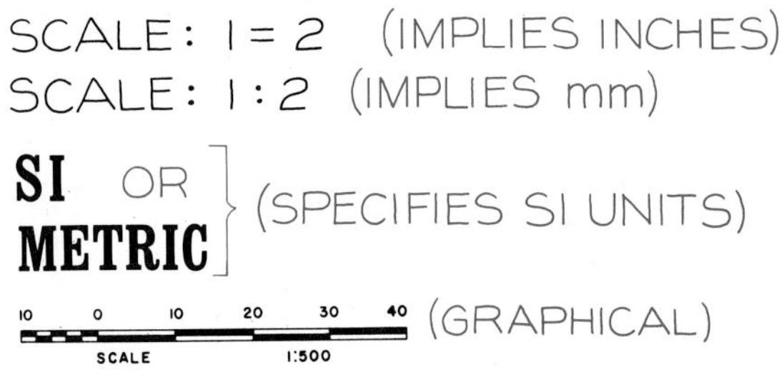

SCALE: 1 = 2 (IMPLIES INCHES)
SCALE: 1 : 2 (IMPLIES mm)

SI OR ⎫
METRIC ⎭ (SPECIFIES SI UNITS)

10   0   10   20   30   40
SCALE   1:500   (GRAPHICAL)

FIG. 16.17 Methods of specifying scales and metric units on working drawings.

Several methods of indicating scales are shown in Fig. 16.17. When the colon is used, such as 1:2, the metric system is implied, whereas the English system is implied when the equal sign is used: 1 = 2. The symbol *SI* or the word *METRIC* on a drawing clearly specifies that the units of measurement are millimeters.

In some cases, a graphical scale is given with calibrations that permit the interpretation of linear units by transferring dimensions from the drawing with your dividers to the scale. This is commonly used for scaling distances on maps.

16.8
Tolerances

General tolerance notes can be given on working drawings to specify the tolerances of dimensions. Such a table of values is shown in Fig. 16.18, where a check can be made to indicate whether the units will be in inches or in millimeters.

FIG. 16.18 General tolerance notes given on working drawings to specify the tolerance permitted on dimensions.

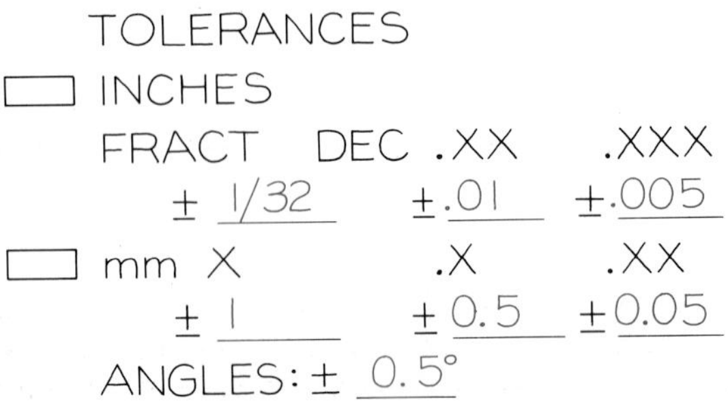

TOLERANCES

☐ INCHES

FRACT   DEC .XX   .XXX
± 1/32   ±.01   ±.005

☐ mm X   .X   .XX
± 1   ± 0.5   ±0.05

ANGLES: ± 0.5°

Plus-or-minus tolerances are given in the blanks under the number of digits under each decimal fraction. For example, this table specifies that each dimension with two-place decimals will have a tolerance of ±0.01″.

Angular tolerances can be given in general notes also. Refer to Chapter 14 for more detailed examples of using general tolerance notes.

16.9
Part names

Each part should be given a name and a number as shown in Fig. 16.19. The letters should be ⅛″ (3 mm) high. The part numbers are placed inside circles, called *balloons*, which are drawn approximately four times the height of the numbers.

NAME OF PART
PLACE NEAR PART
BALLOON 4 X LETTER HEIGHT

FIG. 16.19 Each part of a working drawing should be named and numbered for listing in the parts list.

The part numbers should be placed close to the parts on the working drawing so that it is clear which part they are associated with. Balloons are especially important on assembly drawings since the numbers of the parts refer to the same numbers in the parts list that serves as a cross reference.

16.10
Checking a drawing

All drawings must be checked before they are released for production, since a slight mistake could prove very expensive when many parts are made. The people who check drawings have special qualifications that enable them to suggest revisions and modifications that will result in a better product at less cost. The checker may be a chief drafter who is experienced in this type of work, or the engineer or designer who originated the project. In larger companies, the drawings are reviewed by the various shops involved to determine whether the most efficient methods of production are specified for each particular part.

The checker never checks an original drawing but instead checks a *diazo print* (a blue-line print). The checker marks the print with a colored pencil, making notes and corrections that he or she feels are desirable. The print is returned to the drafter for revision of the original drawing, and another print is made for approval.

In Fig. 16.20, a detail drawing of a special bushing is shown. In this drawing, the various modifications made by checkers are labeled with letters that are circled and placed near the revisions. The changes are listed and dated in the revision record by the drafter. Note that the change numbers are placed in a row below the revisions. This procedure serves as a check on the various revisions to prevent one from being overlooked.

Note that several drafters and checkers were involved in the approval and preparation of the drawing. Tolerances and general information are printed in the title block to ensure uniformity in the production of similar parts.

Checkers are responsible for checking the soundness of the design and its functional characteristics. They are also responsible for the completeness of the drawing, the quality of the drawing, its readability, lettering, drafting techniques, and clarity. A poorly drawn view must be redrawn so that it will reproduce well and be understood by those using it. Quality of

	Max Value	Points Earned
TITLE BLOCK		
Student's name	1	
Checker	1	
Date	1	
Scale	1	
Sheet number	1	
REPRESENTATION OF DETAILS		
Selection of views	5	
Assembly drawings	10	
Positioning of views	5	
DRAFTING PRACTICES		
Line quality	8	
Lettering	8	
Proper dimensioning	10	
Proper use of sections	5	
Proper use of auxiliary views	5	
DESIGN INFORMATION		
Indication of tolerances	5	
General tolerance notes	5	
Fillets and rounds notes	5	
Finish marks	5	
Parts list	8	
Thread notes and symbols	5	
PRESENTATION		
Properly trimmed	2	
Properly folded	2	
Properly stapled	2	
	100	

grade

FIG. 16.21 A checklist for evaluating a working-drawing assignment.

FIG. 16.20 The revisions of this working drawing are noted near the revisions with letters in balloons that are cross-referenced in a table of revisions. (Courtesy of General Motors Corporation.)

lettering is very important. Since working drawings should not be scaled, the shop person must rely on lettered notes and dimensions for the information.

The best method for students to check their drawings is to make a scale drawing of the part from the working drawings. It is easier to find another's mistakes rather than one's own.

A grading scale for checking working drawings prepared by students is given in Fig. 16.21. This general list can be used as an outline for reviewing working drawings to ensure that the major requirements have been met.

16.11
Drafter's log

Drafters should keep a record called a *log* to show all changes that were made during the project. Changes, dates, and the people involved should be recorded for

DRAFTSMAN'S DESIGN LOG

Sheet No. /
of / Sheet

Detailed Description: *Layout and details of new transmission low speed gear and mainshaft combination. Low speed gear to have 10 of splines accurately ground with respect to gear teeth, and mainshaft to have three ground lands for mounting low speed gear on these surfaces.*

Job no. *9344-97*
Job name *Trans. Mainshaft First and Reverse Gear*
Models *1950*
Engineer *Poe*
Job started *3-25-48*
Job finished *4-7-48*
Layout numbers *L-36042*

Job Objective: *To eliminate selective fit on mating parts.*

References: *L-33827*

Progress, Decisions and Authority:
3-30-48 - Messrs. Poe and Poe decided to change from a 22 tooth basic spline with 3 unevenly spaced lands to both a 24 tooth basic spline with 6 evenly spaced lands and a 24 tooth basic spline with 8 evenly spaced lands.
Engineers also requested study of a longer hub for 1st reverse gear to reduce runout.
4-6-48 After preliminary investigation Messrs. Poe and Poe decided to cancel the 8 lands construction and the elongated hub.
4-7-48 Mr. Poe decided to have an additional layout made of a 22 tooth basic spline with alternating teeth and lands.

Calculations and sketches
to be dated and attached.

R. Poe
Signature

FIG. 16.22 A drafter's log should be kept as a record of the project to explain actions taken that might otherwise be forgotten. (Courtesy of General Motors Corporation.)

FIG. 16.23 An assembly drawing is used to explain how the parts of a product such as this Ford tractor are assembled. (Courtesy of Ford Motor Company.)

references as the project progresses, and as a review of the finished project.

An example of a drafter's log is shown in Fig. 16.22. The description of the project and its objectives are given first. Each change and the reason for it are tabulated under "Progress, Decisions and Authority." The people responsible for the changes are mentioned by name.

These notes serve to refresh the memory of anyone who wishes to review the project. Calculations are often made during the process of preparing a drawing. If they are lost or if they are poorly done, it may be necessary to make them again. Consequently, they should be made a permanent part of the log and attached to the log.

16.12
Assembly drawings

When the parts have been made according to the specifications of the working drawings, they will be assembled (Fig. 16.23). This requires a drawing called an *assembly drawing*.

Two general types of assembly drawings are drawn by using (1) pictorial techniques or (2) orthographic projection.

FIG. 16.24 An exploded pictorial assembly. Each part is listed by number in the parts list.

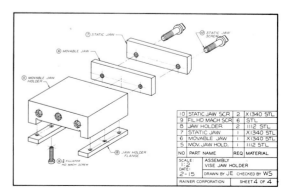

10	STATIC JAW SCR	2	X1340 STL
9	FIL HD MACH SCR	6	STL
8	JAW HOLDER	2	1112 STL
7	STATIC JAW	1	X1340 STL
6	MOVABLE JAW	1	X1340 STL
5	MOV. JAW HOLD.	1	1112 STL
NO	PART NAME	REQ	MATERIAL

SCALE: 1:2
DATE: 2-15
ASSEMBLY VISE JAW HOLDER
DRAWN BY JE CHECKED BY WS
RAINER CORPORATION SHEET 4 OF 4

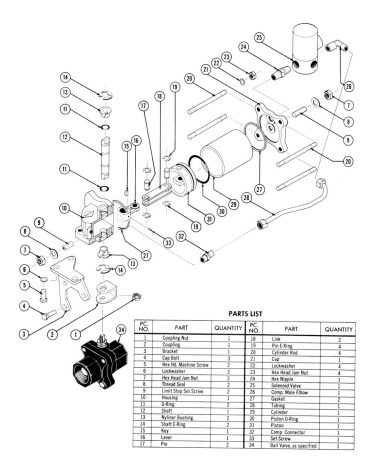

PARTS LIST

PC. NO.	PART	QUANTITY	PC. NO.	PART	QUANTITY
1	Coupling Nut	1	18	Link	2
2	Coupling	1	19	Pin E-Ring	4
3	Bracket	1	20	Cylinder Rod	4
4	Cap Bolt	4	21	Cap	1
5	Hex Hd. Machine Screw	3	22	Lockwasher	4
6	Lockwasher	2	23	Hex Head Jam Nut	4
7	Hex Head Jam Nut	2	24	Hex Nipple	1
8	Thread Seal	2	25	Solenoid Valve	1
9	Limit Stop Set Screw	2	26	Comp. Male Elbow	1
10	Housing	1	27	Gasket	2
11	O-Ring	2	28	Tubing	1
12	Shaft	1	29	Cylinder	1
13	Nyliner Bushing	2	30	Piston O-Ring	1
14	Shaft E-Ring	2	31	Piston	1
15	Key	1	32	Comp. Connector	1
16	Lever	1	33	Set Screw	1
17	Pin	2	34	Ball Valve, as specified	1

FIG. 16.25 An exploded pictorial assembly drawing of the parts of a solenoid control valve. (Courtesy of Jenkins Bros.)

An assembly drawing in pictorial is shown in Fig. 16.24. The parts are numbered and cross-referenced to the parts list, where more information about each part is given. This assembly shows the parts separated, or *exploded* apart, along their center lines of assembly. The pictorial assembly offers the most understandable view of the relationship of the parts. This is more important when the assemblies are complex and difficult to visualize in orthographic views.

Another exploded assembly drawing is shown in Fig. 16.25, where a solenoid control valve is illustrated and its parts list given. Drawings of this type are used in catalogs and maintenance manuals for ease of interpretation.

It is unnecessary to dimension assemblies in most cases, since the parts have been dimensioned individually in the working drawings prior to drawing the assembly.

An example of a dimensioned assembly is shown in Fig. 16.26, where a helicopter frame is drawn as a pictorial. More clarity is provided by this assembly than would be possible in an orthographic assembly.

An orthographic assembly of a special puller is shown in Fig. 16.27. These parts are completely assembled with two end views and a half-section to explain their relationship. A similar orthographic assembly is the subassembly of the gear-cutting fixture in Fig. 16.28. This is a partially exploded assembly.

The outline assembly drawing illustrated in Fig. 16.29 is used to show how various components are connected. Each part is composed of subassemblies that are not shown in detail. The purpose of this type of drawing is to show the overall relationship of the major components with a few key dimensions that would be of importance in its assembly and installation.

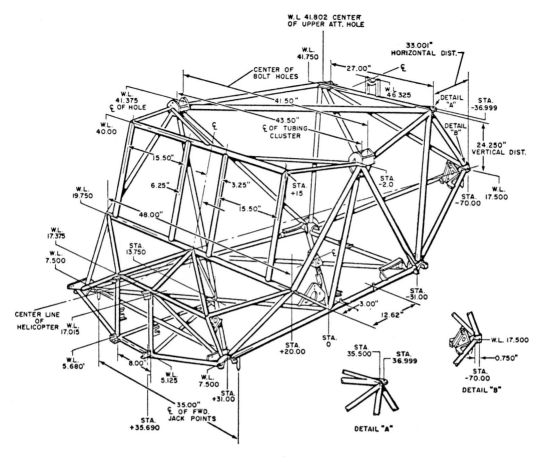

FIG. 16.26 A dimensioned pictorial assembly of a helicopter frame. (Courtesy of Bell Helicopter Company.)

FIG. 16.28 A partially exploded orthographic assembly drawing of an index guide of a cutting fixture.

FIG. 16.27 An assembled orthographic assembly of a special puller. Note that sections can be used to clarify the assembly.

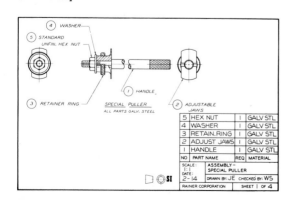

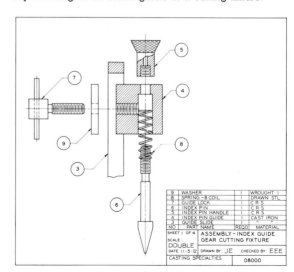

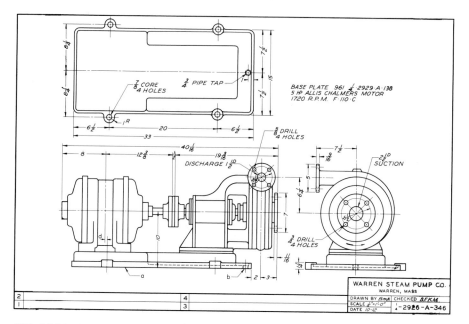

FIG. 16.29 An outline assembly that shows the general relationship of the parts of the assembly and their overall dimensions.

16.13
Tabular dimensions

A single detail drawing can be used effectively to represent similar parts with different dimensions (Fig. 16.30). Each dimension is given a letter that corresponds to a table of dimensions. A particular part—part X, for example—will accommodate a shaft diameter of 62 mm, and the other dimensions are taken from this same horizontal row. You can see that this technique reduces the number of drawings required to illustrate numerous parts.

FIG. 16.30 The same drawings are used in conjuction with the table of values. Parts $V, W, X, Y,$ and Z are dimensioned by using letters on the drawings.

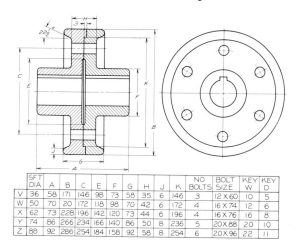

SFT DIA	A	B	C	E	F	G	H	J	K	NO BOLTS	BOLT SIZE	KEY W	KEY D
V	36	58	171	146	98	73	58	35	6	146	3	12 X 60	10	5
W	50	70	20	172	118	98	70	42	6	172	4	16 X 74	12	6
X	62	73	228	196	142	120	73	44	6	196	4	16 X 76	16	8
Y	74	86	266	234	166	140	86	50	8	236	5	20 X 88	20	10
Z	88	92	286	254	184	158	92	58	8	254	6	20 X 96	22	11

16.14
Freehand working drawings

A freehand sketch can serve the same purpose as an instrument drawing, provided the part is simple and the essential dimensions are given (Fig. 16.31). The same principles of working-drawing construction should be followed as when instruments are used.

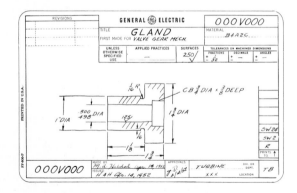

FIG. 16.31 A freehand working drawing with the essential dimensions can be adequate as a detail drawing. (Courtesy of General Electric Company.)

16.15
Castings and forged parts

Two parts are shown in Fig. 16.32 to illustrate the difference between a *forged part* and a machined part. You can see that a forging is a rough form that is made oversize by hammering the metal into shape or by pressing it between two forms (called *dies*) until the general shape is obtained. The forging is then machined to its finished dimensions and tolerances.

FIG. 16.32 The upper part is a "blank" that has been forged. When the forging has been machined, it will appear as shown in the lower photograph.

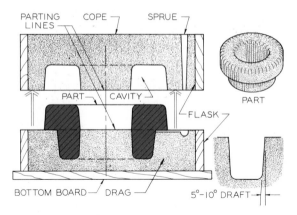

FIG. 16.33 A two-part sand mold is used to produce a casting. The draft of 5° or 10° is necessary to permit withdrawal of the pattern from the sand. The casting must be machined to size it within specified tolerances.

A *casting* is a general shape, much like that of a forging, that must be machined so that it will fit with other parts in an assembly. The casting is formed by pouring molten metal into the void in a mold formed by a pattern that is made slightly larger than the finished part (Fig. 16.33). In order for the part's pattern to be removable from the sand that forms the mold, its sides must be tapered. This taper of 5° to 10° is called the *draft* (Fig. 16.33).

In some industries, casting and forging drawings are made separately from machine drawings. These are more often combined into one drawing, with the understanding that the forgings and castings must be made with additional material to allow for the removal of excess by machining to meet the final design specifications. The body-wheel cylinder drawing in Fig. 16.34 dimensions both the casting and the machining operations in one drawing.

16.16
Sheet metal drawings

Parts that are made of thin metal (sheet metal) are formed by bending or stamping. The flat metal patterns for these parts must be developed graphically.

An example of a sheet metal part is shown in Fig. 16.35. The bend lines are shown on the flat pattern with the angles of bend and the radii of the bends given. The note B.D. means to bend downward, and B.U. means to bend upward.

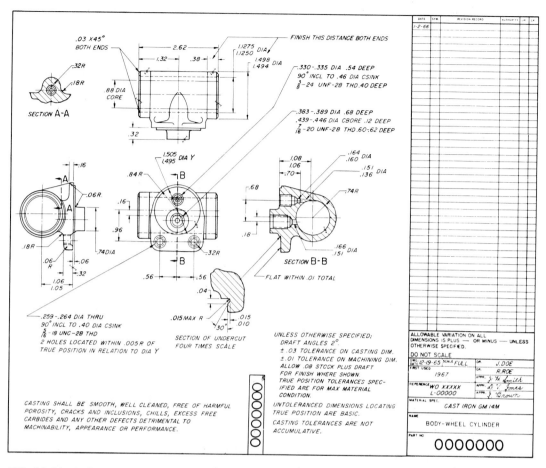

FIG. 16.34 A detail drawing of a casting that shows both the machining operations and the specifications. (Courtesy of General Motors Corporation.)

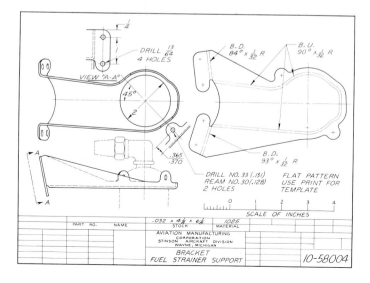

FIG. 16.35 A detail drawing of a sheet metal part. It is shown as a flat pattern and as orthographic views when bent into shape.

Problems

The following problems (Figs. 16.36–16.85) are to be drawn on the sheet sizes assigned or those suggested. Each problem should be drawn with the appropriate dimensions and notes to fully describe the parts and assemblies being drawn.

Working drawings may be made on film or tracing vellum in ink or in pencil. Select a suitable title block and complete it using good lettering practices. Some problems will require more than one sheet to show all the parts properly.

Assemblies should be prepared with a parts list where there are several parts. These may be orthographic or pictorial assemblies, either exploded, assembled, or partially exploded.

The dimensions given in the problems do not always represent good dimensioning practices because of space limitations; but the dimensions given are usually adequate for you to complete the detail drawings. In some cases, there may be omitted dimensions that you must approximate using your own judgment. When making the detail drawings, strive to provide all the necessary information, notes, and dimensions to describe the views completely. Utilize any of the previously covered principles, conventions, and techniques to present the views with the maximum of clarity and simplicity.

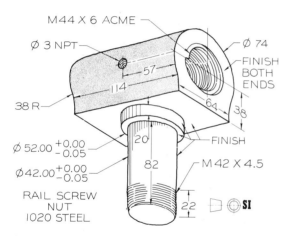

FIG. 16.36 Make a detail drawing on a Size B sheet.

FIG. 16.37 Make a detail drawing on a Size B sheet.

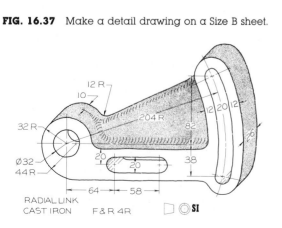

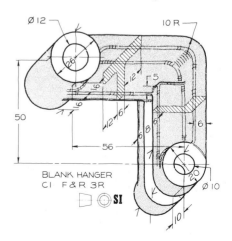

FIG. 16.38 Make a detail drawing on a Size C sheet.

FIG. 16.39 Make a detail drawing on a Size B sheet.

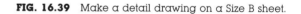

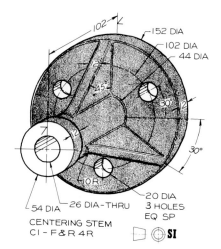

FIG. 16.40 Make a detail drawing on a Size B sheet.

CENTERING STEM
C1 - F & R 4R

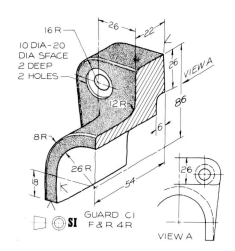

GUARD C1
F & R 4R

VIEW A

FIG. 16.41 Make a detail drawing on a Size B sheet.

FIG. 16.42 Make a detail drawing on a Size B sheet.

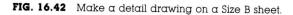
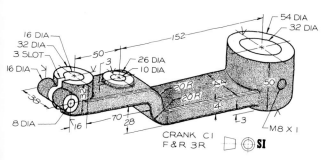

CRANK C1
F & R 3R

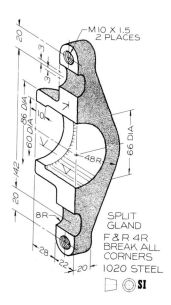

SPLIT
GLAND
F & R 4R
BREAK ALL
CORNERS
1020 STEEL

FIG. 16.43 Make a detail drawing on a Size B sheet.

FIG. 16.44 Make a detail drawing on a Size B sheet.

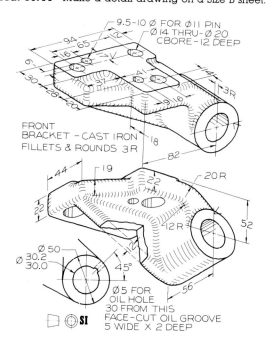

FRONT
BRACKET - CAST IRON
FILLETS & ROUNDS 3R

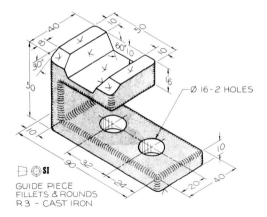

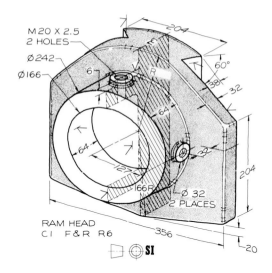

GUIDE PIECE
FILLETS & ROUNDS
R 3 - CAST IRON

Ø 16 - 2 HOLES

FIG. 16.45 Make a detail drawing on a Size B sheet.

M 20 X 2.5
2 HOLES

Ø242

Ø166

Ø 32
2 PLACES

RAM HEAD
CI F & R R6

FIG. 16.48 Make a detail drawing on a Size B sheet.

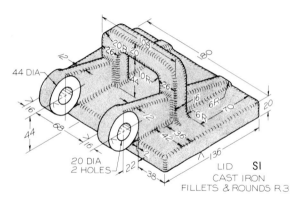

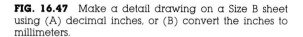

44 DIA

20 DIA
2 HOLES

LID   SI
CAST IRON
FILLETS & ROUNDS R3

FIG. 16.46 Make a detail drawing on a Size B sheet.

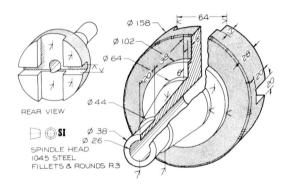

REAR VIEW

SPINDLE HEAD
1045 STEEL
FILLETS & ROUNDS R3

Ø 158
Ø 102
Ø 64
Ø 44
Ø 38
Ø 26

FIG. 16.49 Make a detail drawing on a Size B sheet.

FIG. 16.50 Make a detail drawing on a Size B sheet.

FIG. 16.47 Make a detail drawing on a Size B sheet using (A) decimal inches, or (B) convert the inches to millimeters.

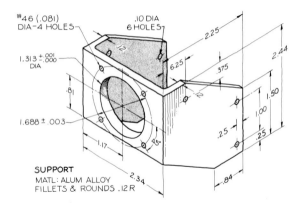

#46 (.081)
DIA-4 HOLES

.10 DIA
6 HOLES

1.313 ±.001
±.000
DIA

1.688 ± .003

SUPPORT
MATL: ALUM ALLOY
FILLETS & ROUNDS .12 R

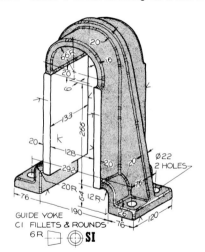

Ø22
2 HOLES

GUIDE YOKE
CI FILLETS & ROUNDS
6R   SI

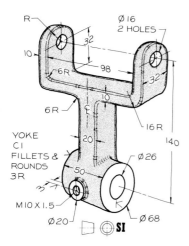

FIG. 16.51 Make a detail drawing on a Size B sheet.

R

Ø 16
2 HOLES

10

32

6R

98

32

6R

10

6R

16 R

20

YOKE
CI
FILLETS &
ROUNDS
3R

Ø 26

140

50

M10 X 1.5

Ø 20

Ø 68

3

SI

Ø 117

45°

92

120°

4.52
4.50

Ø 40

66

ANGULAR TOL ± 0.5°

REAR VIEW

25 R

17 R

Ø 11.04-11.11
3 HOLES EQ SP

39

12

30°

7

5

HUB CRANKSHAFT
PULLEY
CAST IRON GM 232M

2 R

5 R
2 R

22
DIA

Ø 34.92-34.94

11R

3

30°

3R

27.03
26.99

R

6

PARTIAL SECTION

SI

47.31
47.23

37.25-37.40
TO BOTTOM OF SLOT

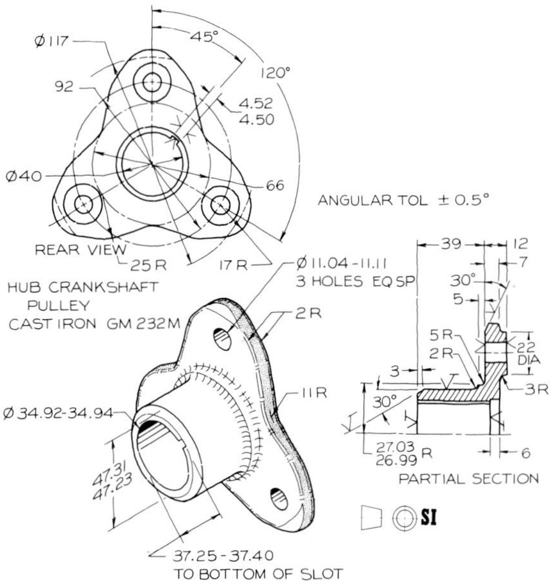

FIG. 16.53 Make a detail drawing on a Size B sheet.

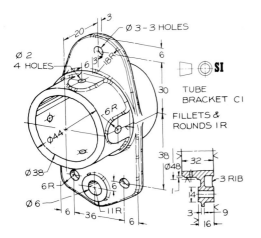

3

20

Ø 3 - 3 HOLES

Ø 2
4 HOLES

6

3

8R

6

SI

30

TUBE
BRACKET CI

6 R

FILLETS &
ROUNDS I R

Ø 44

38

32

Ø 48

6 R

3 RIB

Ø 38

14

6 R

6

Ø 6

3

9

6

36

11R

6

16

FIG. 16.52 Make a detail drawing on a Size B sheet.

FIG. 16.54 Make a detail drawing on a Size B sheet using (A) decimal inches, or (B) convert the inches to millimeters.

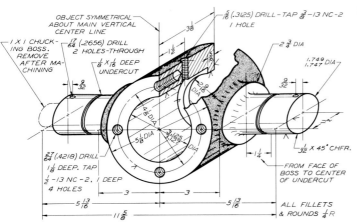

OBJECT SYMMETRICAL
ABOUT MAIN VERTICAL
CENTER LINE

$\frac{5}{16}$ (.3125) DRILL - TAP $\frac{3}{8}$ -13 NC-2
I HOLE

38

I X I CHUCK-
ING BOSS.
REMOVE
AFTER MA-
CHINING

$\frac{17}{64}$ (.2656) DRILL
2 HOLES-THROUGH

$\frac{1}{2}$

$2\frac{3}{4}$ DIA

$\frac{1}{8}$ X $\frac{1}{16}$ DEEP
UNDERCUT

1.749
1.747 DIA

$\frac{9}{32}$

$\frac{9}{32}$

$\frac{27}{64}$ (.4218) DRILL
$1\frac{1}{2}$ DEEP. TAP
$\frac{1}{2}$ -13 NC-2, I DEEP
4 HOLES

$1\frac{1}{4}$

$\frac{1}{32}$ X 45° CHFR.

FROM FACE OF
BOSS TO CENTER
OF UNDERCUT

3

3

ALL FILLETS
& ROUNDS $\frac{1}{4}$ R

$5\frac{13}{16}$

$5\frac{13}{16}$

$11\frac{5}{8}$

CROSSHEAD — COMPRESSOR

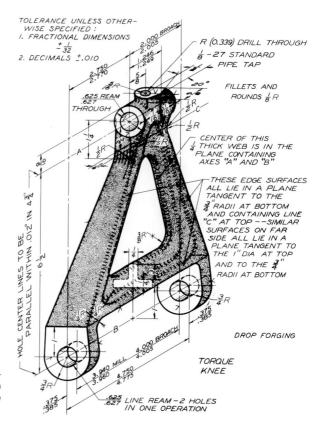

FIG. 16.55 Make a detail drawing on a Size B sheet using (A) decimal inches, or (B) convert the inches to millimeters.

FIG. 16.56 Make a detail drawing on a Size B sheet using (A) decimal inches, or (B) convert the inches to millimeters.

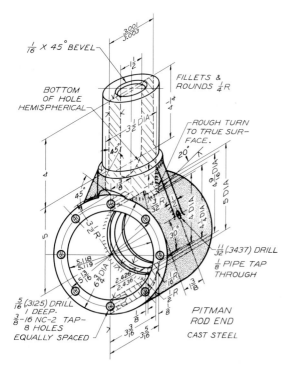

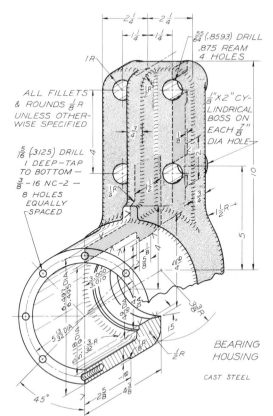

$2\frac{1}{4}$ $2\frac{1}{4}$

$1\frac{1}{4}$ $1\frac{1}{4}$

$\frac{55}{64}$ (.8593) DRILL
.875 REAM
4 HOLES

1R

ALL FILLETS
& ROUNDS $\frac{1}{8}$ R
UNLESS OTHER-
WISE SPECIFIED

$1''$x2"CY-
LINDRICAL
BOSS ON
EACH $\frac{7}{8}''$
DIA HOLE

$\frac{5}{16}$ (.3125) DRILL
I DEEP-TAP
TO BOTTOM —
$\frac{3}{8}$ -16 NC-2 —
8 HOLES
EQUALLY
SPACED

$1\frac{1}{2}$ R

IR

6.497 DIA
6.495

5 $\frac{13}{32}$ DIA

$1\frac{1}{2}$ R

$\frac{3}{8}$ R

15°

$\frac{3}{32}$ R

$\frac{1}{2}$ R

BEARING
HOUSING

CAST STEEL

45°

$2\frac{5}{8}$ $4\frac{3}{8}$

FIG. 16.57 Make a detail drawing on a Size B sheet using (A)
decimal inches, or (B) convert the
inches to millimeters.

FIG. 16.58 Make a detail
drawing on a Size B sheet
using (A) decimal inches,
or (B) convert the inches to
millimeters.

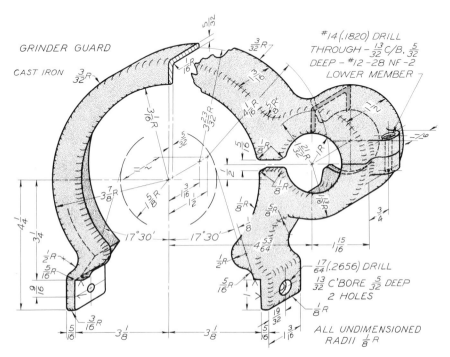

GRINDER GUARD

CAST IRON $\frac{3}{32}$ R

$\frac{5}{32}$

$\frac{3}{32}$ R

$\frac{1}{16}$ R

#14 (.1820) DRILL
THROUGH – $\frac{13}{32}$ C/B, $\frac{5}{32}$
DEEP – #12 -28 NF -2
LOWER MEMBER

$3\frac{1}{16}$ R

$3\frac{23}{32}$ R

$\frac{21}{32}$ R

$\frac{1}{2}$

$\frac{5}{32}$

$\frac{5}{16}$

$3\frac{7}{8}$ R

$\frac{5}{16}$ R

$\frac{3}{16}$ R

$\frac{1}{2}$

$\frac{1}{8}$ R

$\frac{5}{8}$ R

$\frac{3}{4}$

$4\frac{1}{4}$

$3\frac{3}{4}$

$\frac{1}{2}$ R

$\frac{5}{16}$ R

$\frac{9}{16}$

17°30' 17°30'

$4\frac{53}{64}$

$1\frac{15}{16}$

$\frac{1}{2}$ R

$\frac{5}{16}$ R

$\frac{17}{64}$ (.2656) DRILL
$\frac{13}{32}$ C'BORE $\frac{5}{32}$ DEEP
2 HOLES

$\frac{1}{8}$ R

$\frac{3}{16}$ R

$\frac{5}{16}$ $3\frac{1}{8}$ $3\frac{1}{8}$

$\frac{5}{16}$ $1\frac{3}{16}$

$\frac{19}{32}$

ALL UNDIMENSIONED
RADII $\frac{1}{8}$ R

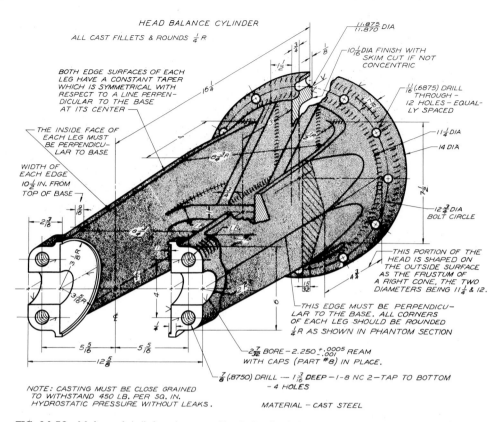

HEAD BALANCE CYLINDER

ALL CAST FILLETS & ROUNDS $\frac{1}{4}R$

11.875 DIA
11.870

$10\frac{1}{16}$ DIA FINISH WITH SKIM CUT IF NOT CONCENTRIC

BOTH EDGE SURFACES OF EACH LEG HAVE A CONSTANT TAPER WHICH IS SYMMETRICAL WITH RESPECT TO A LINE PERPENDICULAR TO THE BASE AT ITS CENTER

$\frac{11}{16}$ (.6875) DRILL THROUGH – 12 HOLES – EQUALLY SPACED

THE INSIDE FACE OF EACH LEG MUST BE PERPENDICULAR TO BASE

$11\frac{1}{4}$ DIA

14 DIA

WIDTH OF EACH EDGE $10\frac{1}{2}$ IN. FROM TOP OF BASE

$12\frac{3}{4}$ DIA BOLT CIRCLE

THIS PORTION OF THE HEAD IS SHAPED ON THE OUTSIDE SURFACE AS THE FRUSTUM OF A RIGHT CONE, THE TWO DIAMETERS BEING $11\frac{1}{4}$ & 12.

THIS EDGE MUST BE PERPENDICULAR TO THE BASE. ALL CORNERS OF EACH LEG SHOULD BE ROUNDED $\frac{1}{4}R$ AS SHOWN IN PHANTOM SECTION

$2\frac{7}{32}$ BORE – 2.250 $^{+.0005}_{-.001}$ REAM WITH CAPS (PART #8) IN PLACE.

$\frac{7}{8}$ (.8750) DRILL — $1\frac{3}{16}$ DEEP – 1-8 NC 2–TAP TO BOTTOM – 4 HOLES

NOTE: CASTING MUST BE CLOSE GRAINED TO WITHSTAND 450 LB. PER SQ. IN. HYDROSTATIC PRESSURE WITHOUT LEAKS.

MATERIAL – CAST STEEL

FIG. 16.59 Make a detail drawing on a Size B sheet using (A) decimal inches, or (B) convert the inches to millimeters.

FIG. 16.60 Make a detail drawing on a Size C sheet using (A) decimal inches, or (B) convert the inches to millimeters.

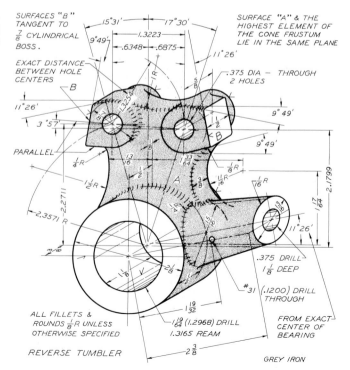

SURFACES "B" TANGENT TO $\frac{7}{8}$ CYLINDRICAL BOSS.

SURFACE "A" & THE HIGHEST ELEMENT OF THE CONE FRUSTUM LIE IN THE SAME PLANE

.375 DIA – THROUGH 2 HOLES

EXACT DISTANCE BETWEEN HOLE CENTERS

.375 DRILL $1\frac{1}{8}$ DEEP

#31 (.1200) DRILL THROUGH

FROM EXACT CENTER OF BEARING

ALL FILLETS & ROUNDS $\frac{1}{8}R$ UNLESS OTHERWISE SPECIFIED

$1\frac{19}{64}$ (1.2968) DRILL 1.3165 REAM

REVERSE TUMBLER

GREY IRON

350

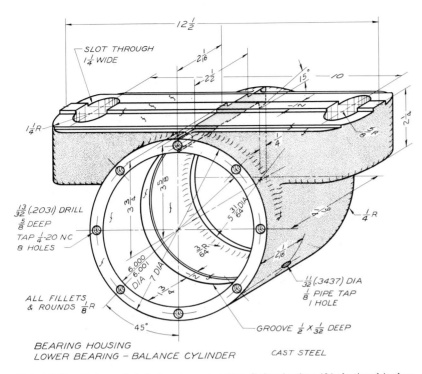

SLOT THROUGH 1¼ WIDE

$12\frac{1}{2}$

$2\frac{1}{16}$

$2\frac{1}{2}$

15°

10

$2\frac{1}{4}$

1¼R

$\frac{5}{8}$R

$\frac{13}{32}$ (.2031) DRILL
$\frac{5}{8}$ DEEP
TAP ¼-20 NC
8 HOLES

3¾
$3\frac{3}{4}$
$3\frac{5}{8}$

$5\frac{31}{64}$ DIA

$\frac{1}{4}$R

ALL FILLETS
& ROUNDS $\frac{1}{8}$R

6.000
6.001
DIA 7 DIA

3¾

3¼

$\frac{3}{64}$R

$2\frac{1}{16}$

$\frac{11}{32}$ (.3437) DIA
$\frac{1}{8}$ PIPE TAP
1 HOLE

45°

GROOVE ½ X $\frac{1}{32}$ DEEP

BEARING HOUSING
LOWER BEARING – BALANCE CYLINDER   CAST STEEL

FIG. 16.61 Make a detail drawing on a Size C sheet using (A) decimal inches, or (B) convert the inches to millimeters.

FIG. 16.62 Make a detail drawing on a Size B sheet using (A) decimal inches or (B) millimeters.

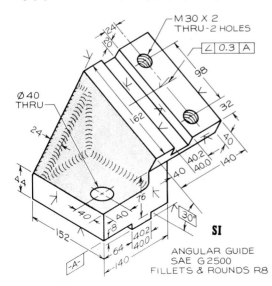

M 30 X 2
THRU-2 HOLES

∠ 0.3 A

98

Ø 40
THRU

32

24

162

140

40.2
40.0

40

24

76

40

40

30°

SI

152

8

64
140

40.2
40.0

-A-

ANGULAR GUIDE
SAE G 2500
FILLETS & ROUNDS R8

FIG. 16.63 Make a detail drawing on a Size B sheet using (A) decimal inches or (B) millimeters.

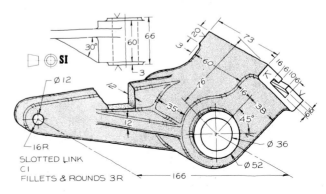

30°

60

66

3

□ ◎ **SI**

20

73

Ø 12

60

76

12

16/16/16

38

45°

35

12

45°

Ø 36

16R

Ø 52

SLOTTED LINK
C1
FILLETS & ROUNDS 3R

166

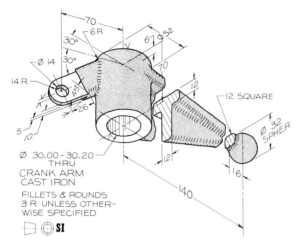

FIG. 16.64 Make a detail drawing on a Size B sheet.

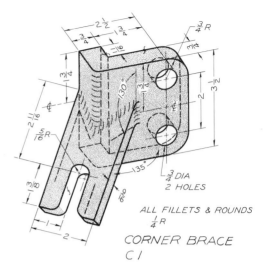

CORNER BRACE
C 1

FIG. 16.66 Make a detail drawing on a Size B sheet. Convert the fractional inches to (A) decimal inches or (B) millimeters.

FIG. 16.65 Make a detail drawing on a Size A sheet.

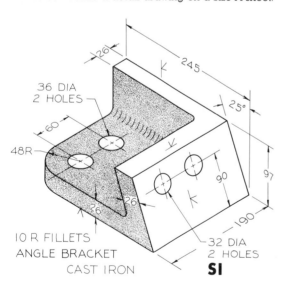

FIG. 16.67 Make a detail drawing on a Size C sheet. Convert the fractional inches to (A) decimal inches or (B) millimeters.

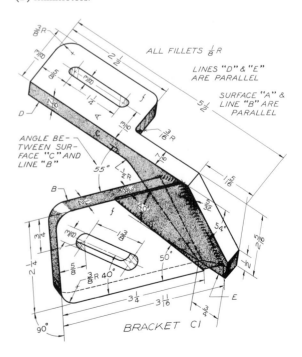

BRACKET C1

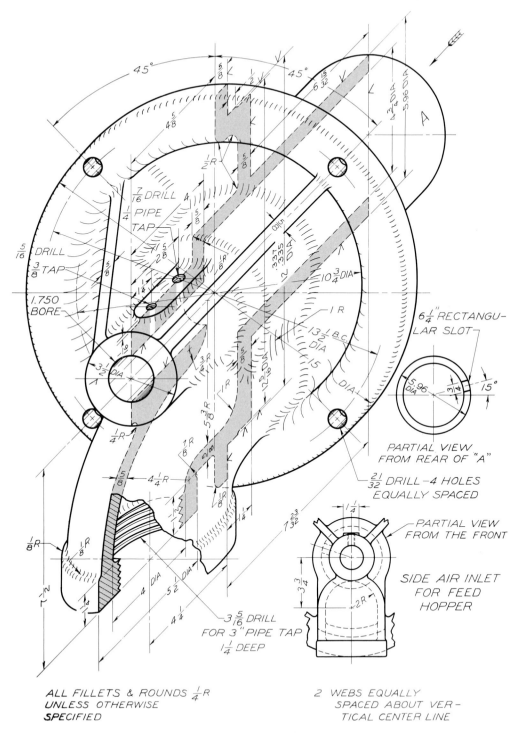

45°

$\frac{5}{8}$ $\frac{1}{2}$ 45° $\frac{15}{32}$ $\frac{3}{4}$ DIA   5.96 DIA   A

$\frac{5}{48}$

$\frac{1}{2}$R $\frac{5}{8}$

$\frac{7}{16}$ DRILL $\frac{1}{4}$ PIPE TAP

$\frac{5}{16}$ DRILL $\frac{3}{8}$ TAP

$\frac{5}{8}$ $\frac{5}{8}$ $1\frac{5}{8}$ $\frac{1}{8}$R $\frac{1}{8}$R   .337 $\frac{.335}{2} DIA$ $10\frac{3}{4}$ DIA

$1\frac{1}{4}$

1.750 BORE

$3\frac{1}{2}$ DIA $\frac{3}{4}$R   1R $\frac{5}{8}$ 1R   13$\frac{1}{2}$ B.C DIA   1R $5\frac{3}{8}$R   .25 DIA   15   DIA

$\frac{1}{4}$R $\frac{1}{8}$R

$\frac{1}{8}$R $\frac{1}{8}$R $\frac{5}{8}$ $4\frac{1}{4}$R $\frac{1}{8}$R $1\frac{1}{4}$

$\frac{1}{8}$R $\frac{1}{8}$R $\frac{1}{4}$

$7\frac{1}{2}$ $1\frac{1}{4}$ 4 DIA $5\frac{1}{4}$ DIA $4\frac{1}{4}$

$6\frac{1}{4}$" RECTANGU-LAR SLOT

5.96 DIA $\frac{3}{4}$ 15°

PARTIAL VIEW FROM REAR OF "A"

$\frac{21}{32}$ DRILL – 4 HOLES EQUALLY SPACED

PARTIAL VIEW FROM THE FRONT

$1\frac{1}{4}$ $1\frac{23}{32}$

$3\frac{3}{4}$ 2R

SIDE AIR INLET FOR FEED HOPPER

$3\frac{5}{16}$ DRILL FOR 3" PIPE TAP $1\frac{1}{4}$ DEEP

ALL FILLETS & ROUNDS $\frac{1}{4}$R UNLESS OTHERWISE SPECIFIED

2 WEBS EQUALLY SPACED ABOUT VER-TICAL CENTER LINE

FIG. 16.68 Make a detail drawing on a Size C sheet. Convert the fractional inches to (A) decimal inches or (B) millimeters.

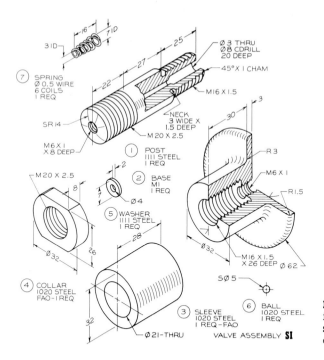

FIG. 16.69 Make a detail drawing of the parts of the valve assembly on Size B sheets. Draw an assembly and provide a parts list.

FIG. 16.70 Make a detail drawing of the parts of the cut-off crank on size B sheets. Convert the fractional dimensions to (A) decimal inches or (B) millimeters. Draw an assembly and provide a parts list.

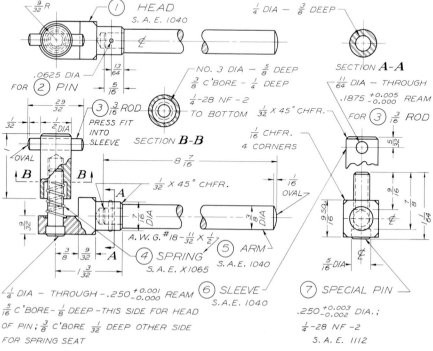

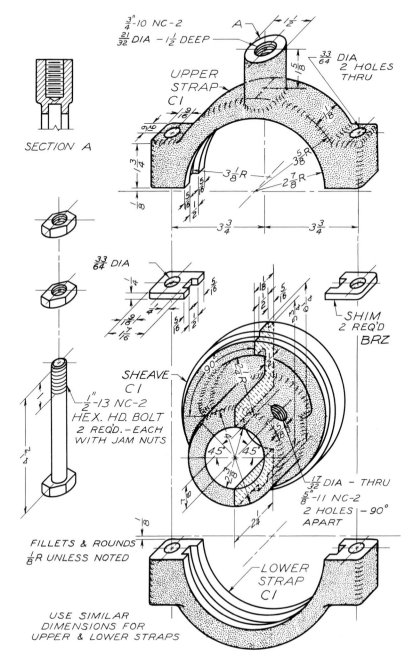

SECTION A

$\frac{3}{4}"$-10 NC-2
$\frac{21}{32}$ DIA – $1\frac{1}{2}$ DEEP

A

$\frac{1}{2}$

$\frac{33}{64}$ DIA
2 HOLES
THRU

UPPER
STRAP
CI

$\frac{5}{8}$

$\frac{9}{16}$

$\frac{9}{16}$

$\frac{3}{4}$

$\frac{1}{8}$

$\frac{1}{8}$

$3\frac{1}{8}R$

$3\frac{5}{8}R$

$2\frac{7}{8}R$

$\frac{5}{16}$ $\frac{5}{16}$

$\frac{1}{2}$

$3\frac{3}{4}$

$3\frac{3}{4}$

$\frac{33}{64}$ DIA

$\frac{1}{4}$

$\frac{5}{16}$

$\frac{9}{16}$

$\frac{1}{4}$

$\frac{7}{16}$

$\frac{5}{16}$ $\frac{1}{2}$

SHIM
2 REQ'D
BRZ

SHEAVE
CI

$\frac{1}{2}"$-13 NC-2
HEX. HD. BOLT
2 REQ'D.—EACH
WITH JAM NUTS

$45°$ $45°$

$\frac{17}{32}$ DIA – THRU
$\frac{5}{8}"$-11 NC-2
2 HOLES – 90°
APART

$\frac{1}{8}$

$2\frac{1}{4}$

FILLETS & ROUNDS
$\frac{1}{8}$R UNLESS NOTED

LOWER
STRAP
CI

USE SIMILAR
DIMENSIONS FOR
UPPER & LOWER STRAPS

FIG. 16.71 Make a detail drawing of the parts of the journal assembly on Size B sheets. Convert the fractional dimensions to (A) decimal inches or (B) millimeters. Draw an assembly and provide a parts list.

FIG. 16.72 Make a detail drawing of the parts of the indicating lever on Size B sheets, and draw an assembly with a parts list.

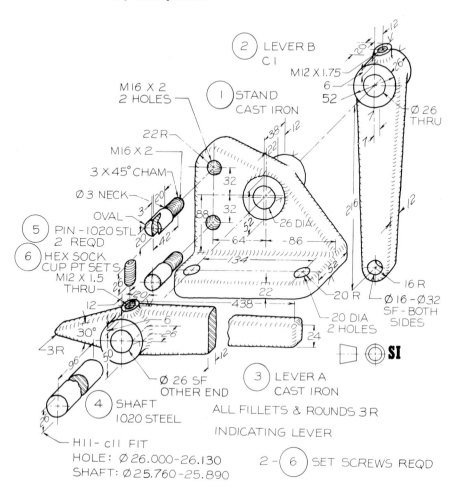

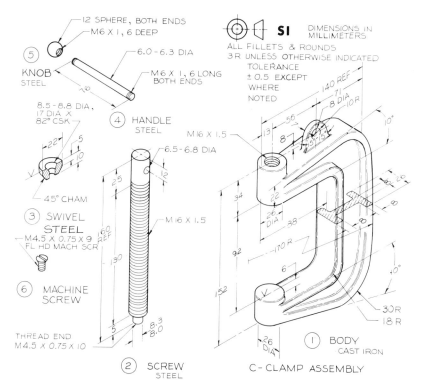

12 SPHERE, BOTH ENDS
M6 X 1, 6 DEEP

(5) KNOB
STEEL

6.0 - 6.3 DIA

M6 X 1, 6 LONG
BOTH ENDS

8.5 - 8.8 DIA,
17 DIA X
82° CSK

(4) HANDLE
STEEL

22   5
10

45° CHAM

(3) SWIVEL
STEEL

M4.5 X 0.75 X 9
FL HD MACH SCR

(6) MACHINE
SCREW

THREAD END
M4.5 X 0.75 X 10

6.5 - 6.8 DIA

25   12

M16 X 1.5

130

5   8.3
8.0

(2) SCREW
STEEL

SI   DIMENSIONS IN
MILLIMETERS
ALL FILLETS & ROUNDS
3 R UNLESS OTHERWISE INDICATED
TOLERANCE
± 0.5 EXCEPT
WHERE
NOTED

M16 X 1.5

140 REF
71
8 DIA   10 R

13   56
8

22

34

26
DIA   38

92   170 R

6

152

30 R
18 R

26
DIA

(1) BODY
CAST IRON

C - CLAMP ASSEMBLY

FIG. 16.73 Make a detail drawing of the parts of the C-clamp assembly on Size B sheets; show an assembly of the parts and provide a parts list.

FIG. 16.74 Make a detail drawing of the parts of the step bearing on Size B sheets, and draw an assembly with a parts list.

FIT SPECIFICATIONS

BEARING & BUSHING
H11/c11 FIT

BUSHING & HANGER
H7/u6 FIT

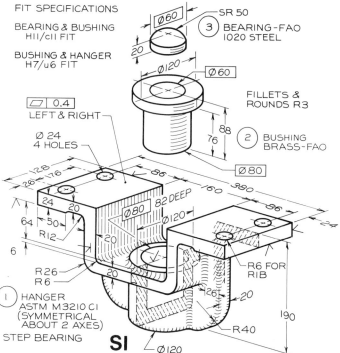

Ø60   SR 50

(3) BEARING-FAO
1020 STEEL

20

Ø120   Ø60

FILLETS &
ROUNDS R3

88

76   (2) BUSHING
BRASS-FAO

Ø80

0.4
LEFT & RIGHT

Ø 24
4 HOLES

128   86
26   176   160   390
24   86

64
50   24
R12
20   Ø80 82 DEEP
Ø120

6

R 26
R 6   20   R6 FOR
RIB

(1) HANGER
ASTM M3210 C1
(SYMMETRICAL
ABOUT 2 AXES)

STEP BEARING   SI

26   20

190

R40

Ø120

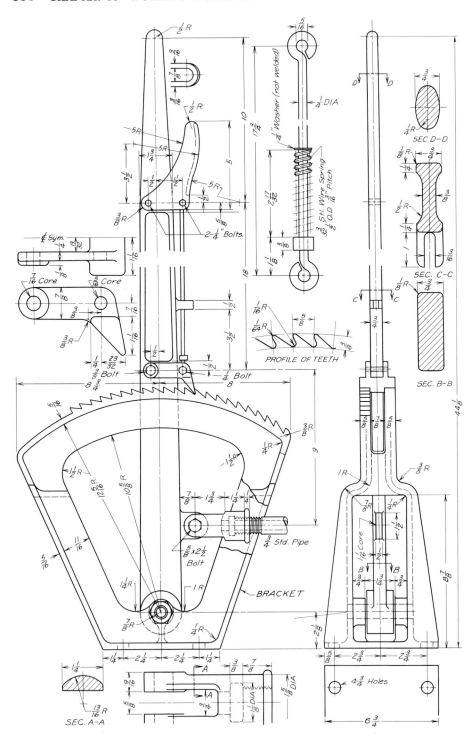

FIG. 16.75 Make detail drawings of the parts of the brake lever on Size B sheets. Convert the fractional inches to (A) decimal inches or (B) millimeters. Draw an assembly of the parts and provide a parts list.

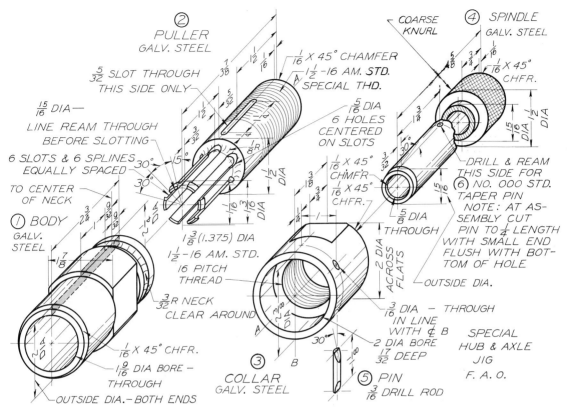

FIG. 16.76 Make detail drawings of the parts of the hub and axle jig on Size B sheets. Convert the fractional inches to (A) decimal inches or (B) millimeters. Draw an assembly of the parts and provide a parts list.

FIG. 16.77 Make a detail drawing of the parts of the special puller on Size B sheets. Convert the fractional inches to (A) decimal inches or (B) millimeters. Draw a pictorial assembly of the parts and give a parts list.

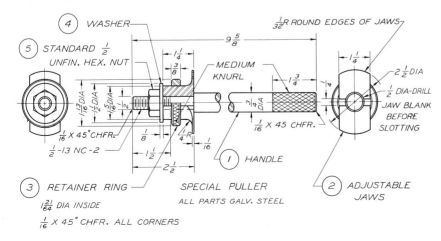

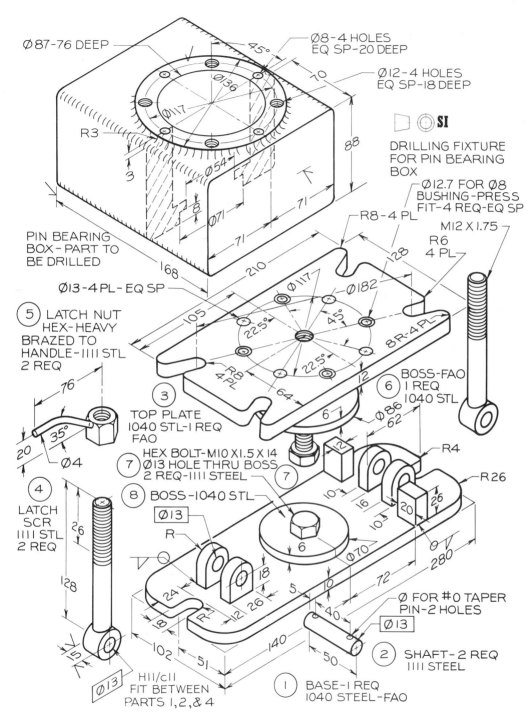

Ø87-76 DEEP

Ø8-4 HOLES
EQ SP-20 DEEP

45°

70

Ø36

Ø117

Ø12-4 HOLES
EQ SP-18 DEEP

88

R3

3

Ø54

8

Ø71

71

71

71

SI

DRILLING FIXTURE
FOR PIN BEARING
BOX

Ø12.7 FOR Ø8
BUSHING-PRESS
FIT-4 REQ-EQ SP

PIN BEARING
BOX-PART TO
BE DRILLED

168

210

R8-4 PL

128

M12 X 1.75

R6
4 PL

Ø13-4 PL-EQ SP

105

Ø117

Ø182

45°

22.5°

8R-4 PL

5 LATCH NUT
HEX-HEAVY
BRAZED TO
HANDLE-1111 STL
2 REQ

22.5°

12

R8
4 PL

64

6

BOSS-FAO
I REQ
1040 STL

76

3 TOP PLATE
1040 STL-1 REQ
FAO

Ø86

62

R4

35°

Ø4

HEX BOLT-M10 X1.5 X 14

7 Ø13 HOLE THRU BOSS
2 REQ-1111 STEEL

7

2

R26

20

R4

4 LATCH
SCR
1111 STL
2 REQ

8 BOSS-1040 STL

Ø13

10

16

20

26

10

R

R

26

18

280

128

24

R

12

26

5

10

Ø70

72

6

Ø FOR #0 TAPER
PIN-2 HOLES

40

Ø13

15

Ø13

102

51

140

50

2 SHAFT- 2 REQ
1111 STEEL

H11/c11
FIT BETWEEN
PARTS 1,2,& 4

1 BASE-I REQ
1040 STEEL-FAO

FIG. 16.78 Make detail drawings of the parts of the drilling jig and crank pin bearing box.
Draw an assembly of the parts and provide a parts list.

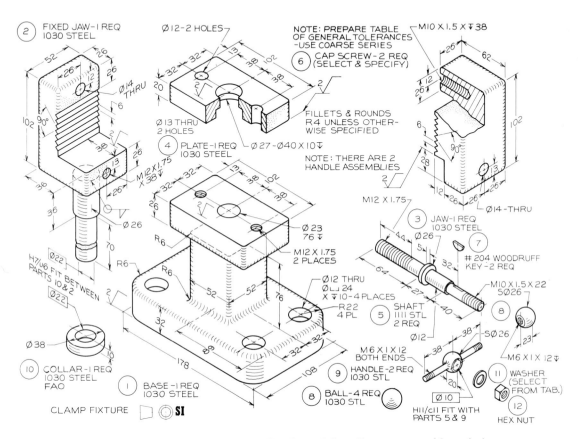

FIG. 16.84 Make detail drawings of the parts of the clamp fixture. Draw an assembly and give a parts list.

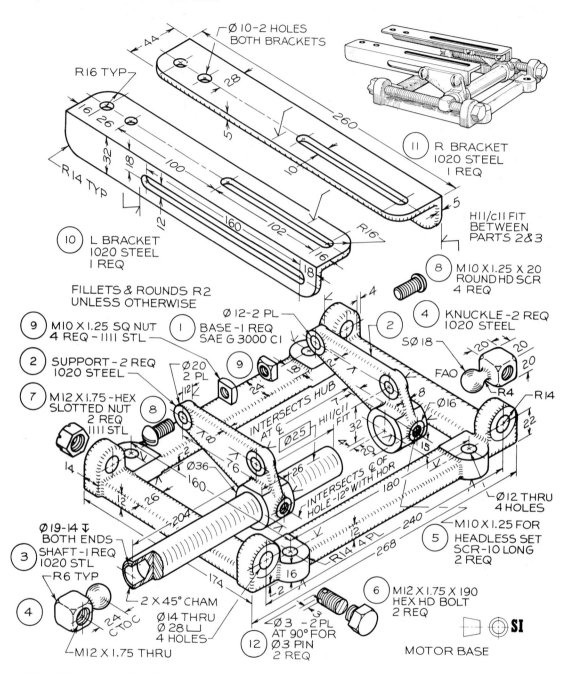

Ø 10-2 HOLES
BOTH BRACKETS

44

R16 TYP

28

260

R14 TYP

16 | 26

32

18

100

5

10

R16

11 R BRACKET
1020 STEEL
1 REQ

5

H11/C11 FIT
BETWEEN
PARTS 2&3

12

160

102

R16

18

6

10 L BRACKET
1020 STEEL
1 REQ

FILLETS & ROUNDS R2
UNLESS OTHERWISE

Ø 12-2 PL

8 M10 X 1.25 X 20
ROUND HD SCR
4 REQ

4 KNUCKLE -2 REQ
1020 STEEL

9 M10 X 1.25 SQ NUT
4 REQ -1111 STL

1 BASE -1 REQ
SAE G 3000 C1

2

2

S Ø 18

FAO

20

20

20

R4

R14

2 SUPPORT - 2 REQ
1020 STEEL

Ø 20
2 PL

9

18

Ø 16

7 M12 X 1.75 -HEX
SLOTTED NUT
2 REQ
1111 STL

12

8

INTERSECTS HUB
AT ℄

8

H11/C11
FIT

Ø 25

32

15

14

Ø 36

Ø 8

2

76

4

26

Ø 12 THRU
4 HOLES

INTERSECTS ℄ OF
HOLE -12° WITH HOR

160

26

12

204

180

240

268

M10 X 1.25 FOR

5 HEADLESS SET
SCR-10 LONG
2 REQ

Ø 19-14
BOTH ENDS

3 SHAFT -1 REQ
1020 STL

R6 TYP

4

24
C TO C

2 X 45° CHAM

Ø 14 THRU
Ø 28
4 HOLES

174

R14 4 PL

12

16

2

6 M12 X 1.75 X 190
HEX HD BOLT
2 REQ

12 Ø 3 -2 PL
AT 90° FOR
Ø 3 PIN
2 REQ

M12 X 1.75 THRU

SI

MOTOR BASE

FIG. 16.85 Make a detail drawing of the parts of the motor base. Draw an assembly and give a parts list.

17

Reproduction Methods and Drawing Shortcuts

17.1
Introduction

So far this text has dealt with the processes of preparing drawings and specifications to communicate ideas in a technical manner. This has progressed through the working-drawing stage, where a detailed drawing is completed on tracing film or paper. Now the drawing must be reproduced, folded, and prepared for filing or transmittal to the users of the drawings. These steps will be discussed in this chapter.

17.2
Reproduction of working drawings

A drawing made by a drafter is of little use in its original form. It would be impractical for the original to be handled by checkers or by workers in the field or in the shop. The drawing would quickly be damaged or soiled, and no copy would be available as a permanent record of the job. Consequently, reproduction of drawings is necessary so that copies can be available for use by the various people concerned. A checker can mark corrections on a work copy without damaging the original drawing. The drafter in turn can make the corrections on the original from the work copy.

Several methods of reproduction are used for making the copies that have traditionally been called "blueprints." This term comes from the original reproduction process that gave a blue background with white lines. The term *blueprint* is still used, although incorrectly, to describe almost all reproduced working drawings regardless of the process.

The processes discussed here are (1) *diazo printing*, (2) *blueprinting*, (3) *microfilming*, (4) *xerography*, and (5) *photostatting*. These are the most often used processes of reproducing engineering drawings.

Diazo printing

The diazo print is more correctly called a "whiteprint" or a "blue-line print" than a blueprint, since it has a white background and blue lines. Other colors of lines are available depending on the type of paper used. The white background makes notes and corrections drawn on the drawing more clearly visible than does the blue background of blueprint.

Both blueprinting and diazo printing require that the original drawing be made on semitransparent tracing paper, cloth, or film that will allow light to pass through the drawing. The paper on which the copy is made, the diazo paper, is chemically treated so that it has a yellow tint on one side. This paper must be stored away from heat and light to prevent spoilage.

The tracing paper or film drawing is placed face up on the yellow side of the diazo paper and is run through the diazo-process machine, which exposes the drawing to a built-in light. The light passes through the tracing paper and burns out the yellow chemical on the diazo paper except where the drawing lines have shielded the paper from the light. After exposure to light, the diazo paper is a duplicate of the original drawing, except that the lines are light yellow. The diazo paper is then passed through the developing unit of the diazo machine, where the yellow lines are developed into permanent blue lines by exposure to ammonia fumes. Diazo printing is a dry process.

A typical diazo printer-developer, which is sometimes called a whiteprinter, is shown in Fig. 17.1. This machine will take sheets up to 42″ wide.

FIG. 17.1 A typical whiteprinter that operates on the diazo process. (Courtesy of Blu-Ray, Inc.)

The speed at which the drawing passes under the light determines the darkness of the copy. A slow speed burns out more of the yellow and produces a clear white background; however, some of the lighter lines of the drawing may be lost. Most diazo copies are made at a somewhat faster speed to give a light tint of blue in the background and stronger lines in the copy. Ink drawings give the best reproductions since the lines are uniform in quality.

It is important to remember that the quality of the diazo print is determined by the quality of the original drawing. A print will not be clear and readable unless the lines of the drawing are dark and dense.

Blueprinting

Blueprints are made with paper that is chemically treated on one side. As in the diazo process, the tracing-paper drawing is placed in contact with the chemically treated side of the paper and exposed to light. The exposed blueprint paper is washed in clear water for a few seconds and is coated with a solution of potassium dichromate. The print is washed again and dried. The wet sheets can be hung on a line to dry or they can be dried by special equipment made for this purpose.

This process is still used, but to a lesser degree than in the past. Since it is a wet process, it requires more time than does the diazo process.

Microfilming

Microfilming is a photographic process that converts large drawings into film copies—either aperture cards or roll film. Drawings must be photographed on either 16-mm or 35-mm film. A camera and copy table are shown in Fig. 17.2.

The roll film or aperture cards can be placed in a microfilm enlarger-printer (Fig. 17.3), where the in-

FIG. 17.2 The Micro-Master 35-mm camera and copy table are used for microfilming engineering drawings. (Courtesy of Keuffel & Esser Company, Morristown, N.J.)

FIG. 17.3 The Bruning 1200 microfilm enlarger-printer makes drawings up to 18″ × 24″ from aperture cards and roll film. (Courtesy of Bruning Company.)

dividual drawings can be viewed on a built-in screen. The selected drawings can then be printed from the film to give standard-size drawings. The range of enlargement varies with the equipment used. Microfilm copies are usually smaller than the original drawings; this saves paper and makes the drawings easier to use.

Microfilming makes it possible to eliminate large, bulky files of drawings, since hundreds of drawings can be stored in miniature size on a small amount of film. The aperture cards shown in Fig. 17.3 are data processing cards that can be catalogued and recalled

FIG. 17.4 This Xerox printer permits the reduction of drawings to less than half size and enlargements to nearly fifty percent larger. (Courtesy of Xerox.)

by a computer to make them accessible with a minimum of effort.

Xerographic reproduction

Xerography is an electrostatic process of duplicating drawings on ordinary, unsensitized paper. This process was developed originally for office duplication uses but has recently been used for the reproduction of engineering drawings.

An advantage of the xerographic process is the possibility of making copies of drawings at a reduced size (Fig. 17.4). The new Xerox 2080 reduces drawings as large as 24″ × 36″ directly from the original to paper sizes ranging from 8″ × 10″ to 14″ × 18″.

Photostatic reproduction

A method of enlarging or reducing drawings by using a camera is called the *photostatic process*. The combination camera and processor shown in Fig. 17.5 is used for photographing drawings and producing high-contrast photographic copies.

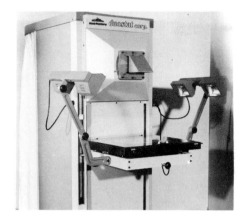

FIG. 17.5 A combination camera-processor for enlarging and reducing drawings to be reproduced as photostats. (Courtesy of the Duostat Corporation.)

The drawing or artwork is placed under the glass of the exposure table (Fig. 17.6), which is lit by built-in lamps. The image can be seen on the glass inside the darkroom, where it is exposed on photographically sensitive paper. The negative paper that has been

FIG. 17.6 The steps in making a photostat with the Duostate camera-processor.

Step 1 The drawing or artwork is placed under the glass on the copyboard.

Step 2 The image is projected inside the darkroom onto photographically sensitive negative paper.

Step 3 The exposed negative paper is placed in contact with the receiver paper, and the two are fed through the developing chemicals.

exposed to the image is placed in contact with receiver paper, and the two are fed through the developing solution to obtain a photostatic copy.

These high-contrast reproductions are often used to prepare artwork that is to be printed by offset printing presses. Additionally, this process can be used to make reproductions on transparent films and for the reproduction of photographs with tones of gray, called halftones.

17.3
Folding the drawing

Once the prints have been finished, the original drawings should be stored in a flat file for future use and updating. The original drawings should not be folded, and handling of any kind should be kept to a minimum.

The printed drawings, on the other hand, are usually folded for transmittal from office to office, often through the mail. The methods of folding Size B, C, D, and E sheets are shown in Fig. 17.7. In each case, the final size after folding is 8½″ × 11″ (or 9″ × 12″). This is the standard modular size that fits most mailing envelopes and file cabinets.

Note that the drawings are folded with the title blocks positioned to be visible at the top and the lower right of the drawing. This is essential in order for a drawing to be retrieved from a file cabinet with ease.

17.4
Overlay drafting techniques

Valuable drafting time can be saved by taking advantage of current processes and materials that utilize a series of overlays to separate parts of a single drawing. For example, engineers and architects often work from a single site plan or floor plan on which a variety of drawings will be made that all utilize the same base plan. The floor plan of a building will be used for the

FIG. 17.7 Standard folds for engineering drawing sheets. The final size in each case is 8½″ × 11″.

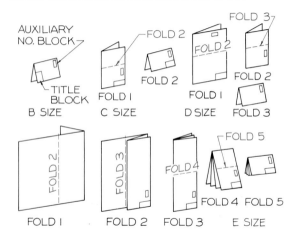

FIG. 17.8 In the pin system, separate overlays are aligned by seven pins mounted on metal strips. (Courtesy of Keuffel & Esser Company, Morristown, N.J.)

electrical plan, furniture arrangement plan, air-conditioning plan, floor materials plan, and so on. You can see that it would be expensive to retrace the plan for each application.

A series of overlays can be used in a system referred to as *pin drafting,* where accurately spaced holes are punched in the polyester drafting film at the top edge of the sheets. These holes are aligned on pins attached to a metal strip that match the holes that were punched in the film (Fig. 17.8). This method ensures accurate alignment or registration of a series of sheets, and the polyester film ensures stability of the material since it does not stretch or sag with changes in humidity.

The steps in using the pin drafting system are shown in Fig. 17.9. The title block can be printed on all drawing film sheets along with the border lines. The overlay sheets need not have borders or a title

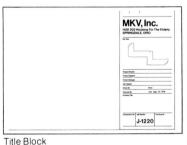

Title Block

Base

FIG. 17.9 Examples of overlays that are overlayed and reproduced to give a combination of several sheets in the final reproduced drawing. (Courtesy of Keuffel & Esser Company, Morristown, N.J.)

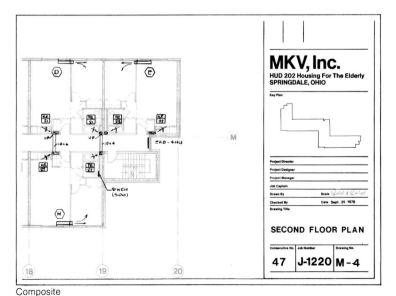

Composite

FIG. 17.10 The process camera that is used to reduce and enlarge engineering drawings is the heart of the pin system. (Courtesy of Keuffel & Esser Company, Morristown, N.J.)

block. The base plan is the sheet that will be common to several drawings.

The composite of the various overlays is shown in the third part of Fig. 17.9. Note that the base plan is printed in gray, which means that it has been screened to the desired percentage of black by a photographic process. This makes the additional information provided by the subcontractor or consultant on the plan more noticeable and easier to read from the base plan.

The set of overlays could be attached by the alignment pins or taped together and run through a diazo machine for full-size prints. Another option of reproduction is the use of a flat-bed process camera (Fig.

17.10) to photograph and reduce the drawings to a standard 8½″ × 11″ size.

You can see that the process of using overlays makes it possible for base plans to be sent to consultants who will overlay them (using the pin registration system) for making their particular additions as overlays. This ensures that the consultants are each working with uniform plans and that they will not accidentally erase or modify any part of the design that was sent to them.

When a large number of prints are needed, reproductions may be printed by offset lithography. This permits them to be printed in multicolors to highlight certain features on drawings.

17.5
Paste-on photos

The engineer who is laying out a manufacturing plant will need to represent many identical machines on a drawing along with other features that are repeated. The architect encounters the same situation when locating furniture or common details such as offices, bathrooms, window details, and the like.

When a number of repetitive drawings are necessary, it is often more economical to use the photographic process to make several reproductions of the drawings on transparent film. These features can be "pasted" into position on the master drawing. (The

FIG. 17.11 When drawings are to be used repetitively on a set of drawings, it may be more economical to photographically reproduce them than to draw them. (Courtesy of Eastman Kodak Company.)

FIG. 17.12 The photographically reproduced drawings are taped in position to complete the overall drawing. (Courtesy of Eastman Kodak Company.)

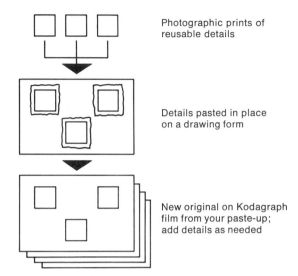

Photographic prints of reusable details

Details pasted in place on a drawing form

New original on Kodagraph film from your paste-up; add details as needed

FIG. 17.13 The steps in photographically revising an engineering drawing. (Courtesy of Eastman Kodak Company.)

term *pasted* is commonly used to describe this attachment, but in reality, tape is used when the drawings are reproduced on transparent film. If the reproductions are made as opaque photostats, then rubber cement can be used.)

The office arrangements shown in Fig. 17.11 are transparencies that have been photographically duplicated from a single drawing. These can be attached to the master drawing to save drawing time. The architect in Fig. 17.12 is composing an entire drawing sheet with paste-on images of repetitive features that were previously drawn.

The steps in preparing final drawings from paste-on photos are schematically shown in Fig. 17.13. Photographic prints of the repetitive details are made and then are pasted into position on the new drawing. In the third step, the drawings are reproduced on film or paper. If reproduced on film, successive prints can be made by the diazo process.

17.6
Photo revisions

If you wish to change the old drawing in Fig. 17.14 to look like the example shown, this can be done by making a clear film reproduction of the parts, cutting them out and taping them in position on a new form.

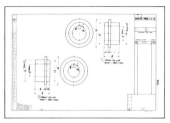

Say you have an existing drawing:

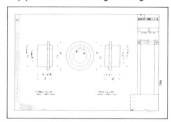

and you want to revise it like this.

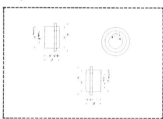

Step 1: First you make a *clear* film reproduction of the original and cut out the elements.

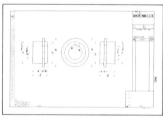

Step 2: Then tape the elements in their new positions on a new form.

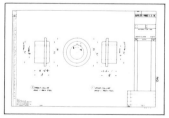

Step 3: Photograph it on film with a matte finish (the tapes and film edges will disappear). Draw in whatever extra detail you want—and you have a new original drawing.

FIG. 17.14 The steps of photographically revising an engineering drawing. (Courtesy of Eastman Kodak Company.)

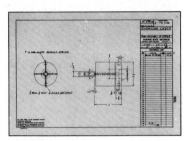

Original drawing

Step 1: Make a reduced-size negative and opaque the area to be revised. (The small size makes it quicker.)

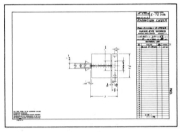

Step 2: Then make a positive on matte film, enlarged to original size.

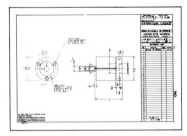

Step 3: Draw in the new detail, and you have a new second original.

FIG. 17.15 The method of modifying a drawing by opaquing the negative. (Courtesy of Eastman Kodak Company.)

The new drawing is then photographed onto a new film on which additional notes and lines can be provided to complete the drawing. This new drawing can be used as a master for making diazo prints.

Opaquing is another method of revising a drawing. A photographic negative is made from the original drawing. The area to be removed is opaqued out by using a brush and an ink like opaquing solution (Fig.

FIG. 17.16 Stick-on lettering can be applied to a drawing by cutting the letters from the plastic sheet, applying them to the drawing in alignment with a guideline, and burnishing the letters to the sheet. (Courtesy of Graphic Products Corporation.)

17.15). You can see that it is advantageous to reduce the negative to a smaller size to lessen the area and effort of opaquing.

The negative is then used to make a positive on polyester drafting film, on which the revision can be drawn and noted in the conventional manner. You now have a new master from which photographic or diazo prints can be made.

17.7
Stick-on materials

A number of companies market stick-on symbols, screens, and lettering that can be applied to drawings to economize on time and improve the appearance of drawings. The three standard types of materials are stick-ons, burnish-ons, and tape-ons.

Stick-on symbols or letters are printed on thin plastic sheets that are cut out with a razor-sharp blade and are transferred to the drawing where the cutout adheres to the drawing. It is burnished permanently to it (Fig. 17.16). The plastic cutout remains as a part of the drawing. This material is available in glossy and matte finishes.

Burnish-on symbols are applied by placing the entire sheet over the drawing and burnishing the desired symbol into place with a rounded-end object, such as the end of a pencil cap. The symbol thus is transferred from the plastic sheet to the drawing surface.

Sheets of symbols can be custom-printed for users who have repetitive needs for trademarks and other often-used symbols (Fig. 17.17). Title blocks are sometimes printed in this manner for application to drawings, to reduce drawing time.

A number of symbols that are used on architectural plans are shown in Fig. 17.18. Plumbing and electrical symbols can be rapidly applied to shorten the drafting time.

FIG. 17.17 Stick-on symbols are available in a wide range of design, and they can be custom-printed to suit the needs of the client. (Courtesy of Graphic Products Corporation.)

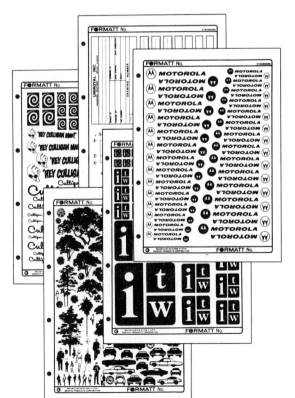

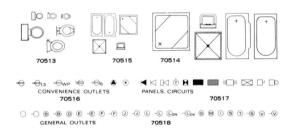

FIG. 17.18 Examples of architectural symbols that can be used on drawings by sticking them to the drawing surface. (Courtesy of Zip-a-Tone Incorporated.)

Colored adhesive tapes in varying widths are available for the preparation of charts and graphs (Fig. 17.19). Tapes are also used to represent wide lines on large drawings. Although tapes can be burnished on tightly, they can be removed for modification when this is desired.

A matte finished sheet is available from several sources that can be typed on with a standard typewriter and then transferred to the drawing where it is attached with its adhesive backing to become a permanent part of the drawing.

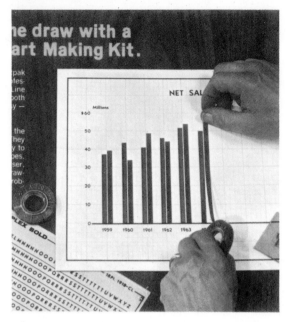

FIG. 17.19 Colored tapes in various widths can be used to speed the process of preparing charts and graphs. (Courtesy of Chartpak Rotex.)

FIG. 17.20 This specially made typewriter can be used by a secretary to type notes on a drawing. (Courtesy of Vari-Typer.)

FIG. 17.21 This assembly is an example of a photodrawing that has saved a considerable amount of drafting time. (Courtesy of LogE-tronics, Inc., and Eastman Kodak Company.)

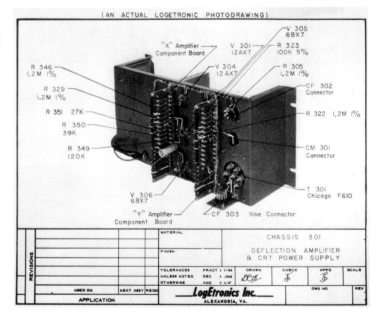

Step 1: Make a halftone print of the photograph.

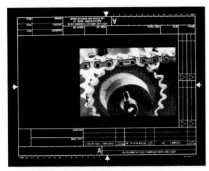

Step 2: Tape the halftone print to a drawing form, and photograph it to produce a negative.

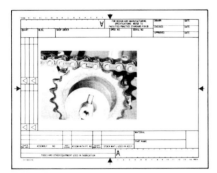

Step 3: Make a positive reproduction on matte film.

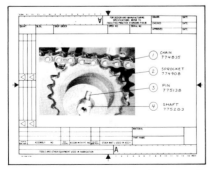

Step 4: Now draw in your callouts—and the job is done.

FIG. 17.22 The steps in making a photodrawing. (Courtesy of Eastman Kodak Company.)

A specially designed typewriter can be used that allows a secretary to assist the drafter by typing notes directly on the surface of the drawing to shorten the time that would be required to hand-letter the notes (Fig. 17.20).

17.8
Photodrawing

Photography can be used as an effective means of economizing when preparing certain types of drawings. This is especially true when existing designs are being modified and where the existing parts can be photographed.

An example of a photodrawing is shown in Fig. 17.21, where an assembly of parts has been noted. In some cases, it would be worthwhile to even build a model for photographing in order to clarify assembly details. This is especially true in the piping industry, where complex refineries are built as models and are then photographed, noted, and reproduced as photodrawings.

The steps of preparing a photodrawing are shown in Fig. 17.22, where it is desired to specify certain parts of a sprocket-and-chain assembly. In Step 1, a halftone print of the photograph is made. (This is the process of screening the photograph, or representing it by a series of dots that give varying tones of gray.) In Step 2, the halftone is taped to a white drawing sheet, which is then photographed to give a negative. The negative is used to produce a positive on polyester drafting film. The notes can be lettered on this drawing film to complete the master drawing (Step 4). The master drawing can then be used to make diazo prints, or it can be microfilmed or reproduced photographically to the desired size.

18

Pictorials

18.1
Introduction

A pictorial is an effective means of communicating an idea. This is especially true if a design is unique or if the person to whom it is being explained has difficulty with the interpretation of multiview drawings.

Pictorials are also helpful in developing a design, since they enable the designer or drafter to work in three dimensions.

Pictorials are sometimes called *technical illustrations*. They are widely used in catalogs, parts manuals, and maintenance publications to describe various products. Practically everyone has used technical illustrations and instructions to assemble a product that was purchased unassembled. Pictorials are used in in-

dustry for the same purpose of putting parts together properly.

This chapter will cover the basic types of pictorial methods: (1) *oblique drawing*, (2) *isometric drawing*, (3) *axonometric projection*, and (4) *perspective projection*. Techniques of rendering and commercial materials will be covered as an introduction to technical illustration.

18.2
Types of pictorials

The three commonly used forms of pictorials are (1) obliques, (2) axonometrics/isometrics, and (3) perspectives. Examples of these are shown in Fig. 18.1.

OBLIQUE PICTORIALS are three-dimensional pictorials made on a plane of paper by projecting from the object with parallel projectors that are *oblique* to the picture plane (Fig. 18.2B).

AXONOMETRIC (ISOMETRIC) projection is a three-dimensional pictorial on a plane of paper drawn by projecting from the object to the picture plane, as illustrated in Fig. 18.2A. The parallel projectors are perpendicular to the picture plane.

FIG. 18.1 The three standard pictorial systems: oblique, isometric, and perspective.

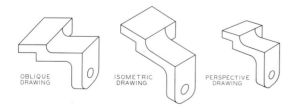

OBLIQUE DRAWING   ISOMETRIC DRAWING   PERSPECTIVE DRAWING

PERSPECTIVE PICTORIALS are drawn with projectors that converge at the viewer's eye and make varying angles with the picture plane (Fig. 18.2C).

18.3
Oblique projections

The system of *oblique projection* is used as the basis of *oblique drawings;* however, oblique projections are seldom used. In Fig. 18.3a, a number of lines of sight are drawn through point 2 of line 1–2. Each line of sight makes a 45° angle with the picture plane, which creates a cone with its apex at 2, and each element on the cone makes a 45° angle with the plane. A variety of projections of line 1–2' on the picture plane can be seen in the front view (b). Each of these projections of 1–2' is equal in length to the true length of 1–2.

The oblique system developed in Fig. 18.3 results in a *cavalier oblique* projection because the projectors make 45° angles with the picture plane, and measurements along the receding axes can be drawn true length in any direction. The term cavalier was used in medieval times to describe a feature of a fortification. Since oblique drawings were used to design fortifications, the term has been used since to describe this type of oblique pictorial.

FIG. 18.2 Types of projection systems for pictorials. (A) Axonometric pictorials are formed by parallel projectors that are perpendicular to the picture plane. (B) Obliques are formed by parallel projectors that are oblique to the picture plane. (C) Perspectives are formed by converging projectors that make varying angles with the picture plane.

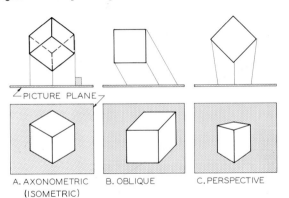

A. AXONOMETRIC (ISOMETRIC)   B. OBLIQUE   C. PERSPECTIVE

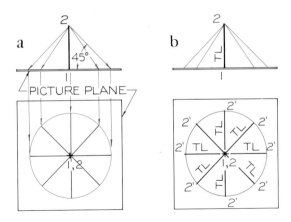

FIG. 18.3 The underlying principle of the cavalier oblique can be seen here where a series of projectors form a cone. Each element makes a 45° angle with the picture plane. Consequently, the projected lengths of 1–2' are true length and are equal in length to line 1–2 that is perpendicular to the picture plane.

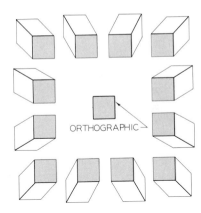

FIG. 18.4 A cavalier oblique is drawn with one surface as a true-size orthographic view. The dimensions along the receding axis are true length, and the axes are drawn at any angle.

Examples of cavalier obliques are shown in Fig. 18.4. The front surface is usually positioned parallel to the picture plane; therefore it will appear true size and as an orthographic view. The measurements along the receding axes are made true length and at any angle.

The top and side views are given as orthographic views, and the front view is drawn as a projection by using the two views of a selected line of sight (Fig. 18.5). Projectors are drawn from the object parallel to

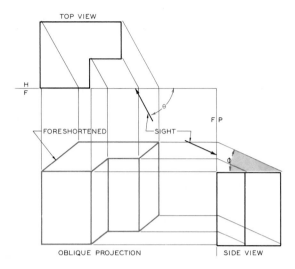

FIG. 18.5 An oblique projection can be drawn at any angle of sight to obtain an oblique pictorial. However, the line of sight should not make an angle less than 45° with the picture plane. This would result in a receding axis longer than true length, thereby distorting the pictorial.

the line of sight to locate their respective points in the oblique view. Surfaces that are parallel to the picture plane will appear true size and shape.

The dimensions along the receding axes are less than true length and greater than half-length; therefore this is called a *general oblique*.

If the angle between the line of sight and the picture plane is less than 45°, the measurements along the receding axes will be greater than true length. This is objectionable since the oblique will be distorted.

18.4
Oblique drawings

Oblique projections are seldom used in the manner just illustrated.

> Instead, three basic types of *oblique drawings* are used that are based on these principles. The three types are (1) *cavalier*, (2) *cabinet*, and (3) *general* (Fig. 18.6).

In each case, the angle of the receding axis can be at any angle between 0° and 90° (Fig. 18.6). Measurements along the receding axes of the *cavalier oblique* are true length (full scale). The *cabinet oblique* has measurements along the receding axes reduced to half-length. The *general oblique* has measurements along the receding axes reduced to between half and full length.

Three examples of cavalier obliques of a cube are shown in Fig. 18.7. Each has a different angle for the receding axes, but the measurements along the receding axes are true length.

A comparison of cavalier and cabinet obliques is given in Fig. 18.8. The cabinet oblique reduces the distortion of an object with a long depth, thereby giving a more pleasing appearance.

FIG. 18.6 Types of obliques

A. The *cavalier oblique* can be drawn with a receding axis at any angle, but measurements along this axis are true length.

B. The *cabinet oblique* can be drawn with a receding axis at any angle, but the measurements along this axis are half-size.

C. The *general oblique* can be drawn with a receding axis at any angle, but the measurements along this axis can vary from half to full size.

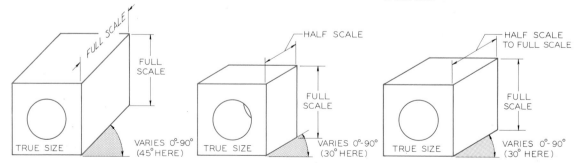

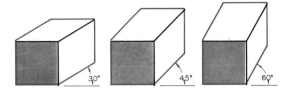

FIG. 18.7 The cavalier oblique is usually drawn with the receding axis at the standard angles of the drafting triangles. Each gives a different view of a box.

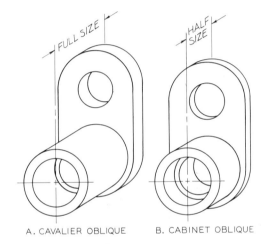

A. CAVALIER OBLIQUE   B. CABINET OBLIQUE

FIG. 18.8 Measurements along the receding axis of a cavalier oblique are full size, and those in a cabinet oblique are half-size.

18.5
Construction of an oblique

An oblique should be drawn by constructing a box using the overall dimensions of height, width, and depth with light construction lines. In Fig. 18.9, the front view is drawn true size in Step 1. In Step 2, the receding axis is drawn at 30° and the depth dimensions are measured true length to give a cavalier oblique. True measurements can be made parallel to the three axes.

The notches are removed from the blocked-in construction box to complete the oblique. These measurements are transferred from the given orthographic views with your dividers.

18.6
Angles in oblique

Angular measurements can be made on the true-size plane of an oblique that is parallel to the picture plane. However, angular measurements will not be true size on the other two planes of the oblique.

To construct an angle in an oblique, coordinates must be used as shown in Fig. 18.10. The sloping surface of 30° must be found by locating the vertex of the angle, H distance from the bottom. The inclination is found by measuring the distance of D along the receding axis to establish the slope. This angle is not equal to the 30° angle that was given in the orthographic view.

FIG. 18.9 Oblique construction

Step 1 The front surface of the oblique is drawn as a true-size plane. The corners are removed.

Step 2 The receding axis is selected, and the true dimensions are measured along this axis.

Step 3 The finished cavalier oblique is strengthened to complete the drawing.

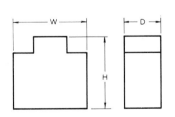

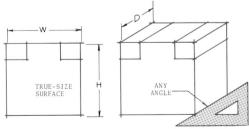

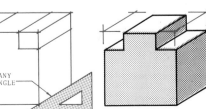

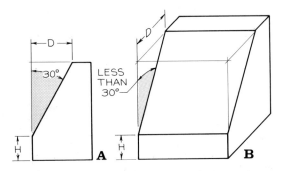

FIG. 18.10 Angles in oblique must be located by using coordinates. They cannot be measured true size except on a true-size plane.

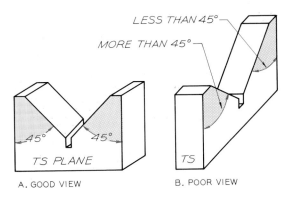

FIG. 18.11 The best view is the view that takes advantage of the ease of construction offered by oblique drawings. The view in (B) is less descriptive and more difficult to construct than the view in (A).

You can see in Fig. 18.11 that a true angle can be measured on a true-size surface. At B, angles along the receding planes are either smaller or larger than their true angles.

It requires less effort and gives a better appearance when obliques are drawn where angles will appear true size as shown in Fig. 18.11A rather than as shown at B.

18.7
Cylinders in oblique

The major advantage of obliques is that circular features can be drawn as true circles when they are parallel to the picture plane. This is illustrated in Fig. 18.12, where an oblique of a cylinder is drawn.

The center lines of the circular end at *A* are drawn, and the receding axis is drawn at any desired angle. The end at *B* is located by measuring along the axis. Circles are drawn at each end using centers *A* and *B*. The circles are connected with tangent lines parallel to the axis. Hidden lines are omitted, and the lines are darkened to complete the oblique.

These same principles are used to construct an object with semicircular features (Fig. 18.13). The oblique is positioned to take advantage of the option of drawing circular features as true circles. Centers C_2 and C_3 are located for drawing two semicircles (Step 3).

FIG. 18.12 A cylinder in oblique

Step 1 The center lines of the circular ends are located at *A* and *B*. The axis is drawn at any angle; since it is true length, this is a cavalier oblique.

Step 2 Circles are drawn by using centers *A* and *B*. These are true circles, since they lie in the orthographic plane of the oblique.

Step 3 Lines are drawn parallel to the axis and tangent to the two end circles. Hidden lines are omitted. Center lines may also be omitted if they are not used for dimensioning purposes.

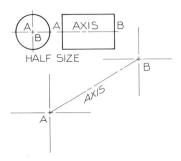
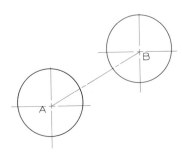
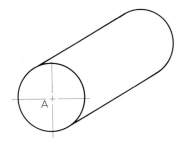

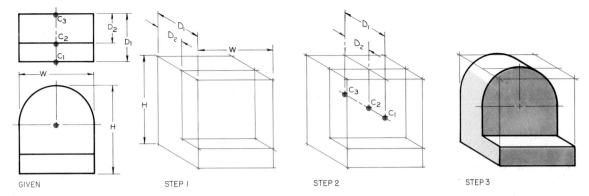

GIVEN | STEP 1 | STEP 2 | STEP 3

FIG. 18.13 Construction of an oblique

Step 1 The overall dimensions are used to block in the oblique pictorial. The notch is removed.

Step 2 The three centers, C_1 C_2, and C_3, are located on each of the planes.

Step 3 The three centers found in Step 2 are used to draw the semicircular features of the oblique. Lines are strengthened.

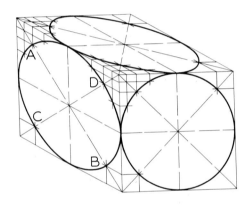

FIG. 18.14 Circular features on the faces of a cavalier oblique of a cube appear as two ellipses and one true circle.

FIG. 18.15 Four-center ellipse in oblique

18.8
Circles in oblique

Although circular features will be true size on a true-size plane of a cavalier oblique pictorial, circular features on the other two planes will appear as ellipses (Fig. 18.14). The ellipses can be found by using coordinates to locate a series of points along the elliptical curves.

A technique of drawing elliptical views in oblique that is used more often is the four-center ellipse technique, which is shown in Fig. 18.15. A rhombus is drawn that would be tangent to the circle at four points. Perpendicular construction lines are drawn

Step 1 The circle that is to be drawn in oblique is blocked in with a square that is tangent to the circle at four points. This square will appear as a rhombus on the oblique plane.

Step 2 Construction lines are drawn perpendicularly from the points of tangency to locate the centers for arcs for drawing two of the four segments of the ellipse.

Step 3 The centers for the two remaining arcs are located.

Step 4 When the four arcs have been drawn, the final result is an approximate ellipse.

GIVEN

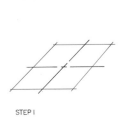

STEP 1

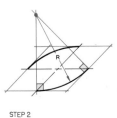

STEP 2

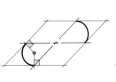

STEP 3

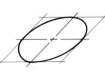

STEP 4

383

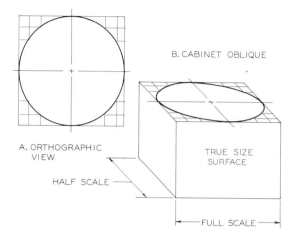

A. ORTHOGRAPHIC VIEW

HALF SCALE

B. CABINET OBLIQUE

TRUE SIZE SURFACE

FULL SCALE

FIG. 18.16 The four-center ellipse technique cannot be used to locate circular shapes on the foreshortened surface of a cabinet oblique. These ellipses must be plotted with coordinates.

from where the center lines cross the sides of the rhombus. This construction is performed as shown in Steps 2 and 3 to locate four centers that are used to draw four arcs that join together to form the ellipse in oblique.

The four-center ellipse method will not work for the cabinet or the general oblique. In these cases, coordinates must be used to locate a series of points on the curve. Coordinates in Fig. 18.16 are shown in orthographic (part A) and then in the cabinet oblique (B). The coordinates that are parallel to the receding axis are reduced to half-size in the oblique. Those in a horizontal direction are drawn full size. The ellipse can be drawn with an irregular curve or with an ellipse template that approximates the plotted points.

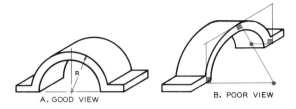

A. GOOD VIEW

B. POOR VIEW

FIG. 18.17 An oblique should be positioned to enable circular features to be drawn with the greatest of ease.

Whenever possible, oblique drawings of objects with circular features should be positioned with circles on true-size planes so they can be drawn as true circles.

In Fig. 18.17, the view at A is better than the one at B since it gives a more descriptive view of the part and is easier to draw. The view at B requires much more construction since the ellipses are drawn by using the four-center ellipse method.

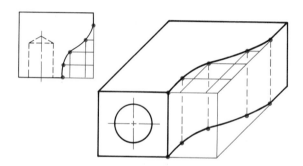

FIG. 18.18 Coordinates are used to establish irregular curves in oblique. The lower curve is found by projecting the points downward a distance equal to the height of the oblique.

FIG. 18.19 The construction of a circular feature on an inclined surface must be found by plotting points using three coordinates, a, b, and c, to locate points 1 and 2 in this example. The plotted points are connected to complete the elliptical feature.

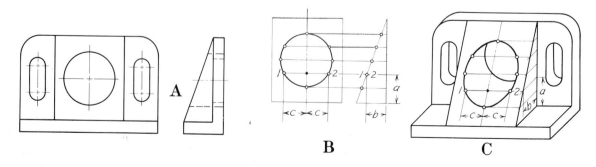

A

B

C

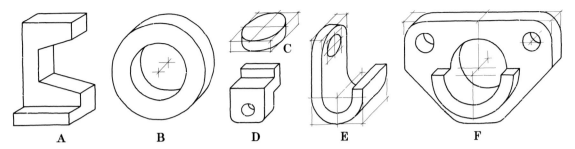

FIG. 18.20 Obliques may be drawn as freehand sketches by utilizing the same principles that were used for instrument pictorials. It is beneficial to use light construction lines to locate the more complex features.

18.9
Curves in oblique

Irregular curves in oblique must be plotted point by point using coordinates (Fig. 18.18). The coordinates are transferred from the orthographic view to the oblique view, and the curve is drawn through these points with an irregular curve.

If the object has a uniform thickness, the lower curve can be found by projecting vertically downward from the upper points a distance equal to the height of the object.

In Fig. 18.19, the elliptical feature on the inclined surface was found by using a series of coordinates to locate the points along its curve. These points are then connected with an irregular curve or by using an ellipse template of approximately the same size.

18.10
Oblique sketching

The ability to make rapid freehand sketches is a valuable asset to the technical person. An understanding of the mechanical principles of oblique construction is essential for sketching obliques.

As shown in Fig. 18.20, the use of light guidelines is helpful in developing a sketch. These guidelines should be drawn lightly so they will not need to be erased when the finished lines of the sketch are darkened.

When sketching on tracing vellum, it is helpful if a printed grid is placed under the vellum to provide guidelines. Grids are available commercially that have a combination of rectangular and angular grids that can be used for drawing oblique sketches.

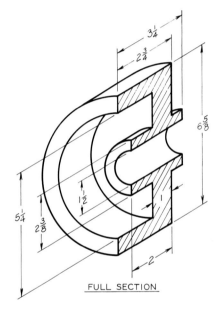

FULL SECTION

FIG. 18.21 Oblique pictorials can be drawn as sections and dimensioned to serve as working drawings.

18.11
Dimensioned obliques

A dimensioned oblique full section is given in Fig. 18.21 where the interior features and the dimensions of the part are shown. This drawing is sufficient for completely describing the part.

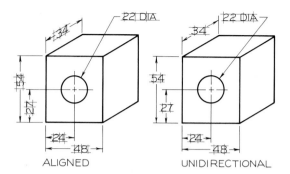

FIG. 18.22 Oblique pictorials can be dimensioned by using either of these methods of applying numerals to the dimension lines.

When dimensions and notes are given in oblique, the numerals and lettering should be applied by using one of the methods shown in Fig. 18.22. In the aligned method, the numerals are aligned with the dimensioned lines. In the unidirectional method, the numerals are all positioned in a single direction regardless of the direction of the dimension lines. Notes that are connected with leaders are positioned horizontally in both methods.

FIG. 18.23 A true isometric projection is found by constructing a view that shows the diagonal of a cube as a point. An isometric drawing is not a true projection since the dimensions are drawn true size rather than reduced in size as in projection.

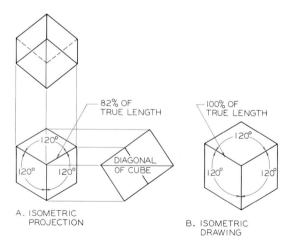

18.12
Isometric pictorials

An *isometric projection* is a type of axonometric projection in which parallel projectors are perpendicular to the picture plane and the diagonal of a cube is seen as a point (Fig. 18.23). The three axes are spaced 120° apart, and the sides are foreshortened to 82% of their true length.

> The term *isometric*, which means "equal measurement," is used to describe this type of pictorial since the planes are equally foreshortened.

An *isometric drawing* is similar to an *isometric projection* except that it is not a true axonometric projection, but an approximate method of drawing a pictorial. Instead of reducing the measurement along the axes to 82%, they are drawn true length (Fig. 18.23B).

A comparison between an isometric projection and an isometric drawing is shown in Fig. 18.24. By using the isometric drawing instead of the isometric projection, pictorials can be measured by using standard scales, the only difference being the 18% increase in size. This is the most commonly used method of drawing isometrics.

The axes of isometric drawings are separated by 120° (Fig. 18.25). Although one of the axes is usually drawn vertically, this is not necessary.

FIG. 18.24 The isometric projection is foreshortened to 82% of full size. The isometric drawing is drawn full size for convenience.

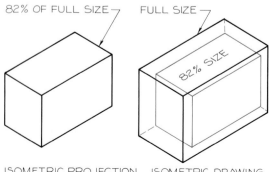

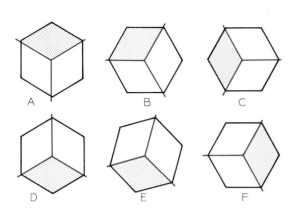

FIG. 18.25 Isometric axes are spaced 120° apart, but they can be revolved into any position.

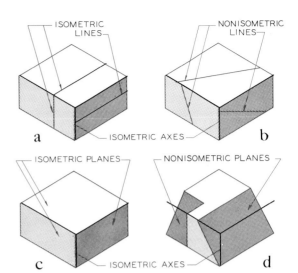

FIG. 18.26 (a) True measurements can be made along isometric lines (parallel to the three axes). (b) Nonisometric lines cannot be measured true length. (c) The three isometric planes are indicated; there are no nonisometric planes in this drawing. (d) Nonisometric planes are planes that are inclined to any of the three isometric planes of a cube.

Construction of an isometric drawing

An isometric drawing is begun by drawing three axes 120° apart. Lines that are parallel to these axes are called *isometric lines* (Fig. 18.26a). True measurements can be made along isometric lines. Measurements cannot be made along nonisometric lines (part b).

The three surfaces of a cube in isometric are called *isometric planes* (Fig. 18.26c). Planes that are parallel to

these planes are isometric planes, and planes that are not parallel to them are called nonisometric planes.

To draw an isometric, you will need a scale and a 30°–60° triangle (Fig. 18.27). Begin by constructing a plane of the isometric using the dimensions of height

FIG. 18.27 Construction of an isometric drawing

Step 1 A 30°–60° triangle is used in combination with a horizontal straightedge to begin the isometric. The dimensions of W, D, and H from the orthographic views were doubled in this case.

Step 2 The depth dimension, D, in the left side view is laid off in isometric. Parallel lines appear parallel in isometric drawings, as well as in orthographic views.

Step 3 The final sides of the isometric are drawn and the lines are strengthened. The 30°–60° triangle automatically separates the axes by 120°.

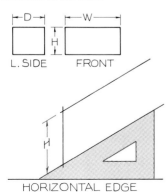

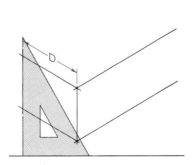

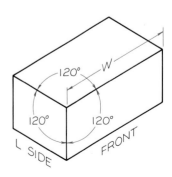

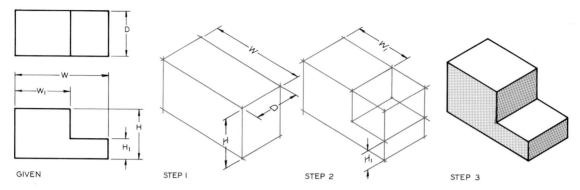

FIG. 18.28   Layout of an isometric drawing (simple)

Step 1   The object is blocked in using the overall dimensions. The notch is removed.

Step 2   The notch is located by establishing its end points.

Step 3   The lines are strengthened to complete the drawing.

(H) and depth (D). In Step 3, the third dimension, width (W), is used to complete the isometric drawing.

It is recommended that all isometric drawings be blocked in using light guidelines, as shown in Fig. 18.28, and overall dimensions of W, D, and H. Other dimensions can be taken from the given views and measured along the isometric axes to locate notches and portions that are removed from the "blocked-in" drawing.

A more complex isometric is shown in Fig. 18.29. Again, the object is blocked in using H, W, and D, and portions of the block are removed to complete the isometric drawing.

18.13
Angles in isometric

Angles cannot be measured true size in an isometric drawing since the surfaces of an isometric are not true size. Angles must be located by using coordinates measured along isometric lines as shown in Fig. 18.30. Lines AB and BC are equal in length in the orthographic view, but they are shorter and longer than true length in the isometric drawing.

FIG. 18.29   Layout of an isometric drawing (complex)

Step 1   The overall dimensions of the object are used to lightly block in the object. One notch is removed by using dimensions taken from the given views.

Step 2   The second notch is removed using dimension H_1.

Step 3   The final lines of the isometric are strengthened.

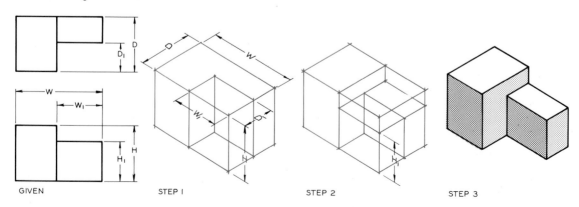

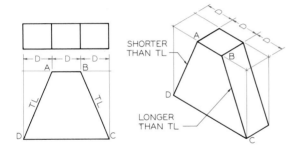

FIG. 18.30 Inclined surfaces must be found by using coordinates measured along the isometric axes. The lengths of angular lines will not be true length in isometric.

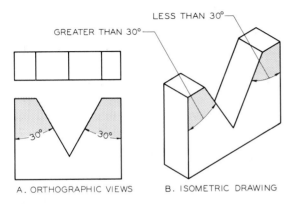

FIG. 18.31 Angles in isometric must be found by using coordinates. Angles will not appear true size in isometric.

A similar example can be seen in Fig. 18.31 where two angles are drawn in isometric. The equal-size angles in orthographic are less and greater than true size in the isometric drawing.

An isometric drawing of an object with an inclined surface is drawn in three steps in Fig. 18.32. The object is blocked in pictorially, and portions are removed. The extreme ends of the inclined plane are located along the isometric axes, and these points are connected to complete the pictorial.

18.14
Circles in isometric

The drawing of circular features in isometric is the most difficult construction problem in this type of pictorial.

Three methods of constructing circles in isometric drawings are (1) *point plotting*, (2) *four-center ellipse construction*, and (3) *ellipse templates*.

Circles: point plotting

A series of points on a circle can be located in an isometric drawing by using two dimensions that are parallel to the isometric axes. These two dimensions are called *coordinates* (Fig. 18.33).

FIG. 18.32 Construction of an isometric drawing with an inclined plane

Step 1 The object is blocked in using the overall dimensions. The notch is removed.

Step 2 The inclined plane is located by establishing its end points.

Step 3 The lines are strengthened to complete the drawing.

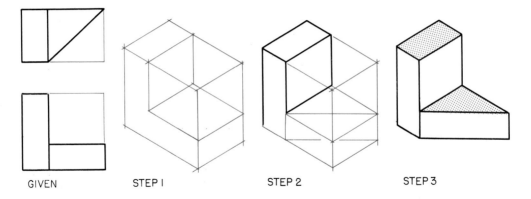

GIVEN   STEP I   STEP 2   STEP 3

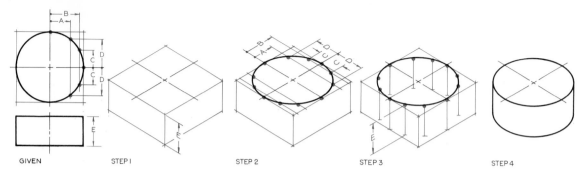

FIG. 18.33 Plotting circles

Step 1 The cylinder is blocked in using the overall dimensions. The center lines locate the points of tangency of the ellipse.

Step 2 Coordinates are used to locate points on the circumference of the circle.

Step 3 The lower ellipse is found by dropping each point a distance equal to the height of the cylinder, E.

Step 4 The two ellipses can be drawn with an irregular curve and connected with tangent lines to complete the cylinder.

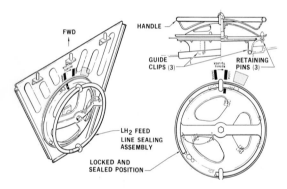

FIG. 18.34 An example of parts that have been drawn by using ellipses in isometric to represent circles. This is a handwheel that was proposed for use in an orbital workshop to be launched into space in the future. (Courtesy of NASA.)

The cylinder is blocked in and drawn pictorially, with the center lines added in Step 1. Coordinates A, B, C, and D are used in Step 2 to locate points along the ellipse and are connected with an irregular curve.

The lower ellipse located on the bottom plane of the cylinder can be found by using a second set of coordinates. The most efficient method is by measuring the distance, E, vertically beneath each point that was located on the upper ellipse (Step 3). A plotted ellipse is a true ellipse, and it is equivalent to a 35° ellipse on an isometric plane.

An example of a design composed of circular features that were drawn in isometric is the handwheel shown in Fig. 18.34.

FIG. 18.35 The four-center ellipse

Step 1 The diameter of the circle is used to construct a rhombus that is tangent to the ellipse.

Step 2 Perpendicular lines are drawn from the midpoints of the sides of the rhombus to locate four centers.

Step 3 Using the four centers and two radii, the four-center ellipse is drawn tangent to the rhombus.

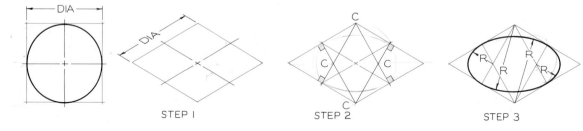

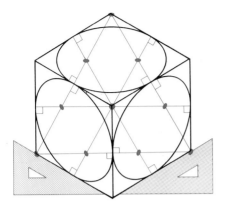

FIG. 18.36 Four-center ellipses can be drawn on all three surfaces of an isometric drawing.

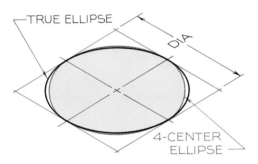

FIG. 18.37 The four-center ellipse is not a true ellipse, but an approximate ellipse.

Circles: four-center ellipse construction

The four-center ellipse method can be used to construct an approximate ellipse in isometric by using four arcs that are drawn with a compass (Fig. 18.35).

The four-center ellipse is drawn by blocking in the orthographic view of the circle with a square that is tangent to the circle at four points. This square is drawn in isometric as a rhombus (Step 1). The four centers are found by constructing perpendiculars to the sides of the rhombus at the midpoints of the sides (Step 2). The four arcs are drawn to give the completed four-center ellipse (Step 3). This method can be used to draw ellipses on any of the three isometric planes since each is equally foreshortened, as illustrated in Fig. 18.36.

You can see in Fig. 18.37 that the four-center ellipse is only an approximate ellipse when it is compared with a true ellipse.

Circles: ellipse templates

A specially designed ellipse template can be purchased for drawing ellipses in isometric. A typical example is shown in Fig. 18.38.

The diameters of the ellipses on the template are measured along the direction of the isometric lines, since this is how diameters are measured in an iso-

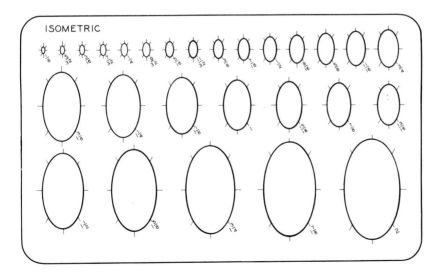

FIG. 18.38 The diameter of a circle in isometric is measured along the direction of the isometric axes. Therefore the major diameter of an isometric ellipse is greater than the measured diameter. The minor diameter is perpendicular to the major diameter.

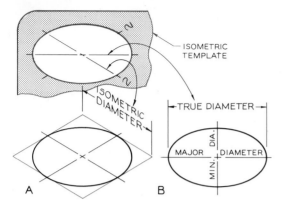

FIG. 18.39 The isometric ellipse template is a special template designed to reduce drafting time. Note that the isometric diameters of the circles are not the major diameters of the ellipses but are parallel to the isometric axes.

metric drawing (Fig. 18.39). The maximum diameter that can be measured across the ellipse is the *major diameter,* which is a true diameter. Consequently the size of the diameter marked on the template is less than the ellipses' major diameter, the true diameter.

The isometric ellipse template can be used to draw an ellipse by constructing the center lines of the ellipse in isometric and aligning the ellipse template with these isometric lines (Fig. 18.39A).

FIG. 18.40 Cylinder: four-center method

Step 1 A rhombus is drawn in isometric at each end of the cylinder's axis.

Step 2 A four-center ellipse is drawn within each rhombus.

Step 3 Lines are drawn tangent to each rhombus to complete the isometric drawing.

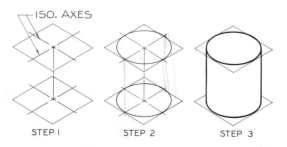

18.15
Cylinders in isometric

A cylinder can be drawn in isometric by using the four-center ellipse method as shown in Fig. 18.40. A rhombus is drawn at each end of the cylinder's axis, with the center lines drawn as isometric lines (Step 1). The ellipses are drawn using the four-center ellipse method at each end (Step 2). The ellipses are connected with tangent lines and the lines are darkened in Step 3.

A cylinder can be drawn by using the ellipse template as illustrated in Fig. 18.41. The axis of the cylinder is drawn and perpendiculars are constructed at each end (Step 1). Since the axis of a right cylinder is perpendicular to the major diameter of its elliptical end, the ellipse template is positioned as shown in Step 2. The size of the ellipse is marked near the elliptical hole on the template. The ellipses are drawn at each end and are connected with tangent lines (Step 3).

Internal cylinders (holes) are drawn by using the principles used for cylinders that were just covered. To construct a cylindrical hole in the block (Fig. 18.42), begin by locating the center of the hole on the isometric plane. The axis of the cylinder is drawn parallel to

FIG. 18.41 Cylinders: ellipse template

Step 1 The axis of the cylinder is drawn to its proper length, and perpendiculars are drawn at each end.

Step 2 The elliptical ends are drawn by aligning the major diameter with the perpendiculars at the ends of the axis. The isometric diameter of the isometric ellipse template is given along the isometric axis.

Step 3 The ellipses are connected with tangent lines to complete the isometric drawing. Hidden lines are omitted.

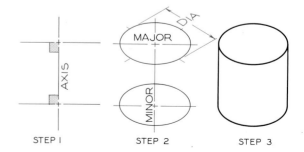

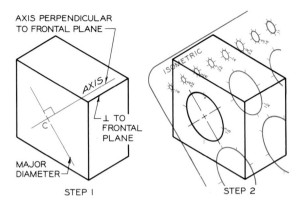

FIG. 18.42 Cylinders in isometric

Step 1 The center of the hole with a given diameter is located on a face of the isometric drawing. The axis of the cylinder is drawn from the center parallel to the isometric axis that is perpendicular to the plane of the circle. The major diameter is drawn perpendicular to the axis.

Step 2 The 1⅜″ ellipse template is used to draw the ellipse by aligning the major and minor diameters with the guidelines on the template.

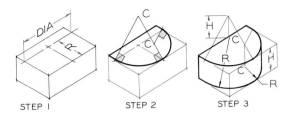

FIG. 18.43 Semicircular features

Step 1 Objects with semicircular features can be drawn by blocking in the objects as if they had square ends. The center lines are drawn to locate the centers and tangent points.

Step 2 Perpendiculars are drawn from each point of tangency to locate two centers. These are used to draw half of a four-center ellipse.

Step 3 The lower surface can be drawn by lowering the centers by the distance of H, the thickness of the part. The same radii are used with these new centers to complete the isometric.

the isometric axis that is perpendicular to this plane through its center (Step 1). The ellipse template is aligned with the major and minor diameters to complete the elliptical view of the cylindrical hole (Step 2).

18.16
Partial circular features

When an object has a semicircular end as in Fig. 18.43, the four-center ellipse method can be used with only two centers to draw half the circle (Step 2). To draw the lower ellipse at the bottom of the object, the centers are projected downward a distance of H, which is equal to the height of the object. These centers are used with the same radii that were used on the upper surface to draw the arcs.

In Fig. 18.44, an object with rounded corners is blocked in and the center lines are located at each rounded corner (Step 1). An ellipse template is used

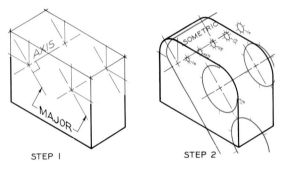

FIG. 18.44 To construct rounded corners on an object, the center lines of the ellipse are drawn at Step 1. The elliptical corners are drawn with an ellipse template at Step 2.

to construct the rounded corners at Step 2. The rounded corners could have been constructed by the four-center ellipse method or by plotting points on the arcs.

A similar drawing involving the construction of ellipses is the conical shape in Fig. 18.45. The ellipses are blocked in at the top and bottom surfaces (Step 1), and the half ellipses are drawn in Step 2 by using a template or the four-center method. The oblique drawing is shown in Step 3.

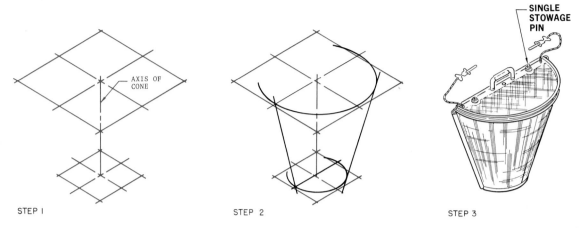

STEP I   STEP 2   STEP 3

FIG. 18.45 Construction of a cone in isometric

Step 1 The axis of the cone is constructed. Each circular end of the cone is blocked in.

Step 2 An ellipse guide is used for constructing the circles in isometric at each end. These ends are connected to give the outline of the object.

Step 3 The remaining details of the screen storage provisions are added to complete the isometric. (Courtesy of the National Aeronautics and Space Administration.)

18.17
Measuring angles

Angles in isometric may be located by coordinates, as shown in Fig. 18.46, since angles will not appear true in an isometric.

A second method of measuring and locating angles is the ellipse template method shown in Fig. 18.47. Since the hinge line is perpendicular to the path of revolution, an ellipse is drawn in Step 1 with the major diameter perpendicular to the hinge line. A true circle is drawn with a diameter that is equal to the major diameter of the ellipse.

In Step 2, point A is located on the ellipse and is projected to the circle. This locates the direction of a horizontal line. From this line, the angle of revolution of the hinged part can be measured true size, 120° in this example (Step 3). Point B is projected to the ellipse to locate a line at 120° in isometric.

The thickness of the revolved part is found by drawing line C perpendicular to line A and projecting back to the ellipse. A smaller ellipse through point D is drawn to locate the thickness of the revolved part in

FIG. 18.46 Inclined surfaces in isometric must be located by using coordinates laid off parallel to the isometric axes. True angles cannot be measured in isometric drawings.

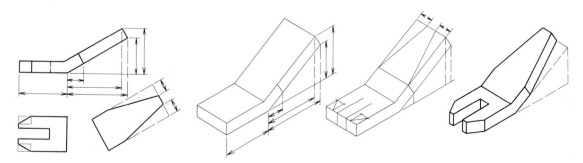

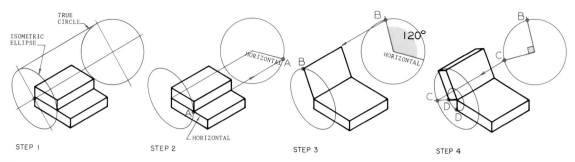

FIG. 18.47 Measuring angles with an ellipse template

Step 1 An ellipse is drawn with the major diameter perpendicular to the hinge line of the two parts. Any size of ellipse could be used. A true circle is drawn with its center on the projection of the hinge line and with a diameter equal to the major diameter of the ellipse.

Step 2 Point *A* is projected to the circle to locate the direction of the horizontal in the circular view.

Step 3 The position of rotation is measured 120° from the horizontal to locate point *B*, which is then projected to the ellipse to locate the position of the revolved surface.

Step 4 To locate the perpendicular to the surface, a 90° angle is drawn in the circular view, and point *C* is projected to the ellipse; a line is drawn from point *C* on the ellipse to the center of the ellipse. A smaller ellipse is drawn to pass through point *D* on the lower part of the object. The point where this ellipse intersects the line from *C* to the center of the ellipse establishes the thickness of the revolved part.

the isometric. The remaining lines are drawn parallel to these key lines.

18.18
Curves in isometric

Irregular curves must be plotted point by point using coordinates to locate each point. Points *A* through *F*

are located in the orthographic view with coordinates of width and depth (Fig. 18.48). These coordinates are transferred to the isometric view of the blocked-in part (Step 1). The plotted points are connected with an irregular curve.

Each point on the upper curve is projected downward for a distance of *H*, which is equal to the height of the part, to locate points on the lower curve (Step 2). The points are connected with an irregular curve to complete the isometric.

FIG. 18.48 Plotting irregular curves

Step 1 Coordinates are established in the orthographic views.

Step 2 One set of coordinates is transferred to the isometric view.

Step 3 The second set of coordinates is transferred to the isometric to establish points on the ellipse.

Step 4 The plotted points are connected with an elliptical curve. An ellipse template can usually serve as a guide for connecting the points.

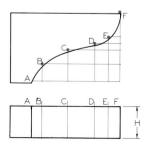
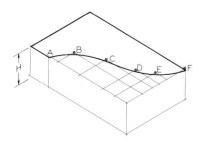
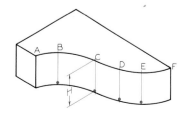

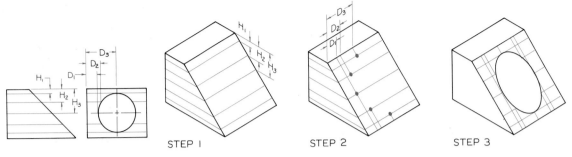

FIG. 18.49 Construction of ellipses on an inclined plane

Step 1 Draw two coordinates to locate a series of points on the irregular curve. These coordinates must be parallel to the standard W, D, and H dimensions.

Step 2 Block in the shape using overall dimensions. Locate points on the irregular curve using the coordinates from the orthographic views.

Step 3 Since the object has a uniform thickness, the lower curve can be found by projecting downward the distance H from the upper points. Connect the points with an irregular curve.

18.19
Ellipses on nonisometric planes

When an ellipse lies on a nonisometric plane such as the one shown in Fig. 18.49, points on the ellipse can be plotted to locate the ellipse.

Coordinates are located in the orthographic views and then transferred to the isometric as shown in Steps 1 and 2. The plotted points can be connected with an irregular curve, or an ellipse template can be selected that will approximate the plotted points (Step 3). The isometric ellipse template cannot be used for this purpose.

18.20
Surfaces of revolution

A surface of revolution is a solid form made by revolving a plane about an axis of revolution. Two examples of surfaces of revolution are shown in Fig. 18.50. To draw these in isometric, the axes were drawn through point 0 to point 4. Circular cross sections were located

FIG. 18.50 Surfaces of revolution must be constructed by drawing a series of cross sections that are connected to give their overall shapes. The cross sections in these examples are circles that are drawn as ellipses in isometric.

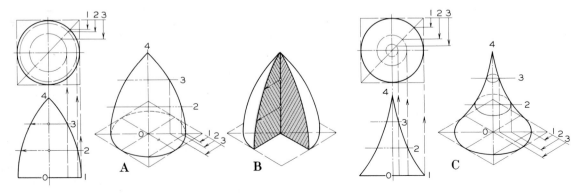

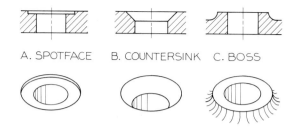

A. SPOTFACE   B. COUNTERSINK   C. BOSS

FIG. 18.51 Examples of circular features drawn in isometric. These can be drawn by using ellipse templates.

along this axis by using the correct elliptical diameters taken from the orthographic views for each.

The elliptical cross sections are connected with tangent lines to find the outlines of the objects in isometrics. The more cross sections that are used, the more accurate will be the location of the tangent lines.

18.21
Machine parts in isometric

Orthographic views of a spotface, countersink, and boss are shown in Fig. 18.51. The isometric pictorials of each are shown also. The isometric drawings of these features could be drawn by point-plotting the circular features, by the four-center method, or by us-

FIG. 18.52 Threads in isometric

Step 1 Draw the cylinder that is to be threaded by using an ellipse template.

Step 2 Lay off perpendiculars that are spaced by a distance equal to the pitch of the thread, P.

Step 3 Draw a series of ellipses to represent the threads. The chamfered end is drawn by using an ellipse whose major diameter is equal to the root diameter of the threads.

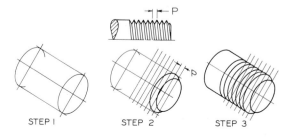

ing an ellipse template. The template is by far the easiest method.

A threaded shaft can be drawn in isometric as shown in Fig. 18.52 by first drawing the cylinder that is to be threaded in isometric (Step 1). In Step 2, the major diameters of the crest lines of the thread are drawn separated at a distance of P, the pitch of the thread. In Step 3, ellipses are drawn by aligning the major diameter of the ellipse template with the perpendiculars to the cylinder's axis. Note that the 45° chamfered end is drawn by using a smaller ellipse at the end.

A hexagon head nut (Fig. 18.53) is drawn in three steps by using an ellipse template. The nut is blocked in and an ellipse drawn tangent to the rhombus. The hexagon is constructed by locating the distance across a flat, W, parallel to the isometric axes.

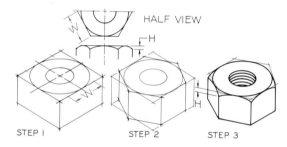

FIG. 18.53 Construction of a nut

Step 1 The overall dimensions of the nut are used to block in the nut.

Step 2 The hexagonal sides are constructed at the top and bottom.

Step 3 The chamfer is drawn with an irregular curve. Threads are drawn to complete the isometric.

The other sides of the hexagon are found in Step 2 by drawing lines tangent to the ellipse. The distance H is laid off at each corner to establish the amount of chamfer at each corner (Step 3).

A hexagon head bolt is drawn in two positions in Fig. 18.54. The washer face can be seen on the lower side of the head, and the chamfer on the upper side of the bolt head.

Many machine parts are composed of spherical shapes. The sphere in isometric is constructed as shown in Fig. 18.55. Three ellipses are drawn as isometric planes with a common center. The center is used to construct a circle that will be tangent to each

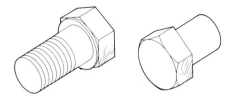

FIG. 18.54 Isometric drawings of the upper and lower sides of a hexagon-head bolt.

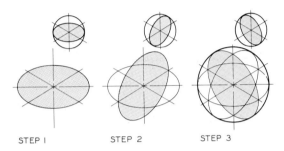

STEP 1   STEP 2   STEP 3

FIG. 18.55 An isometric sphere

Step 1 The three intersecting isometric axes are drawn. The ellipse template is used to draw the horizontal elliptical section.

Step 2 The isometric ellipse template is used to draw one of the vertical elliptical sections.

Step 3 The third vertical elliptical section is drawn, and the center is used to draw a circle tangent to the three ellipses.

FIG. 18.56 Spherical features

Step 1 An isometric ellipse template is used to draw the elliptical features of a round-head screw.

Step 2 The slot in the head is drawn, and the lines are darkened to complete the isometric of the head.

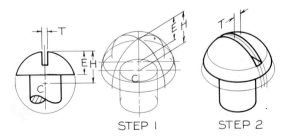

STEP 1   STEP 2

ellipse as shown in Step 3. The sphere is larger in isometric than in a true axonometric projection.

A portion of a sphere is used to draw the round-head screw in Fig. 18.56. A hemisphere is constructed in Step 1. The center line of the slot is located along one of the isometric planes. The thickness of the head is measured as distance E from the highest point on the sphere. The slot is drawn in Step 3 to complete the isometric drawing.

18.22
Isometric sections

A full section can be drawn in isometric to clarify internal details that might otherwise be overlooked (Fig. 18.57). Half-sections can also be used as illustrated in Fig. 18.58. The same part shown as both a full section and a half-section is given in Fig. 18.59.

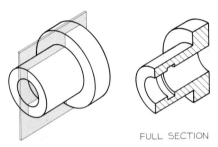

FULL SECTION

FIG. 18.57 Parts can be shown in isometric sections to clarify internal features, such as this full section.

FIG. 18.58 An isometric drawing of a half-section can be constructed.

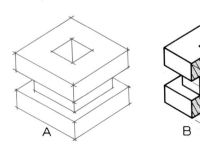

A   B

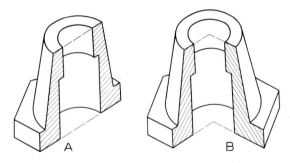

FIG. 18.59 A comparison of isometric full and half-sections of the same part.

18.23
Dimensioned isometrics

When it is advantageous to dimension and note a part shown in isometric, either the aligned or the unidirectional method illustrated in Fig. 18.60 can be used to apply the notes. In both cases, notes connected with leaders are positioned horizontally. Always use guidelines for your lettering and numerals.

18.24
Fillets and rounds

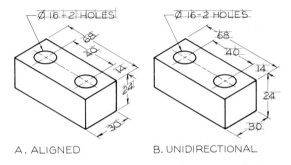

A. ALIGNED   B. UNIDIRECTIONAL

FIG. 18.60 Dimensions can be placed on isometric drawings by using either of the techniques shown here. Guidelines should always be used for the lettering.

Fillets and rounds in isometric can be represented by either of the methods shown in Fig. 18.61 to give added realism to a pictorial drawing. The examples at A and B show how intersecting guidelines are drawn equal in length to the radii of the fillets and rounds, and arcs are drawn tangent to these lines. These arcs can be drawn freehand or with an ellipse template. The method in part C utilizes freehand lines drawn parallel or concentric with the directions of the fillets and rounds.

An example of these two methods is shown in Fig. 18.62. The stipple shading was applied by using an adhesive overlay film that can be purchased.

FIG. 18.61 Representation of fillets and rounds

A. Fillets and rounds can be represented by segments of an isometric ellipse if guidelines are constructed at intervals.

B. Fillets and rounds can be represented by elliptical arcs by constructing radial guidelines.

C. Fillets and rounds can be represented by lines that run parallel to the fillets and rounds.

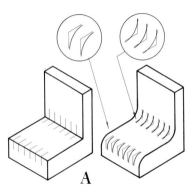

A

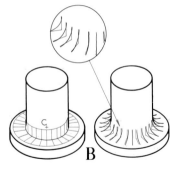

B

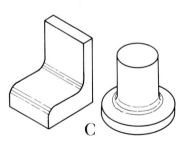

C

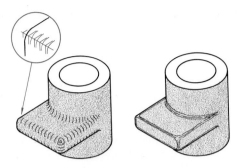

FIG. 18.62 Two methods of representing fillets and rounds on a part.

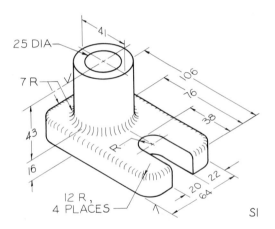

FIG. 18.63 An isometric drawing with complete dimensions and fillets and rounds represented.

FIG. 18.64 Part numbers should be applied to assemblies as shown at B and by avoiding the errors made at A.

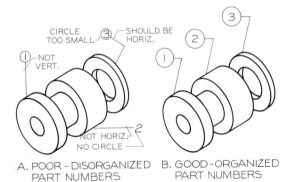

When fillets and rounds are illustrated as shown in Fig. 18.63 and dimensions are applied, it is much easier to understand the features of the part than when it is represented by orthographic views. The dimensions in this example are shown using the aligned method.

18.25
Isometric assemblies

Assemblies are used to explain how a series of parts is assembled. The more common mistakes in applying leaders to an assembly are shown in Fig. 18.64A. The more acceptable techniques are shown in part B. The numbers in the circles ("balloons") refer to the number given to each part that appears in the parts list.

An assembly used in a parts manual is shown in Fig. 18.65, along with its parts list. This assembly is "exploded" apart so the parts are separated, but it is clear how they would be assembled. The assembly in

FIG. 18.65 An industrial example of an assembly.

ELECTRIC OPERATORS for Jenkins Ball Valves
PARTS LIST

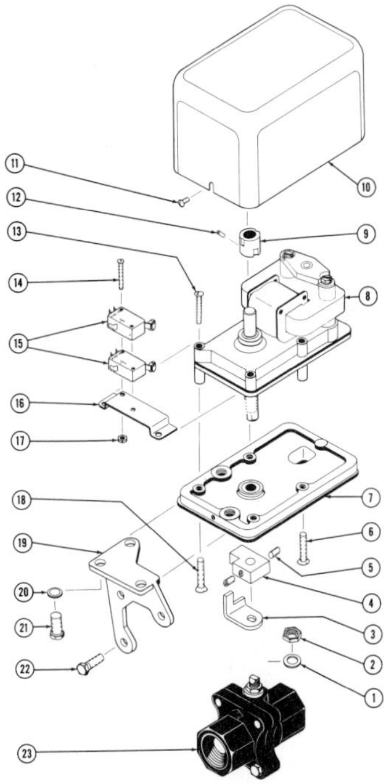

PARTS LIST

PC. NO.	PART	QUANTITY
1	Coupling Lockwasher	1
2	Coupling Nut	1
3	Coupling—Driven Half	1
4	Coupling—Driving Half	1
5	Coupling—Set Screw	2
6	Motor Mounting Screw	2
7	Motor Mounting Plate	1
8	Gear Motor	1
9	Cam	1
10	Cover	1
11	Cover Screw	2
12	Cam—Set Screw	1
13	Bracket Mounting Screw	2
14	Switch Mounting Screw	2
15	Switches	2
16	Switch Bracket	1
17	Switch Mounting Nut	2
18	Motor Mounting Screw	2
19	Bracket	1
20	Lockwasher	2
21	Operator Mounting Screw	2
22	Cap Bolt	3
23	Ball Valve as specified	1

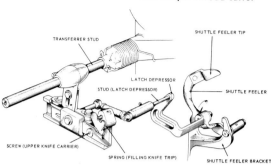

*25 Draper thread cutter

FIG. 18.66 An assembly drawing with parts assembled and named. (Courtesy of Draper Corporation.)

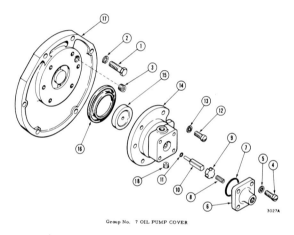

Group No. 7 OIL PUMP COVER

FIG. 18.67 An exploded isometric assembly.

Fig. 18.66 is totally assembled, and each part is named by note. An exploded isometric assembly is given in Fig. 18.67.

18.26
Piping systems in isometric

Piping layouts are often drawn as isometrics to clarify complex systems that are hard to interpret when shown in orthographic views. Some of the more often used piping connectors are shown in Fig. 18.68. The components are shown both in orthographic and in isometric views. Ellipse templates are used to represent circular arcs in isometric. These are single-line representations since the pipes are represented by single, heavy lines.

A portion of a pipe system is shown in Fig. 18.69. Additional pipe symbols are given in Appendix 9. All symbols are drawn in isometric using the same principles presented in this section.

18.27
Axonometric projection

An *axonometric projection* is a form of orthographic projection in which the pictorial view is projected perpendicularly onto the picture plane with parallel projectors. The object is positioned in an angular position

FIG. 18.68 A comparison of orthographic views and isometric pictorials of single-line representations of piping symbols. Note that the isometric template is used for constructing rounded corners in the isometric drawings.

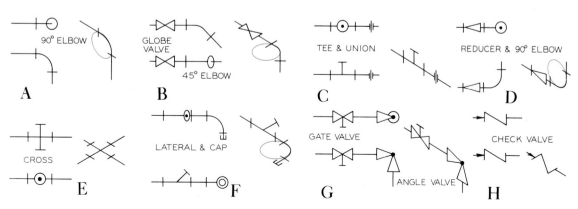

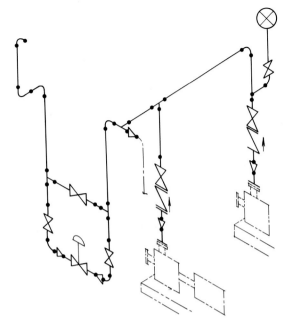

FIG. 18.69 A typical piping system drawn in isometric using the appropriate symbols.

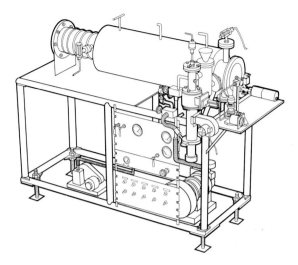

FIG. 18.71 A trimetric projection of a cold diffusion pump. (Courtesy of Aro, Inc.)

with the picture plane so that its pictorial projection will be a three-dimensional view rather than a two-dimensional view as an orthographic view.

Three types of axonometric pictorials are possible: (1) isometric, (2) dimetric, or (3) trimetric. The *isometric projection* is the view where the diagonal of a cube is viewed as a point. The planes will be equally foreshortened and axes equally spaced 120° apart (Fig. 18.70). The measurements along the three axes will be equal, but less than true length since this is a true projection.

A *dimetric projection* is an axonometric projection of a cube in which two planes are equally foreshort-

FIG. 18.70 The three types of axonometric projection.

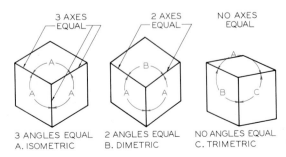

ened and two of the axes are separated by equal angles (part B). The measurements along two of the axes are equal, but less than true length.

A *trimetric projection* is an axonometric projection of a cube where all three planes are unequally foreshortened and the angles between the three axes and their lengths are different (Fig. 18.70C).

Figure 18.71 is a trimetric projection of a cold diffusion pump. Note that the angles between the three axes are unequal; this identifies it as a trimetric.

18.28
Axonometric construction

All axonometric constructions can be made in the same manner as the trimetric constructed in Fig. 18.72.

A cube should always be used instead of the object to be drawn as an axonometric. By using a cube, you can construct axonometric scales that can be used to draw trimetrics of many objects, not just a single part.

The trimetric scales found in Step 3 can be lengthened to accommodate any size of part. Using axonometric scales eliminates the need for duplicating the construction when trimetrics of similar angles are drawn.

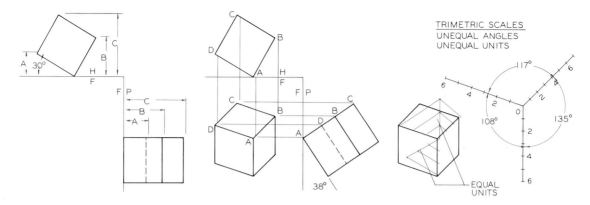

FIG. 18.72 Trimetric-scale construction

Step 1 Revolve the top view 30° clockwise. Find the side view by transferring dimensions *A, B,* and *C* from the top view. If the revolution had been 45° in the top view, the resulting projection would be either a dimetric or an isometric.

Step 2 Tilt the side view 34°. This will change the projection of the top view but not the width dimension; consequently it is unnecessary to change the top view. Determine the trimetric projection of the cube by projecting from the top and side views as in orthographic projection.

Step 3 All sides of a cube are equal; therefore divide each axis (or edge) into an equal number of units by proportional divisions as shown, even though the three axes in a trimetric have different lengths. The three axes can be extended and scaled into as many units as desired.

Trimetric scales are not complete unless the ellipse angles are found for each plane so that an ellipse template can be used. In a trimetric, a different ellipse angle will be used for each plane since each plane is foreshortened differently.

When a trimetric is given (Fig. 18.73), a true-size plane can be found by drawing lines 1–2, 2–3, and 3–1 perpendicular to the three axes (Step 1). The side view of this true-size plane is found as a vertical edge in Step 2. Using these two views, the angles between

FIG. 18.73 Ellipse template angles for a trimetric scale

Step 1 The planes of a cube are mutually perpendicular; therefore line *OA* is perpendicular to plane *OCDE,* and line *OE* is perpendicular to plane *ABCO.* On each plane, construct true-length lines, which are located perpendicular to the axis lines *AO, CO,* and *EO.* The true-length lines are 1–2, 2–3, and 3–1. These lines will intersect at points on the axes forming triangle 1–2–3.

Step 2 Since plane 1–2–3 is composed of true-length lines, it is true size in the trimetric projection. Determine the side view of the plane, which is a frontal plane by projection. Find the 90° angle of the cube in the side view by constructing a semicircle, using the edge view of plane 1–2–3 as the diameter. Project point *O* to the semicircle where the 90° angle is inscribed.

Step 3 Two views permit auxiliary views to be found. Determine the edge view of each principal plane by locating the point views of lines 1–2, 2–3, and 3–1. The ellipse guide angle is the angle between the edge views of 1–0–2, 2–0–3, and 1–0–3 and the line of sight. Position the ellipse guides on each plane so that the major diameter is parallel to the true-length lines on that plane.

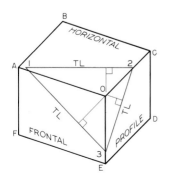
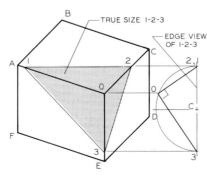
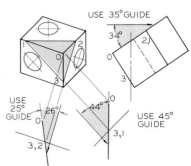

403

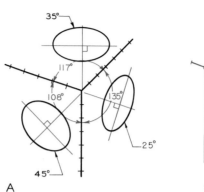

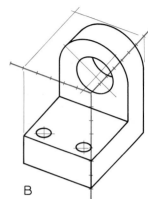

FIG. 18.74 The complete trimetric scales showing the units along the axes and the ellipse template angles. The object at B was drawn by using the axonometric scales.

A   B

the lines of sight and each plane can be found in Step 3. These angles are the ellipse angles for each surface.

The ellipses are positioned on each plane with their major diameters perpendicular to the axes intersecting each plane. This gives you a set of trimetric scales with calibrations and the ellipse angles for each plane (Fig. 18.74A). These scales can be used to construct a trimetric by overlaying the scales with tracing vellum, as shown in Fig. 18.74B.

A second technique of constructing an axonometric is shown in Fig. 18.75. The three axes are located in a convenient position, and plane 1–2–3 is constructed with three true-length lines in Step 1. Semicircles are drawn with diameters equal to these true-length lines. Angles are inscribed inside each to give a 90° angle at point 0 for each semicircle (Step 2). The top, front, and side views of the object are located at point 0 for each view, and each view is projected back where the three projectors converge at common points.

FIG. 18.75 Axonometric projection

Step 1 The three axes of an axonometric projection can be selected and drawn at the angle of your choice. True-length perpendiculars are drawn so they intersect at 1, 2, and 3.

Step 2 Semicircles are projected from each true-length line. Right angles are inscribed in the semicircle, 2–0–3 for example.

Step 3 The three orthographic views of the object are located with the same corner placed at the right angles found in Step 2. These views are projected back to intersect and form a trimetric in this case.

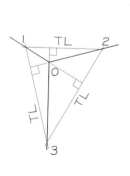

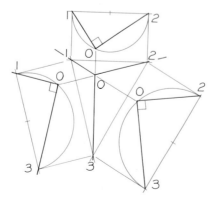

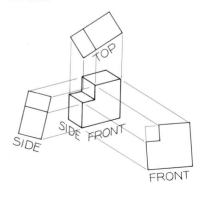

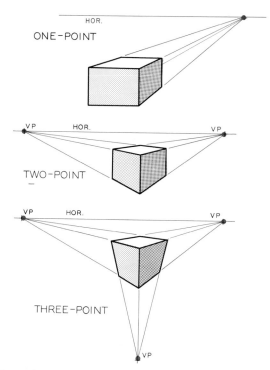

HOR.

ONE-POINT

VP HOR. VP

TWO-POINT

VP HOR. VP

THREE-POINT

VP

FIG. 18.76 A comparison of one-point, two-point, and three-point perspectives.

18.29
Perspectives

A *perspective* is a view that is normally seen by the eye or camera and is the most realistic form of pictorial. All parallel lines converge at infinite vanishing points as they recede from the observer.

> The three types of perspectives (Fig. 18.76) are (1) *one-point*, (2) *two-point*, and (3) *three-point*, depending on the number of vanishing points used in their construction.

The *one-point perspective* has one surface of the object that is parallel to the picture plane; therefore it is true shape. The other sides vanish to a single point on the horizon, called a vanishing point.

A *two-point perspective* is a pictorial that is positioned with two sides at an angle to the picture plane; this requires two vanishing points (Fig. 18.76). All horizontal lines converge at the vanishing points, but vertical lines remain vertical and have no vanishing point.

The *three-point perspective* utilizes three vanishing points since the object is positioned so that all sides of it are at an angle with the picture plane (Fig. 18.76). The three-point perspective is used in drawing larger objects, such as buildings.

18.30
One-point perspectives

The steps of drawing a one-point perspective are shown in Fig. 18.77. Here are given the top and side view of the object, the picture plane, the station point (S.P.), the horizon, and the ground line.

THE PICTURE PLANE is the plane on which the perspective is projected. It appears as an edge in the top view.

THE STATION POINT is the location of the observer's eye in the plan view. The front view of the station point will always lie on the horizon.

THE HORIZON is a horizontal line in the front view that represents an infinite horizontal, such as the surface of the ocean.

THE GROUND LINE is an infinite horizontal line in the front view that passes through the base of the object being drawn.

THE CENTER OF VISION (C.V.) is a point that lies on the picture plane in the top view and on the horizon in the front view. In both cases, it is on the line from the station point that is perpendicular to the picture plane.

When drawing any perspective, the station point (S.P.) should be located far enough away from the object so that the perspective can be contained in a cone of vision that is not more than 30° (Fig. 18.78). If a larger cone of vision is used, the perspective will be distorted.

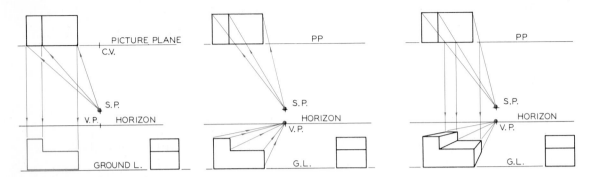

FIG. 18.77 Construction of a one-point perspective

Step 1 Since the object is parallel to the picture plane, there will be only one vanishing point, which will be located on the horizon below the station point. Projections from the top and side views establish the front plane. This surface is true size, since it lies in the picture plane.

Step 2 Draw projectors from the station point to the rear points of the object in the top view and from the front view to the vanishing point on the horizon. In a one-point perspective, the vanishing point is the front view of the station point.

Step 3 Construct vertical projectors from the top view to the front view from the points where the projectors cross the picture plane. These projectors intersect the lines leading to the vanishing point. This is called a one-point perspective, since the lines converge at a single point.

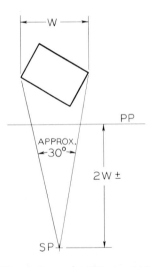

FIG. 18.78 The station point (SP) should be placed far enough away from the object to permit the cone of vision to be less than 30° to reduce distortion.

Measuring points for one-point perspective

A measuring point is an additional vanishing point used to locate measurements along the receding lines that vanish to the horizon. In Fig. 18.79, the measur-

ing point of a one-point perspective is found by revolving line 0–2 into the picture plane to 0–2′. The measuring point is found by drawing a construction line from the station point to the picture plane parallel to 2–2′. The measuring point is located on the horizon by projection from the picture plane (Step 1).

Since the distance 0–2 is equal to 0–2′, depth dimensions can be laid off along the ground line and then projected to the measuring point. This locates the rear corner of the one-point perspective (Step 2). The depth from the picture plane to the front of the object is found in the same manner.

The use of the measuring-point method eliminates the need for placing the top view in the customary top-view position. Instead the dimensions can be transferred to the ground line in a more convenient manner.

18.31
Two-point perspectives

If two surfaces of an object are positioned at an angle to the picture plane, two vanishing points will be required to draw it as a perspective. Different views can be obtained by changing the relationship between the horizontal and the ground line, as illustrated in Fig. 18.80.

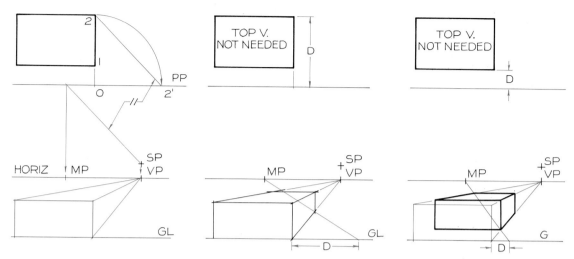

FIG. 18.79 One-point perspective—measuring points

Step 1 Line 0–2 is revolved into the picture plane to locate point 2'. A line is drawn parallel to 2'–2 through SP to the PP. The measuring point is located on the horizon.

Step 2 Distance D is laid off along the ground line from the front corner of the perspective. This distance is projected to the measuring point to locate the rear corner of the perspective.

Step 3 The front of the object is located by laying off distance D from the corner of the perspective and projecting to the measuring point. This locates the front surface of the object behind the picture plane.

FIG. 18.80 Different perspectives can be obtained by locating the horizontal over, under, and through the object.

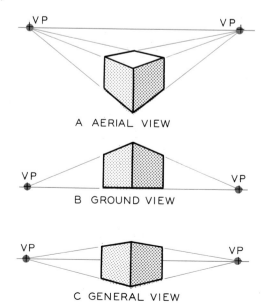

A AERIAL VIEW

B GROUND VIEW

C GENERAL VIEW

An *aerial view* will be obtained when the horizon is placed above the ground line and the top of the object in the front view. When the ground line and the horizon coincide in the front view, a *ground-level view* will be obtained. This gives the view that would be seen if your eye were looking from the ground. A *general view* is one where the horizon is placed above the ground line and through the object, usually at a height equal to the height of a person (part C).

The steps of constructing a two-point perspective are shown in Fig. 18.81. The horizon has been placed above the object to give a slight aerial view. The vanishing points are found by drawing construction lines parallel to the sides of the object from the station point in the top view.

Since line *AB* lies in the picture plane, it will be true length in the perspective. All height dimensions must originate at this vertical line because it is the only true-length line in the perspective. Points *C* and *D* are found by projecting to *AB,* and then projecting toward the vanishing points.

A typical two-point perspective is shown in Fig. 18.82, where the construction has not been separated into steps. By referring to Fig. 18.80, you will be able

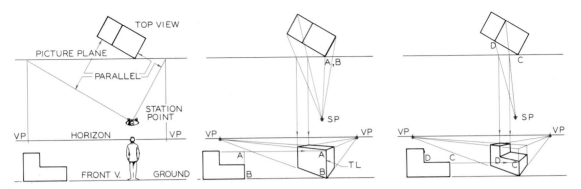

FIG. 18.81   Construction of a two-point perspective

Step 1 Construct projectors that extend from the top view of the station point to the picture plane parallel to the forward edges of the object. Project these points vertically to a horizontal line in the front view to locate vanishing points. Locate the horizon in a convenient position. Draw the ground line below the horizon and construct the side view on the ground line.

Step 2 Since all lines in the picture plane are true length, line *AB* is true length. Consequently, line *AB* is projected from the side view to determine its height. Then project each end of *AB* to the vanishing points to determine two perspective planes. Draw projectors from the station point to the exterior edges of the top view. Project the intersections of these projectors with the picture plane to the front view to determine the limits of the object.

Step 3 The box obtained in Step 2 must have a notch removed. Determine point *C* in the front view by projecting from the side view to the true-length line *AB*. Draw a projector from point *C* to the left vanishing point. Point *D* will lie on this projector beneath the point where a projector from the station point to the top view of point *D* crosses the picture plane. Complete the notch by projecting to the respective vanishing points.

FIG. 18.82   A two-point perspective of an object.

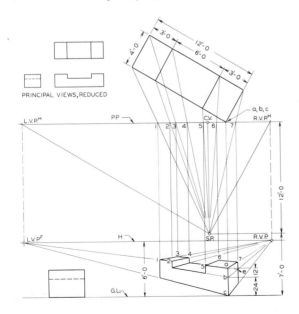

to understand the development of the construction used.

The object in Fig. 18.83 does not make contact with the picture plane in the top view as in the previous examples. To draw a perspective of this object, the lines of the object must be extended to intersect the picture plane. The height is measured, at this point, and the infinite plane is drawn to the vanishing point.

The corner of the object can be located on this infinite plane by projecting it to the picture plane in the top view with a projector from the station point. This corner does not contact the picture plane, and none of the height dimensions will be true length.

Two-point perspectives: measuring points

Measuring points are found in Fig. 18.84 to aid in the construction of two-point perspectives. The use of measuring points eliminates the need to have the top

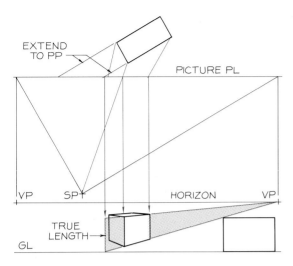

FIG. 18.83 A two-point perspective of an object that is not in contact with the picture plane.

view in the top-view position after the vanishing points have been found.

Two vanishing points are found for two-point perspectives. These are used to locate dimensions along the receding planes that vanish to the horizon, as shown in Step 2. Depth dimensions are laid off along the ground line from point B, and secondary planes are passed from these points to their measuring points (Step 3).

Measuring points are used in Fig. 18.85 to construct a two-point perspective. The same principles were used in this construction as in the previous example.

Arcs in perspective

Arcs in two-point perspectives must be found by using coordinates to locate points along the curves in perspective (Fig. 18.86). Points 1 through 7 are located along the semicircular arc in the orthographic view. These same points are found in the perspective by projecting coordinates from the top and side views.

FIG. 18.84 Construction of measuring points

Step 1 In the top view, revolve lines AB and BC into the picture plane using point B as the center of revolution. Draw construction lines through points A–A' and points C–C'. These lines represent edge views of vertical planes passing through the corner points A and C.

Step 2 Draw lines from the station point parallel to lines A–A' and C–C' to the picture plane. Project these points of intersection to the horizon to locate two measuring points. Distances AB and BC can be laid off true length on the ground line (GL), since they have been revolved into the picture plane in the top view.

Step 3 Extend imaginary planes from points A and C on the ground line to their respective measuring points. These planes intersect the infinite planes, which are extended to the vanishing points, to locate the two corners of the block. True measurements can to be laid off on the ground line and projected to measuring points to find corner points.

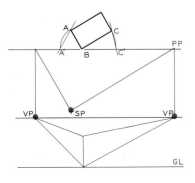 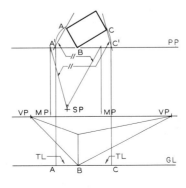 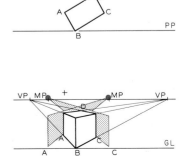

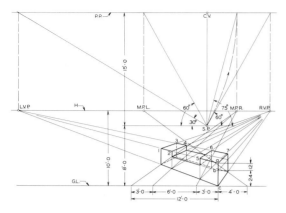

FIG. 18.85 A two-point perspective drawn with the use of measuring points.

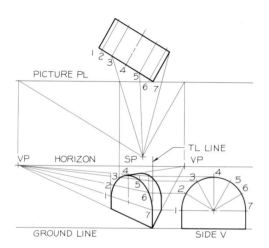

FIG. 18.86 A two-point perspective of an object with semicircular features.

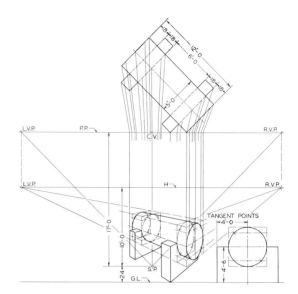

FIG. 18.87 A two-point perspective of an object with circular features.

The points do not form a true ellipse, but a slightly egg-shaped oval. An irregular curve is used to connect the points.

A two-point perspective of a cylindrical object is constructed in Fig. 18.87. Again, it is necessary to locate points along the circle in perspective. Eight points are located in this example to approximate the perspective of the circular ends.

Sloping planes

A sloping plane that is not horizontal will not have its vanishing point on the horizon, but below or above the horizon. In Fig. 18.88, a roof slopes 25° with the horizontal in two directions. The vanishing points of these two planes are found in Step 2 by revolving *AB* into the picture plane to *AB'*, and *B'* is projected to the horizon. From this point, two lines are drawn at 25° with the horizon to the vertical projector through the vanishing point found in Step 1. This locates the vanishing points of the two planes of the roof.

The planes of the roof are located in the perspective and are projected to their respective vanishing points (Step 3). Only horizontal planes vanish to the horizon.

18.32
Three-point perspectives

A three-point perspective can be constructed by positioning the object so that all its sides make angles with the picture plane. In Fig. 18.89, the top and side views are orthographic views that have been tilted with respect to the picture plane. The station point is the same distance from the picture plane in the side view.

Three vanishing points are located by constructing lines from the two views of the station point that are

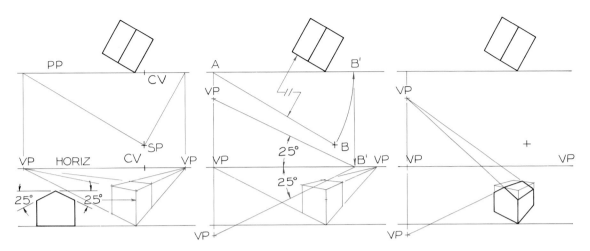

FIG. 18.88 Perspectives of sloping surfaces

Step 1 The vanishing points are located and the perspective of the object is drawn as if it had no sloping planes. The point where the sloping plane begins its slope is found on the true-length vertical line by projecting from the side view.

Step 2 Line *AB* is rotated in the top view to locate *B'* on PP which is then projected to the horizon. Since both planes slope 25°, their vanishing points are drawn at 25° with the horizon from *B'* to a vertical line that passes through the left VP.

Step 3 Each sloping plane is found by using the two vanishing points found in Step 2 and projecting from points found on the vertical lines where the slopes begin.

FIG. 18.89 The construction of a three-point perspective.

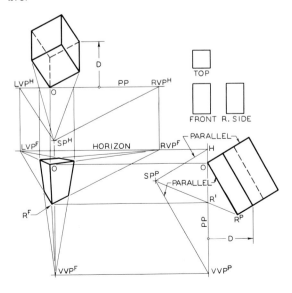

parallel to the sides of object. The corner points in perspective are found by projecting to the picture plane and then to the perspective view where the lines vanish to their respective vanishing points.

In laying out a three-point perspective, you must begin by positioning the top and side views. The vanishing point for vertical lines will lie along a vertical construction line through the station point (S.P.).

A shortcut method of drawing a three-point perspective is shown in Fig. 18.90. An equilateral triangle with 60° angles is drawn, and a convenient point, *O*, is located within the triangle. The vertexes of the triangle are used as the three vanishing points.

Width and depth are laid off on either side of point *O*. These dimensions converge at the two horizontal vanishing points to locate the top view of the box. Height is laid off along a construction line through *O* that is parallel to one of the other sides of the triangle. Height is projected back to the vertical line from *O* to the lower vanishing point. Now that the three dimensions have been located, the three-point perspective can be completed.

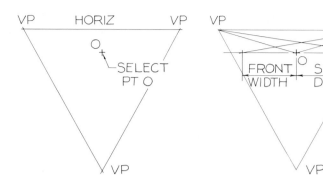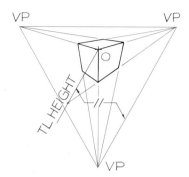

FIG. 18.90 Three-point perspective

Step 1 Draw an equilateral triangle. The vanishing points lie at each vertex. Locate point O anywhere within this triangle, and it will be the beginning point of the perspective.

Step 2 Draw a horizontal through point O. The width and depth can be laid off from point O, and projectors are drawn to the two upper vanishing points. This locates the top surface.

Step 3 The height is laid off along a line that is parallel to one of the triangle's sides. The height is projected back to where it intersects a line from point O to the lower VP.

18.33
Perspective charts

Perspective charts are time-saving grids that eliminate the need for constructing vanishing points and other tedious constructions (Fig. 18.91). The grid is drawn to scale so that it can be overlaid by tracing vellum and perspectives drawn by using the printed grid. Scales can be assigned to the grid for varying sizes of perspectives.

FIG. 18.91 Use of a perspective grid. (Courtesy of Graphicraft.)

Perspective charts come in a variety of angles and views to suit most perspective needs. Perspective grids are excellent for preliminary sketches, since they give a general idea of the appearance of the finished perspective in the shortest time.

18.34
Shades and shadows
in orthographic

Shades and shadows are added to drawings to increase the realism of their appearance. A surface is in shade when the light does not strike it. For example, if a block were facing the sunlight, its backside would not be in the direct rays of light but would be in *shade*.

When an object shields another surface from the sunlight, the dark area that is cast on this surface is called *shadow*.

The triangular prism in Fig. 18.92A is drawn in orthographic projection. The rays of light are chosen to be 45° backward and downward in the two given views. The shadow of the object on the frontal plane is found by projecting to the vertical plane in the top view and downward to the projectors from the front view. One surface is in shade since it is away from the light rays. The shadow of the object helps you instinctively visualize the shape of the object.

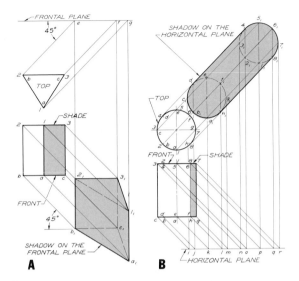

FIG. 18.92 A. The construction of the shadow of a triangular prism in orthographic on a vertical plane. B. The construction of the shadow of the cylinder in orthographic on a horizontal plane.

The shades and shadows of a cylinder are found in Fig. 18.92B. Points are located around the circular ends of the cylinder. These are projected to the horizontal plane and upward to the projectors from the top view. The shadow will be an edge in the front view, but it can be seen true size in the top view. Since

the circular ends of the cylinder are parallel to the horizontal, their shadows will be true circles that are connected with tangent lines.

Since the shadow does not begin at the edge of the cylinder in the top view, you have the impression that the object is suspended above the horizontal surface on which the shadow is cast.

18.35
Shades and shadows—a block in isometric

A light ray is established in Fig. 18.93 that is parallel to the picture plane since the vertical and horizontal legs of the triangle are perpendicular and true length. The shadows of two vertical lines are found in Step 2 when the rays are passed through points A and B. The shadows are found at A' and B' where horizontal lines from the vertical lines intersect the light rays. The process is continued to find the shadows A', B', and C', which are connected to complete the shadow, and the shaded area is indicated.

These same principles are used to find the shades and shadows of the more complex object shown in Fig. 18.94

FIG. 18.93 Shades and shadows of a cube

Step 1 The light-source triangle is constructed for a ray of light that is parallel to the drawing paper.

Step 2 The shadows cast from vertical corners passing through points A and B are found. Note that the projectors from each end of the vertical corners are parallel to the sides of the light-source triangle.

Step 3 The shadow of the vertical line through point C is found even though it is hidden. Line $C'B'$ is drawn to complete the outline of the shadow.

Step 4 The object is shaded to indicate the surfaces in shade and the shadow.

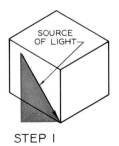

STEP 1

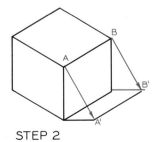

STEP 2

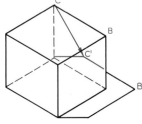

STEP 3

STEP 4

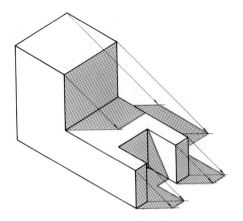

FIG. 18.94 Shades and shadows constructed for an object in isometric.

18.36
Shades and shadows—cylinders in isometric

The shadow of a cylinder is found by locating the shadows of a number of its elements as if they were individual lines (Fig. 18.95).

The light source is parallel to the picture plane in this example. The shadows of the point along the upper circle are connected to complete the shadow. The shaded area is located where the light source is tangent to the upper surface of the cylinder.

When a cylinder is in a horizontal position, the circular ends are blocked in and coordinates are used

FIG. 18.95 Construction of the shades and shadows of a vertical cylinder. The shadows of each element are found one at a time. The light is parallel to the picture plane.

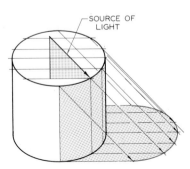

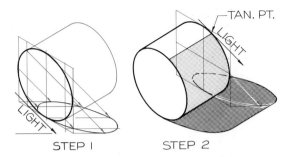

FIG. 18.96 The construction of shades and shadows of a horizontal cylinder. The light is parallel to the picture plane.

to project points to the horizontal surface (Fig. 18.96). This gives an elliptical shadow for each end. The elliptical shadows of both ends are found in the same manner to complete the shadow. The area of shade is located where the ray of light is tangent to the circular end of the cylinder.

18.37
Shades and shadows—angular light

It is unnecessary for the ray of light to be parallel to the picture plane. It can be in any direction. A triangle and source of light is established in Fig. 18.97, Step 1. The ray of light passes through the end of a vertical line at A and intersects with the projector that passes through the bottom of the line, parallel to the triangle (Step 2). This locates the shadow A'.

Note that horizontal lines and their shadows are parallel and equal in length.

FIG. 18.97 The construction of the shadow of an isometric with a source of light that is oblique to the picture plane.

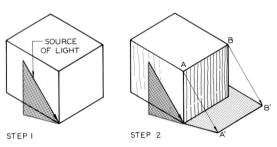

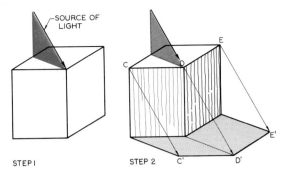

FIG. 18.98 The construction of the shadow of an oblique drawing of a cube with an angular light source.

18.38
Shades and shadows—obliques

An oblique drawing can have shades and shadows applied in the same manner used for isometrics. A triangle is formed to establish the ray of light in Step 1 of Fig. 18.98. The shadows of the vertical corner lines are found. Points C', B', and E' are connected to complete the outline of the shadow. Both the vertical surfaces of the block that are visible are in shade.

18.39
Shades and shadows—perspectives

To find shades and shadows of perspectives, begin by establishing the ray of light. Since the coordinates of the ray of light in Fig. 18.99 intersect at 90°, the light is parallel to the picture plane.

FIG. 18.99 The construction of shades and shadows of a perspective. The light source is parallel to the picture plane.

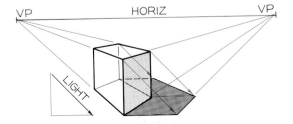

The shadows of the vertical corners of the object are found by passing projectors through the upper corners that are parallel to the light source. They intersect on the horizontal plane where the projectors from the bottom of vertical lines are extended parallel to the horizontal leg of the light triangle.

The shadows of each upper point are found in this manner and are connected to form the outline of the shadow. Note that horizontal lines and their shadows converge to the same vanishing points since they are parallel in reality.

The shades and shadows of an object are shown in Fig. 18.100 where the source of light is at an angle with the picture plane. The light ray is drawn in the top and front views at selected angles. The vanishing points of horizontal and angular shadows are found on the horizontal and below the ground line.

The shadows of the vertical line project the VPL (vanishing point, light) where they intersect the shadows that converge at VPH (vanishing point, horizontal). This is repeated for each corner to complete the outline of the shadow.

18.40
Rendering techniques

Rendering or shading an object is a technique of adding realism to a pictorial. Examples of freehand and instrument rendering are shown on several basic shapes in Fig. 18.101. These illustrations were made by using India ink, but a similar effect can be achieved by using black pencil lines.

A stipple shading technique was used to render the components of a jet turbo engine in Fig. 18.102. This rendered drawing is more realistic than it would be without surface shading.

18.41
Overlay film

Overlay film is an acetate film on which is printed a pattern that can be applied to a drawing to shade an illustration. It is applied as shown in Fig. 18.103. These films have adhesive backings and can be burnished permanently to the surface by a firm rubbing pressure. They are trimmed to size by using a pointed stylus or razor blade.

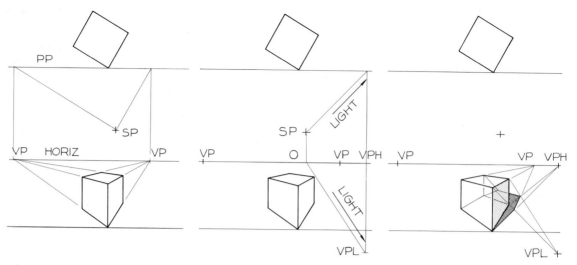

FIG. 18.100 Shades and shadows—oblique light source

Step 1 The perspective of the object is drawn in the conventional manner.

Step 2 The top and front views of the ray of light are established as desired. The SP is projected to the horizon to locate point O. The vanishing point of the light ray (VPL) is found by drawing a construction line parallel to it in the front view from point O.

Step 3 A shadow of a vertical line is found by locating the intersection between the projectors to VPL and VPH.

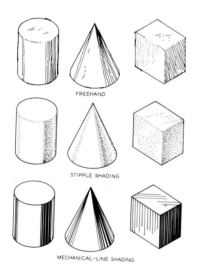

FREEHAND

STIPPLE SHADING

MECHANICAL-LINE SHADING

FIG. 18.101 Line and dot shading add realism to a pictorial.

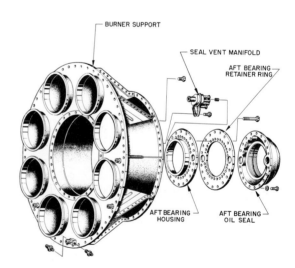

BURNER SUPPORT

SEAL VENT MANIFOLD

AFT BEARING RETAINER RING

AFT BEARING HOUSING

AFT BEARING OIL SEAL

FIG. 18.102 A pictorial assembly that was drawn by using a stippling technique with ink. (Courtesy of General Motors Corporation.)

FIG. 18.103 The steps of applying overlay film to shade an area. (Courtesy of Artype Incorporated.)

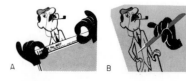

A B C

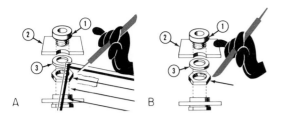

FIG. 18.104 The steps of applying leaders and numbers to an assembly drawing. (Courtesy of Artype Incorporated.)

Patterns, symbols, letters, numbers, arrowheads, and the like are available on film. Arrowheads are applied to an assembly in Fig. 18.104. Overlay films are available with both dull and glossy surfaces.

Examples of various patterns and symbols are shown in Fig. 18.105. Film is also available in percentage screens that are labeled in percentage and lines per inch (Fig. 18.106). The percentage indicates the percentage of solid black; the lines per inch represent the number of rows of dots per inch. A 27.5-line screen is an open screen, whereas a 55-line screen is composed of smaller dots that are twice as close together.

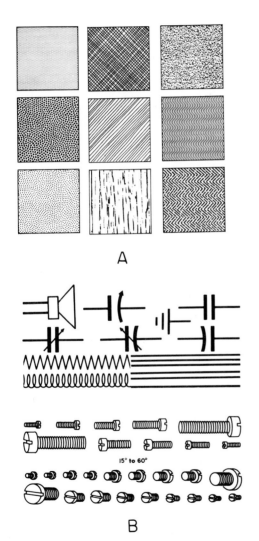

A

B

FIG. 18.105 Examples of a few of the (A) patterns and (B) symbols that are available in 9″ × 12″ sheets of overlay film. (Courtesy of Artype Incorporated.)

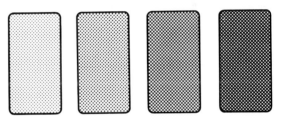

FIG. 18.106 Overlay film is available in screens at 10-percent increments. A 10-percent screen is 10 percent solid black, and a 70-percent screen is 70 percent solid black.

18.42
Photographic illustrations

Technical illustrations can be made from photographs as shown in Fig. 18.107. Drawings can emphasize certain aspects of the photograph that might not otherwise be clear. The photograph can be converted to a drawing by projecting it onto the drawing surface by an opaque projector where it can be outlined and then rendered.

Photographs of a large size can be overlaid with tracing vellum or polyester film and the drawing made by tracing from the photograph. A bold illustration is required if it is to be reduced to a much smaller size for publication.

Photographs can be used to illustrate parts such as the assembly in Fig. 18.108. These parts were positioned and photographed, and the background was masked (whited out) to eliminate the props used to support them while they were being photographed.

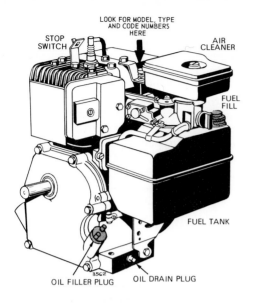

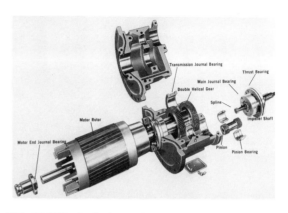

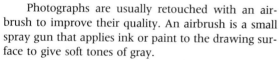

FIG. 18.108 A photographic assembly that has been retouched with an airbrush. (Courtesy of Carrier Air Conditioning Company.)

FIG. 18.107 An example of a line drawing made from a photograph to improve its reproduction when reduced to a small size. (Courtesy of Briggs & Stratton Corporation.)

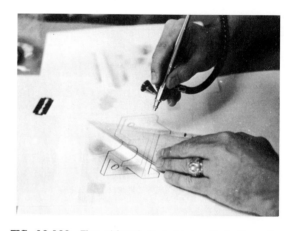

FIG. 18.109 The airbrush is a device that allows the illustrator to apply a light spray of ink or paint on an illustration to obtain gradual tones.

FIG. 18.110 This drawing was rendered by using a multimedia method with pencil, ink, and liquid-tipped markers. (Courtesy of Ford Motor Company.)

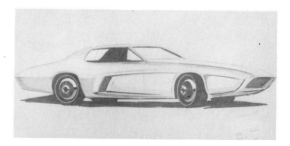

Photographs are usually retouched with an airbrush to improve their quality. An airbrush is a small spray gun that applies ink or paint to the drawing surface to give soft tones of gray.

An airbrush is shown in Fig. 18.109 as it is used to shade a drawing. The illustrator who uses an airbrush must use numerous masks to shield the portion of the drawing surface where shading is unwanted. The masks are called *friskets*. They are applied to the surface, and windows are cut from them to open the areas to be sprayed.

18.43
Multimedia illustrations

The technical illustrator may use many media to achieve the desired effect when making a technical il-lustration. These can include pencil, ink, charcoal, overlay film, airbrush, and liquid-tipped markers.

The freehand sketch of the automobile in Fig. 18.110 was made with a combination of pencil and liquid-tipped markers. The result is a pleasing sketch that enables the designer to develop and communicate styling ideas while working.

Problems

The following problems are to be solved on Size A or B sheets as assigned by your teacher. Select the appropriate scale that will best take advantage of the space available on each sheet.

Obliques

1–31. Construct either cavalier, cabinet, or general obliques of the objects assigned.

Isometrics

1–31. Construct isometric drawings of the objects assigned.

Axonometrics

1–31. Construct axonometric scales and find the ellipse template angles for each surface by using a cube rotated into positions of your choice. Calibrate the scales to represent ½″ intervals. Overlay the scales and draw axonometrics of the objects assigned.

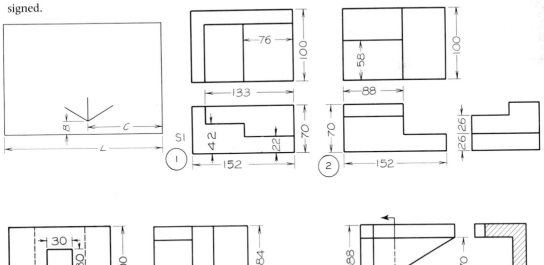

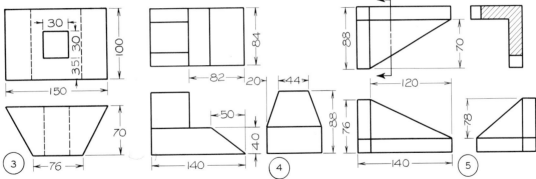

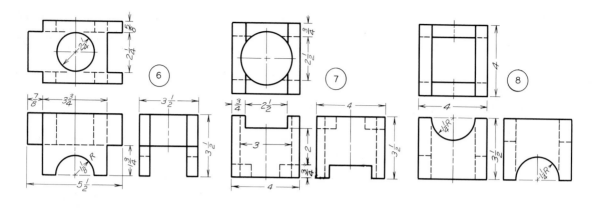

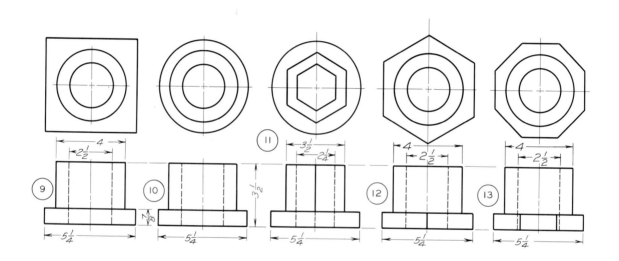

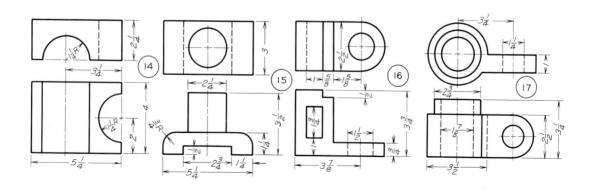

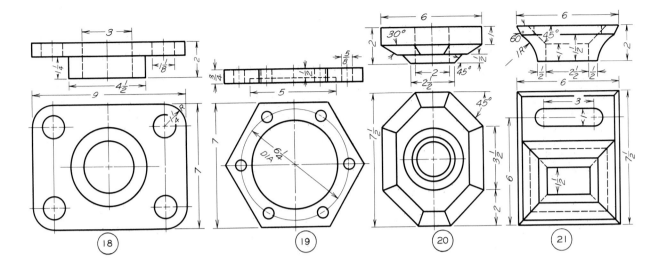

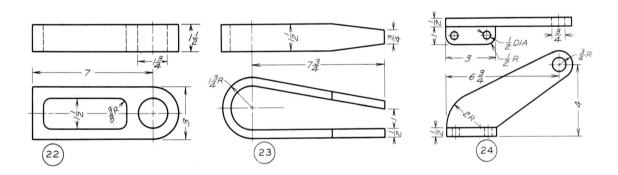

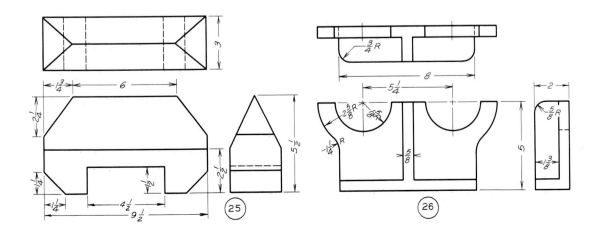

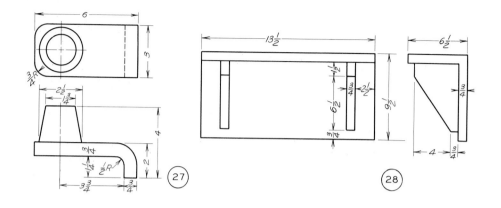

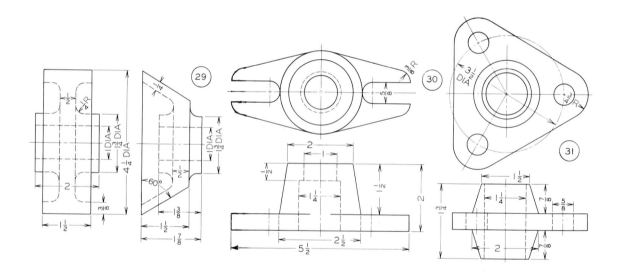

Perspectives

32–37. Lay out these perspective problems on Size B sheets and complete the perspectives as assigned.

38–42. Construct three-point perspectives of parts 1 through 5 as assigned. Select the most appropriate scale to take best advantage of the space available.

Shades and Shadows

43–55. Construct pictorials (any type as assigned) of parts 1–13 and find the shades and shadows of each. Establish the source of light that will best enhance the pictorial.

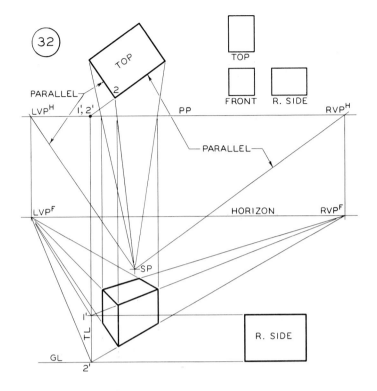

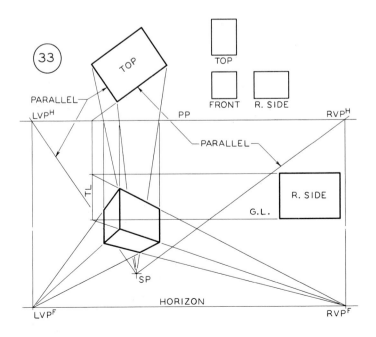

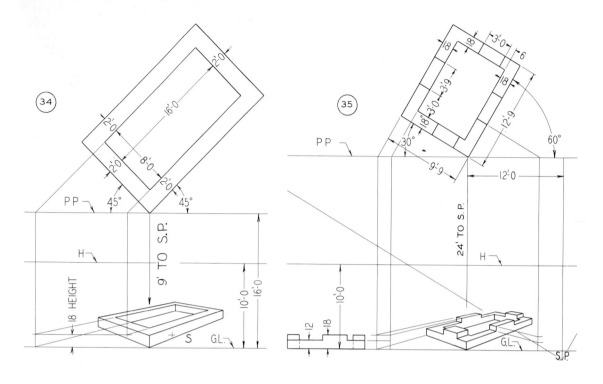

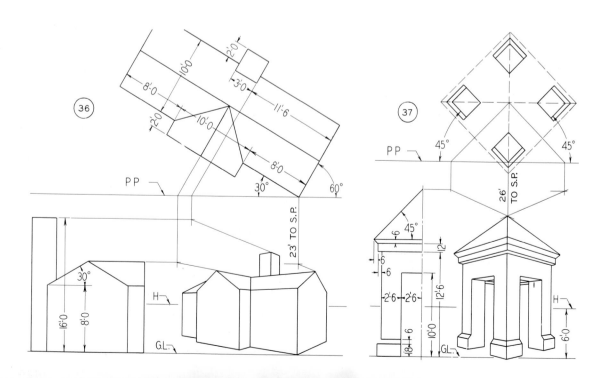

19 Graphs

19.1
Introduction

Data and information expressed as numbers and words are difficult to analyze or evaluate unless they are transcribed into graphical form. Drawings of data shown graphically are called *graphs*. They are sometimes called *charts,* an acceptable term, but one that is more appropriate when referring to maps, which are specialized forms of graphs.

Graphs are helpful in the communication of data to others (Fig. 19.1); consequently this is a popular means of briefing other people on trends that would otherwise be difficult to communicate. The trends of a plotted curve on a graph can be compared to the

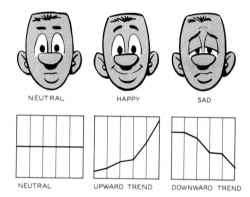

FIG. 19.2 Curves on a graph are similar to expressions on a face.

expressions on a person's face, which is a graph of sorts that reveals a person's feelings (Fig. 19.2). For example, a flat curve shows no change, while an upwardly inclined curve indicates a positive increase. A downward curve, on the other hand, represents a downward trend and a negative result.

This chapter will deal with the more commonly used graphs. The basic types are:

1. Pie graphs
2. Bar graphs
3. Linear coordinate graphs
4. Logarithmic coordinate graphs
5. Semilogarithmic coordinate graphs
6. Schematics and diagrams

FIG. 19.1 Graphs are helpful in presenting technical data to one's associates.

FIG. 19.3 When graphs are drawn to be photographed, they must be laid out at a proportion that will match the proportion of the film in the camera.

19.2
Size proportions of graphs

Graphs may be used to illustrate technical reports that are reproduced in quantity, and they may be used for projection by slide or overhead projectors. In all cases, the proportion of the graph must be determined so that it will match the proportion of the space or the format of the visual aid.

FIG. 19.4 This diagonal-line method can be used for constructing areas whose sides are proportional to those of a 35-mm slide.

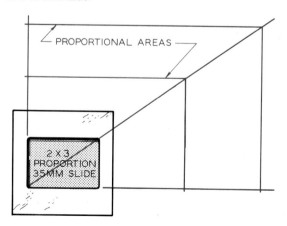

If a graph is to be photographed by a 35-mm camera (Fig. 19.3), the graph must conform to the standard size of the 35-mm film that is used, approximately 2 × 3, as shown in Fig. 19.4.

The proportions of the area in which the graph is to be drawn can be enlarged or reduced by using the diagonal-line method, as illustrated in Fig. 19.4. The horizontal dimension of the slide opening is extended to the right, and the left edge is extended upward. Any point on either of these extended lines is projected to the diagonal and then to the other extended line, to give an area of equal proportion.

19.3
Pie graphs

Pie graphs compare the relationship of parts to a whole when there are only a few parts. Figure 19.5 shows the distribution of skilled workers employed in industry; this graph gives a good visual comparison of these groups.

The method of drawing a pie graph is shown in Fig. 19.6. Note that the given data do not give as good an impression of the comparisons as does the pie graph, even though the data are quite simple.

To facilitate lettering within narrow spaces, the thin sectors should be placed as nearly horizontal as

FIG. 19.5 A pie graph shows the relationship of the parts to a whole. It is effective when there are only a few parts.

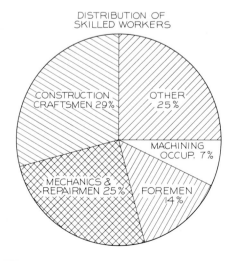

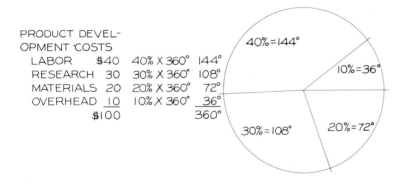

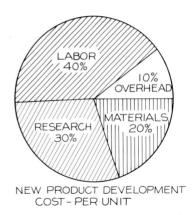

NEW PRODUCT DEVELOPMENT
COST – PER UNIT

FIG. 19.6 Drawing pie graphs

Step 1 The total sum of the parts is found, and the percentage of each is found. Each percentage is multiplied by 360° to find the angle of each sector of the pie graph.

Step 2 The circle is drawn to represent the pie graph. Each sector is drawn using the degrees found in Step 1. The smaller sectors should be placed as nearly horizontal as possible.

Step 3 The sectors are labeled with their proper names and percentages. In some cases, it might be desirable to include the exact numbers in each sector as well.

possible. This provides more room for the label and the percentage. The actual percentage should be given in all cases, and it may be desirable to give the actual numbers or values as well in each sector.

19.4
Bar graphs

Bar graphs are effective for comparing values, especially since they are well understood by the general public. For example, the production of timber for var-

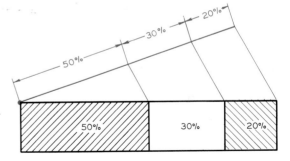

FIG. 19.8 The method of constructing a single bar where the sum of all the parts will be 100%.

ious uses is compared in Fig. 19.7. In this example, the bars show not only the overall production (total lengths of the bars), but the portions of the total devoted to the three uses of the timber.

A bar graph can be composed of a single bar (Fig. 19.8). The total length of the bar is 100%, and the bar is divided into lengths that are proportional to the percentages represented by each of the three parts of the bar.

The method of constructing a bar graph is given in Fig. 19.9. In this case, the title of the graph is placed inside the graph, where space is available. The title could have been placed under or over the graph.

For bar graphs, the data should be sorted by arranging the bars in ascending or descending order,

FIG. 19.7 In this example, each bar represents 100% of the total amount, and each bar represents different totals.

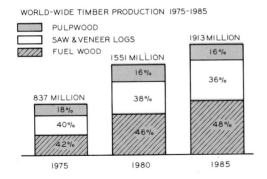

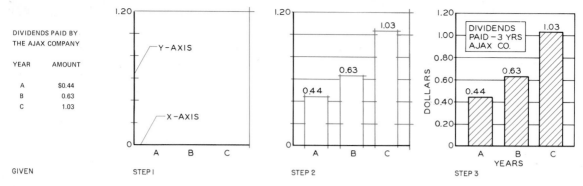

FIG. 19.9 Construction of a bar graph

Given These data are to be plotted as a bar graph.

Step 1 Lay off the vertical and horizontal axes so that the data will fit on the grid. Make the bars begin at zero.

Step 2 Construct and label the bars. The width of the bars should be different from the space between them. Horizontal grid lines should not pass through the bars.

Step 3 Strengthen lines, title the graph, label the axes, and crosshatch the bars.

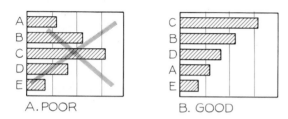

FIG. 19.10 The bars at A are arranged alphabetically. The resulting graph is not as easy to evaluate as the one at B, where the bars have been sorted and arranged in descending order.

FIG. 19.11 A horizontal bar graph that is arranged in descending order to show where engineers are employed.

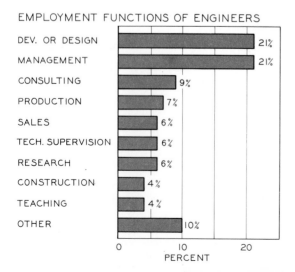

EMPLOYMENT FUNCTIONS OF ENGINEERS

DEV. OR DESIGN	21%
MANAGEMENT	21%
CONSULTING	9%
PRODUCTION	7%
SALES	6%
TECH. SUPERVISION	6%
RESEARCH	6%
CONSTRUCTION	4%
TEACHING	4%
OTHER	10%

PERCENT

since it is desirable to know how the data ranks from category to category (Fig. 19.10). An arbitrary arrangement of bars, alphabetically or numerically, results in a graph that is more difficult to evaluate than the descending arrangement at B.

If the data are sequential and involve time, such as sales per month, it would be less effective to rank the data in ascending order because it is more important to see variations related to periods of time.

Bars in a bar graph may be horizontal, as shown in Fig. 19.11, or vertical, as shown in Fig. 19.12. In both of these cases, the bars are arranged in order of length for ease of comparison and ranking of the data. It is desirable for the bars of a graph to begin at zero to show a true comparison in the data.

FIG. 19.12 This bar graph has the bars arranged in descending order to compare several sources of pollution. (Courtesy of Boeing Company.)

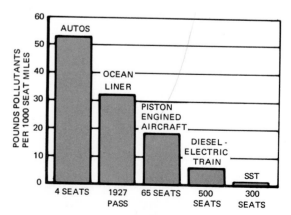

19.5
Linear coordinate graphs

A typical coordinate graph is given in Fig. 19.13 with the accompanying notes that explain its important features. When divisions along an axis of the graph are equal, the scale is linear. The other axis is also divided into equal units and therefore is also a linear scale.

Points are plotted on the grid by using two measurements, called *coordinates*, made along each axis. The plotted points are indicated by using easy-to-draw symbols, such as circles, that can be drawn with a template.

The horizontal axis of the graph is called the *abscissa* or *x*-axis, and the vertical scale at the left is called the *ordinate* or *y*-axis. In mathematics, data are plotted in terms of *x*- and *y*-coordinates on a graph.

Once the points have been plotted, the curve is drawn from point to point. (The line that is drawn to represent the plotted points is called a curve whether it is a smooth or a broken line.) The curve should not close up the plotted points but should leave them as open circles or symbols (see Fig. 19.15).

The curve must be drawn as a heavy, prominent line, since it is the most important part of the graph and shows the data. In Fig. 19.13, there are two curves; therefore it is helpful if they are drawn as different types of lines to distinguish between them. Each is labeled with a note and a leader.

The title of the graph is placed inside the graph in a box to explain the graph. It is a good rule to give enough information on a graph to make it understandable without additional text.

Units are labeled along the *x*- and *y*-axes with labels that designate the units that the graph is comparing.

Broken-line graphs

The steps of drawing a linear coordinate graph are shown in Fig. 19.14. In this case, the points are connected with a broken-line curve since the data points are ten years apart on the *x*-axis. Thus it is impossible to assume that the change in the data is a smooth, continual progression from point to point.

For the best appearance, the plotted points should not be crossed by the curve or the grid lines of the graph (Fig. 19.15). Each circle or symbol used to plot points should be about ⅛″ (5 mm) in diameter.

Different symbols can be used to plot points, along with distinctively different lines to represent the curves. Several approved symbols and lines are shown in Fig. 19.16.

The title of a graph can be placed in any of the three positions shown in Fig. 19.17. The title should never be one as meaningless as "Graph" or "Coordinate Graph." Instead, it should explain the graph by giving important information, such as the company, date, source of data, and general comparisons being shown.

The example in Fig. 19.18A shows a properly labeled axis. You can see that the axis in part B has too many grid lines and too many units labeled along the

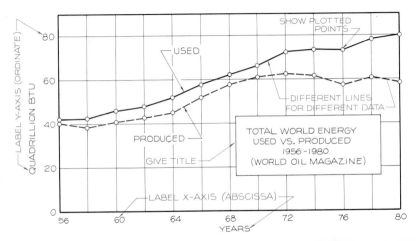

FIG. 19.13 The basic linear coordinate graph with the important features identified.

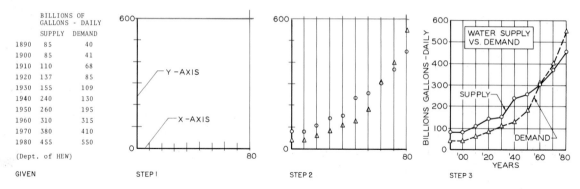

BILLIONS OF GALLONS - DAILY		
	SUPPLY	DEMAND
1890	85	40
1900	85	41
1910	110	68
1920	137	85
1930	155	109
1940	240	130
1950	260	195
1960	310	315
1970	380	410
1980	455	550

(Dept. of HEW)

GIVEN   STEP I   STEP 2   STEP 3

FIG. 19.14   Construction of a broken-line graph

Given   A record of water supply and water demand since 1890 plotted as a line graph.

Step 1   The vertical and horizontal axes are laid off to provide space for the largest values.

Step 2   The points are plotted directly over the respective years. Different symbols are used for each curve.

Step 3   The data points are connected with straight lines, the axes are labeled, the graph is titled, and the lines are strengthened.

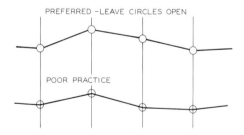

FIG. 19.15   The curve of a graph should be drawn from point to point, but it should not close up the symbols used to locate the plotted points.

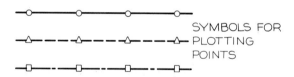

FIG. 19.16   Any of these symbols or lines can be used effectively to represent different curves on a single graph. The symbols are about ⅛" (4 mm) in diameter.

FIG. 19.17   Title placement on a graph

A.   The title of a graph can be lettered inside a box placed within the area of the graph. The perimeter lines of the box should not coincide with grid lines within the graph.

B.   The title can be placed over the graph. The title should be drawn in ⅛" letters or slightly larger.

C.   The title can be placed under the graph. It is good practice to be consistent when a series of graphs is used in the same report.

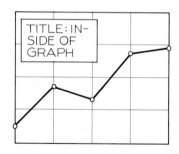

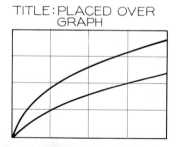

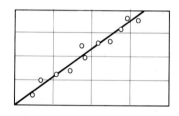

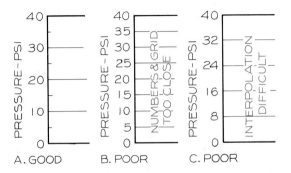

FIG. 19.18 The scale at A is the best. It has about the right number of grid lines and divisions, and the numbers are given in well-spaced, easy-to-interpolate form. The numbers at B are too close, and there are too many grid lines. The units at C are given in units that make interpolation difficult by eye.

axis; this clutters the graph without adding to its value. On the other hand, the units selected at C make it difficult to interpolate easily between the labeled values. For example, it is more difficult to locate a value such as 22 by eye on this scale than on the one at A.

Smooth-line graphs

In order to determine whether the points on a graph will be connected with a broken-line curve as previ-

FIG. 19.19 When the process that is graphed involves gradual, continuous changes in relationships, the curve should be drawn as a smooth line.

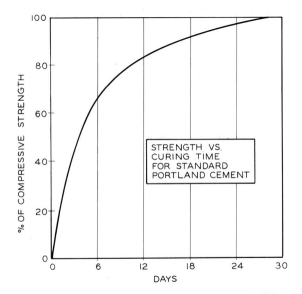

ously shown or a smooth-line curve as shown in Fig. 19.19, you must have an understanding of the data. You instinctively understand that the strength of cement and its curing time will result in a smooth, continuous relationship that should be connected by a smooth curve. Even if the data points do not plot to lie on the curve, you know that the deviation of the points from this curve is due to errors of measurement or the methods used in collecting the data.

Similarly, the strength of clay tile, as related to its absorption characteristics, is an example of data that yield a smooth curve (Fig. 19.20). Note that the plotted data do not lie on the curve. Since you know that the relationship is a continuous one that should be connected with a smooth curve, the *best curve* is drawn to interpret the data to give an average representation of the points.

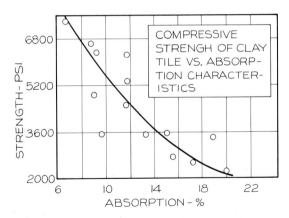

FIG. 19.20 If it is known that a relationship plotted on a graph should yield a smooth gradual curve, a smooth-line "best" curve is drawn to represent the average of the plotted points. You must use your judgment and knowledge of the data in cases of this type.

For the same reasons, you know that there is a smooth-line curve relationship between miles per gallon and the speed at which a car is driven. Two engines are compared in Fig. 19.21 with two smooth-line curves. The effect of speed on several automotive characteristics is compared in Fig. 19.22.

When a smooth-line curve is used to connect data points, there is the implication that you can *interpolate* between the plotted points to estimate other values. Points connected by a broken-line curve imply that you cannot interpolate between the plotted points.

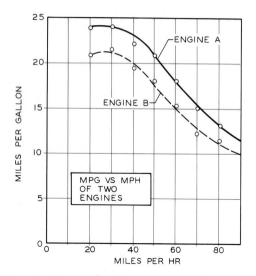

FIG. 19.21 These are "best" curves that approximate the data without necessarily passing through each data point. Inspection of the data tells you that this curve should be a smooth-line curve rather than a broken-line curve.

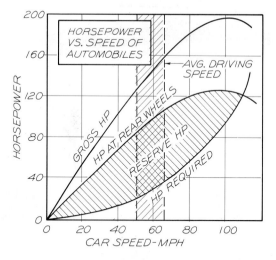

FIG. 19.22 A linear coordinate graph is used here to analyze data affecting the design of an automobile's power system.

Straight-line graphs

Some graphs have neither broken-line curves nor smooth-line curves, but straight-line curves as shown in the example in Fig. 19.23. You can determine a third value from the two given values by using this

graph. For example, if you are driving 70 miles per hour and it takes 5 seconds to react to apply your brakes, you will have traveled 500 feet in this interval of time.

Two-scale coordinate graphs

Graphs can be drawn with different scales in combination, such as the one shown in Fig. 19.24. The ver-

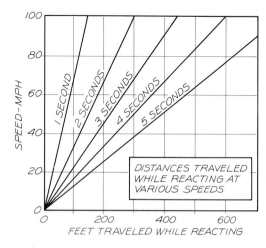

FIG. 19.23 A graph can be used to determine a third value when two variables are known. Taking this information from a graph is easier than computing each answer separately.

FIG. 19.24 This is a composite graph with different scales along each y-axis. The curves are labeled so reference can be made to the applicable scale.

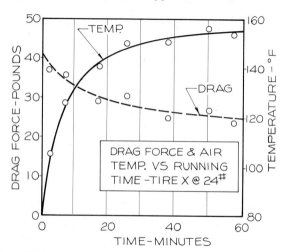

tical scale at the left is in units of pounds, and the one at the right is in units of degrees of temperature. Both curves are drawn using their respective *y*-axes, and each curve is labeled.

Care must be taken to avoid confusing the reader of a graph of this type. However, these graphs are effective when comparing related variables such as the drag force and air temperature of a type of automobile tire, as shown in this example.

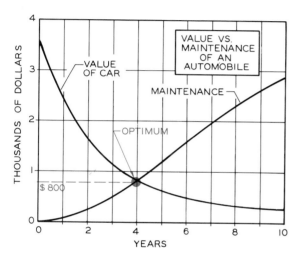

FIG. 19.25 This graph shows the optimum time to sell a car, based on the intersection of two curves that represent the depreciation of the car's value and its increasing maintenance costs.

FIG. 19.26 Optimization graphs

Step 1 Lay out the graph and plot the given curves.

Step 2 Add the two curves to find a third curve. Distance A is shown transferred to locate a point on the third curve. The lowest point of the "total" curve is the optimum point of 11,000 units.

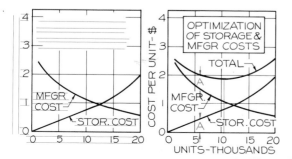

Optimization graphs

Graphs can be used effectively to optimize data. For example, the optimization of the depreciation of an automobile and its increase in maintenance costs is shown in Fig. 19.25. These two sets of data are plotted, and the curves cross at an *x*-axis value of four years. At this time, the cost of maintenance is equal to the value of the car, which indicates that this might be a desirable time to exchange it for a new one.

Another optimization graph is constructed in two steps in Fig. 19.26. The manufacturing cost per unit is reduced as more units are made, but the warehousing cost increases. A third curve is found in Step 2 by adding the two curves. Value *A* is shown to illustrate how this value is transferred with your dividers to add it to the manufacturing cost curve. The "total" curve tells you that the optimum number to manufacture at a time is about 11,000 units. When more or fewer are manufactured, the expense per unit is greater.

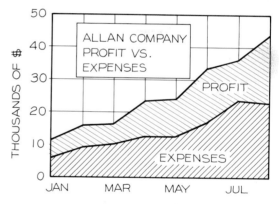

FIG. 19.27 This graph is a combination of a coordinate graph and an area graph. The upper curve represents the total of two values plotted, one above the other.

Composite graphs

The graph in Fig. 19.27 is a composite between an area graph and a coordinate graph. The lower curve is plotted first. The upper curve is found by adding the values to the lower curve so that the two areas represent the data. The upper curve is equal to the sum of the two *y*-values.

The graph in Fig. 19.28 is a combination of a coordinate graph and a bar graph that is used to show

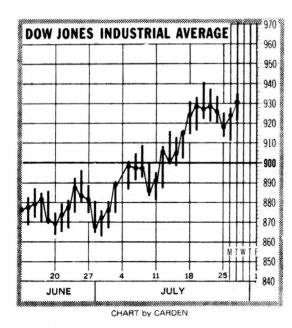

CHART by CARDEN

FIG. 19.28 This graph is a combination of a coordinate graph and a bar graph. The bars represent the ranges of selling during a day, and the broken-line curve connects the points at which the market closed each day.

the Dow-Jones industrial stock average. The bars represent the daily ranges in the index. The broken-line curve connects the points where the market closed for each day.

Break-even graphs

Coordinate graphs are helpful in evaluating marketing and manufacturing costs that are used to determine the selling cost for a product. The break-even graph in Fig. 19.29 is drawn to reveal that 10,000 units must be sold at $3.50 each to cover the manufacturing and development costs. Sales in excess of 10,000 result in profit.

A second type of break-even graph (Fig. 19.30) uses the cost of manufacturing per unit versus the number of units produced. In this example, the development costs must be incorporated into the unit costs. The manufacturer can determine how many units must be sold to break even at a given price, or the price per unit if a given number is selected. In this example, a sales price of 80¢ requires that 8400 units be sold to break even.

FIG. 19.29 Break-even graph

Step 1 The graph is drawn to show the cost ($20,000 in this case) of developing the product. It is determined that each unit would cost $1.50 to manufacture. This is a total investment of $35,000 for 10,000 units.

Step 2 In order for the manufacturer to break even at 10,000, the units must be sold for $3.50 each. Draw a line from zero through the break-even point for $35,000.

Step 3 The loss is $20,000 at zero units and becomes progressively less until the break-even point is reached. The profit is the difference between the cost and income to the right of the break-even point.

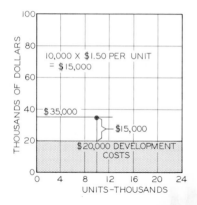

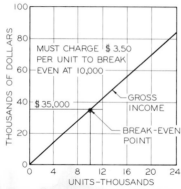

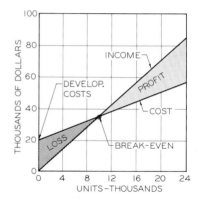

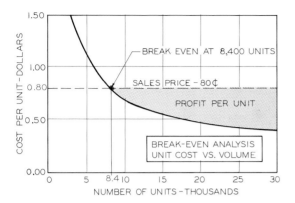

FIG. 19.30 The break-even point can be found on a graph that shows the relationship between the cost per unit, which includes the development cost, and the number of units produced. The sales price is a fixed price. The break-even point is reached when 8400 units have been sold at 80¢ each.

19.6
Logarithmic coordinate graphs

Both scales of a logarithmic grid are calibrated into divisions that are equal to the logarithms of the units represented. Commercially printed logarithmic grid paper is available in many variations that can be used for graphing data.

The graph in Fig. 19.31 has a logarithmic grid and shows the geometry of standard railroad cars as they relate to the tracks so that there will not be more than a maximum projection width of 12 feet around curves. Extremely large values can be shown on logarithmic grids since the lengths are considerably compressed.

19.7
Semilogarithmic coordinate graphs

Semilogarithmic graphs are referred to as ratio graphs because they give graphical representations of ratios. One scale, usually the vertical scale, is logarithmic, and the other is linear (divided into equal divisions). Two curves that are parallel on a semilogarithmic graph have equal percentage increases.

In Fig. 19.32, the same data are shown plotted on a linear grid and on a semilogarithmic grid. The semilogarithmic graph reveals that the percent of change from 0 to 5 is greater for curve B than for curve A, since curve B is steeper. This comparison was not apparent in the plot on the linear grid.

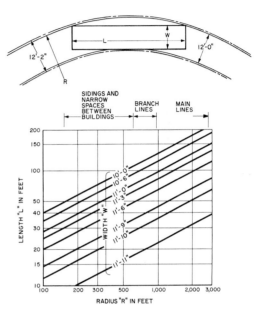

FIG. 19.31 This logarithmic graph shows the maximum load projection of 12 feet in relation to the length of a railroad car and the radius of the curve. (Courtesy of *Plant Engineering.*)

FIG. 19.32 When plotted on a standard grid, curve A appears to be increasing at a greater rate than curve B. However, the true rate of increase can be seen when the same data are plotted on a semilogarithmic graph in part B.

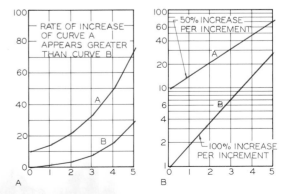

The relationship between the linear scale and the logarithmic scale is shown in Fig. 19.33. Equal divisions along the linear scale have unequal ratios, and equal divisions along the log scale have equal ratios.

Log scales can be drawn to have one or many cycles. Each cycle increases by a factor of 10. For example, the scale in Fig. 19.34A is a three-cycle scale, and the one at B is a two-cycle scale. When these must be drawn to a special length, commercially printed log

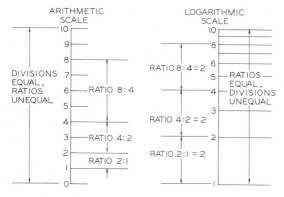

FIG. 19.33 The spacings on an arithmetic scale are equal, with unequal ratios between points. The spacings on logarithmic scales are unequal, but equal spaces represent equal ratios.

FIG. 19.34 Logarithmic paper can be purchased or drawn using several cycles. Three-, two-, and one-cycle scales are shown here. Calibrations can be drawn on a scale of any length by projecting from a printed scale as shown in part C.

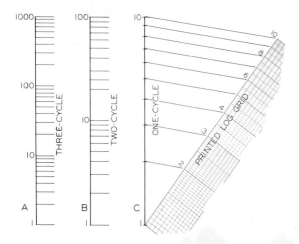

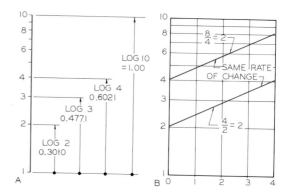

FIG. 19.35 A number's logarithm is used to locate its position on a log scale (A). This makes it possible to see the true rate of change at any location on a semilogarithmic graph (B).

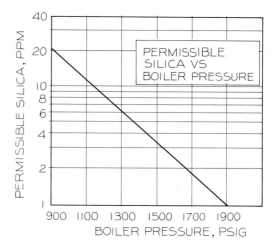

FIG. 19.36 A semilogarithmic graph is used to compare the permissible silica (parts per million) in relation to the boiler pressure.

scales can be used to transfer graphically the calibrations to the scale being drawn (Fig. 19.34C).

In Fig. 19.35, the calibrations along the log scale are separated by the difference in their logarithms. The logarithms are laid off by using a convenient scale that is calibrated in decimal divisions.

It can be seen in Fig. 19.35B that parallel straight-line curves yield equal ratios of increase. Figure 19.36 is an example of a semilogarithmic graph used to present industrial data.

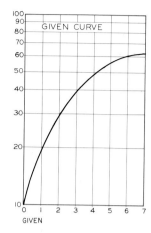 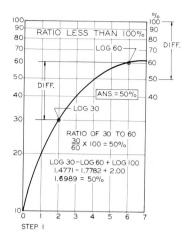 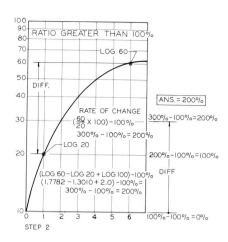

FIG. 19.37 Percentage graphs

Given The data are plotted on a semilogarithmic graph to enable you to determine percentages and ratios in much the same manner that you use a slide rule.

Step 1 In finding the percent that a smaller number is of a larger number, you know that the percent will be less than 100%. The log of 30 is subtracted from the log of 60 with dividers and is transferred to the percent scale at the right, where 30 is found to be 50% of 60.

Step 2 To find the percent of increase, a smaller number is divided into a larger number to give a value greater than 100%. The difference between the logs of 60 and 20 is found with dividers and is measured upward from 100% at the right, to find that the percent of increase is 200%.

"Semilog" graphs have several disadvantages. The most critical one is that they are misunderstood by many people who do not recognize them as being different from linear coordinate graphs. Also, zero values cannot be shown on log scales.

Percentage graphs

The percent that one number is of another, or the percent increase of one number that is greater than the other, can be determined by using a semilogarithmic graph. Examples of both are shown in Fig. 19.37.

Data plotted in Step 1 are used to find the percent that 30 is of 60, two points on the curve. The vertical distance between the two points is equal to the difference of their logarithms. This distance is subtracted from the log of 100 at the right of the graph, to give a value of 50% as a direct reading.

In Step 2, the percent of increase between two points is transferred from the grid to the lower end of the log scale and measured upward. It is measured upward since the increase is greater than zero. The percent of increase is measured directly from scale. These methods can be used to find percent increases or decreases of any sets of points on the grid.

19.8
Polar graphs

Polar graphs are drawn with a series of concentric circles with the origin at the center. Lines are drawn from the center toward the perimeter of the graph, where the data can be plotted through 360° by measuring values from the origin. The illumination of a lamp is shown in Fig. 19.38, where the maximum lighting of the lamp is 550 lumens at 35° from the vertical, for example.

This type of graph is used to plot the areas of illumination of all types of lighting fixtures. Polar graph paper is available commercially for drawing graphs of this type.

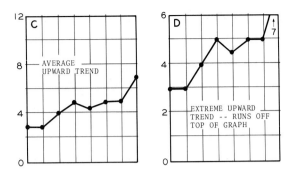

FIG. 19.45 The upward trend of the data at D appears greater than the curve at C. Both graphs show the same data.

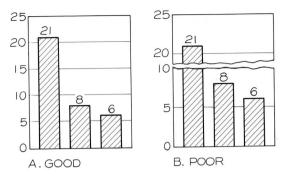

FIG. 19.46 This bar graph distorts the data at B because a portion of the graph has been removed, negating the pictorial value of the graph.

Problems

The problems below are to be drawn on Size A sheets (8½″ × 11″) in pencil or ink as specified. Follow the techniques that were covered in this chapter and the examples that are given as you solve the problems.

Pie graphs

1. Draw a pie graph that compares the employment of male youth between the ages of 16 and 21: Operators—25%; Craftsmen—9%; Professionals, technicians, and managers—6%; Clerical and sales—17%; Service—11%; Farm workers—13%; and Laborers—19%.

2. Draw a pie graph that shows the proportional relationship between the following members of the technological team: Engineers—985,000; Technicians—932,000; and Scientists—410,000.

3. Construct a pie graph of the following percentages of the employment status of graduates of two-year technician programs one year after graduation: Employed—63%; Continuing full-time study—23%; Considering job offers—6%; Military—6%; and Other—2%.

4. Construct a pie graph that shows the relationship between the types of degrees held by engineers in aeronautical engineering: Bachelor—65%; Master—29%; and Ph.D.—6%.

5. Draw a pie graph for the following average annual expenditures of a state on public roads: Construction—$13,600,000; Maintenance—$7,100,000; Purchase and upkeep of equipment—$2,900,000; Bonds—$5,600,000; and Engineering and administration—$1,600,000.

6. Draw a pie graph that shows the data given in Problem 9.

7. Draw a pie graph that shows the data given in Problem 10.

Bar graphs

8. Draw a bar graph that depicts the unemployment rate of high-school graduates and dropouts in various age categories. The age groups and the percent of unemployment of each group are given in the following table:

Ages	Percent of Labor Force	
	Graduates	Dropouts
16–17	18	22
18–19	12.5	17.5
20–21	8	13
22–24	5	9

9. Draw a single bar that represents 100% of a die casting alloy. The proportional parts of the alloy are: Tin—16%; Lead—24%; Zinc—38.8%; Aluminum—16.4%; and Copper—4.8%.

10. Draw a bar graph that compares the number of skilled workers employed in various occupations. Arrange the graph for ease of interpretation and comparison of occupations. Use the following data: Carpenters—82,000; All-round machinists—310,000; Plumbers—350,000; Bricklayers—200,000; Appliance servicers—185,000; Automotive mechanics—760,000; Electricians—380,000; and Painters—400,000.

11. Draw a bar graph that represents the flow of a river in cubic feet per second (cfs) as shown in the following table. Show bars that represent the data for ten days in the first month only. Omit the second month.

Day of Month	Rate of Flow in 1000 cfs	
	1st Month	2nd Month
1	19	19
2	130	70
3	228	79
4	129	33
5	65	19
6	32	14
7	17	15
8	13	11
9	22	19
10	32	27

12. Draw a bar graph that shows the airline distances in statute miles from New York to the cities listed in the table below. Arrange in ascending or descending order.

- Berlin 3965
- Buenos Aires 5300
- Honolulu 4960
- London. 3465
- Manila 8510
- Mexico City 2090
- Moscow 4665
- Paris. 3634
- Tokyo 6740

13. Draw a bar graph that compares the corrosion resistance of the materials listed in the table below:

	Loss in Weight %	
	In Atmosphere	In Sea Water
Common steel	100	100
10% Nickel steel	70	80
25% Nickel steel	20	55

14. Draw a bar graph using the data in Problem 1.

15. Draw a bar graph using the data in Problem 2.

16. Draw a bar graph using the data in Problem 3.

17. Construct a rectangular grid graph to show the accident experience of Company A. Plot the numbers of disabling accidents per million person-hours of work on the y-axis. Years will be plotted on the x-axis. Data: 1970—1.21; 1971—0.97; 1972—0.86; 1973—0.63; 1974—0.76; 1975—0.99; 1976—0.95; 1977—0.55; 1978—0.76; 1979—0.68; 1980—0.55; 1981—0.73; 1982—0.52; 1983—0.46.

Linear coordinate graphs

18. Using the data given in Table 19.1, draw a linear coordinate graph that compares the supply and demand of water in the United States from 1890 to 1980. Supply and demand are given in billions of gallons of water per day.

19. Present the data in Table 19.2 in a linear coordinate graph to decide which lamps should be selected to provide economical lighting for an industrial plant. The table gives the candlepower directly under the lamps (0°) and at various angles from the vertical when the lamps are mounted at a height of 25 feet.

20. Construct a linear coordinate graph that shows the relationship in energy costs (mills per kilowatt-hour) and the percent capacity of two types of power plants. Plot energy costs along the y-axis, and the capacity factor along the x-axis. The plotted curve will compare the costs of a nuclear plant with a gas- or oil-fired plant. Data for a gas-fired plant: 17 mills, 10%; 12 mills, 20%; 8 mills, 40%; 7 mills, 60%; 6 mills, 80%; 5.8 mills, 100%. Nuclear plant data: 24 mills, 10%; 14 mills, 20%;

Semilogarithmic graphs

36. Construct a semilogarithmic graph with the y-axis as a two-cycle log scale from 1 to 100 and the x-axis as a linear scale from 1 to 7. Plot the data below to show the survivability of a shelter at varying distances from a one-megaton air burst. The data consist of overpressure in psi along the y-axis, and distance from ground zero in miles along the x-axis. The data points represent an 80% chance of survival of the shelter. Data: 1 mile, 55 psi; 2 miles, 11 psi; 3 miles, 4.5 psi; 4 miles, 2.5 psi; 5 miles, 2.0 psi; 6 miles, 1.3 psi.

37. The growth of two divisions of a company, Division A and Division B, is given in the data below. Plot the data on a rectilinear graph and on a semilog graph. The semilog graph should have a one-cycle log scale on the y-axis for sales in thousands of dollars, and a linear scale on the x-axis showing years for a six-year period. Data in dollars: 1 yr, A = \$11,700 and B = \$44,000; 2 yr, A = \$19,500 and B = \$50,000; 3 yr, A = \$25,000 and B = \$55,000; 4 yr, A = \$32,000 and B = \$64,000; 5 yr, A = \$42,000 and B = \$66,000; 6 yr, A = \$48,000 and B = \$75,000. Which division has the better growth rate?

38. Draw a semilog chart showing probable engineering progress throughout history. Use the following indices: 40,000 B.C. = 21; 30,000 B.C. = 21.5; 20,000 B.C. = 22; 16,000 B.C. = 23; 10,000 B.C. = 27; 6000 B.C. = 34; 4000 B.C. = 39; 2000 B.C. = 49; 500 B.C. = 60; A.D. 1900 = 100. Horizontal scale 1″ = 10,000 years. Height of cycle = about 5″. Two-cycle printed paper may be used if available.

39. Plot the data from Problem 24 as a semilogarithmic graph.

40. Plot the data from Problem 26 as a semilogarithmic graph.

Percentage graphs

41. Plot the data given in Problem 18 on a semilog graph. What is the percent of increase in the demand for water from 1890 to 1920? What percent of the demand is the supply for the following years: 1900, 1930, and 1970?

42. Using the graph plotted in Problem 37, determine the percent of increase of Division A and Division B from year 1 to year 4. Also, what percent of sales of Division A are the sales of Division B at the end of year 2? At the end of year 6?

43. On semilog paper, plot two values from Problem 26—water horsepower and electric horsepower—compared with gallons per minute along the x-axis. What percent is water horsepower of the electric horsepower when 75 gallons per minute are being pumped? What is the percent increase of the electric horsepower from 0 to 185 gallons per minute?

Organizational charts

44. Draw an organizational chart for a city government organized as follows: The electorate elects school board, city council, and municipal court officers. The city council is responsible for the civil service commission, city manager, and city planning board. The city manager's duties cover finance, public safety, public works, health and welfare, and law.

45. Draw an organizational chart for a manufacturing plant. The sales manager, chief engineer, treasurer, and general manager are responsible to the president. Each of these officers has a department force. The general manager has three department heads: master mechanic, plant superintendent, and purchasing agent. The plant superintendent has charge of the shop foremen, under whom are the working forces, and also has direct charge of the shipping, tool and die, inspection, order, and stores and supplies departments.

Polar Graphs

46. Construct a polar graph of the data given in Problem 19.

47. Construct a polar graph of the following illumination, in lumens at various angles, emitted from a luminaire. The zero-degrees position is vertically under the overhead lamp. Data: 0°, 12,000; 10°, 15,000; 20°, 10,000; 30°, 8000; 40°, 4200; 50°, 2500; 60°, 1000; 70°, 0. The illumination is symmetrical about the vertical.

20
Pipe Drafting

20.1
Introduction

An understanding of pipe drafting begins with a familiarity of the types of pipe that are available. The commonly used types of pipe are (1) steel pipe, (2) cast-iron pipe, (3) copper, brass, and bronze pipe and tubing, and (4) plastic pipe.

The standards for the grades and weights for pipe and pipe fittings are specified by several organizations to ensure uniformity of size and strength of interchangeable components. Among these organizations are the American National Standards Institute (ANSI), the American Society for Testing and Materials (ASTM), the American Petroleum Institute (API), and the Manufacturers Standardization Society (MSS).

20.2
Welded and seamless steel pipe

Traditionally, steel pipe has been specified in three weights—standard (STD), extra strong (XS), and double extra strong (XXS). These designations are still in use, and their specifications are listed in the ANSI B 36.10-1979 standards. However, additional designations for pipe called *schedules* have been introduced to provide the pipe designer with a wider selection of pipe to cover more applications.

The ten schedules are Schedules 10, 20, 30, 40, 60, 80, 100, 120, 140, and 160. The wall thicknesses of the pipes vary from the thinnest, in Schedule 10, to the thickest, in Schedule 160. The outside diameters are of a constant size for pipes of the same nominal size in all schedules.

Schedule designations correspond to STD, XS, and XXS specifications in some cases, as is shown partially in Table 20.1. This table has been abbreviated from the ANSI B 36.10-1979 tables by omitting a number of the pipe sizes and schedules. The most often used schedules are 40, 80, and 120.

You will note in the table that pipes from the smallest size up to and including 12″ pipes are specified by their inside diameter (ID), which means that the outside diameter (OD) is larger than the specified size. The inside diameters are the same size as the nominal sizes of the pipe for standard-weight pipe. For XS and XXS pipe, the inside diameters are slightly different in size from the nominal size. Beginning with the 14″ diameter pipes, the nominal sizes represent the outside diameters of the pipe.

The standard lengths for steel pipe are 20 feet and 40 feet. Seamless steel (SMLS STL) pipe is a smooth pipe with no weld seams along its length. Welded pipe is formed into a cylinder and is butt-welded (BW) at the seam, or it is joined with an electric resistance weld (ERW).

Welded and seamless pipe are the most widely used types of pipe for process and chemical refining plants. The type and grade of pipe with the proper characteristics must be selected to resist a wide range of pressures and corrosive agents.

TABLE 20.1
DIMENSIONS AND WEIGHTS OF WELDED AND SEAMLESS STEEL PIPE
(ANSI B 36.10-1979)

	Inch Units			Identification			SI Units		
Inch Nominal Size	O.D. DIA. (in.)	Wall Thk. (in.)	Weight (lbs/ft)	*STD XS XXS	Sch. No.		O.D. DIA. (mm)	Wall Thk. (mm)	Weight (kg/m)
½	0.84	0.11	0.85	STD	40		21.3	2.8	1.3
1	1.32	0.13	1.68	STD	40		33.4	3.4	2.5
1	1.30	0.18	2.17	XS	80		33.4	4.6	3.2
1	1.30	0.36	3.66	XXS			33.4	9.1	5.5
2	2.38	0.22	3.65	STD	40		60.3	3.9	5.4
2	2.38	0.22	5.02	XS	80		60.3	5.5	7.5
2	2.38	0.44	9.03	XXS			60.3	11.1	13.4
4	4.50	0.23	10.79	STD	40		114.3	6.0	16.1
4	4.50	0.34	14.98	XS	80		114.3	8.6	42.6
4	4.50	0.67	27.54	XXS			114.3	17.1	41.0
8	8.63	0.32	28.55	STD	40		219.1	8.2	42.6
8	8.63	0.50	43.39	XS	80		219.1	12.7	64.6
8	8.63	0.88	74.40	XXS			219.1	22.2	107.9
12	12.75	0.38	49.56	STD			323.0	9.5	67.9
12	12.75	0.50	65.42	XS			323.0	12.7	97.5
12	12.75	1.00	125.40	XXS	120		133.9	25.4	187.0
14	†14.00	0.38	54.57	STD	30		355.6	9.5	87.3
14	14.00	0.50	72.08	XS			355.6	12.7	107.4
18	18.00	0.38	70.59	STD			457.0	9.5	106.2
18	18.00	0.50	93.45	XS			457.0	12.7	139.2
24	24.00	0.38	94.62	STD	20		610.0	9.5	141.1
24	24.00	0.50	125.49	XS			610.0	12.7	187.1
30	30.00	0.38	118.65	STD			762.0	9.5	176.8
30	30.00	0.50	157.53	XS	20		762.0	12.7	234.7
40	40.00	0.38	158.70	STD			1016.0	9.5	236.5
40	40.00	0.50	210.90	XS			1016.0	12.7	314.2

*Standard (STD)
  Extra strong (XS)
  Double extra strong (XXS)
†Beginning with 14″ DIA pipe, the nominal size represents the outside diameter (O.D.).
This table has been compressed by omitting many of the available pipe sizes. The nominal sizes of pipes that are listed in the complete table are: ⅛″, ¼″, ⅜″, ½″, ¾″, 1″, 1¼″, 1½″, 2″, 2½″, 3″, 3½″, 4″, 5″, 6″, 8″, 10″, 12″, 14″, 16″, 18″, 20″, 22″ . . . (at 2″ increments up to 60″).

20.3
Cast-iron pipe

Cast-iron pipe is used for the transportation of liquids, water, gas, and sewerage. When used as a sewerage pipe, cast-iron pipe is referred to as "soil pipe." Cast-iron pipe is available in diameter sizes from 3 inches to 60 inches.

The standard lengths of this type of pipe are 5 feet and 10 feet. Cast iron is more brittle and more subject to cracking when it is loaded than is steel pipe. Therefore it should not be used where high pressures of weights will be applied to it.

20.4
Copper, brass, and bronze piping

Copper, brass, and bronze are used to manufacture piping and tubing for use in applications where there must be a high resistance to corrosive elements, such as acidic soils and chemicals that are transmitted through the pipes. Copper pipe is used for gas and water transmission when the pipes are placed within or under concrete slab foundations of buildings. This ensures that the pipes will not have to be replaced because of corrosion. The standard length of pipes made of these nonferrous materials is 12 feet.

Tubing is a smaller-size pipe that can be easily bent when it is made of copper, brass, or bronze. An example of a system of tubing is shown in Fig. 20.1. The term *piping* is considered to apply to rigid pipes that are larger than tubes, usually in excess of 2″ in diameter.

20.5
Miscellaneous pipes

Other materials that are used to manufacture pipes are aluminum, asbestos-cement, concrete, polyvinyl chloride (PVC), and various other plastics. Each of these materials has its special characteristics that make it desirable or economical for certain applications. The

method of designing and detailing piping systems by the pipe drafter is essentially the same regardless of the piping material used.

20.6
Pipe joints

The basic connection in a pipe system is the joint where two straight sections of pipe fit together. Three types of joints are illustrated in Fig. 20.2: *screwed*, *welded*, and *flanged*.

SCREWED JOINTS are joined by pipe threads of the type covered in Chapter 10. A table of pipe-thread specifications is given in Appendix 10. Pipe threads are tapered at a ratio of 1 to 16 on the radius (Fig.

FIG. 20.1 The small pipes in this illustration are called tubes because they are less than 2″ in diameter and can be bent to form angles.

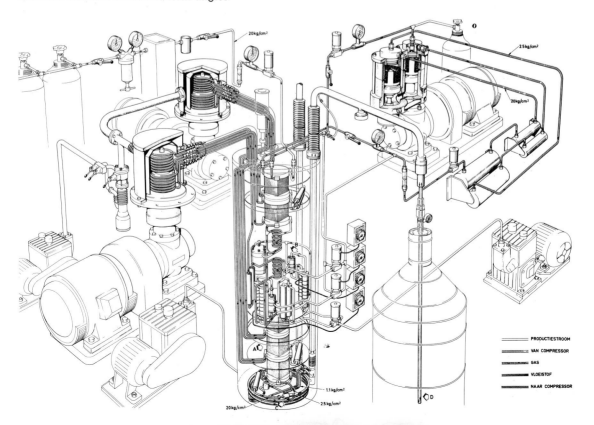

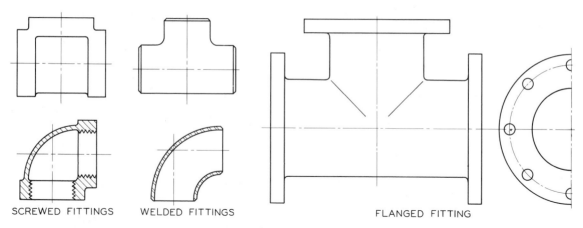

FIG. 20.2 The three basic types of joints are screwed, welded, and flanged joints.

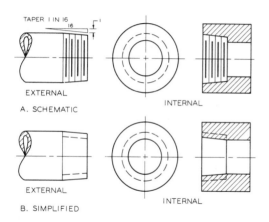

FIG. 20.3 A screwed pipe utilizes a type of threading called pipe threads that binds tightly as they are screwed together.

20.3). As the pipes are screwed together, the threads bind to form a snug, locking fit. A cementing compound is applied to the threads before joining, to improve the seal.

FLANGED JOINTS, shown in Fig. 20.4, are welded to the straight sections of pipe, which are then bolted together with a series of bolts around the perimeter of the flanges. Flanged joints form strong, rigid joints that can withstand high pressure and permit disassembly of the joints when needed. Several types of flange faces are shown in Fig. 20.5 and in Appendix 13.

WELDED JOINTS are joined by welded seams around the perimeter of the pipe to form butt welds. Welded joints are used extensively in "big-inch" pipelines that are used for transporting petroleum products cross-country.

FIG. 20.4 Types of flanged joints and the methods of attaching the flanges to the pipes. (Courtesy of Vogt Machine Co.)

WELDING NECK FLANGES   LAP JOINT FLANGES   SOCKET WELDING FLANGES   SLIP-ON WELDING FLANGES   THREADED FLANGES

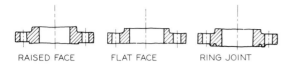

RAISED FACE   FLAT FACE   RING JOINT

FIG. 20.5 Three types of flange faces are the raised face (RF), flat face (FF), and ring joint (RJ).

BELL-AND-SPIGOT (B&S) JOINTS are used to join cast-iron pipes, as illustrated in Fig. 20.6. The spigot is placed inside the bell, and the two are sealed with molten lead or a commercially available sealing ring that snaps into position to form a sealed joint.

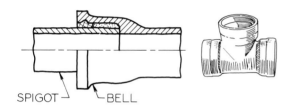

SPIGOT   BELL

FIG. 20.6 A bell-and-spigot joint (B&S) is used to connect cast-iron pipes.

SOLDERING is used to connect smaller pipes and tubular connections. Soldering is usually limited to nonferrous tubing. However, screwed fittings are available to connect tubing, as shown in Fig. 20.7. In this example, the joint is sealed more firmly as the fitting is tightened.

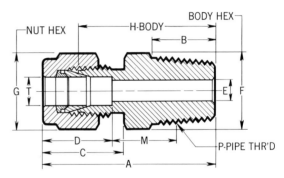

FIG. 20.7 This screwed fitting is used to attach small tubing. (Courtesy of Crawford Fitting Co.)

20.7
Screwed fittings

A number of standard fittings used to assemble a piping system are shown in Fig. 20.8. The two types of graphical symbols that are used to represent fittings and pipe are *double-line* and *single-line symbols*.

Double-line symbols are pictorially more representative of the fittings and pipes since they are drawn to scale with double lines. Single-line symbols are more symbolic since the size of the pipe is drawn with a single line and the fittings are drawn as single-line schematic symbols. Either type of representation can be used to present a piping system in orthographic and pictorial views.

Fittings are available in three weights: *standard* (STD), *extra strong* (XS), and *double extra strong* (XXS). These weights match the standard weights of the pipes to which they will be connected. Other weights of fittings are available from manufacturers, but these three weights are stocked by practically all suppliers.

A piping system of screwed fittings is shown in Fig. 20.9 with single-line symbols in a single-line system to call attention to them. These could just as well have been drawn with the double-line symbols. You should become familiar with the names and methods of specifying the various fittings by type and size.

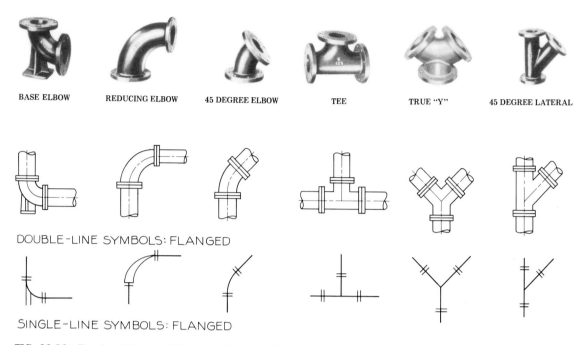

FIG. 20.10 Standard flanged fittings are shown in this example along with the single-line and double-line symbols that are used to represent them.

FIG. 20.11 Standard fittings, welded, are shown in this example along with the single-line and double-line symbols that are used to represent them.

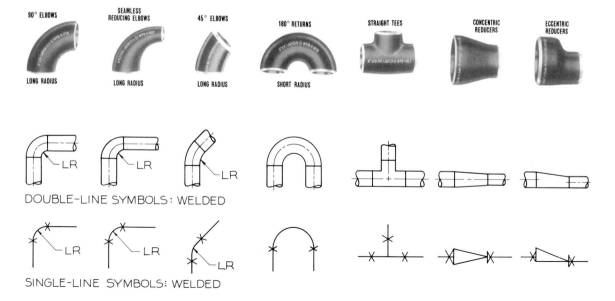

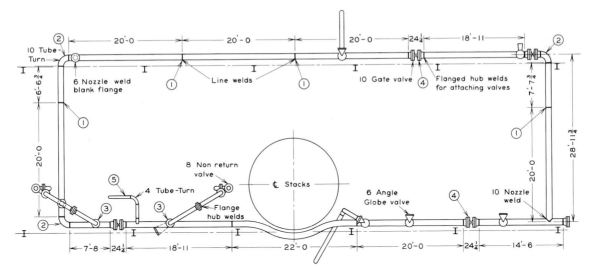

FIG. 20.12 A piping layout that uses a series of welded and flanged joints. This system is drawn with double-line symbols.

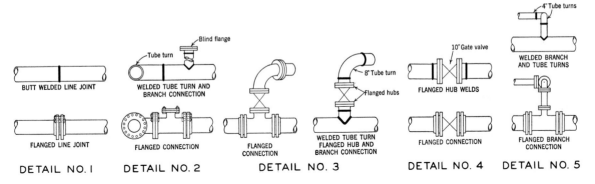

FIG. 20.13 A comparison of flanged and welded joints drawn with double-line symbols.

relief valves, to name a few. The three basic types—*gate, globe,* and *check valves*—are shown in Fig. 20.14 with single-line symbols.

GATE VALVES are used to turn the flow within a pipe on or off with the least restriction of flow through the valve. These valves are not meant to be used to regulate the degree of flow.

GLOBE VALVES are used not only to turn the flow on and off, but also to regulate the flow to a desired level by adjusting the handwheel control.

ANGLE VALVES are types of globe valves that turn at 90° angles at bends in the piping system. They have the same controlling features as the straight globe valves.

CHECK VALVES permit the flow in the pipe in only one direction. A backward flow is checked by either a movable piston or a swinging washer that will be activated by a change of flow to close the opening (Fig. 20.14).

Symbols for the other types of valves can be found in Fig. 20.15. Examples of double-line symbols depicting valves and fittings are shown in Fig. 20.16.

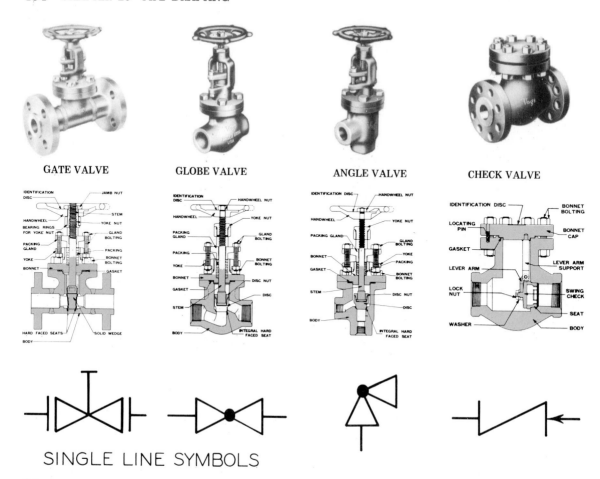

GATE VALVE GLOBE VALVE ANGLE VALVE CHECK VALVE

SINGLE LINE SYMBOLS

FIG. 20.14 The three basic types of valves are gate, globe, and check valves. (Photographs courtesy of Vogt Machine Co.)

FIG. 20.15 Examples of types of valves and fittings.

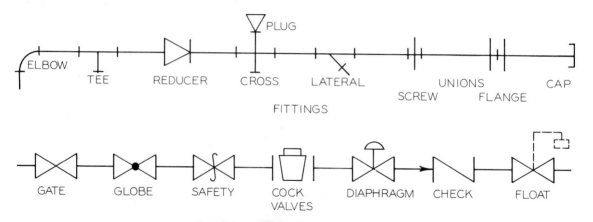

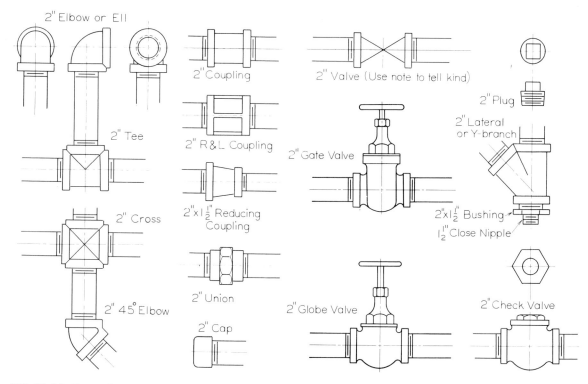

FIG. 20.16 Examples of valves and fittings drawn with double-line symbols.

FIG. 20.17 Fittings and valves drawn here with single-line symbols are drawn differently in the various orthographic views. Observe these differences and how the various views are drawn.

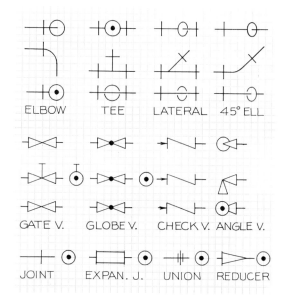

20.11
Fittings in orthographic views

Fittings and valves must be shown from any view in orthographic projection; consequently you must be familiar with the methods of representing different views of these components. Two and three views of typical fittings and valves are shown in Fig. 20.17. These fittings are drawn as single-line screwed fittings, but the same general principles can be used to represent other types of joints as double-line drawings. Several views of fittings are given with double-line symbols in Fig. 20.16.

Observe the various views of the fittings and notice how the direction of an elbow can be shown by a slight variation in the different views. The same techniques are used to represent tees and laterals.

A piping system is shown in a single orthographic view in Fig. 20.18, where a combination of double-line and single-line symbols is drawn. Note that arrows are used to give the direction of flow in the system. Different joints are screwed, welded, and flanged.

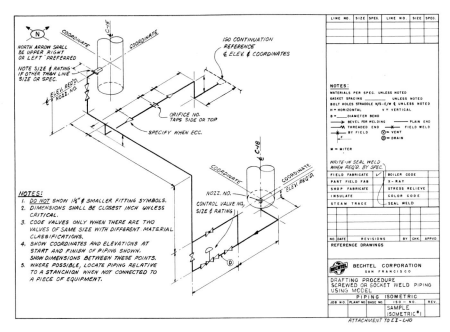

FIG. 20.24 A suggested format for preparing spool drawings by the Bechtel Corporation is shown here. (Courtesy of the Bechtel Corporation.)

FIG. 20.25 A detail drawing of a cylindrical vessel that is connected into a pipe system.

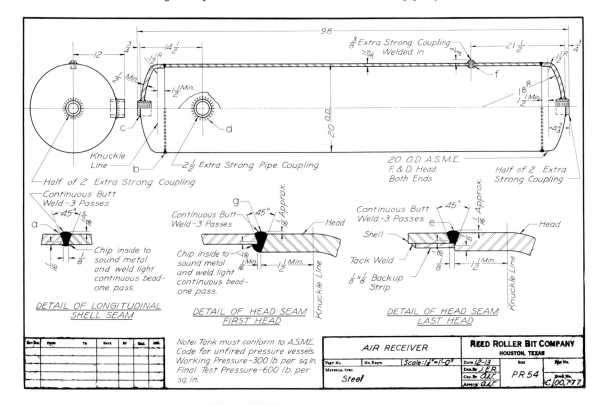

Problems

1. On a Size A sheet, draw five orthographic views of the fittings listed below. The views should include the front, top, bottom, left, and right views. Draw two fittings per page. Refer to Fig. 20.17 and Appendix 9 as guides in making these drawings. Use single-line symbols to draw the following screwed fittings: 90° ell, 45° ell, tee, lateral, cap reducing ell, cross, concentric reducer, check valve, union, globe valve, gate valve, and bushing.

2. Same as Problem 1, but draw the fittings as flanged fittings.

3. Same as Problem 1, but draw the fittings as welded fittings.

4. Same as Problem 1, but draw the fittings as double-line screwed fittings.

5. Same as Problem 1, but draw the fittings as double-line flanged fittings.

6. Same as Problem 1, but draw the fittings as double-line welded fittings.

7. Convert the single-line sketch in Fig. 20.26 into a double-line system that will fit on a Size A sheet.

FIG. 20.26 Problem 7.

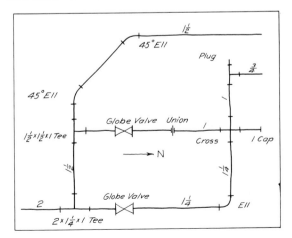

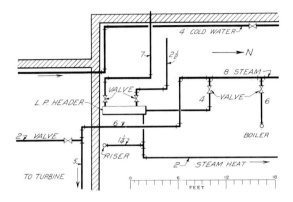

FIG. 20.27 Problem 8.

8. Convert the single-line pipe system in Fig. 20.27 into a double-line drawing that will fit on a Size A sheet, using the graphical scale given in the drawing to select the best scale for the system.

9. Convert the piping system shown in isometric in Fig. 20.20 into a double-line isometric drawing that will fit on a Size B sheet.

10. Convert the piping system given in Fig. 20.20 into a two-view orthographic drawing with single-line symbols.

11. Convert the isometric drawing of the piping system in Fig. 20.24 into a two-view orthographic drawing that will fit on a Size B sheet. Take the measurements from the given drawing and select a convenient scale.

12. Convert the isometric of the piping system in Fig. 20.19 into a double-line isometric that will fit on a Size B sheet.

13. Convert the orthographic piping system in Fig. 20.9 into a single-line isometric drawing that will fit on a Size B sheet. Estimate the dimensions.

14. Convert the orthographic piping system in Fig. 20.9 into a double-line orthographic view that will fit on a Size B sheet.

21

Electric/
Electronics
Drafting

21.1
Introduction

Electric/electronics drafting is a specialty area in the field of drafting technology. Electrical drafting is related to the transmission of electrical power used in large quantities in the home and industry for lighting, heating, and equipment operation. Electronics drafting deals with circuits in which electronic tubes or transistors are used, where power is used in smaller quantities. Examples of electronic equipment are radios, televisions, computers, and similar products.

FIG. 21.1 The recommended line weights for drawing electronics diagrams.

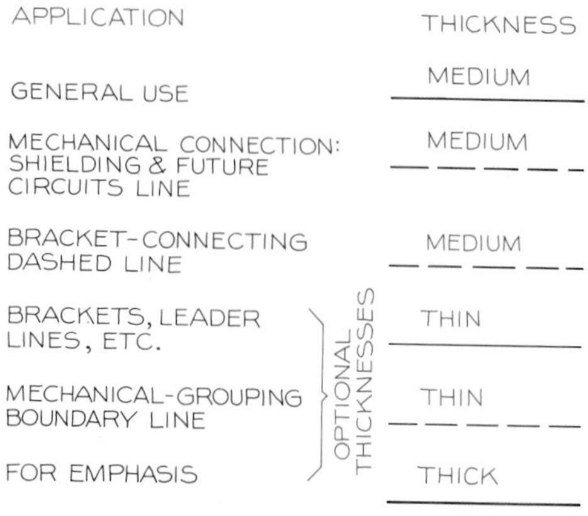

APPLICATION		THICKNESS
GENERAL USE		MEDIUM
MECHANICAL CONNECTION: SHIELDING & FUTURE CIRCUITS LINE		MEDIUM
BRACKET-CONNECTING DASHED LINE		MEDIUM
BRACKETS, LEADER LINES, ETC.	OPTIONAL THICKNESSES	THIN
MECHANICAL-GROUPING BOUNDARY LINE		THIN
FOR EMPHASIS		THICK

Electronics drafters are responsible for the preparation of drawings that will be used in fabricating the circuit and thereby bringing the product into being. They will work from sketches and specifications developed by the engineer or electronics technologist. Although it is not necessary that they be knowledgeable in the design of a circuit, drafters must be familiar with standard methods of representing various parts within the electrical/electronic circuit.

A major portion of the text in this chapter has been extracted from ANSI Y14.15, *Electrical and Electronics Diagrams,* the standard that regulates the drafting techniques used in this area. The symbols used were taken from ANSI Y32.2, *Graphic Symbols for Electrical and Electronics Diagrams.*

21.2
Types of diagrams

Electronic circuits are classified and drawn in the format of one of the following types of diagrams:

1. Single-line diagrams
2. Schematic diagrams
3. Connection diagrams

The suggested line weights for drawing these diagrams are shown in Fig. 21.1.

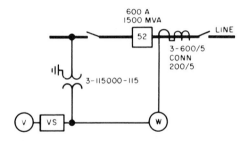

FIG. 21.2 A portion of a single-line diagram where heavy lines represent the primary circuits, and medium lines represent the connections to the current and potential sources. (Courtesy of ANSI.)

Single-line diagrams

Single-line diagrams use single lines and graphic symbols to show the course of an electric circuit or system of circuits and the parts and devices within it. A single-line diagram is used to represent both AC and DC systems, as illustrated in Fig. 21.2. An example of a single-line diagram of an audio system is shown in Fig. 21.3. Primary circuits are indicated by thick connecting lines, and medium lines are used to represent connections to the current and potential sources.

Single-line diagrams show the connections of meters, major equipment, and instruments. In addition, ratings are often given to supplement the graphic symbols to provide such information as kilowatt ratings, voltages, cycles and revolutions per minute, and generator ratings. An example of a diagram of this type with the ratings added is shown in Fig. 21.4.

Schematic diagrams

Schematic diagrams are graphic symbols showing the electrical connections and functions of a specific circuit arrangement. Although the schematic diagram enables one to trace the circuit and its functions, the physical sizes, shapes, and locations of various components are not given. A schematic diagram is illustrated in Fig. 21.5 (see pages 466–467); this diagram should be referred to for applications of the principles covered in this chapter.

Connection diagrams

Connection diagrams show the connections and installations of the parts and devices of the system. In addition to showing the internal connections, external

FIG. 21.3 A typical single-line diagram for illustrating electronics and communications circuits. (Courtesy of ANSI.)

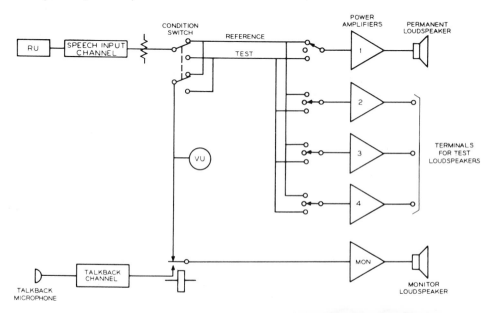

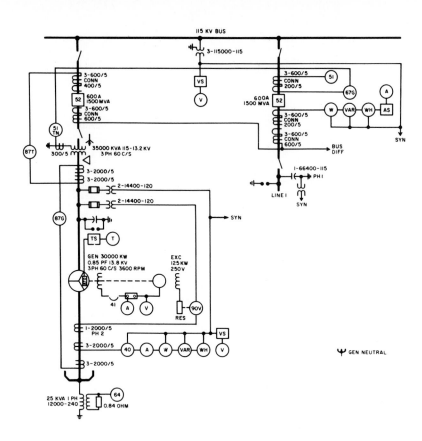

FIG. 21.4 A single-line diagram that is used to illustrate a power switchgear complete with the device designations noted on the diagram. (Courtesy of ANSI.)

FIG. 21.7 Connections should be shown with single-point junctions as shown at (A). Dots may be used to call attention to connections as shown at (B); and dots *must* be used when there are multiples of the type shown at (C).

FIG. 21.6 A three-dimensional connection diagram that shows the circuit and its components with the necessary dimensions to explain how it is connected or installed. (Courtesy of the General Motors Corporation.)

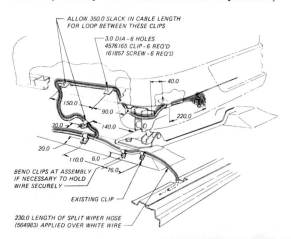

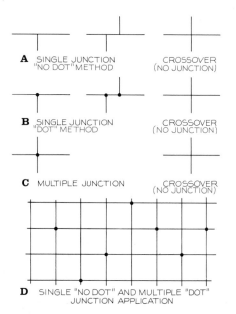

connections, or both, the connection diagram shows the physical arrangement of the parts. It can be described as an installation diagram such as the one shown in Fig. 21.6.

21.3
Schematic diagram connecting symbols

Of the symbols used to communicate the composition of a circuit graphically, the most basic symbols are those used to represent connections of parts within the circuit. The use of dots to show connections is optional; it is preferable to omit them if clarity is not sacrificed by so doing (Fig. 21.7A). The dots are used to distinguish between connecting lines and those that simply pass over each other (B). It is preferred that connecting wires have single junctions wherever possible.

When the layout of a circuit does not permit the use of single junctions, and lines within the circuit must cross, then dots must be used to distinguish between crossing and connecting lines (C, D).

INTERRUPTED PATHS are breaks in lines within a schematic diagram that are interrupted to conserve space when this can be done without confusion. For example, the circuit in Fig. 21.8 has been interrupted since the lines do not connect the left and the right sides of the circuit. Instead, the ends of the lines are

FIG. 21.8 Circuits may be interrupted and connections not shown by lines if they are properly labeled to clarify their relationship to the part of the circuit that has been removed. The connections above are labeled to match those on the left and right sides of the illustration. (Courtesy of ANSI.)

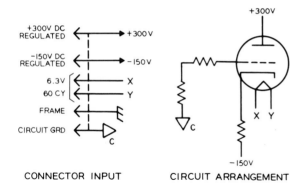

CONNECTOR INPUT   CIRCUIT ARRANGEMENT

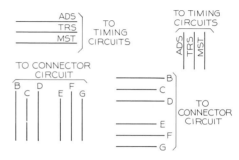

FIG. 21.9 Brackets and notes may be used to specify the destinations of interrupted circuits, as shown in this illustration.

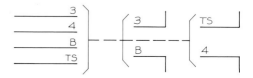

FIG. 21.10 The connections of interrupted circuits can be indicated by using brackets and a dashed line in addition to labeling the lines. The dashed line should not be drawn to appear as an extension of one of the lines in the circuit.

labeled to correspond to the matching notes at the other side of the interrupted circuit.

There will be occasions where sets of lines in a horizontal or vertical direction will be interrupted (Fig. 21.9). Brackets will be used to interrupt the circuit, and notes will be placed outside the brackets to indicate the destinations of the wires or their connections.

In some cases, a dashed line is used to connect brackets that interrupt circuits (Fig. 21.10). The dashed line should be drawn so that it will not be mistaken as a continuation of one of the lines within the bracket.

MECHANICAL LINKAGES that are closely related to electronic functions may be shown as part of a schematic diagram (Fig. 21.11). An arrangement of this type helps clarify the relationship of the electronics circuit to the mechanical components.

21.4
Graphic symbols

The symbols covered in this section are extracted from the ANSI Y32.2 standard, and they are adequate for practically all diagrams. However, when a highly spe-

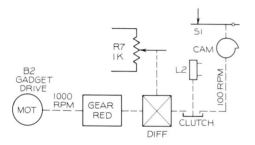

FIG. 21.11 If mechanical functions are closely related to electrical functions, it may be desirable to link the mechanical components within the schematic diagram.

cialized part needs to be shown and a symbol for it is not provided in these standards, it is permissible for the drafter to develop his or her own symbol provided it is properly labeled and the meaning is clearly understood.

The symbols that are presented in Figs. 21.12 through 21.16 are drawn on a grid composed of 5-mm (0.20 inch) squares that have been reduced. The actual dimensions of the symbols can be approximated by using the grid and equating each square to its full-size measurement of 5 mm. Symbols may be drawn larger or smaller to fit the size of your layout provided the relative proportions of the symbols are kept about the same.

The symbols in Figs. 21.12 through 21.16 are but a few of the more commonly used symbols. There are

FIG. 21.12 These graphic symbols that are used to represent parts within a schematic diagram are drawn on a 5-mm (0.20-inch) grid. The suggested sizes of the symbols can be found by taking the dimensions from the grid to draw the symbols full size.

FIG. 21.13 The six upper symbols are used to represent often-used types of electron tubes. The interpretation of the parts that make up each symbol is given in the lower half of the figure.

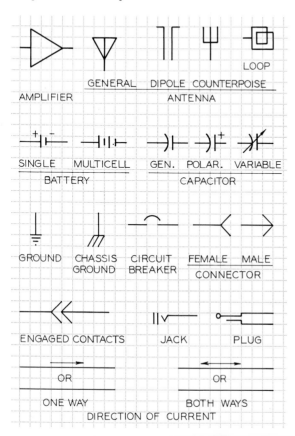

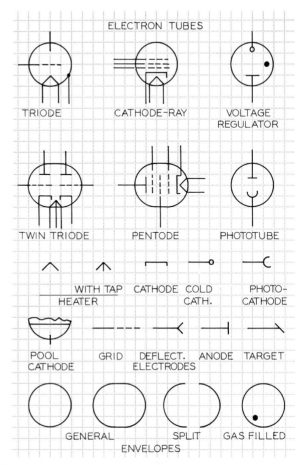

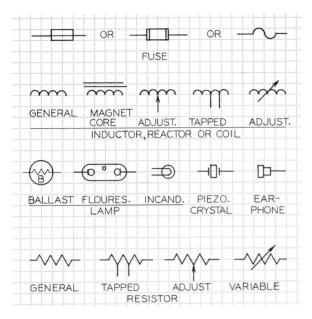

FIG. 21.14 Graphics symbols of standard circuit components.

between five and six hundred different symbols in the ANSI standard for variations of the basic electrical/ electronics symbols.

Electronics symbols can be drawn with conventional drawing equipment and instruments by approximating the proportions of the symbols; drafting templates of electronics symbols are available if you wish to save time and effort. An example of an electronics-symbol template is shown in Fig. 21.17.

Some of the symbols presented as examples have been noted to provide designations of their sizes or ratings. The need for this additional information depends on the requirements and the usage of the schematic diagrams in which the designations appear. Additional information on the preparation of identification information is covered in Section 21.7.

21.5
Terminals

Terminals are the ends of devices that are attached in a circuit with connecting wires. Examples of devices with terminals that are specified in circuit diagrams are switches, relays, and transformers. The graphic

symbol for a terminal is an open circle that is the same size as the solid circle used to indicate a connection.

SWITCHES are used to turn a circuit on or off, or else to actuate a certain part of it while turning another off. Examples of graphical techniques for labeling switches are shown in the schematic diagram in Fig. 21.5. In this case, a table is used with notes to clarify the switching connections of the terminals.

When a group of parts is enclosed or shielded (drawn enclosed with dashed lines) and the terminal circles have been omitted, the terminal markings should be placed immediately outside the enclosure, as shown in Fig. 21.5 at $T2$, $T3$, $T4$, $T5$ and $T8$. The terminal identifications should be added to the graphic symbols that correspond to the actual physical mark-

FIG. 21.15 Graphics symbols for semiconductor devices and transistors. The arrows in the middle of the figure illustrate the meanings of the arrows used in the transistor symbols shown below them.

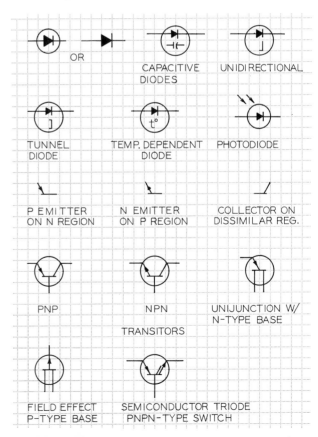

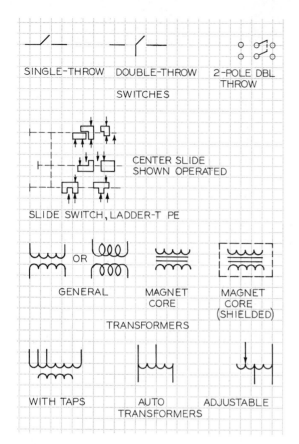

SINGLE-THROW   DOUBLE-THROW   2-POLE DBL THROW

SWITCHES

CENTER SLIDE SHOWN OPERATED

SLIDE SWITCH, LADDER-T PE

GENERAL   MAGNET CORE   MAGNET CORE (SHIELDED)

TRANSFORMERS

WITH TAPS   AUTO   ADJUSTABLE

TRANSFORMERS

FIG. 21.16 Graphic symbols for representing switches and transformers.

FIG. 21.17 Various types of template are available for drawing the graphic symbols used in schematic diagrams. (Courtesy of Frederick Post Company.)

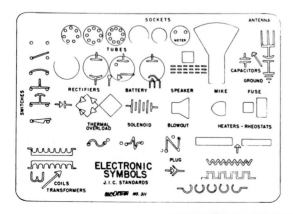

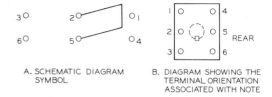

A. SCHEMATIC DIAGRAM SYMBOL

B. DIAGRAM SHOWING THE TERMINAL ORIENTATION ASSOCIATED WITH NOTE

FIG. 21.18 An example of a method of labeling the terminals of a toggle switch on a schematic diagram (A), and a diagram that illustrates the toggle switch when it is viewed from its rear (B).

ings that appear on or near the terminals of the part, such as 10.7 *MC* for transformer *T*3 in Fig. 21.5. When terminal parts are not marked with identifying numbers, notes should be assigned arbitrarily by the drafter or engineer. Several examples of notes and symbols that explain the parts of a diagram are shown in Figs. 21.18, 21.19, and 21.20.

Colored wires or symbols are often used to identify the various leads that connect to terminals. When colored wires need to be identified on a diagram, the colors are lettered on the drawing, as was done for the transformer *T*10 of Fig. 21.5. An example of the identification of a capacitor, *C*40, that is marked with geometric symbols is shown in Fig. 21.5.

ROTARY TERMINALS are used to regulate the resistance in some circuits, and the direction of rotation of the dial is indicated on the schematic diagram. The abbreviations *CW* (clockwise) or *CCW* (counterclockwise) are placed adjacent to the movable contact when it is in its extreme clockwise position, as shown in Fig. 21.21A. The movable contact has an arrow at its end.

FIG. 21.19 An example of a rotary switch as it would appear on a schematic diagram (left), and a diagram that shows the numbered terminals of the switch when viewed from its rear. (Courtesy of ANSI.)

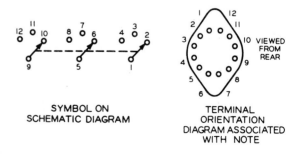

SYMBOL ON SCHEMATIC DIAGRAM

TERMINAL ORIENTATION DIAGRAM ASSOCIATED WITH NOTE

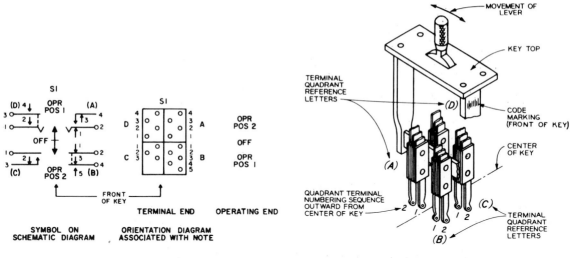

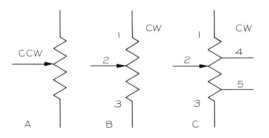

FIG. 21.20 An example of a typical lever switch as it would appear on a schematic diagram (left, part A), and an orientation diagram that shows the numbered terminals of the switch when viewed from its operating end (right, part A). A pictorial of the lever switch and its four quadrants is given in part B of the figure. (Courtesy of ANSI.)

If the device terminals are not marked, numbers may be used with the resistor symbol and the number 2 assigned to the adjustable contact (Fig. 21.21B). Other fixed taps may be sequentially numbered and added as shown in part C.

The position of a switch as it relates to the function of a circuit should be indicated on the schematic diagram. A method of showing functions of a variable switch is shown in Fig. 21.22. The arrow represents the movable end of the switch that can be positioned

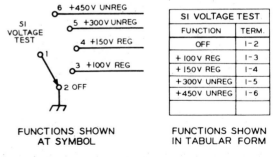

SI VOLTAGE TEST	
FUNCTION	TERM.
OFF	1-2
+ 100 V REG	1-3
+ 150 V REG	1-4
+300 V UNREG	1-5
+450 V UNREG	1-6

FUNCTIONS SHOWN AT SYMBOL   FUNCTIONS SHOWN IN TABULAR FORM

FIG. 21.22 For more complex switches, position-to-position function relations may be shown by using symbols on the schematic diagram or by a table of values located elsewhere on the diagram. (Courtesy of ANSI.)

FIG. 21.21 To indicate the direction of rotation of rotary switches on a schematic diagram, the abbreviations *CW* (clockwise) and *CCW* (counterclockwise) are placed near the movable contact (part A). If the device terminals are not marked, numbers may be used with the resistor symbols and the number 2 assigned to the adjustable contact (part B). Additional contacts may be labeled as shown at C.

to connect with several circuits. The different functional positions of the rotary switch are shown both by symbol and by table in this illustration.

Another method of representing a rotary switch is shown in Fig. 21.23 by symbol and by table. The tabular form is preferred owing to the complexity of this particular switch. The dashes between the numbers in the table indicate that the numbers have been connected. For example, when the switch is in position 2, the following terminals are connected: 1 and 3, 5 and 7, and 9 and 11. A table of this type is used at the bottom of Fig. 21.5.

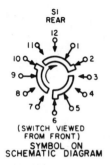

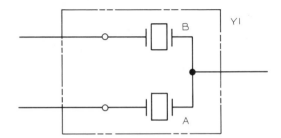

FIG. 21.23 A rotary switch may be shown on a schematic diagram with its terminals labeled as shown at the left, or its functions can be given in a table placed elsewhere on the drawing as shown at the right. Dashes are used to indicate the linkage of the numbered terminals. For example, 1–2 means that terminals 1 and 2 are connected in the "off" position. (Courtesy of ANSI.)

SI (REAR)		
POS	FUNCTION	TERM.
1	OFF(SHOWN)	1-2,5-6,9-10
2	STANDBY	1-3,5-7,9-11
3	OPERATE	1-4,5-8,9-12

(SWITCH VIEWED FROM FRONT)
SYMBOL ON SCHEMATIC DIAGRAM

FUNCTIONS SHOWN IN TABULAR FORM

Electron tubes have pins that fit into sockets that have terminals connecting into circuits. Pins are labeled with numbers placed outside the symbol used to represent the tube, as shown in Fig. 21.24, and are numbered in a clockwise direction with the tube viewed from its bottom.

21.6
Separation of parts

In complex circuits, it is often advantageous to separate elements of a multielement part with portions of the graphic symbols drawn in different locations on the drawing. An example of this method of separation is the switch labeled *S1A*, *S1B*, *S1C*, and so on, in Fig. 21.5. The switch is labeled *S1*, and the letters that follow, called suffixes, are used to designate different parts of the same switch. Suffix letters may also be used to label subdivisions of an enclosed unit that is

FIG. 21.24 Tube pin numbers should be placed outside the tube envelope and adjacent to the connecting lines. (Courtesy of ANSI.)

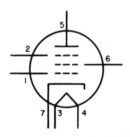

FIG. 21.25 As subdivisions within the complete part, crystals *A* and *B* are referred to as *Y1A* and *Y1B*.

made up of a series of internal parts, such as the crystal unit shown in Fig. 21.25. These crystals are referred to as *Y1A* and *Y1B*.

Rotary switches of the type shown in Fig. 21.26 are designated as *S1A*, *S1B*, and so on. The suffix letters *A*, *B*, etc., are labeled in sequence beginning with the knob and working away from it. Each end of the various sections of the switch should be viewed from the same end. When the rear and front of the switches need to be used, the words *FRONT* and *REAR* are added to the designations.

FIG. 21.26 Parts of rotary switches are designated with suffix letters *A*, *B*, *C*, etc., and are referred to as *S1A*, *S1B*, *S1C*, etc. The words *FRONT* and *REAR* are added to these designations when both sides of the switch are used. (Courtesy of ANSI.)

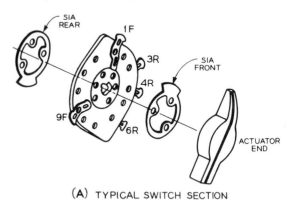

(A) TYPICAL SWITCH SECTION

(B) GRAPHIC SYMBOL

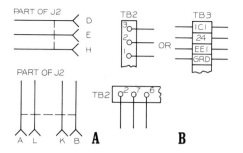

FIG. 21.27 The portions of connectors or terminal boards are functionally separated on a diagram; the words *PART OF* may precede the reference designation of the entire portion. Or conventional breaks can be used to indicate graphically that the part drawn is only a portion of the whole.

Portions of items such as terminal boards, connectors, or rotary switches may be separated on a diagram provided this is noted for clarity. The words *PART OF* may precede the identification of the portion of the circuit of which it is a part, as shown in Fig. 21.27A. A second method of showing a part of a system is to use conventional break lines that make the note *PART OF* unnecessary.

21.7
Reference designations

A combination of letters and numbers that identify items on a schematic diagram are called reference designations. These designations are used to identify the

components not only on the drawing but also in the related documents that refer to them. Reference designations should be placed close to the symbols that represent the replaceable items of a circuit on a drawing. Items that are not separately replaceable may be identified if this is considered necessary. Mounting devices for electron tubes, lamps, fuses, and the like are seldom identified on schematic diagrams.

It is standard practice to begin each reference designation with an uppercase letter that may be followed by a numeral with no hyphen between them. The number usually represents a portion of the part being represented. The lowest number of a designation should be assigned to begin at the upper left of the schematic diagram and proceed consecutively from left to right and top to bottom throughout the drawing.

Some of the standard abbreviations used to designate parts of an assembly are: amplifier—A, battery—BT, capacitor—C, connector—J, piezo-electric crystal—Y, fuse—F, electron tube—V, generator—G, rectifier—CR, resistor—R, transformer—T, and transistor—Q.

As the circuit is being designed, some of the numbered elements may be deleted from the circuit drawing. The numbered elements that remain should not be renumbered even though there is a missing element within the sequence of numbers used to label the parts. Instead, a table of the type shown in Fig. 21.28 can be used to list the parts that have been omitted from the circuit. The highest designations are also given in the table as a check to be sure that all parts are considered.

Electron tubes are labeled not only with reference designations but with type designation and circuit function, as shown in Fig. 21.29. This information is labeled in three lines, such as V5/35C5/OUTPUT, which are located adjacent to the symbol.

FIG. 21.28 Reference designations are used to identify parts of a circuit. They are labeled in a numerical sequence from left to right beginning at the upper left of the diagram. If parts are later deleted from the system, the ones deleted should be listed in a table, along with the highest reference number designations.

HIGHEST REFERENCE DESIGNATIONS	
R 72	C 40
REFERENCE DESIGNATIONS NOT USED	
R8, R10, R61 R64, R70	C12, C15, C17 C20, C22

FIG. 21.29 Three lines of notes can be used with electron tubes to specify reference designations, type designation, and function. This information should be located adjacent to the symbol, and preferably above it. (Courtesy of ANSI.)

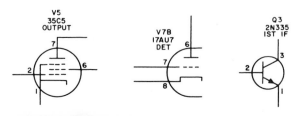

		Symbol	
Multiplier	Prefix	Method 1	Method 2
10^{12}	tera	T	T
10^9	giga	G	G
10^6 (1,000,000)	mega	M	M
10^3 (1000)	kilo	k	K
10^{-3} (.001)	milli	m	MILLI
10^{-6} (.000001)	micro	μ	U
10^{-9}	nano	n	N
10^{-12}	pico	p	P
10^{-15}	femto	f	F
10^{-18}	atto	a	A

A. MULTIPLIERS

Range in Ohms	Express as	Example
Less than 1000	ohms	.031 470
1000 to 99,999	ohms or kilohms	1800 15,853 10k 82k
100,000 to 999,999	kilohms or megohms	220k .22M
1,000,000 or more	megohms	3.3M

B. RESISTANCE

Range in Picofarads	Express as	Example
Less than 10,000	picofarads	152.4 pF 4700 pF
10,000 or more	microfarads	.015 μF 30 μF

C. CAPACITANCE

FIG. 21.30 Multipliers should be used to reduce the number of zeros in a number (part A). Examples of expressing units of capacitance and resistance are shown at B and C. (Courtesy of ANSI.)

21.8
Numerical units of function

Functional units such as the values of resistance, capacitance, inductance, and voltage should be specified with the fewest number of zeros by using the multipliers in Fig. 21.30A as prefixes. Examples using this method of expression are shown in parts B and C of Fig. 21.30, where units of resistance and capacitance are given. When four-digit numbers are given, the commas should be omitted. One thousand should be written as 1000, not as 1,000. You should recognize and use the lowercase or uppercase prefixes as indicated in the table of Fig. 21.30.

A general note can be used where certain units are repeated on a drawing to reduce time and effort.

The following is recommended form for a note of this type:

> UNLESS OTHERWISE SPECIFIED: RESISTANCE VALUES ARE IN OHMS. CAPACITANCE VALUES ARE IN MICROFARADS.

or

> CAPACITANCE VALUES ARE IN PICOFARADS.

A note for specifying capacitance values is:

> CAPACITANCE VALUES SHOWN AS NUMBERS EQUAL TO OR GREATER THAN UNITY ARE IN pF AND NUMBERS LESS THAN UNITY ARE IN μF.

Examples of the placement of the reference designations and the numerical values of resistors are shown in Fig. 21.31.

21.9
Functional identification of parts

The readability of a circuit is improved if parts are labeled to indicate their functions. Test points are labeled on drawings with the letters *Tp* and their suffix

FIG. 21.31 Methods of labeling the units of resistance on a schematic diagram.

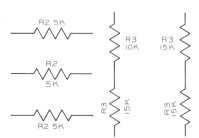

FIG. 21.32 The result of the miniaturization of electronic devices can be seen here where a solid state logic technology (SLT) chip the size of a dot gives the same function as its predecessors, the transistor and the vacuum tube. (Courtesy of IBM.)

numbers. The sequence of the suffix numbers should be the same as the sequence of troubleshooting the circuit when it is defective. As an alternative, the test function can be indicated on the diagram below the reference designation.

Additional information may be included on a schematic diagram to aid in the maintenance of the system. Examples of additional information are:

- DC resistance of windings and coils.
- Critical input and output impedance values.

FIG. 21.33 This is a magnified view of a printed circuit where the circuit has actually been printed and etched on a circuit board, and the devices are then soldered into position. (Courtesy of Bishop Industries Corporation.)

- Wave shapes (voltage or current) at significant points.
- Wiring requirements for critical ground points, shielding, pairing, and the like.
- Power or voltage ratings of parts.
- Caution notation for electrical hazards at maintenance points.
- Circuit voltage values at significant points (tube pins, test points, terminal boards, etc.).
- Zones (grid system) on complex schematics.
- Signal flow direction in main signal paths.

21.10
Printed circuits

Printed circuits are universally used for miniature electronic components and computer systems. The degree of miniaturization that has occurred in the electronics industry can be seen in Fig. 21.32, where the function of a vacuum tube has been replaced by a chip transistor the size of a dot.

In this method, the drawings of the circuits are drawn up to four times or more the size that the circuit will ultimately be printed. The drawings for the printed circuit are usually drawn in black India ink on acetate film and are then photographically reduced to the desired size. The circuit is "printed" onto an insulated board made of plastic or ceramics, and the devices within the circuit are connected and soldered (Fig. 21.33).

The steps of fabricating a solid state logic technology (SLT) module that was used in the IBM 360 computer is shown in Fig. 21.34. The tiny module (about one-half inch square) had its chip transistor positioned and attached in the last step of assembly before it was encased in its protective metal shell (Fig. 21.35).

Some printed circuits are printed on both sides of the circuit board; this requires two photographic negatives, as shown in Fig. 21.36, that were made from positive drawings (black lines on a white background). Each drawing for each side can be made on separate sheets of acetate that are laid over each other when the second diagram is drawn. However, a more efficient method uses red and blue tape that can be used for making a single drawing (Fig. 21.37) from which two negatives are photographically made. Filters are used on the process camera to drop out the red for one negative and a different filter for dropping

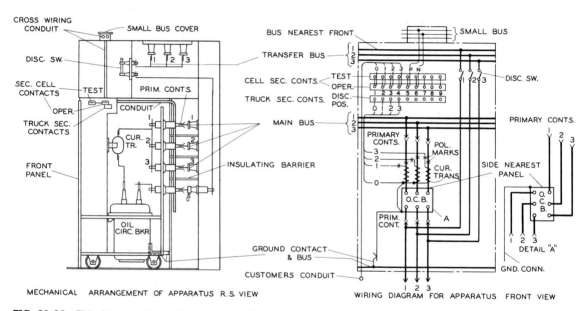

MECHANICAL ARRANGEMENT OF APPARATUS R.S. VIEW

WIRING DIAGRAM FOR APPARATUS FRONT VIEW

FIG. 21.39 This drawing shows views of a metal-enclosed switchgear to describe the arrangement of the apparatus and also gives the wiring diagram for the unit.

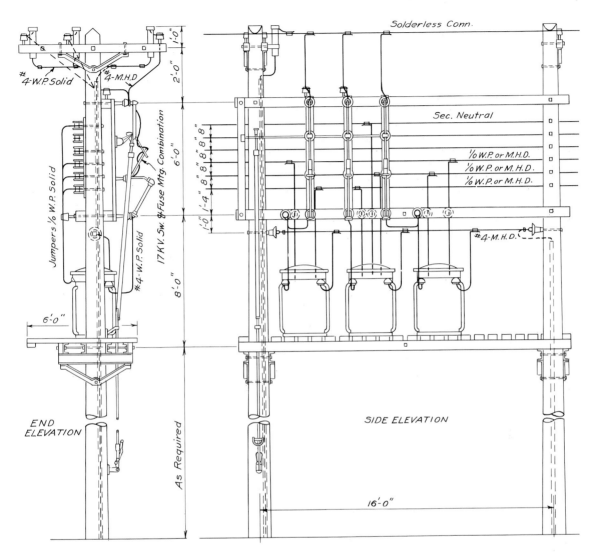

FIG. 21.40 This installation drawing of an electrical substation is sufficient to explain its installation to the contractor who must assemble and construct it. (Courtesy of Railway and Industrial Engineering Co., Greensburg, Pennsylvania.)

Problems

1. On a Size A sheet, make a schematic diagram of the circuit shown in Fig. 21.41.

2. On a Size A sheet, make a schematic diagram of the circuit shown in Fig. 21.42.

3. On a Size A sheet, make a schematic diagram of the circuit shown in Fig. 21.43.

4. On a Size A sheet, make a schematic diagram of the circuit shown in Fig. 21.44.

5. On a Size A sheet, make a schematic diagram of the circuit shown in Fig. 21.45.

FIG. 21.41 Problem 1: A low-pass inductive-input filter. (Courtesy of NASA.)

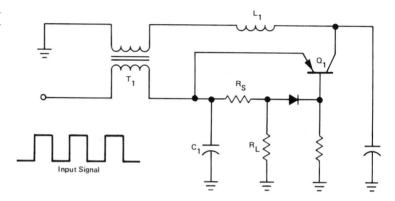

FIG. 21.42 Problem 2: A quadruple-sampling processor. (Courtesy of NASA.)

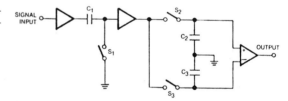

FIG. 21.43 Problem 3: A temperature-compensating DC restorer circuit designed to condition the video circuit of a Fairchild area-image sensor. (Courtesy of NASA.)

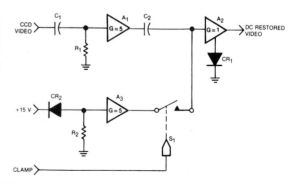

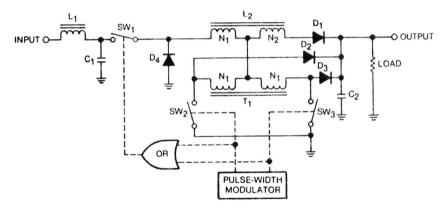

FIG. 21.44 Problem 4: A "buck/boost" voltage regulator. (Courtesy of NASA.)

FIG. 21.45 Problem 5: An improved power-factor controller. (Courtesy of NASA.)

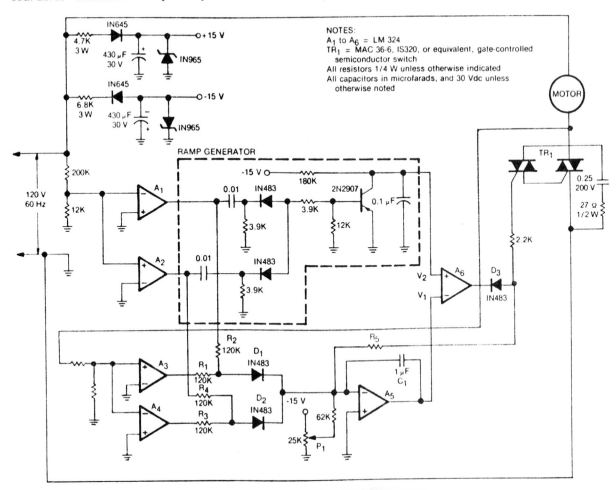

6. On a Size B sheet, make a schematic diagram of the circuit shown in Fig. 21.46.

7. On a Size C sheet, make a schematic diagram of the circuit shown in Fig. 21.47.

8. On a Size C sheet, make a schematic diagram of the circuit shown in Fig. 21.48 and give the parts list.

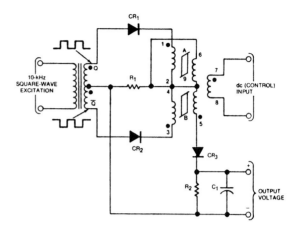

FIG. 21.46 Problem 6: A magnetic-amplified DC transducer. (Courtesy of NASA.)

FIG. 21.47 Problem 7: An electrocardiograph (EKG) signal conditioner. (Courtesy of NASA.)

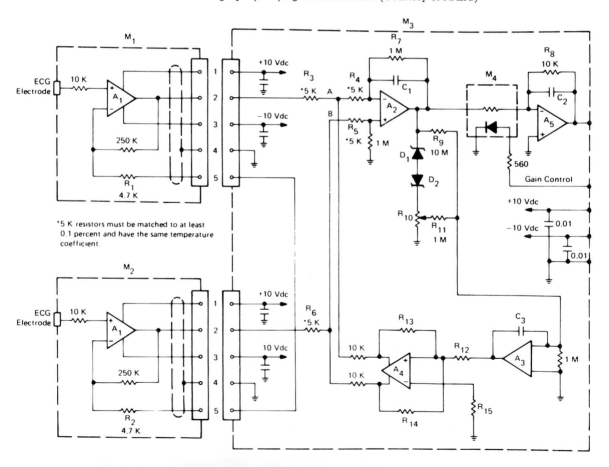

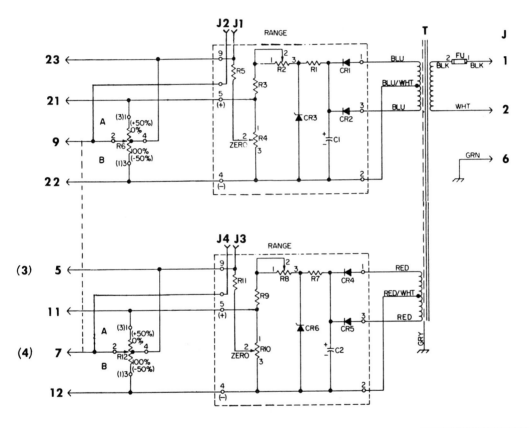

COMPONENT	DESCRIPTION	
CR1, CR2, CR3, CR4	DIODE	1N3193
CR3 & CR6	ZENER DIODE	24V 3W ± .5V TOL
R1 & R7	RESISTOR	560 Ω 5W 5%
R3 & R9	RESISTOR	56 Ω
R5 & R11	RESISTOR	2.7K 1/2W 5%
R2 & R8	TRIM POT	50 Ω
R4 & R10	TRIM POT	5K
R6 & R12	POT	500 Ω 2W DUAL
C1 & C2	CAPACITOR	60 MFD 60V
T	TRANSFORMER	
FU	FUSE	1 AMP
M	EDGEWISE METER	
J	CONNECTOR	
J1, J2, J3, J4	TEST JACK	

TYPE	R6 SCALE		M SCALE	
	A	B	BOTTOM	TOP
RS1100C	0%	100%	0%	100%
RS1110C	0%	100%	PER ENGINEERING DATA	
RS3100C	+50%	-50%	-50%	+50%
RS4100C	0%	100%	0%	100%
RS2120C	100%	0%	OMIT	
RS2100C	100%	0%	0%	100%

NOTE: OUTPUT WIRED TO TERMINALS 3 AND 4 ON
TYPES RS1100C, RS1110C, AND RS3100C.

OUTPUT WIRED TO TERMINALS 5 AND 7 ON
TYPES RS2100C, RS2120C, AND RS4100C.

FIG. 21.48   Problem 8: A schematic diagram of a dual output manual station and its parts list. (Courtesy of NASA.)

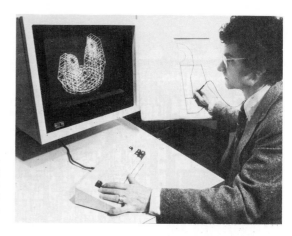

FIG. 22.3 This engineer has constructed a 3-D finite element model of a gun mount on the Applicon display by digitizing a multiview engineering drawing. (Courtesy of Applicon.)

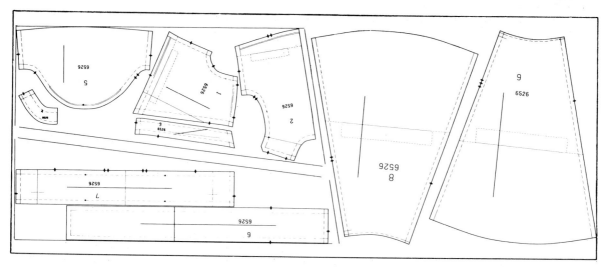

FIG. 22.4 This clothing pattern was designed, plotted, and cut out by utilizing computer graphics on a Versatec plotter. Programs have been written that will plot this same pattern for graduated sizes of clothing. (Courtesy of Versatec.)

FIG. 22.5 This Producer drafting system is composed of a keyboard, CRT, digitizer, computer, and plotter. It has three work stations. (Courtesy of Bausch and Lomb.)

son in observing a three-dimensional design. Most 3-D computer graphics systems permit the observer to view a design from any angle as it is revolved on the screen.

22.3
Hardware systems

A basic computer graphics system consists of a *computer*, a *terminal*, a *plotter*, and a *printer*. Additional devices such as *digitizers* and *light pens* may be used for direct input of graphics information (Fig. 22.5).

Computer

Computers are the devices that receive the input of the programmer, execute the programs, and then produce some form of useful output. The largest computers, called *mainframes*, are big, fast, powerful, and expensive. The smallest computers, called *microcomputers*, have recently come into widespread use for personal and small-business applications. Microcomputers are much cheaper than mainframes; they are excellent for graphical applications where massive data storage and high speed are not essential, and low price and small size are important.

The *minicomputer* is a medium-sized computer that lies between the mainframe and the microcomputer in performance and cost.

Terminal

The terminal is the device by which the user communicates with the computer. It usually consists of a keyboard with some type of output device, such as a typewriter or television screen (Fig. 22.6). Most graphics terminals use one of three types of television screens or CRTs (cathode-ray tubes). One type is *raster scanned*, which means that the picture display is being refreshed or scanned from left to right and top to bottom at a rate of about 60 times per second.

A second type is the *storage tube*, where the image is drawn on the screen much like a drawing on an erasable blackboard.

The third and most powerful type of screen is *vector refreshed*, which means that each line in the picture is being continuously redrawn by the computer. The

FIG. 22.6 An engineer is shown working at a computer graphics system's terminal where he has access to a keyboard and a menu tablet. (Courtesy of Bausch & Lomb.)

vector refresh type requires more computer power than is available from the smaller minicomputers and microcomputers.

Plotters

The plotter is the machine that is directed by the computer and the program to make a drawing. The two basic types of plotters are *flatbed* and *drum*. The flatbed plotter is a large flat bed to which the drawing paper is attached. A pen is moved about over the paper in a raised or lowered position to complete the drawing (Fig. 22.7). This type of plotter is well suited to plotting on a variety of surfaces in addition to paper. The drum plotter uses a special type of paper that is held on a spool and rolled over a rotating drum (Fig. 22.8). The drum rotates in two directions as the pen that is suspended above the surface of the paper moves left or right along the drum.

Printers

The printer can be thought of as a typewriter that is operated by the program and the computer. Many varieties are available in a wide range of prices. The speed and type quality offered by the printer are what usually determine the purchase price. One of the most important applications of the printer in a computer system is the printing of a copy of a computer program

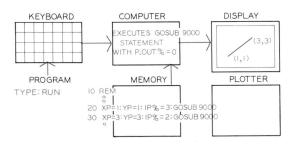

FIG. 22.13 A graphics display can be obtained on the display screen when a program has the statement P.OUT% = 0. The command RUN is typed into the keyboard, and the computer commands the display to plot the graphical output from the program.

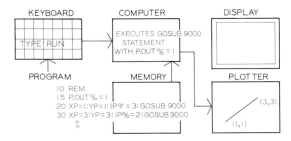

FIG. 22.14 A graphics plot can be obtained at the plotter when the program has the statement P.OUT% = 1. The command RUN is typed into the keyboard, and the computer directs the plotter to plot the graphical output.

FIG. 22.15 Data from a drawing can be fed into the computer by means of a digitizer board by using a pen to digitize points on the drawing. The digitized points are transferred to the computer, and these data are input into the existing program by the statement GOSUB 9500. The command RUN, input at the keyboard, runs the program.

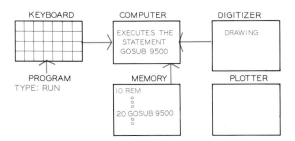

will not be compatible with all subroutines and systems.

GETTING DATA FROM THE DIGITIZER Data are obtained from the digitizer by executing the subroutine GOSUB 9500, which defines the variables DIGI.X, DIGI.Y, and DIGI.P%. The values of DIGI.X and DIGI.Y are the coordinates of the digitizer pen, and DIGI.P% has the value of either 2 or 3 for pen down and up commands. As shown in Fig. 22.15, the RUN command causes the computer to execute the program statements. Executing the subroutine GOSUB 9500 causes the coordinates and pen value to be accepted from the digitizer.

22.5
Basic programming rules

The example programs given below were written in a language called *BASIC*, which is different from, but has many similarities to, the *FORTRAN* language. Some elementary aspects of BASIC programming are covered in this section.

A *program* is a sequence of instructions and specifications that will be received by a computer, causing it to perform the operations desired by the programmer. Each statement must be given a number that is an integer between 0 and 65535. For example, a statement may be written as 30 X = X + 1, where 30 is the statement number and X = X + 1 is the statement.

The following rules apply to a statement that is entered after a line number:

1. If a line number of a newly entered statement is the same as a previously entered statement, then the new line will replace the old line with the same line number.

2. If a line containing only a line number and no statement is entered, then any previously entered statement with the same line number will be deleted from the program.

3. If the line number of a newly entered statement is different from any other line in the program, the statement will be added to the program. When the statements are listed by the computer, they will be arranged in ascending order (that is, 10, 20, 30, etc.).

Some BASIC *commands* should not be given line numbers and will be executed immediately by BASIC. These commands cause BASIC to operate on the current program or to perform some system function. Listed below are some of the commonly used commands and their functions:

LIST—causes the current program to be listed on the terminal device (usually a CRT).

NEW—causes the current program to be erased.

SAVE "PROGRAM NAME HERE"—saves the program under the given name that is enclosed in quotations. Example: SAVE "PROG1"

LOAD "PROGRAM NAME HERE"—transfers a program by name, in quotations, from the disc into memory and makes it the current program. Example: LOAD "PROG1"

KILL "PROGRAM NAME HERE"—removes the named program, in quotations, from the disc. Example: KILL "PROG1"

FILES—produces a list of the programs that are stored on a particular disc.

RUN—begins the execution of the current program.

All BASIC programs are executed (set into operation) by the command RUN, which is typed by using the keyboard of the terminal. The program will continue to run until one of the following events occurs:

1. The statement STOP is executed;
2. An invalid statement is executed;
3. The program runs out of sequential line numbers to execute.

VARIABLE NAMES Alphabetic or numeric characters up to 40 characters long, provided there are no spaces, can be used for variable names. Periods can be used effectively to separate words, such as: PROG.ONE. Variable names cannot be identical to commands such as LIST, RUN, and the like.

ARRAYS Subscripted variables can be assigned any valid variable name followed by a number in parentheses. The letter, or letter and number combination, is the name of the array, and the number in parentheses locates the element within the array and is called a subscript. For example, the tenth element in array B would be B(10), and the sixth element in array F3 would be F3(6).

To allocate space in the computer's memory for an array, the DIM (dimensioning) statement is used at the beginning of a BASIC program. A statement such as 10 DIM X(50), Y(50), X9(10), P(2,10), would allocate 50 elements for the array of X, 50 elements for the array Y, 10 elements for the variable X9, and two groups of 10 elements each for the variable P.

REMARK Remark statements are used to document a program within the program for future reference, but they have no effect on the function of the program when it is run. The REMARK statement can be shortened to REM when written in the program. You will note in example programs in the following sections that REM statements are essential to understanding the logic of the programmer as he or she writes the program.

Examples of REM statements are given below:

10 REM this is an example of
20 REM basic remark statements
30 REM that are ignored during
40 REM execution
50 REM of the basic program

MATHEMATICAL FUNCTIONS The version of BASIC that is presented in this chapter has eight mathematical functions that can be used in numerical expressions. These functions are listed below:

- ABS (numerical values here)—Returns the absolute value of the numerical expression. *Example:* ABS(3) = 3, ABS(−3) = 3, and ABS(0) = 0.

- SGN (numerical expression here)—Returns 1, 0, or −1, which indicates whether the numerical expression is positive, zero-valued, or negative, respectively. *Example:* SGN (10) = 1, SGN (0) = 0, and SGN (−3.2) = −1.

- INT (numerical expression here)—Returns the greatest integer value that is less than or equal to the value of the numerical expression. *Example:* INT (3) = 3, INT (3.9) = 3, and INT (−3.5) = −4.

- LOG (numerical expression here)—Returns an approximation of the natural logarithm of the value of the numerical expression. If LOG is specified with a value less than or equal to zero, a program error will occur. *Example:* LOG (1) = 0, LOG (7) = 1.945901, and LOG (0.1) = −2.3025851.

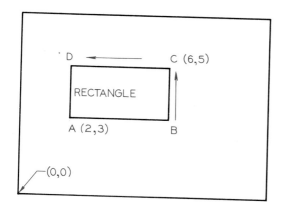

```
100 REM PROGRAM TO DRAW A RECTANGLE
110 REM GIVEN COORDINATES OF TWO CORNERS
120 REM POINT A AND POINT C
130 REM ==================================
140 REM POINT A ---> XA,YA
150 REM POINT B ---> XC,YA
160 REM POINT C ---> XC,YC
170 REM POINT D ---> XA,YC
180 XA = 2 : YA = 3
190 XC = 6 : YC = 5
200 IP%=0 : GOSUB 9000
210 XP=0 : YP=0 : IP%=3 : GOSUB 9000
220 REM PLOT TO POINT A
230 XP=XA : YP=YA : IP%=3 : GOSUB 9000
240 REM PLOT TO POINT B
250 XP=XC : YP=YA : IP%=2 : GOSUB 9000
260 REM PLOT TO POINT C
270 XP=XC : YP=YC : IP%=2 : GOSUB 9000
280 REM PLOT TO POINT D
290 XP=XA : YP=YC : IP%=2 : GOSUB 9000
300 REM PLOT TO POINT A
310 XP=XA : YP=YA : IP%=2 : GOSUB 9000
320 REM RETURN TO ORIGIN
330 XP=0 : YP=0 : IP%=3 : GOSUB 9000
340 STOP
350 END
```

FIG. 22.22 A program and a display of its output for drawing a rectangle when the coordinates of each corner are known.

After the pen is initialized at 0, 0, it is directed to point *A* (XA, YA) with the pen up in line 230. Statements 250, 270, 290, and 310 successively move the pen to points *B*, *C*, *D*, and back to *A*. You can see that a variety of different rectangles can be drawn by changing the numerical values of the variables in statements 180 and 190.

Inclined rectangles can be constructed when the following are known: the coordinates of one corner, the height and width, and the angle of inclination from 0° to 360°. In the example in Fig. 22.23, these known values are given in statements 190 through 220.

The angle is converted to radians in line 240. The coordinates at corner *B* are XB and YB, and these are computed in lines 260 and 270 as the cosine and sine of the 30° angle multiplied by the width of the rectan-

gle. Similarly, the coordinates of corner *C* are found by using the sine and cosine of the height of the rectangle in statements 290 and 300. The coordinates of corner *D* are defined in lines 320 and 330.

The variables for each corner are sequentially inserted into the plot statement in statements 370

FIG. 22.23 A program and a display of its output for drawing a rectangle by expressing its dimensions in variables that can be easily changed for other variations of size and rotation.

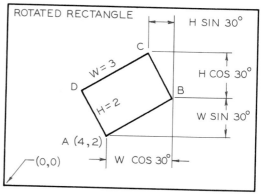

```
100 REM PROGRAM TO DRAW A RECTANGLE
110 REM GIVEN THE COORDINATE OF THE POINT A,
120 REM THE HEIGHT, THE WIDTH, AND THE ROTATION
130 REM ABOUT THE POINT A IN DEGREES
140 REM =========================================
150 REM POINT A ---------------------> XA,YA
160 REM HEIGHT ---------------------> R.HEIGHT
170 REM WIDTH ----------------------> R.WIDTH
180 REM ROTATION ABOUT POINT A -----> R.ANGLE
190 XA = 4 : YA = 2
200 R.HEIGHT = 2
210 R.WIDTH = 3
220 R.ANGLE = 30
230 REM CONVERT R.ANGLE TO RADIANS
240 R.ANGLE = R.ANGLE * 3.1416 / 180
250 REM COMPUTE POINT B
260 XB = XA + R.WIDTH * COS(R.ANGLE)
270 YB = YA + R.WIDTH * SIN(R.ANGLE)
280 REM COMPUTE POINT C
290 XC = XB - R.HEIGHT * SIN(R.ANGLE)
300 YC = YB + R.HEIGHT * COS(R.ANGLE)
310 REM COMPUTE POINT D
320 XD = XA - R.HEIGHT * SIN(R.ANGLE)
330 YD = YA + R.HEIGHT * COS(R.ANGLE)
340 IP%=0 : GOSUB 9000
350 XP=0 : YP=0 : IP%=3 : GOSUB 9000
360 REM PLOT TO POINT A
370 XP=XA : YP=YA : IP%=3 : GOSUB 9000
380 REM PLOT TO POINT B
390 XP=XB : YP=YB : IP%=2 : GOSUB 9000
400 REM PLOT TO POINT C
410 XP=XC : YP=YC : IP%=2 : GOSUB 9000
420 REM PLOT TO POINT D
430 XP=XD : YP=YD : IP%=2 : GOSUB 9000
440 REM PLOT TO POINT A
450 XP=XA : YP=YA : IP%=2 : GOSUB 9000
460 REM RETURN TO ORIGIN
470 XP=0 : YP=0 : IP%=3 : GOSUB 9000
480 STOP
490 END
```

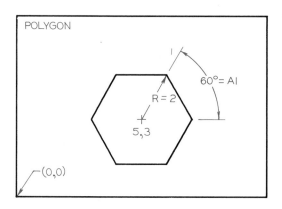

```
100 REM PROGRAM TO DRAW A POLYGON
110 REM GIVEN CENTER, RADIUS,
120 REM AND NUMBER OF SIDES.
130 REM ==========================
140 REM CENTER ---> XCENTER , YCENTER
150 REM RADIUS ---> RADIUS
160 REM #SIDES ---> NSIDES
170 XCENTER=5 : YCENTER=3
180 RADIUS = 2
190 NSIDES = 6
200 IP%=0 : GOSUB 9000
210 XP=0 : YP=0 : IP%=3 : GOSUB 9000
220 FOR ANGLE = 0 TO 360 STEP 360/NSIDES
230    REM CONVERT ANGLE TO RADIANS
240    ANGLE.R = ANGLE * 3.1416 / 180
250    REM COMPUTE COORDINATES OF POINT X,Y
260    XP = XCENTER + RADIUS * COS(ANGLE.R)
270    YP = YCENTER + RADIUS * SIN(ANGLE.R)
280    REM MOVE PEN UP IF ANGLE = 0
290    IF ANGLE = 0 THEN IP%=3 ELSE IP%=2
300    GOSUB 9000
310 NEXT ANGLE
320 XP=0 : YP=0 : IP%=3 : GOSUB 9000
330 STOP
340 END
```

FIG. 22.24 A program and its resulting display for drawing regular polygons.

22.10
Plotting polygons

A regular polygon, having a number of sides each of the same length, can be constructed by using the program shown in Fig. 22.24 when the following are given: the coordinates of the center, the distance from the center to each corner (RADIUS), and the number of sides (NSIDES). This information is given in lines 170 through 190.

The coordinates of corner 1 are found in statements 260 and 270, where the RADIUS is multiplied by the sine and cosine of the angle between each corner point and each is added to the coordinates of the center of the polygon. An IF statement is used in line 290 that tells the plotter to lift the pen when the angle is at zero degrees (horizontal) and to move to the first calculated point. A FOR NEXT loop is used between statements 220 and 310 where the steps of angle ANGLE are specified to be 360/NSIDES, which is 30° in this example.

The values of each set of X- and Y-coordinates are computed and plotted for each successive 30° angle through 360° and back to the point of beginning, where the pen is then lifted and the loop is ended. The pen is then directed by statement 320 to return to the origin, 0, 0.

22.11
Circles

The program for drawing a circle is identical to the one used to draw regular polygons. The only exception is the smallness of the size of the steps of the angle from corner to corner of the polygon. The circle is really no more than a polygon with many small sides.

Like the polygon, the following must be given: the coordinates of the center (XCENTER and YCENTER) and the radius, R, as given in statements 170 and 180 of the program in Fig. 22.25. The angular steps are given in the FOR NEXT loop as 5° increments, and the resulting polygon approximates a circle.

As the circle is drawn larger, the angular steps can be specified in smaller increments. Small circles can be drawn with slightly larger increments, and plotting time can be reduced without affecting the appearance of the plotted circle.

22.12
Programs with data statements

An alternative method of programming the coordinates of points is to use DATA statements as given in lines 160 through 200 of Fig. 22.26. The three numbers in each DATA statement represent X, Y, and pen

through 450 to drive the pen to each of the four corners. The use of variables makes it possible to change any or all of the variables in statements 190 through 220 and thereby obtain a different rectangle with a minimum of programming modification.

497

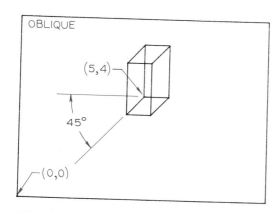

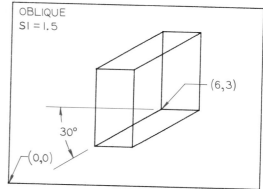

```
100 REM PROGRAM TO DRAW A FIGURE GIVEN THE
110 REM COORDINATES FOR ITS POINTS ON THE
120 REM X,Y, & Z AXIS.   THESE VALUES ARE
130 REM TRANSLATED ONTO AN OBLIQUE AXIS
140 REM BEFORE THEY ARE PLOTTED.   THE
150 REM COORDINATES ARE DEFINED WITH "DATA"
160 REM STATEMENTS CONTAINING X,Y,Z & PEN
170 REM VALUES.   X.ORG AND Y.ORG ARE THE
180 REM X & Y LOCATION OF THE ORIGIN OF THE
190 REM FIGURE.   Z.SCALE IS THE Z-AXIS
200 REM SCALE FACTOR.   SCALE IS THE OVERALL
210 REM SCALE FACTOR. ANGLE IS THE Z-AXIS
220 REM ANGLE.  THE LAST DATA VALUE IS
230 REM ALWAYS 999.
240 REM DEFINE ORIGIN, SCALE AND ROTATION
250 X.ORG=5 : Y.ORG=4
260 Z.SCALE=.5 : SCALE=1 : ANGLE = 45
270 DATA 0,0,0,3,1,0,0,2,1,2,0,2,1,2,2,2
280 DATA 1,0,2,2,0,0,2,2,0,0,0,2,0,2,0,2
290 DATA 0,2,2,2,0,0,2,2,1,0,2,3,1,0,0,2
300 DATA 0,2,2,3,1,2,2,2,0,2,0,3,1,2,0,2
310 DATA 0,0,0,3,999
320 IP%=0 : GOSUB 9000
330 REM CONVERT ANGLE TO RADIANS
340 ANGLE = ANGLE * 3.1416 /180
350 REM ------ READ LOOP ------
360 READ XP
370 IF XP = 999 THEN STOP
380 READ YP
390 READ ZP
400 READ IP%
410 XP=X.ORG+((XP-(ZP*COS(ANGLE)*Z.SCALE)))*SCALE
420 YP=Y.ORG+((YP-(ZP*SIN(ANGLE)*Z.SCALE)))*SCALE
430 REM PLOT XP,YP & IP%
440 GOSUB 9000
450 GOTO 360
460 REM ------ END OF LOOP ------
470 END
```

FIG. 22.33 A program and its plot that converts given data points into an oblique drawing.

```
100 REM PROGRAM TO DRAW A FIGURE GIVEN THE
110 REM COORDINATES FOR ITS POINTS ON THE
120 REM X,Y, & Z AXIS.   THESE VALUES ARE
130 REM TRANSLATED ONTO AN OBLIQUE AXIS
140 REM BEFORE THEY ARE PLOTTED.   THE
150 REM COORDINATES ARE DEFINED WITH "DATA"
160 REM STATEMENTS CONTAINING X,Y,Z & PEN
170 REM VALUES.   X.ORG AND Y.ORG ARE THE
180 REM X & Y LOCATION OF THE ORIGIN OF THE
190 REM FIGURE.   Z.SCALE IS THE Z-AXIS
200 REM SCALE FACTOR.   SCALE IS THE OVERALL
210 REM SCALE FACTOR. ANGLE IS THE Z-AXIS
220 REM ANGLE.  THE LAST DATA VALUE IS
230 REM ALWAYS 999.
240 REM DEFINE ORIGIN, SCALE AND ROTATION
250 X.ORG=6 : Y.ORG=3
260 Z.SCALE=1 : SCALE=1.5 : ANGLE = 30
270 DATA 0,0,0,3,1,0,0,2,1,2,0,2,1,2,2,2
280 DATA 1,0,2,2,0,0,2,2,0,0,0,2,0,2,0,2
290 DATA 0,2,2,2,0,0,2,2,1,0,2,3,1,0,0,2
300 DATA 0,2,2,3,1,2,2,2,0,2,0,3,1,2,0,2
310 DATA 0,0,0,3,999
320 IP%=0 : GOSUB 9000
330 REM CONVERT ANGLE TO RADIANS
340 ANGLE = ANGLE * 3.1416 /180
350 REM ------ READ LOOP ------
360 READ XP
370 IF XP = 999 THEN STOP
380 READ YP
390 READ ZP
400 READ IP%
410 XP=X.ORG+((XP-(ZP*COS(ANGLE)*Z.SCALE)))*SCALE
420 YP=Y.ORG+((YP-(ZP*SIN(ANGLE)*Z.SCALE)))*SCALE
430 REM PLOT XP,YP & IP%
440 GOSUB 9000
450 GOTO 360
460 REM ------ END OF LOOP ------
470 END
```

FIG. 22.34 A cavalier oblique has been plotted by the accompanying program. Note that the drawing has been enlarged since a scale factor (SCALE) of 1.5 was given.

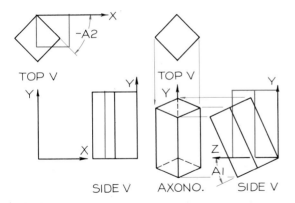

FIG. 22.35 An axonometric view of an object is found by revolving the top view about the Y-axis and the side view about the X-axis. The resulting front view is an axonometric projection.

22.15
Axonometrics and isometrics

An axonometric pictorial can be programmed and plotted by revolving the views of the part about the Y-axis and then the X-axis by the desired angles, as shown orthographically in Fig. 22.35. The resulting front view is the axonometric pictorial of the object. Notice the assignments of positive and negative directions of rotation in each view.

A program and the plot of an axonometric that has been revolved $-30°$ about the Y-axis in a negative direction, and revolved $30°$ in a positive direction about the X-axis, are shown in Fig. 22.36. In line 270, the coordinates of the origin of the figure (X.ORG, Y.ORG), the two angles of rotation (X.ANGLE and Y.ANGLE), and the scale factor (SCALE) are given.

The data statements, lines 270 through 310, give the coordinates of the points of the pictorial that are to be drawn. The equations in lines 410, 420, and 430 determine the axonometric coordinates of the points that are to be plotted. Each of the data sets is looped through these equations until all have been calculated and plotted.

Since the three types of axonometrics are isometrics, dimetrics, and trimetrics, you can see that this

converted by these formulas. A loop is used between lines 360 and 450 for reading, calculating, and plotting the points.

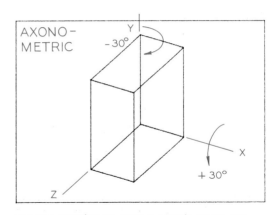

```
100 REM PROGRAM TO DRAW A FIGURE GIVEN THE
110 REM COORDINATES FOR ITS POINTS ON THE
120 REM X,Y, & Z AXIS.  THESE VALUES ARE
130 REM TRANSLATED ONTO AN AXONOMETRIC AXIS
140 REM BEFORE THEY ARE PLOTTED.  THE
150 REM COORDINATES ARE DEFINED WITH "DATA"
160 REM STATEMENTS CONTAINING X,Y,Z & PEN
170 REM VALUES.  X.ORG AND Y.ORG ARE THE
180 REM X & Y LOCATION OF THE ORIGIN OF THE
190 REM FIGURE.  SCALE IS THE OVERALL SCALE
200 REM FACTOR.  X.ANGLE IS THE ROTATION
210 REM ABOUT THE X AXIS.  Y.ANGLE IS THE
220 REM ROTATION ABOUT THE Y AXIS.
230 REM ===================================
240 REM DEFINE ORIGIN, SCALE AND ROTATION
250 X.ORG=5 : Y.ORG=3 : SCALE=2
260 X.ANGLE=30 : Y.ANGLE=-30
270 DATA 0,0,0,3,1,0,0,2,1,2,0,2,1,2,2,2
280 DATA 1,0,2,2,0,0,2,2,0,0,0,2,0,2,0,2
290 DATA 0,2,2,2,0,0,2,2,1,0,2,3,1,0,0,2
300 DATA 0,2,2,3,1,2,2,2,0,2,0,3,1,2,0,2
310 DATA 0,0,0,3,999
320 REM PLOT DATA
330 IP%=0 : GOSUB 9000
340 REM CONVERT ANGLE TO RADIANS
350 X.ANGLE = X.ANGLE * 3.1416 / 180
360 Y.ANGLE = Y.ANGLE * 3.1416 / 180
370 REM ------ READ LOOP ------
380 READ XP
390 IF XP = 999 THEN STOP
400 READ YP,ZP,IP%
410 YP=XP*SIN(Y.ANGLE)*SIN(X.ANGLE)+YP*COS(X.ANGLE)
420 YP=(YP-ZP*COS(Y.ANGLE)*SIN(X.ANGLE))*SCALE+Y.ORG
430 XP=(XP*COS(Y.ANGLE)+ZP*SIN(Y.ANGLE))*SCALE+X.ORG
440 REM PLOT XP,YP & IP%
450 GOSUB 9000
460 GOTO 380
470 REM ------ END OF LOOP ------
480 END
```

FIG. 22.36 This program rotates the top and side views by 30° to give this axonometric pictorial.

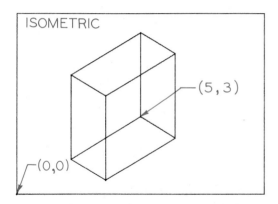

```
100 REM PROGRAM TO DRAW A FIGURE GIVEN THE
110 REM COORDINATES FOR ITS POINTS ON THE
120 REM X,Y, & Z AXIS.  THESE VALUES ARE
130 REM TRANSLATED ONTO AN ISOMETRIC AXIS
140 REM BEFORE THEY ARE PLOTTED.  THE
150 REM COORDINATES ARE DEFINED WITH "DATA"
160 REM STATEMENTS CONTAINING X,Y,Z & PEN
170 REM VALUES.  X.ORG AND Y.ORG ARE THE
180 REM X & Y LOCATION OF THE ORIGIN OF THE
190 REM FIGURE.  SCALE IS THE OVERALL SCALE
200 REM FACTOR.  X.ANGLE IS THE ROTATION
210 REM ABOUT THE X AXIS(-45).  Y.ANGLE IS
220 REM THE ROTATION ABOUT THE Y AXIS(35.3).
230 REM ====================================
240 REM DEFINE ORIGIN, SCALE AND ROTATION
250 X.ORG=5 : Y.ORG=3 : SCALE=2
260 X.ANGLE=35.3 : Y.ANGLE=-45
270 DATA 0,0,0,3,1,0,0,2,1,2,0,2,1,2,2,2
280 DATA 1,0,2,2,0,0,2,2,0,0,0,2,0,2,0,2
290 DATA 0,2,2,2,0,0,2,2,1,0,2,3,1,0,0,2
300 DATA 0,2,2,3,1,2,2,2,0,2,0,3,1,2,0,2
310 DATA 0,0,0,3,999
320 REM PLOT DATA
330 IP%=0 : GOSUB 9000
340 REM CONVERT ANGLE TO RADIANS
350 X.ANGLE = X.ANGLE * 3.1416 / 180
360 Y.ANGLE = Y.ANGLE * 3.1416 / 180
370 REM ------ READ LOOP ------
380 READ XP
390 IF XP = 999 THEN STOP
400 READ YP,ZP,IP%
410 YP=XP*SIN(Y.ANGLE)*SIN(X.ANGLE)+YP*COS(X.ANGLE)
420 YP=(YP-ZP*COS(Y.ANGLE)*SIN(X.ANGLE))*SCALE+Y.ORG
430 XP=(XP*COS(Y.ANGLE)+ZP*SIN(Y.ANGLE))*SCALE+X.ORG
440 REM PLOT XP,YP & IP%
450 GOSUB 9000
460 GOTO 380
470 REM ------ END OF LOOP ------
480 END
```

FIG. 22.37 The axonometric program can be used to draw an isometric pictorial by revolving the top view and the side view so the point view of a cube's diagonal will be found in the axonometric view.

program can be used for a variety of axonometric pictorials. An isometric projection is shown plotted in Fig. 22.37. The program that was used for this pictorial was the one used to draw the dimetric in Fig. 22.36. The only difference was the two angles of rotation.

22.16
The basic character-plotting function

Much like the basic plotting subroutine located at line number 9000 and called by using GOSUB 9000, characters are plotted with a subroutine located at 9300 that is called by using GOSUB 9300 (Fig. 22.38). However, before the character subroutine can be called, an initialization subroutine must be called by using GOSUB 9400.

The initialization subroutine performs certain functions needed before the character subroutine can work properly. The initializing subroutine, called GOSUB 9400, should be executed only once per program.

Similar to the plotting subroutine, certain key variables must be assigned values before calling the character-plotting subroutine. These are CHAR$, which should be assigned the string of characters to be plotted; CHAR.X and CHAR.Y, which should be assigned the X- and Y-coordinate locations of the lower left corner of the first character string; CHAR.H, which should be assigned a value representing the height of the characters in inches; and CHAR.A, which should

FIG. 22.38 This plot statement is used for writing text or characters. GOSUB 9400 is the *character initialization subroutine*, which loads the characters and is called only once per program. CHAR.X and CHAR.Y are the X- and Y-coordinates of the lower left corner of a string of characters that are to be plotted. CHAR.H is the desired height of the characters in inches. CHAR.A is the angle of rotation of a string of characters in a counterclockwise direction from the horizontal. CHAR$ is assigned the string of characters that are to be plotted. The character plotting subroutine is called with GOSUB 9300.

```
                  ┌ CHARACTER INITIALIZATION
                  │     SUBROUTINE
                  ↓
100  GOSUB 9400
                    X-COORD          Y-COORD
110  CHAR.X= 5 :CHAR.Y= 3
     CHAR HEIGHT ┐    CHAR STRING ANGLE ┐
120  CHAR.H= .2 :CHAR.A= 45
     ┌ STRING VARIABLE
130  CHAR$ = " CHAR STRING "
                    ┌ CHARACTER PLOTTING
140  GOSUB 9300
```

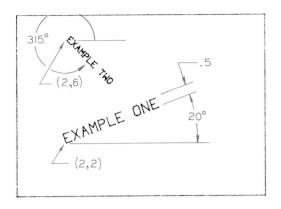

```
100 REM PROGRAM TO SHOW HOW CHARACTERS ARE
110 REM PLOTTED USING THE CHARACTER PLOTTING
120 REM SUBROUTINE.
130 REM ==================================
140 REM INITIALIZE
150 IP%=0 : GOSUB 9000 : GOSUB 9400
160 REM NOW DEFINE THE CHARACTER STRING TO BE
170 REM PLOTTED, IT'S ORIGIN, HEIGHT
180 REM AND ROTATION.
190 CHAR$="EXAMPLE ONE"
200 REM DEFINE ORIGIN
210 CHAR.X=2 : CHAR.Y=2
220 REM DEFINE DEFINE ROTATION AND HEIGHT.
230 CHAR.A=20 : CHAR.H=.5
240 REM NOW PLOT THE STRING
250 GOSUB 9300
260 REM NOW PLOT EXAMPLE STRING 2
270 CHAR.X=2 : CHAR.Y=6 : CHAR.H=.3
280 CHAR.A=315 : CHAR$="EXAMPLE TWO"
290 GOSUB 9300
300 REM RETURN TO ORIGIN
310 XP=0 : YP=0 : IP%=3 : GOSUB 9000
320 STOP
330 END
```

FIG. 22.39 An example program and its output on a display screen are shown where the words, *EXAMPLE ONE* are plotted.

be assigned the degrees of rotation of the string about the lower left corner of the first character of the string (Fig. 22.39).

22.17
Advanced applications

Using the concepts of computer graphics learned thus far, one might imagine how the following more advanced applications were created. All these plotted examples are the output of programs using the ideas of circles, rectangles, or simple lines and points.

BAR GRAPH The bar graph in Fig. 22.40 was generated from information about the number of bars, the maximum bar value, X-axis, and graph titles. Looking at the graph in parts rather than as a whole, it can be seen that the bars are just simple rectangles. The tick marks along the X-axis are single lines, and the text was plotted by using the FNC character function. Putting these parts together in the right places created the bar graph in Fig. 22.40.

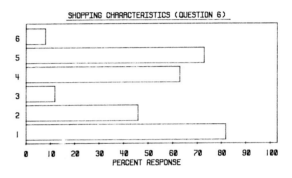

1. IS FOR DISCOUNT STORES.
2. IS FOR T.V. SPECIALITY STORES.
3. IS FOR WHOLESALE OUTLETS.
4. IS FOR DEPARTMENT STORES.
5. IS FOR RETAIL OUTLETS.
6. IS FOR MAIL ORDER CATALOGUE OR ADVERTISEMENTS.

FIG. 22.40 A bar graph is the result of using a number of plot statements and rectangle subroutines in addition to the text-plotting statement.

PLATE CAM The plate cam example in Fig. 22.41 is a very good example of computer-aided design of different shapes of cams. The one shown, designed for a knife-edge follower, was developed from the given specifications. This cam design was plotted by a computer program that used only the specifications shown.

The program is based on a circle routine, with the radius of the circle defined according to the step angle and a parabolic function of the lift and the rise and fall angles. This program used an axis-drawing subroutine to plot the border and tick-marks and the character-plotting functions to draw the text.

A more advanced example of computer graphics is the pictorial in Fig. 22.42, which illustrates an assembly of parts. This drawing was developed by Texas Instruments.

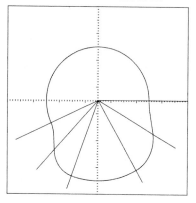

PLATE CAM CAD EXAMPLE

SPECIFICATIONS
Minimum Radius = 2in.
Lift   = 1in.
High Dwell   = 50deg.
Rise Angle   = 45deg.
Fall Angle   = 60deg.

FIG. 22.41 This cam profile is an example of an advanced computer graphics plotting problem.

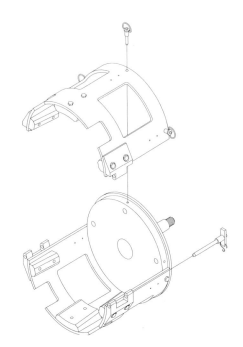

FIG. 22.42 A computer-drawn pictorial of an assembly of several parts. (Courtesy of Texas Instruments.)

Problems

The problems below can be assigned to be written, programmed, and run on a computer graphics system. When a system is not available, the programs can be written and then plotted by hand on a 7.5″ × 10″ grid to simulate a display screen. When plotted by hand, it is suggested that the programs and plots be done on a Size AV sheet or sheets.

1. Write a program that will plot the lines given below. Obtain a printout of your program and a plot of the lines when your program has run.

Problem	Line	
A	A(2,2)	B (6,6)
B	B(1,6)	C (9,2)
C	E(3,5)	F (8,3)
D	G(8,5)	H (3,3)

2. Write a program that will plot the horizontally positioned rectangles from the information given below. Obtain a printout of your program and a plot of the rectangles when your program has run.

Problem	Lower Left Corner	Height	Width
A	(2,2)	4″	5″
B	(1,1)	6″	5″
C	(3,1)	2″	3″
D	(2,1)	5″	4″

3. Write a program that will plot the angular lines below by using trigonometric functions of the angle of inclination, and the length of the line. Obtain a printout of your program and plot the lines when your program has run.

Problem	Lower Left Point	Length	Angle with Horizontal
A	(1,1)	5″	30°
B	(3,1)	4″	45°
C	(7,2)	3″	135°
D	(5,2)	4″	195°

4. Write a program that will plot the rectangles that incline with the horizontal at the angles given in the table below. Use variables similar to those illustrated in Fig. 22.23. Obtain a printout of your program and plot the rectangles when your program has run.

Problem	Lower Left Corner	Height	Width	Angle
A	(2,0.5)	4″	5″	15°
B	(2,1)	3″	4″	35°
C	(3,1)	2″	3.4″	40°
D	(2,1)	4.5″	4″	10°

5. Write a program that will plot regular polygons using the specifications given in the table below for each. Refer to Fig. 22.24. Obtain a printout of your program and plot the solutions when your program has run.

Problem	No. of Sides	Center	Radius
A	4	(5,4)	2.5″
B	6	(5,4)	2.75″
C	8	(5,4)	2.00″
D	10	(5,4)	2.25″

6. Write a program that will plot circles using the specifications given in Problem 5 by increasing the number of sides of the polygons to approximate the circles. Obtain a printout of your program and plot the solutions when your program has run.

7. Write a program that will use data points to plot the front views of any of the assigned objects shown in Fig. 22.43. Place the lower left corner at (2,2). Obtain a printout of your program and plot the solutions when your program has run.

8. Write a main program and a subroutine for plotting the rectangles assigned in Problem 4. Refer to Fig. 22.27. Obtain a printout of your program and a plot of the solutions when the program has run.

9. Write a main program and a subroutine for plotting the circles assigned in Problem 6. Refer to Fig. 22.29. Obtain a printout of your program and a plot of the solutions when the program has run.

10. By referring to Fig. 22.33, write a similar program to plot one of the oblique pictorials below by using the orthographic views of the objects in Fig. 22.43 and the supplementary data in Table 22.2. Obtain a printout of your program and a plot of the solutions when the program has run.

11. By referring to Fig. 22.36, write a similar program to plot one of the axonometric pictorials using the

TABLE 22.2

Part No.	Origin	Overall Scale Factor	Z-axis Scale Factor	Angle
1	(5,3)	0.50	0.50	30°
2	(5,3)	0.65	1.00	45°
3	(5,3)	0.40	0.75	36°
4	(5,3)	0.70	0.60	25°

FIG. 22.43 Orthographic views of problems where each square is equal to 0.50 inches. The same corner point in the top and front views is the origin for each problem.

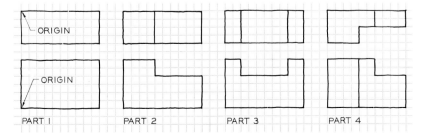

TABLE 22.3

Part	Origin	Overall Scale Factor	Rotation about Y-axis	Rotation about X-axis
1	(5,3)	0.40	−20°	40°
2	(5,3)	0.50	−45°	35.5°
3	(5,3)	0.60	−50°	15°
4	(5,3)	0.37	−15°	20°

orthographic views of the objects in Fig. 22.43 and the supplementary data in Table 22.3. Obtain a printout of your program and a plot of the solutions when the program has run.

12. Write a program similar to the one given in Fig. 22.38 that will plot three lines of text (of your choice) that will be plotted at different angles and will not overlap.

APPENDIXES
Contents

APPENDIX 1
ABBREVIATIONS (ANSI Z 32.13)

Word	Abbreviation	Word	Abbreviation	Word	Abbreviation
Abbreviate	ABBR	Bench mark	BM	Centigrade	C
Absolute	ABS	Between	BET.	Centigram	CG
Account	ACCT	Between centers	BC	Centimeter	cm
Actual	ACT.	Between		Chain	CH
Adapter	ADPT	perpendiculars	BP	Chamfer	CHAM
Addendum	ADD.	Bevel	BEV	Change notice	CN
Adjust	ADJ	Bill of material	B/M	Change order	CO
Advance	ADV	Birmingham wire gage	BWG	Channel	CHAN
After	AFT.	Blueprint	BP	Check	CHK
Aggregate	AGGR	Board	BD	Check valve	CV
Air condition	AIR COND	Boiler	BLR	Chemical	CHEM
Airport	AP	Bolt circle	BC	Chord	CHD
Airplane	APL	Both sides	BS	Circle	CIR
Allowance	ALLOW	Bottom	BOT.	Circuit	CKT
Alloy	ALY	Bottom chord	BC	Circular	CIR
Alteration	ALT	Boundary	BDY	Circular pitch	CP
Alternate	ALT	Bracket	BRKT	Circumference	CIRC
Alternating current	AC	Brake horsepower	BHP	Clockwise	CW
Altitude	ALT	Brass	BRS	Coated	CTD
Aluminum	AL	Brazing	BRZG	Cold drawn	CD
American National		Break	BRK	Cold drawn steel	CDS
Standard	AMER NATL STD	Breaker	BKR	Cold finish	CF
American wire gage	AWG	Bridge	BRDG	Cold punched	CP
Ammeter	AM	Brinnell hardness	BH	Cold rolled	CR
Amount	AMT	British Standard	BR STD	Cold rolled steel	CRS
Ampere	AMP	British Thermal Units	BTU	Column	COL
Anneal	ANL	Broach	BRO	Combination	COMB.
Antenna	ANT.	Bronze	BRZ	Combustion	COMB
Apparatus	APP	Brown & Sharp (Wire gage,		Commutator	COMM
Appendix	APPX	same as AWG)	B&S	Company	CO
Approved	APPD	Building	BLDG	Concentric	CONC
Approximate	APPROX	Bulkhead	BHD	Concrete	CONC
Arc weld	ARC/W	Bureau	BU	Condition	COND
Area	A	Bureau of Standards	BU STD	Connect	CONN
Armature	ARM.	Bushing	BUSH.	Contact	CONT
Asbestos	ASB	Button	BUT.	Cord	CD
Asphalt	ASPH	Buzzer	BUZ	Corporation	CORP
Assembly	ASSY	By-pass	BYP	Corrugate	CORR
Association	ASSN			Cotter	COT
Atomic	AT	Cabinet	CAB.	Counterclockwise	CCW
Authorized	AUTH	Cadmium plate	CD PL	Counterbore	CBORE
Auxiliary	AUX	Calculate	CALC	Counterdrill	CDRILL
Avenue	AVE	Calibrate	CAL	Counterpunch	CPUNCH
Average	AVG	Calorie	CAL	Countersink	CSK
Avoirdupois	AVDP	Capacitor	CAP	Coupling	CPLG
Azimuth	AZ	Cap screw	CAP SCR	Crank	CRK
		Case harden	CH	Cross section	XSECT
Babbitt	BAB	Cast iron	CI	Cubic	CU
Back pressure	BP	Cast steel	CS	Cubic centimeter	cc
Balance	BAL	Casting	CSTG	Cubic feet per minute	CFM
Ball bearing	BB	Castle nut	CAS NUT	Cubic feet per second	CFS
Barometer	BAR	Catalogue	CAT.	Cubic foot	CU FT
Barrel	BBL	Cement	CEM	Cubic inch	CU IN.
Base line	BL	Center	CTR	Cubic meter	CU M
Base plate	BP	Centerline	CL	Cubic yard	CU YD
Battery	BAT.	Center of gravity	CG	Current	CUR
Bearing	BRG	Center to center	C to C	Cylinder	CYL

Cont.

APPENDIX 1
ABBREVIATIONS (ANSI Z 32.13) (Cont.)

Word	Abbreviation	Word	Abbreviation	Word	Abbreviation
Decimal	DEC	Fillister	FIL	Illustrate	ILLUS
Dedendum	DED	Filter	FLT	Inch	(") IN.
Degree	(°) DEG	Finish	FIN.	Inches per second	IPS
Department	DEPT	Finish all over	FAO	Include	INCL
Design	DSGN	Flange	FLG	Industrial	IND
Detail	DET	Flat head	FH	Information	INFO
Develop	DEV	Fluid	FL	Inside diameter	ID
Diagonal	DIAG	Focus	FOC	Instrument	INST
Diagram	DIAG	Foot	(') FT	Insulate	INS
Diameter	DIA	Forging	FORG	Interior	INT
Diametrical pitch	DP	Forward	FWD	Internal	INT
Dimension	DIM.	Foundation	FDN	Intersect	INT
Direct current	DC	Foundry	FDRY	Iron	I
Discharge	DISCH	Frequency	FREQ	Irregular	IRREG
Distance	DIST	Front	FR		
District	DIST			Jack	J
Ditto	DO.			Joint	JT
Dovetail	DVTL	Gage	GA	Junction	JCT
Dowel	DWL	Gallon	GAL	Junction box	JB
Down	DN	Galvanize	GALV		
Dozen	DOZ	Galvanized iron	GI	Key	K
Drafting	DFTG	Galvanized steel	GS	Keyseat	KST
Draftsman	DFTSMN	Gasket	GSKT	Keyway	KWY
Drawing	DWG	General	GEN	Kiln-dried	KD
Drill	DR	Government	GOVT	Kip (1000 lb)	K
Drive	DR	Governor	GOV	Knots	KN
Drop forge	DF	Grade	GR		
		Grade line	GL	Laboratory	LAB
		Gram	G	Lateral	LAT
Each	EA	Gravity	G	Latitude	LAT
East	E	Grind	GRD	Left	L
Eccentric	ECC	Groove	GRV	Left hand	LH
Effective	EFF	Ground	GRD	Length	LG
Elbow	ELL	Gypsum	GYP	Letter	LTR
Electric	ELEC			Light	LT
Elevation	ELEV			Line	L
Engineer	ENGR	Half-round	½ RD	Logarithm	LOG.
Equal	EQ	Handle	HDL	Lubricate	LUB
Equipment	EQUIP.	Hanger	HGR	Lumber	LBR
Equivalent	EQUIV	Hard	H		
Estimate	EST	Hard-drawn	HD	Machine	MACH
Exterior	EXT	Harden	HDN	Malleable	MALL
Extra heavy	X HVY	Hardware	HDW	Malleable iron	MI
Extra strong	X STR	Head	HD	Manhole	MH
		Headless	HDLS	Manual	MAN.
		Headquarters	HQ	Manufacture	MFR
Fabricate	FAB	Heat	HT	Material	MATL
Face to face	F to F	Heat treat	HT TR	Maximum	MAX
Fahrenheit	F	Hexagon	HEX	Mechanical	MECH
Fairing	FAIR.	High-pressure	HP	Mechanism	MECH
Far side	FS	High-speed	HS	Median	MED
Federal	FED.	Horizontal	HOR	Metal	MET.
Feet	(') FT	Horsepower	HP	Meter (Instrument or	
Feet per minute	FPM	Hot rolled	HR	measure of length)	M
Feet per second	FPS	Hot rolled steel	HRS	Miles	MI
Field	FLD	Hour	HR	Miles per gallon	MPG
Figure	FIG.	Hundredweight	CWT	Miles per hour	MPH
Fillet	FIL	Hydraulic	HYD		

APPENDIX 2
CONVERSION TABLES (Cont.)

Power conversions

Kilogrammeters/second	$\times$ 9.806 65	= watts
Kilowatts	$\times$ 3.414 43 $\times$ 10^3	= BTU/hour
	$\times$ 2.655 22 $\times$ 10^6	= footpounds/hour
	$\times$ 4.425 37 $\times$ 10^4	= footpounds/minute
	$\times$ 737.562	= footpounds/second
	$\times$ 1.019 726 $\times$ 10^7	= gramcentimeters/second
	$\times$ 1.341 02	= horsepower
	$\times$ 3.6 $\times$ 10^6	= joules/hour
	$\times$ 10^3	= joules/second
	$\times$ 3.671 01 $\times$ 10^5	= kilogrammeters/hour
	$\times$ 999.835	= international watt
Watts	$\times$ 44.253 7	= footpounds/minute
	$\times$ 1.341 02 $\times$ 10^{-3}	= horsepower
	$\times$ 1	= joules/second

Time conversions

(No attempt has been made in this brief treatment to correlate solar, mean solar, sidereal, and mean sidereal days.)

Mean solar days	$\times$ 24	= mean solar hours
Mean solar hours	$\times$ 3.600 $\times$ 10^3	= mean solar seconds
	$\times$ 60	= mean solar minutes

Angle conversions

Degrees	$\times$ 60	= minutes
	$\times$ 1.745 329 3 $\times$ 10^{-2}	= radians
Degrees/foot	$\times$ 5.726 145 $\times$ 10^{-4}	= radians/centimeter
Degrees/minute	$\times$ 2.908 8 $\times$ 10^{-4}	= radians/second
	$\times$ 4.629 629 $\times$ 10^{-5}	= revolutions/second
Degrees/second	$\times$ 1.745 329 3 $\times$ 10^{-2}	= radians/second
	$\times$ 0.166	= revolutions/minute
	$\times$ 2.77 $\times$ 10^{-3}	= revolutions/second
Minutes	$\times$ 1.667 $\times$ 10^{-2}	= degrees
	$\times$ 2.908 8 $\times$ 10^{-4}	= radians
	$\times$ 60	= seconds
Radians	$\times$ 0.159 154	= circumferences
	$\times$ 57.295 77	= degrees
	$\times$ 3.437 746 $\times$ 10^3	= minutes
Seconds	$\times$ 2.777 $\times$ 10^{-4}	= degrees
	$\times$ 1.667 $\times$ 10^{-2}	= minutes
	$\times$ 4.848 136 8 $\times$ 10^{-6}	= radians
Steradians	$\times$ 0.159 154 9	= hemispheres
	$\times$ 7.957 74 $\times$ 10^{-2}	= spheres
	$\times$ 0.636 619 7	= spherical right angles

Mass conversions

Grains	$\times$ 6.479 8 $\times$ 10^{-2}	= grams
	$\times$ 2.285 71 $\times$ 10^{-3}	= avoirdupois ounces

Cont.

Mass conversions

Grams	$\times$ 15.432 358	= grains
	$\times$ 3.527 396 $\times$ 10^{-2}	= avoirdupois ounces
	$\times$ 2.204 62 $\times$ 10^{-3}	= avoirdupois pounds
Kilograms	$\times$ 564.383 4	= avoirdupois drams
	$\times$ 2.204 622 6	= avoirdupois pounds
	$\times$ 2.2	= pounds
	$\times$ 9.842 065 $\times$ 10^{-4}	= long tons
	$\times$ 10^{-3}	= metric tons
	$\times$ 1.102 31 $\times$ 10^{-3}	= short tons
Avoirdupois ounces	$\times$ 28.349 5	= grams
	$\times$ 6.25 $\times$ 10^{-2}	= avoirdupois pounds
	$\times$ 0.911 458	= troy ounces
Avoirdupois pounds	$\times$ 256	= drams
	$\times$ 4.535 923 7 $\times$ 10^2	= grams
	$\times$ 0.453 592 4	= kilograms
	$\times$ 16	= ounces
Long tons	$\times$ 2.24 $\times$ 10^3	= avoirdupois pounds
	$\times$ 1.106 046 9	= metric tons
	$\times$ 1.12	= short tons
Metric tons	$\times$ 10^3	= kilograms
	$\times$ 2.204 622 $\times$ 10^3	= avoirdupois pounds
Short tons	$\times$ 2 $\times$ 10^3	= avoirdupois pounds
	$\times$ 907.184 74	= kilograms

Force conversions

Dynes	$\times$ 10^{-5}	= newtons
Newtons	$\times$ 10^5	= dynes
	$\times$ 0.224 808	= pounds-force
Pounds	$\times$ 4.448 22	= newtons

Energy conversions

British Thermal Units (thermochemical)	$\times$ 1.054 35 $\times$ 10^3	= joules
	$\times$ 2.928 27 $\times$ 10^{-4}	= kilowatthours
	$\times$ 1.054 35 $\times$ 10^3	= wattseconds
Foot-pound-force	$\times$ 1.355 818 0	= joules
	$\times$ 0.138 255	= kilogramforce-meters
	$\times$ 3.766 16 $\times$ 10^{-7}	= kilowatthours
	$\times$ 1.355 818 0	= newtonmeters
Joules	$\times$ 9.484 5 $\times$ 10^{-4}	= British Thermal Units
	$\times$ 0.737 562	= foot-pounds-force
	$\times$ 0.101 971 6	= kilogramforce-meters
	$\times$ 2.777 7 $\times$ 10^{-7}	= kilowatthours
	$\times$ 1	= wattseconds
Kilogramforce-meters	$\times$ 9.287 7 $\times$ 10^{-3}	= British Thermal Units
	$\times$ 7.233 01	= foot-pounds-force
	$\times$ 9.806 65	= joules
	$\times$ 9.806 65	= newtonmeters
	$\times$ 2.724 0 $\times$ 10^{-3}	= watthours

APPENDIX 2
CONVERSION TABLES (Cont.)

Energy conversions

Kilowatthours	$\times$ 3.409 52 $\times$ 10^3	= British Thermal Units
	$\times$ 2.655 22 $\times$ 10^6	= foot-pounds-force
	$\times$ 1.341 02	= horsepowerhours
	$\times$ 3.6 $\times$ 10^6	= joules
	$\times$ 3.670 98 $\times$ 10^5	= kilogramforce-meters
Newtonmeters	$\times$ 0.101 971	= kilogramforce-meters
	$\times$ 0.737 562	= poundforce-feet
Watthours	$\times$ 3.414 43	= British Thermal Units
	$\times$ 2.655 22 $\times$ 10^3	= foot-pounds-force
	$\times$ 3.6 $\times$ 10^3	= joules
	$\times$ 3.670 98 $\times$ 10^2	= kilogramforce-meters

Pressure conversions

Atmospheres	$\times$ 1.013 25	= bars
	$\times$ 1.033 23 $\times$ 10^3	= grams/square centimeter
	$\times$ 1.033 23 $\times$ 10^7	= grams/square meter
	$\times$ 14.696 0	= pounds/square inch
	$\times$ 760	= torrs
	$\times$ 101	= kilopascals
Bars	$\times$ 0.986 923	= atmospheres
	$\times$ 10^6	= baryes
	$\times$ 1.019 716 $\times$ 10^7	= grams/square meter
	$\times$ 1.019 716 $\times$ 10^4	= kilogramsforce/square meter
	$\times$ 14.503 8	= poundsforce/square inch
Baryes	$\times$ 10^{-6}	= bars
Inches of mercury	$\times$ 3.386 4 $\times$ 10^{-2}	= bars
	$\times$ 345.316	= kilogramsforce/square meter
	$\times$ 70.726 2	= poundsforce/square foot
Pascal	$\times$ 1	= newton/square meter

APPENDIX 3
LOGARITHMS OF NUMBERS

N	0	1	2	3	4	5	6	7	8	9
1.0	.0000	.0043	.0086	.0128	.0170	.0212	.0253	.0294	.0334	.0374
1.1	.0414	.0453	.0492	.0531	.0569	.0607	.0645	.0682	.0719	.0755
1.2	.0792	.0828	.0864	.0899	.0934	.0969	.1004	.1038	.1072	.1106
1.3	.1139	.1173	.1206	.1239	.1271	.1303	.1335	.1367	.1399	.1430
1.4	.1461	.1492	.1523	.1553	.1584	.1614	.1644	.1673	.1703	.1732
1.5	.1761	.1790	.1818	.1847	.1875	.1903	.1931	.1959	.1987	.2014
1.6	.2041	.2068	.2095	.2122	.2148	.2175	.2201	.2227	.2253	.2279
1.7	.2304	.2330	.2355	.2380	.2405	.2430	.2455	.2480	.2504	.2529
1.8	.2553	.2577	.2601	.2625	.2648	.2672	.2695	.2718	.2742	.2765
1.9	.2788	.2810	.2833	.2856	.2878	.2900	.2923	.2945	.2967	.2989
2.0	.3010	.3032	.3054	.3075	.3096	.3118	.3139	.3160	.3181	.3201
2.1	.3222	.3243	.3263	.3284	.3304	.3324	.3345	.3365	.3385	.3404
2.2	.3424	.3444	.3464	.3483	.3502	.3522	.3541	.3560	.3579	.3598
2.3	.3617	.3636	.3655	.3674	.3692	.3711	.3729	.3747	.3766	.3784
2.4	.3802	.3820	.3838	.3856	.3874	.3892	.3909	.3927	.3945	.3962
2.5	.3979	.3997	.4014	.4031	.4048	.4065	.4082	.4099	.4116	.4133
2.6	.4150	.4166	.4183	.4200	.4216	.4232	.4249	.4265	.4281	.4298
2.7	.4314	.4330	.4346	.4362	.4378	.4393	.4409	.4425	.4440	.4456
2.8	.4472	.4487	.4502	.4518	.4533	.4548	.4564	.4579	.4594	.4609
2.9	.4624	.4639	.4654	.4669	.4683	.4698	.4713	.4728	.4742	.4757
3.0	.4771	.4786	.4800	.4814	.4829	.4843	.4857	.4871	.4886	.4900
3.1	.4914	.4928	.4942	.4955	.4969	.4983	.4997	.5011	.5024	.5038
3.2	.5051	.5065	.5079	.5092	.5105	.5119	.5132	.5145	.5159	.5172
3.3	.5185	.5198	.5211	.5224	.5237	.5250	.5263	.5276	.5289	.5302
3.4	.5315	.5328	.5340	.5353	.5366	.5378	.5391	.5403	.5416	.5428
3.5	.5441	.5453	.5465	.5478	.5490	.5502	.5514	.5527	.5539	.5551
3.6	.5563	.5575	.5587	.5599	.5611	.5623	.5635	.5647	.5658	.5670
3.7	.5682	.5694	.5705	.5717	.5729	.5740	.5752	.5763	.5775	.5786
3.8	.5798	.5809	.5821	.5832	.5843	.5855	.5866	.5877	.5888	.5899
3.9	.5911	.5922	.5933	.5944	.5955	.5966	.5977	.5988	.5999	.6010
4.0	.6021	.6031	.6042	.6053	.6064	.6075	.6085	.6096	.6107	.6117
4.1	.6128	.6138	.6149	.6160	.6170	.6180	.6191	.6201	.6212	.6222
4.2	.6232	.6243	.6253	.6263	.6274	.6284	.6294	.6304	.6314	.6325
4.3	.6335	.6345	.6355	.6365	.6375	.6385	.6395	.6405	.6415	.6425
4.4	.6435	.6444	.6454	.6464	.6474	.6484	.6493	.6503	.6513	.6522
4.5	.6532	.6542	.6551	.6561	.6571	.6580	.6590	.6599	.6609	.6618
4.6	.6628	.6637	.6646	.6656	.6665	.6675	.6684	.6693	.6702	.6712
4.7	.6721	.6730	.6739	.6749	.6758	.6767	.6776	.6785	.6794	.6803
4.8	.6812	.6821	.6830	.6839	.6848	.6857	.6866	.6875	.6884	.6893
4.9	.6902	.6911	.6920	.6928	.6937	.6946	.6955	.6964	.6972	.6981
5.0	.6990	.6998	.7007	.7016	.7024	.7033	.7042	.7050	.7059	.7067
5.1	.7076	.7084	.7093	.7101	.7110	.7118	.7126	.7135	.7143	.7152
5.2	.7160	.7168	.7177	.7185	.7193	.7202	.7210	.7218	.7226	.7235
5.3	.7243	.7251	.7259	.7267	.7275	.7284	.7292	.7300	.7308	7316
5.4	.7324	.7332	.7340	.7348	.7356	.7364	.7372	.7380	.7388	.7396
N	0	1	2	3	4	5	6	7	8	9

APPENDIX 5
WEIGHTS AND MEASURES

UNITED STATES SYSTEM

LINEAR MEASURE

Inches	Feet	Yards	Rods	Furlongs	Miles
1.0 =	.08333	= .02778 =	.0050505 =	.00012626 =	.00001578
12.0 =	1.0	= .33333 =	.0606061 =	.00151515 =	.00018939
36.0 =	3.0	= 1.0 =	.1818182 =	.00454545 =	.00056818
198.0 =	16.5	= 5.5 =	1.0 =	.025 =	.003125
7920.0 =	660.0	= 220.0 =	40.0 =	1.0 =	.125
63360.0 =	5280.0	= 1760.0 =	320.0 =	8.0 =	1.0

SQUARE AND LAND MEASURE

Sq. Inches	Square Feet	Square Yards	Sq. Rods	Acres	Sq. Miles
1.0 =	.006944 =	.000772			
144.0 =	1.0 =	.111111			
1296.0 =	9.0 =	1.0 =	.03306 =	.000207	
39204.0 =	272.25 =	30.25 =	1.0 =	.00625 =	.0000098
	43560.0 =	4840.0 =	160.0 =	1.0 =	.0015625
		3097600.0 =	102400.0 =	640.0 =	1.0

AVOIRDUPOIS WEIGHTS

Grains	Drams	Ounces	Pounds	Tons
1.0 =	.03657 =	.002286 =	.000143 =	.0000000714
27.34375 =	1.0 =	.0625 =	.003906 =	.00000195
437.5 =	16.0 =	1.0 =	.0625 =	.00003125
7000.0 =	256.0 =	16.0 =	1.0 =	.0005
14000000.0 =	512000.0 =	32000.0 =	2000.0 =	1.0

DRY MEASURE

Pints	Quarts	Pecks	Cubic Feet	Bushels
1.0 =	.5 =	.0625 =	.01945 =	.01563
2.0 =	1.0 =	.125 =	.03891 =	.03125
16.0 =	8.0 =	1.0 =	.31112 =	.25
51.42627 =	25.71314 =	3.21414 =	1.0 =	.80354
64.0 =	32.0 =	4.0 =	1.2445 =	1.0

LIQUID MEASURE

Gills	Pints	Quarts	U. S. Gallons	Cubic Feet
1.0 =	.25 =	.125 =	.03125 =	.00418
4.0 =	1.0 =	.5 =	.125 =	.01671
8.0 =	2.0 =	1.0 =	.250 =	.03342
32.0 =	8.0 =	4.0 =	1.0 =	.1337
			7.48052 =	1.0

METRIC SYSTEM

UNITS
Length—Meter : Mass—Gram : Capacity—Liter

for pure water at 4°C. (39.2°F.)

1 cubic decimeter or 1 liter = 1 kilogram

1000 Milli $\begin{cases} meters \text{ (mm)} \\ grams \text{ (mg)} \\ liters \text{ (ml)} \end{cases}$ = 100 Centi $\begin{cases} meters \text{ (cm)} \\ grams \text{ (cg)} \\ liters \text{ (cl)} \end{cases}$ = 10 Deci $\begin{cases} meters \text{ (dm)} \\ grams \text{ (dg)} \\ liters \text{ (dl)} \end{cases}$ = 1 $\begin{cases} meter \\ gram \\ liter \end{cases}$

1000 $\begin{cases} meters \\ grams \\ liters \end{cases}$ = 100 Deka $\begin{cases} meters \text{ (dkm)} \\ grams \text{ (dkg)} \\ liters \text{ (dkl)} \end{cases}$ = 10 Hecto $\begin{cases} meters \text{ (hm)} \\ grams \text{ (hg)} \\ liters \text{ (hl)} \end{cases}$ = 1 Kilo $\begin{cases} meter \text{ (km)} \\ gram \text{ (kg)} \\ liter \text{ (kl)} \end{cases}$

1 Metric Ton	= 1000 Kilograms
100 Square Meters	= 1 Are
100 Ares	= 1 Hectare
100 Hectares	= 1 Square Kilometer

APPENDIX 6
DECIMAL EQUIVALENTS AND TEMPERATURE CONVERSION

DECIMAL EQUIVALENTS—INCH-MILLIMETER CONVERSION TABLE

1/2	1/4	1/8	1/16	1/32	1/64	Decimals	Millimeters
					1	.015625	.396875
				1		.031250	.793750
					3	.046875	1.190625
			1			.062500	1.587500
					5	.078125	1.984375
				3		.093750	2.381250
					7	.109375	2.778125
		1				.125000	3.175000
					9	.140625	3.571875
				5		.156250	3.968750
					11	.171875	4.365625
			3			.187500	4.762500
					13	.203125	5.159375
				7		.218750	5.556250
					15	.234375	5.953125
	1					.250000	6.350000
					17	.265625	6.746875
				9		.281250	7.143750
					19	.296875	7.540625
			5			.312500	7.937500
					21	.328125	8.334375
				11		.343750	8.731250
					23	.359375	9.128125
		3				.375000	9.525000
					25	.390625	9.921875
				13		.406250	10.318750
					27	.421875	10.715625
			7			.437500	11.112500
					29	.453125	11.509375
				15		.468750	11.906250
					31	.484375	12.303125
1						.500000	12.700000

1/2	1/4	1/8	1/16	1/32	1/64	Decimals	Millimeters
					33	.515625	13.096875
				17		.531250	13.493750
					35	.546875	13.890625
			9			.562500	14.287500
					37	.578125	14.684375
				19		.593750	15.081250
					39	.609375	15.478125
		5				.625000	15.875000
					41	.640625	16.271875
				21		.656250	16.668750
					43	.671875	17.065625
			11			.687500	17.462500
					45	.703125	17.859375
				23		.718750	18.256250
					47	.734375	18.653125
	3					.750000	19.050000
					49	.765625	19.446875
				25		.781250	19.843750
					51	.796875	20.240625
			13			.812500	20.637500
					53	.828125	21.034375
				27		.843750	21.431250
					55	.859375	21.828125
		7				.875000	22.225000
					57	.890625	22.621875
				29		.906250	23.018750
					59	.921875	23.415625
			15			.937500	23.812500
					61	.953125	24.209375
				31		.968750	24.606250
					63	.984375	25.003125
2	4	8	16	32	64	1.000000	25.400000

APPENDIX 6

DECIMAL EQUIVALENTS AND TEMPERATURE CONVERSION (Cont.)

TEMPERATURE CONVERSION

-210 to 0

C.	C. or F.	F.
-134	-210	-346
-129	-200	-328
-123	-190	-310
-118	-180	-292
-112	-170	-274
-107	-160	-256
-101	-150	-238
-95.6	-140	-220
-90.0	-130	-202
-84.4	-120	-184
-78.9	-110	-166
-73.3	-100	-148
-67.8	-90	-130
-62.2	-80	-112
-56.7	-70	-94
-51.1	-60	-76
-45.6	-50	-58
-40.0	-40	-40
-34.4	-30	-22
-28.9	-20	-4
-23.3	-10	14
-17.8	0	32

1 to 25

C.	C. or F.	F.
-17.2	1	33.8
-16.7	2	35.6
-16.1	3	37.4
-15.6	4	39.2
-15.0	5	41.0
-14.4	6	42.8
-13.9	7	44.6
-13.3	8	46.4
-12.8	9	48.2
-12.2	10	50.0
-11.7	11	51.8
-11.1	12	53.6
-10.6	13	55.4
-10.0	14	57.2
-9.44	15	59.0
-8.89	16	60.8
-8.33	17	62.6
-7.78	18	64.4
-7.22	19	66.2
-6.67	20	68.0
-6.11	21	69.8
-5.56	22	71.6
-5.00	23	73.4
-4.44	24	75.2
-3.89	25	77.0

26 to 50

C.	C. or F.	F.
-3.33	26	78.8
-2.78	27	80.6
-2.22	28	82.4
-1.67	29	84.2
-1.11	30	86.0
-0.56	31	87.8
0	32	89.6
0.56	33	91.4
1.11	34	93.2
1.67	35	95.0
2.22	36	96.8
2.78	37	98.6
3.33	38	100.4
3.89	39	102.2
4.44	40	104.0
5.00	41	105.8
5.56	42	107.6
6.11	43	109.4
6.67	44	111.2
7.22	45	113.0
7.78	46	114.8
8.33	47	116.6
8.89	48	118.4
9.44	49	120.2
10.0	50	122.0

51 to 75

C.	C. or F.	F.
10.6	51	123.8
11.1	52	125.6
11.7	53	127.4
12.2	54	129.2
12.8	55	131.0
13.3	56	132.8
13.9	57	134.6
14.4	58	136.4
15.0	59	138.2
15.6	60	140.0
16.1	61	141.8
16.7	62	143.6
17.2	63	145.4
17.8	64	147.2
18.3	65	149.0
18.9	66	150.8
19.4	67	152.6
20.0	68	154.4
20.6	69	156.2
21.1	70	158.0
21.7	71	159.8
22.2	72	161.6
22.8	73	163.4
23.3	74	165.2
23.9	75	167.0

76 to 100

C.	C. or F.	F.
24.4	76	168.8
25.0	77	170.6
25.6	78	172.4
26.1	79	174.2
26.7	80	176.0
27.2	81	177.8
27.8	82	179.6
28.3	83	181.4
28.9	84	183.2
29.4	85	185.0
30.0	86	186.8
30.6	87	188.6
31.1	88	190.4
31.7	89	192.2
32.2	90	194.0
32.8	91	195.8
33.3	92	197.6
33.9	93	199.4
34.4	94	201.2
35.0	95	203.0
35.6	96	204.8
36.1	97	206.6
36.7	98	208.4
37.2	99	210.2
37.8	100	212.0

101 to 340

C.	C. or F.	F.
43	110	230
49	120	248
54	130	266
60	140	284
66	150	302
71	160	320
77	170	338
82	180	356
88	190	374
93	200	392
99	210	410
100	212	413
104	220	428
110	230	446
116	240	464
121	250	482
127	260	500
132	270	518
138	280	536
143	290	554
149	300	572
154	310	590
160	320	608
166	330	626
171	340	644

341 to 490

C.	C. or F.	F.
177	350	662
182	360	680
188	370	698
193	380	716
199	390	734
204	400	752
210	410	770
216	420	788
221	430	806
227	440	824
232	450	842
238	460	860
243	470	878
249	480	896
254	490	914

491 to 750

C.	C. or F.	F.
260	500	932
266	510	950
271	520	968
277	530	986
282	540	1004
288	550	1022
293	560	1040
299	570	1058
304	580	1076
310	590	1094
316	600	1112
321	610	1130
327	620	1148
332	630	1166
338	640	1184
343	650	1202
349	660	1220
354	670	1238
360	680	1256
366	690	1274
371	700	1292
377	710	1310
382	720	1328
388	730	1346
393	740	1364
399	750	1382

INTERPOLATION FACTORS

C.		F.	C.		F.
0.56	1	1.8	3.33	6	10.8
1.11	2	3.6	3.89	7	12.6
1.67	3	5.4	4.44	8	14.4
2.22	4	7.2	5.00	9	16.2
2.78	5	9.0	5.56	10	18.0

NOTE:—The numbers in bold face type refer to the temperature either in degrees Centigrade or Fahrenheit which it is desired to convert into the other scale. If converting from Fahrenheit degrees to Centigrade the equivalent temperature will be found in the left column, while if converting from degrees Centigrade to degrees Fahrenheit, the answer will be found in the column on the right.

$$°F = \frac{9}{5}\,(°C) + 32$$

$$°C = \frac{5}{9}\,(°F - 32)$$

APPENDIX 7
WEIGHTS AND SPECIFIC GRAVITIES

Substance	Weight Lb. per Cu. Ft.	Specific Gravity	Substance	Weight Lb. per Cu. Ft.	Specific Gravity
METALS, ALLOYS, ORES			**TIMBER, U. S. SEASONED**		
			Moisture Content by Weight:		
Aluminum, cast, hammered	165	2.55-2.75	Seasoned timber 15 to 20%		
Brass, cast, rolled	534	8.4-8.7	Green timber up to 50%		
Bronze, 7.9 to 14% Sn	509	7.4-8.9	Ash, white, red	40	0.62-0.65
Bronze, aluminum	481	7.7	Cedar, white, red	22	0.32-0.38
Copper, cast, rolled	556	8.8-9.0	Chestnut	41	0.66
Copper ore, pyrites	262	4.1-4.3	Cypress	30	0.48
Gold, cast, hammered	1205	19.25-19.3	Fir, Douglas spruce	32	0.51
Iron, cast, pig	450	7.2	Fir, eastern	25	0.40
Iron, wrought	485	7.6-7.9	Elm, white	45	0.72
Iron, spiegel-eisen	468	7.5	Hemlock	29	0.42-0.52
Iron, ferro-silicon	437	6.7-7.3	Hickory	49	0.74-0.84
Iron ore, hematite	325	5.2	Locust	46	0.73
Iron ore, hematite in bank	160-180	....	Maple, hard	43	0.68
Iron ore, hematite loose	130-160	....	Maple, white	33	0.53
Iron ore, limonite	237	3.6-4.0	Oak, chestnut	54	0.86
Iron ore, magnetite	315	4.9-5.2	Oak, live	59	0.95
Iron slag	172	2.5-3.0	Oak, red, black	41	0.65
Lead	710	11.37	Oak, white	46	0.74
Lead ore, galena	465	7.3-7.6	Pine, Oregon	32	0.51
Magnesium, alloys	112	1.74-1.83	Pine, red	30	0.48
Manganese	475	7.2-8.0	Pine, white	26	0.41
Manganese ore, pyrolusite	259	3.7-4.6	Pine, yellow, long-leaf	44	0.70
Mercury	849	13.6	Pine, yellow, short-leaf	38	0.61
Monel Metal	556	8.8-9.0	Poplar	30	0.48
Nickel	565	8.9-9.2	Redwood, California	26	0.42
Platinum, cast, hammered	1330	21.1-21.5	Spruce, white, black	27	0.40-0.46
Silver, cast, hammered	656	10.4-10.6	Walnut, black	38	0.61
Steel, rolled	490	7.85			
Tin, cast, hammered	459	7.2-7.5			
Tin ore, cassiterite	418	6.4-7.0			
Zinc, cast, rolled	440	6.9-7.2			
Zinc ore, blende	253	3.9-4.2	**VARIOUS LIQUIDS**		
			Alcohol, 100%	49	0.79
			Acids, muriatic 40%	75	1.20
			Acids, nitric 91%	94	1.50
			Acids, sulphuric 87%	112	1.80
VARIOUS SOLIDS			Lye, soda 66%	106	1.70
			Oils, vegetable	58	0.91-0.94
Cereals, oats............bulk	32	....	Oils, mineral, lubricants	57	0.90-0.93
Cereals, barley............bulk	39	....	Water, 4°C. max. density	62.428	1.0
Cereals, corn, rye........bulk	48	....	Water, 100°C	59.830	0.9584
Cereals, wheat............bulk	48	....	Water, ice	56	0.88-0.92
Hay and Straw............bales	20	....	Water, snow, fresh fallen	8	.125
Cotton, Flax, Hemp	93	1.47-1.50	Water, sea water	64	1.02-1.03
Fats	58	0.90-0.97			
Flour, loose	28	0.40-0.50			
Flour, pressed	47	0.70-0.80			
Glass, common	156	2.40-2.60	**GASES**		
Glass, plate or crown	161	2.45-2.72			
Glass, crystal	184	2.90-3.00	Air, 0°C. 760 mm.	.08071	1.0
Leather	59	0.86-1.02	Ammonia	.0478	0.5920
Paper	58	0.70-1.15	Carbon dioxide	.1234	1.5291
Potatoes, piled	42		Carbon monoxide	.0781	0.9673
Rubber, caoutchouc	59	0.92-0.96	Gas, illuminating	.028-.036	0.35-0.45
Rubber goods	94	1.0-2.0	Gas, natural	.038-.039	0.47-0.48
Salt, granulated, piled	48	....	Hydrogen	.00559	0.0693
Saltpeter	67	....	Nitrogen	.0784	0.9714
Starch	96	1.53	Oxygen	.0892	1.1056
Sulphur	125	1.93-2.07			
Wool	82	1.32			

The specific gravities of solids and liquids refer to water at 4°C., those of gases to air at 0°C. and 760 mm. pressure. The weights per cubic foot are derived from average specific gravities, except where stated that weights are for bulk, heaped or loose material, etc.

(Courtesy of the American Institute of Steel Construction.)

APPENDIX 7
WEIGHTS AND SPECIFIC GRAVITIES (Cont.)

Substance	Weight Lb. per Cu. Ft.	Specific Gravity	Substance	Weight Lb. per Cu. Ft.	Specific Gravity
ASHLAR MASONRY			**MINERALS**		
Granite, syenite, gneiss	165	2.3-3.0	Asbestos	153	2.1-2.8
Limestone, marble	160	2.3-2.8	Barytes	281	4.50
Sandstone, bluestone	140	2.1-2.4	Basalt	184	2.7-3.2
			Bauxite	159	2.55
MORTAR RUBBLE			Borax	109	1.7-1.8
MASONRY			Chalk	137	1.8-2.6
Granite, syenite, gneiss	155	2.2-2.8	Clay, marl	137	1.8-2.6
Limestone, marble	150	2.2-2.6	Dolomite	181	2.9
Sandstone, bluestone	130	2.0-2.2	Feldspar, orthoclase	159	2.5-2.6
			Gneiss, serpentine	159	2.4-2.7
DRY RUBBLE MASONRY			Granite, syenite	175	2.5-3.1
Granite, syenite, gneiss	130	1.9-2.3	Greenstone, trap	187	2.8-3.2
Limestone, marble	125	1.9-2.1	Gypsum, alabaster	159	2.3-2.8
Sandstone, bluestone	110	1.8-1.9	Hornblende	187	3.0
			Limestone, marble	165	2.5-2.8
BRICK MASONRY			Magnesite	187	3.0
Pressed brick	140	2.2-2.3	Phosphate rock, apatite	200	3.2
Common brick	120	1.8-2.0	Porphyry	172	2.6-2.9
Soft brick	100	1.5-1.7	Pumice, natural	40	0.37-0.90
			Quartz, flint	165	2.5-2.8
CONCRETE MASONRY			Sandstone, bluestone	147	2.2-2.5
Cement, stone, sand	144	2.2-2.4	Shale, slate	175	2.7-2.9
Cement, slag, etc.	130	1.9-2.3	Soapstone, talc	169	2.6-2.8
Cement, cinder, etc.	100	1.5-1.7			
VARIOUS BUILDING			**STONE, QUARRIED, PILED**		
MATERIALS			Basalt, granite, gneiss	96	....
Ashes, cinders	40-45	....	Limestone, marble, quartz	95	....
Cement, portland, loose	90	....	Sandstone	82	....
Cement, portland, set	183	2.7-3.2	Shale	92	....
Lime, gypsum, loose	53-64	....	Greenstone, hornblende	107	....
Mortar, set	103	1.4-1.9			
Slags, bank slag	67-72	....			
Slags, bank screenings	98-117	....			
Slags, machine slag	96	....	**BITUMINOUS SUBSTANCES**		
Slags, slag sand	49-55	....	Asphaltum	81	1.1-1.5
			Coal, anthracite	97	1.4-1.7
EARTH, ETC., EXCAVATED			Coal, bituminous	84	1.2-1.5
Clay, dry	63	....	Coal, lignite	78	1.1-1.4
Clay, damp, plastic	110	....	Coal, peat, turf, dry	47	0.65-0.85
Clay and gravel, dry	100	....	Coal, charcoal, pine	23	0.28-0.44
Earth, dry, loose	76	....	Coal, charcoal, oak	33	0.47-0.57
Earth, dry, packed	95	....	Coal, coke	75	1.0-1.4
Earth, moist, loose	78	....	Graphite	131	1.9-2.3
Earth, moist, packed	96	....	Paraffine	56	0.87-0.91
Earth, mud, flowing	108	....	Petroleum	54	0.87
Earth, mud, packed	115	....	Petroleum, refined	50	0.79-0.82
Riprap, limestone	80-85	....	Petroleum, benzine	46	0.73-0.75
Riprap, sandstone	90	....	Petroleum, gasoline	42	0.66-0.69
Riprap, shale	105	....	Pitch	69	1.07-1.15
Sand, gravel, dry, loose	90-105	....	Tar, bituminous	75	1.20
Sand, gravel, dry, packed	100-120	....			
Sand, gravel, dry, wet	118-120	....			
EXCAVATIONS IN WATER			**COAL AND COKE, PILED**		
Sand or gravel	60	....	Coal, anthracite	47-58	....
Sand or gravel and clay	65	....	Coal, bituminous, lignite	40-54	....
Clay	80	....	Coal, peat, turf	20-26	....
River mud	90	....	Coal, charcoal	10-14	....
Soil	70	....	Coal, coke	23-32	....
Stone riprap	65	....			

The specific gravities of solids and liquids refer to water at 4°C., those of gases to air at 0°C. and 760 mm. pressure. The weights per cubic foot are derived from average specific gravities, except where stated that weights are for bulk, heaped or loose material, etc.

APPENDIX 8
WIRE AND SHEET METAL GAGES

WIRE AND SHEET METAL GAGES
IN DECIMALS OF AN INCH

Name of Gage	United States Standard Gage*		The United States Steel Wire Gage	American or Brown & Sharpe Wire Gage	New Birmingham Standard Sheet & Hoop Gage	British Imperial or English Legal Standard Wire Gage	Birmingham or Stubs Iron Wire Gage	Name of Gage
Principal Use	Uncoated Steel Sheets and Light Plates		Steel Wire except Music Wire	Non-Ferrous Sheets and Wire	Iron and Steel Sheets and Hoops	Wire	Strips, Bands, Hoops and Wire	Principal Use
Gage No.	Weight Oz. per Sq. Ft.	Approx. Thickness Inches	Thickness, Inches					Gage No.
7/0's			.4900		.6666	.500		7/0's
6/0's			.4615	.5800	.625	.464		6/0's
5/0's			.4305	.5165	.5883	.432	.500	5/0's
4/0's			.3938	.4600	.5416	.400	.454	4/0's
3/0's			.3625	.4096	.500	.372	.425	3/0's
2/0's			.3310	.3648	.4452	.348	.380	2/0's
0			.3065	.3249	.3964	.324	.340	0
1			.2830	.2893	.3532	.300	.300	1
2			.2625	.2576	.3147	.276	.284	2
3	160	.2391	.2437	.2294	.2804	.252	.259	3
4	150	.2242	.2253	.2043	.250	.232	.238	4
5	140	.2092	.2070	.1819	.2225	.212	.220	5
6	130	.1943	.1920	.1620	.1981	.192	.203	6
7	120	.1793	.1770	.1443	.1764	.176	.180	7
8	110	.1644	.1620	.1285	.1570	.160	.165	8
9	100	.1495	.1483	.1144	.1398	.144	.148	9
10	90	.1345	.1350	.1019	.1250	.128	.134	10
11	80	.1196	.1205	.0907	.1113	.116	.120	11
12	70	.1046	.1055	.0808	.0991	.104	.109	12
13	60	.0897	.0915	.0720	.0882	.092	.095	13
14	50	0747	.0800	.0641	.0785	.080	.083	14
15	45	.0673	.0720	.0571	.0699	.072	.072	15
16	40	.0598	.0625	.0508	.0625	.064	.065	16
17	36	.0538	.0540	.0453	.0556	.056	.058	17
18	32	.0478	.0475	.0403	.0495	.048	.049	18
19	28	.0418	.0410	.0359	.0440	.040	.042	19
20	24	.0359	.0348	.0320	.0392	.036	.035	20
21	22	.0329	.0318	.0285	.0349	.032	.032	21
22	20	.0299	.0286	.0253	.0313	.028	.028	22
23	18	.0269	.0258	.0226	.0278	.024	.025	23
24	16	.0239	.0230	.0201	.0248	.022	.022	24
25	14	.0209	.0204	.0179	.0220	.020	.020	25
26	12	.0179	.0181	.0159	.0196	.018	.018	26
27	11	.0164	.0173	.0142	.0175	.0164	.016	27
28	10	.0149	.0162	.0126	.0156	.0148	.014	28
29	9	.0135	.0150	.0113	.0139	.0136	.013	29
30	8	.0120	.0140	.0100	.0123	.0124	.012	30
31	7	.0105	.0132	.0089	.0110	.0116	.010	31
32	6.5	.0097	.0128	.0080	.0098	.0108	.009	32
33	6	.0090	.0118	.0071	.0087	.0100	.008	33
34	5.5	.0082	.0104	.0063	.0077	.0092	.007	34
35	5	.0075	.0095	.0056	.0069	.0084	.005	35
36	4.5	.0067	.0090	.0050	.0061	.0076	.004	36
37	4.25	.0064	.0085	.0045	.0054	.0068		37
38	4	.0060	.0080	.0040	.0048	.0060		38
39			.0075	.0035	.0043	.0052		39
40			.0070	.0031	.0039	.0048		40

* U. S. Standard Gage is officially a weight gage, in oz. per sq. ft. as tabulated. The Approx. Thickness shown is the "Manufacturers' Standard" of the American Iron and Steel Institute, based on steel as weighing 501.81 lbs. per cu. ft. (489.6 true weight plus 2.5 percent for average over-run in area and thickness). The A.I.S.I. standard nomenclature for flat rolled carbon steel is as follows:

Widths, Inches	Thicknesses, Inch							
	0.2500 and thicker	0.2499 to 0.2031	0.2030 to 0.1875	0.1874 to 0.0568	0.0567 to 0.0344	0.0343 to 0.0255	0.0254 to 0.0142	0.0141 and thinner
To 3½ incl.	Bar	Bar	Strip	Strip	Strip	Strip	Sheet	Sheet
Over 3½ to 6 incl.	Bar	Bar	Strip	Strip	Strip	Sheet	Sheet	Sheet
" 6 to 12 "	Plate	Strip	Strip	Strip	Sheet	Sheet	Sheet	Sheet
" 12 to 32 "	Plate	Sheet	Sheet	Sheet	Sheet	Sheet	Sheet	Black Plate
" 32 to 48 "	Plate	Sheet	Sheet	Sheet	Sheet	Sheet	Sheet	Sheet
" 48	Plate	Plate	Plate	Sheet	Sheet	Sheet	Sheet	———

APPENDIX 9
PIPING SYMBOLS

	TYPE OF FITTING	DOUBLE LINE CONVENTION					SINGLE LINE CONVENTION					FLOW DIAGRAM
		FLANGED	SCREWED	B & S	WELDED	SOLDERED	FLANGED	SCREWED	B & S	WELDED	SOLDERED	
1	Joint											
2	Joint - Expansion											
3	Union											
4	Sleeve											
5	Reducer											
6	Reducer - Eccentric											
7	Reducing Flange											
8	Bushing											
9	Elbow - 45°											
10	Elbow - 90°											
11	Elbow - Long radius											
12	Elbow - (turned up)											
13	Elbow - (turned down)											
14	Elbow - Side outlet (outlet up)											
15	Elbow - Side outlet (outlet down)											
16	Elbow - Base											
17	Elbow - Double branch											
18	Elbow - Reducing											
19	Lateral											
20	Tee											
21	Tee - Single sweep											

Cont.

APPENDIX 9
PIPING SYMBOLS (Cont.)

TYPE OF FITTING		DOUBLE LINE CONVENTION					SINGLE LINE CONVENTION					FLOW DIAGRAM
		FLANGED	SCREWED	B & S	WELDED	SOLDERED	FLANGED	SCREWED	B & S	WELDED	SOLDERED	
22	Tee - Double sweep											
23	Tee - (outlet up)											
24	Tee - (outlet down)											
25	Tee - Side outlet (outlet up)											
26	Tee - Side outlet (outlet down)											
27	Cross											
28	Valve - Globe											
29	Valve - Angle											
30	Valve - Motor operated globe											Motor operated
31	Valve - Gate											
32	Valve - Angle gate											
33	Valve - Motor operated gate											Motor operated
34	Valve - Check											
35	Valve - Angle check											
36	Valve - Safety											
37	Valve - Angle safety											
38	Valve - Quick opening											
39	Valve - Float operating											
40	Stop Cock											

APPENDIX 16

TAP DRILL SIZES FOR AMERICAN NATIONAL AND UNIFIED COARSE AND FINE THREADS

$$p = \text{pitch} = \frac{1}{\text{No. thrd. per in.}}$$

$$d = \text{depth} = p \times .649519$$

$$f = \text{flat} = \frac{p}{8}$$

$$\text{pitch diameter} = d - \frac{.6495}{N}$$

For Nos. 575 and 585 Screw Thread Micrometers

Size	Threads per inch NC UNC	Threads per inch NF UNF	Outside Diameter Inches	Pitch Diameter Inches	Root Diameter Inches	Tap Drill Approx. 75% Full Thread	Decimal Equiv. of Tap Drill
0	..	80	.0600	.0519	.0438	3/64	.0469
1	64	..	.0730	.0629	.0527	53	.0595
1	..	72	.0730	.0640	.0550	53	.0595
2	56	..	.0860	.0744	.0628	50	.0700
2	..	64	.0860	.0759	.0657	50	.0700
3	48	..	.0990	.0855	.0719	47	.0785
3	..	56	.0990	.0874	.0758	46	.0810
4	40	..	.1120	.0958	.0795	43	.0890
4	..	48	.1120	.0985	.0849	42	.0935
5	40	..	.1250	.1088	.0925	38	.1015
5	..	44	.1250	.1102	.0955	37	.1040
6	32	..	.1380	.1177	.0974	36	.1065
6	..	40	.1380	.1218	.1055	33	.1130
8	32	..	.1640	.1437	.1234	29	.1360
8	..	36	.1640	.1460	.1279	29	.1360
10	24	..	.1900	.1629	.1359	26	.1470
10	..	32	.1900	.1697	.1494	21	.1590
12	24	..	.2160	.1889	.1619	16	.1770
12	..	28	.2160	.1928	.1696	15	.1800
1/4	20	..	.2500	.2175	.1850	7	.2010
1/4	..	28	.2500	.2268	.2036	3	.2130
5/16	18	..	.3125	.2764	.2403	F	.2570
5/16	..	24	.3125	.2854	.2584	I	.2720
3/8	16	..	.3750	.3344	.2938	5/16	.3125
3/8	..	24	.3750	.3479	.3209	Q	.3320
7/16	14	..	.4375	.3911	.3447	U	.3680
7/16	..	20	.4375	.4050	.3726	25/64	.3906
1/2	13	..	.5000	.4500	.4001	27/64	.4219
1/2	..	20	.5000	.4675	.4351	29/64	.4531
9/16	12	..	.5625	.5084	.4542	31/64	.4844
9/16	..	18	.5625	.5264	.4903	33/64	.5156
5/8	11	..	.6250	.5660	.5069	17/32	.5312
5/8	..	18	.6250	.5889	.5528	37/64	.5781
3/4	10	..	.7500	.6850	.6201	21/32	.6562
3/4	..	16	.7500	.7094	.6688	11/16	.6875
7/8	9	..	.8750	.8028	.7307	49/64	.7656
7/8	..	14	.8750	.8286	.7822	13/16	.8125

Cont.

APPENDIX 16

TAP DRILL SIZES FOR AMERICAN NATIONAL AND UNIFIED COARSE AND FINE THREADS (Cont.)

Size	Threads per inch NC UNC	Threads per inch NF UNF	Outside Diameter Inches	Pitch Diameter Inches	Root Diameter Inches	Tap Drill Approx. 75% Full Thread	Decimal Equiv. of Tap Drill
1	8	..	1.0000	.9188	.8376	7/8	.8750
1	..	12	1.0000	.9459	.8917	59/64	.9219
1⅛	7	..	1.1250	1.0322	.9394	63/64	.9844
1⅛	..	12	1.1250	1.0709	1.0168	1 3/64	1.0469
1¼	7	..	1.2500	1.1572	1.0644	1 7/64	1.1094
1¼	..	12	1.2500	1.1959	1.1418	1 11/64	1.1719
1⅜	6	..	1.3750	1.2667	1.1585	1 7/32	1.2187
1⅜	..	12	1.3750	1.3209	1.2668	1 19/64	1.2969
1½	6	..	1.5000	1.3917	1.2835	1 11/32	1.3437
1½	..	12	1.5000	1.4459	1.3918	1 27/64	1.4219
1¾	5	..	1.7500	1.6201	1.4902	1 9/16	1.5625
2	4½	..	2.0000	1.8557	1.7113	1 25/32	1.7812
2¼	4½	..	2.2500	2.1057	1.9613	2 1/32	2.0313
2½	4	..	2.5000	2.3376	2.1752	2¼	2.2500
2¾	4	..	2.7500	2.5876	2.4252	2½	2.5000
3	4	..	3.0000	3.8376	2.6752	2¾	2.7500
3¼	4	..	3.2500	3.0876	2.9252	3	3.0000
3½	4	..	3.5000	3.3376	3.1752	3¼	3.2500
3¾	4	..	3.7500	3.5876	3.4252	3½	3.5000
4	4	..	4.0000	3.3786	3.6752	3¾	3.7500

(Courtesy of the L. S. Starrett Company.)

APPENDIX 21

AMERICAN STANDARD HEXAGON HEAD BOLTS AND NUTS

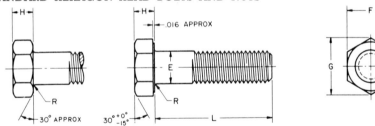

Dimensions of Hex Cap Screws (Finished Hex Bolts)

Nominal Size or Basic Product Dia		Body Dia E		Width Across Flats F			Width Across Corners G		Height H			Radius of Fillet R	
		Max	Min	Basic	Max	Min	Max	Min	Basic	Max	Min	Max	Min
1/4	0.2500	0.2500	0.2450	7/16	0.4375	0.428	0.505	0.488	5/32	0.163	0.150	0.025	0.015
5/16	0.3125	0.3125	0.3065	1/2	0.5000	0.489	0.577	0.557	13/64	0.211	0.195	0.025	0.015
3/8	0.3750	0.3750	0.3690	9/16	0.5625	0.551	0.650	0.628	15/64	0.243	0.226	0.025	0.015
7/16	0.4375	0.4375	0.4305	5/8	0.6250	0.612	0.722	0.698	9/32	0.291	0.272	0.025	0.015
1/2	0.5000	0.5000	0.4930	3/4	0.7500	0.736	0.866	0.840	5/16	0.323	0.302	0.025	0.015
9/16	0.5625	0.5625	0.5545	13/16	0.8125	0.798	0.938	0.910	23/64	0.371	0.348	0.045	0.020
5/8	0.6250	0.6250	0.6170	15/16	0.9375	0.922	1.083	1.051	25/64	0.403	0.378	0.045	0.020
3/4	0.7500	0.7500	0.7410	1 1/8	1.1250	1.100	1.299	1.254	15/32	0.483	0.455	0.045	0.020
7/8	0.8750	0.8750	0.8660	1 5/16	1.3125	1.285	1.516	1.465	35/64	0.563	0.531	0.065	0.040
1	1.0000	1.0000	0.9900	1 1/2	1.5000	1.469	1.732	1.675	39/64	0.627	0.591	0.095	0.060
1 1/8	1.1250	1.1250	1.1140	1 11/16	1.6875	1.631	1.949	1.859	11/16	0.718	0.658	0.095	0.060
1 1/4	1.2500	1.2500	1.2390	1 7/8	1.8750	1.812	2.165	2.066	25/32	0.813	0.749	0.095	0.060
1 3/8	1.3750	1.3750	1.3630	2 1/16	2.0625	1.994	2.382	2.273	27/32	0.878	0.810	0.095	0.060
1 1/2	1.5000	1.5000	1.4880	2 1/4	2.2500	2.175	2.598	2.480	15/16	0.974	0.902	0.095	0.060
1 3/4	1.7500	1.7500	1.7380	2 5/8	2.6250	2.538	3.031	2.893	1 3/32	1.134	1.054	0.095	0.060
2	2.0000	2.0000	1.9880	3	3.0000	2.900	3.464	3.306	1 7/32	1.263	1.175	0.095	0.060
2 1/4	2.2500	2.2500	2.2380	3 3/8	3.3750	3.262	3.897	3.719	1 3/8	1.423	1.327	0.095	0.060
2 1/2	2.5000	2.5000	2.4880	3 3/4	3.7500	3.625	4.330	4.133	1 17/32	1.583	1.479	0.095	0.060
2 3/4	2.7500	2.7500	2.7380	4 1/8	4.1250	3.988	4.763	4.546	1 11/16	1.744	1.632	0.095	0.060
3	3.0000	3.0000	2.9880	4 1/2	4.5000	4.350	5.196	4.959	1 7/8	1.935	1.815	0.095	0.060

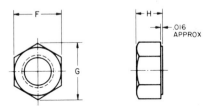
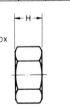
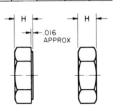

Dimensions of Hex Nuts and Hex Jam Nuts

Nominal Size or Basic Major Dia of Thread		Width Across Flats F			Width Across Corners G		Thickness Hex Nuts H			Thickness Hex Jam Nuts H		
		Basic	Max	Min	Max	Min	Basic	Max	Min	Basic	Max	Min
1/4	0.2500	7/16	0.4375	0.428	0.505	0.488	7/32	0.226	0.212	5/32	0.163	0.150
5/16	0.3125	1/2	0.5000	0.489	0.577	0.557	17/64	0.273	0.258	3/16	0.195	0.180
3/8	0.3750	9/16	0.5625	0.551	0.650	0.628	21/64	0.337	0.320	7/32	0.227	0.210
7/16	0.4375	11/16	0.6875	0.675	0.794	0.768	3/8	0.385	0.365	1/4	0.260	0.240
1/2	0.5000	3/4	0.7500	0.736	0.866	0.840	7/16	0.448	0.427	5/16	0.323	0.302
9/16	0.5625	7/8	0.8750	0.861	1.010	0.982	31/64	0.496	0.473	5/16	0.324	0.301
5/8	0.6250	15/16	0.9375	0.922	1.083	1.051	35/64	0.559	0.535	3/8	0.387	0.363
3/4	0.7500	1 1/8	1.1250	1.088	1.299	1.240	41/64	0.665	0.617	27/64	0.446	0.398
7/8	0.8750	1 5/16	1.3125	1.269	1.516	1.447	3/4	0.776	0.724	31/64	0.510	0.458
1	1.0000	1 1/2	1.5000	1.450	1.732	1.653	55/64	0.887	0.831	35/64	0.575	0.519
1 1/8	1.1250	1 11/16	1.6875	1.631	1.949	1.859	31/32	0.999	0.939	39/64	0.639	0.579
1 1/4	1.2500	1 7/8	1.8750	1.812	2.165	2.066	1 1/16	1.094	1.030	23/32	0.751	0.687
1 3/8	1.3750	2 1/16	2.0625	1.994	2.382	2.273	1 11/64	1.206	1.138	25/32	0.815	0.747
1 1/2	1.5000	2 1/4	2.2500	2.175	2.598	2.480	1 9/32	1.317	1.245	27/32	0.880	0.808

(Courtesy of ANSI; B18.2.1–1965 and ANSI; B18.2.2–1965.)

APPENDIX 22
FILLISTER HEAD AND ROUND HEAD CAP SCREWS

Fillister Head Cap Screws

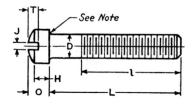

Nominal Size	D Body Diameter		A Head Diameter		H Height of Head		O Total Height of Head		J Width of Slot		T Depth of Slot	
	Max	Min	Max	Min	Max	Min	Max	Min	Max	Min	Max	Min
1/4	0.250	0.245	0.375	0.363	0.172	0.157	0.216	0.194	0.075	0.064	0.097	0.077
5/16	0.3125	0.307	0.437	0.424	0.203	0.186	0.253	0.230	0.084	0.072	0.115	0.090
3/8	0.375	0.369	0.562	0.547	0.250	0.229	0.314	0.284	0.094	0.081	0.142	0.112
7/16	0.4375	0.431	0.625	0.608	0.297	0.274	0.368	0.336	0.094	0.081	0.168	0.133
1/2	0.500	0.493	0.750	0.731	0.328	0.301	0.413	0.376	0.106	0.091	0.193	0.153
9/16	0.5625	0.555	0.812	0.792	0.375	0.346	0.467	0.427	0.118	0.102	0.213	0.168
5/8	0.625	0.617	0.875	0.853	0.422	0.391	0.521	0.478	0.133	0.116	0.239	0.189
3/4	0.750	0.742	1.000	0.976	0.500	0.466	0.612	0.566	0.149	0.131	0.283	0.223
7/8	0.875	0.866	1.125	1.098	0.594	0.556	0.720	0.668	0.167	0.147	0.334	0.264
1	1.000	0.990	1.312	1.282	0.656	0.612	0.803	0.743	0.188	0.166	0.371	0.291

All dimensions are given in inches.

The radius of the fillet at the base of the head:
    For sizes 1/4 to 3/8 in. incl. is 0.016 min and 0.031 max,
        7/16 to 9/16 in. incl. is 0.016 min and 0.047 max,
        5/8 to 1 in. incl. is 0.031 min and 0.062 max.

Round Head Cap Screws

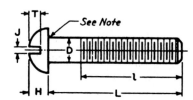

Nominal Size	D Body Diameter		A Head Diameter		H Height of Head		J Width of Slot		T Depth of Slot	
	Max	Min	Max	Min	Max	Min	Max	Min	Max	Min
1/4	0.250	0.245	0.437	0.418	0.191	0.175	0.075	0.064	0.117	0.097
5/16	0.3125	0.307	0.562	0.540	0.245	0.226	0.084	0.072	0.151	0.126
3/8	0.375	0.369	0.625	0.603	0.273	0.252	0.094	0.081	0.168	0.138
7/16	0.4375	0.431	0.750	0.725	0.328	0.302	0.094	0.081	0.202	0.167
1/2	0.500	0.493	0.812	0.786	0.354	0.327	0.106	0.091	0.218	0.178
9/16	0.5625	0.555	0.937	0.909	0.409	0.378	0.118	0.102	0.252	0.207
5/8	0.625	0.617	1.000	0.970	0.437	0.405	0.133	0.116	0.270	0.220
3/4	0.750	0.742	1.250	1.215	0.546	0.507	0.149	0.131	0.338	0.278

All dimensions are given in inches.

Radius of the fillet at the base of the head:
    For sizes 1/4 to 3/8 in. incl. is 0.016 min and 0.031 max,
        7/16 to 9/16 in. incl. is 0.016 min and 0.047 max,
        5/8 to 1 in. incl. is 0.031 min and 0.062 max.

(Courtesy of ANSI; B18.6.2–1956.)

APPENDIX 32
STANDARD KEYS AND KEYWAYS

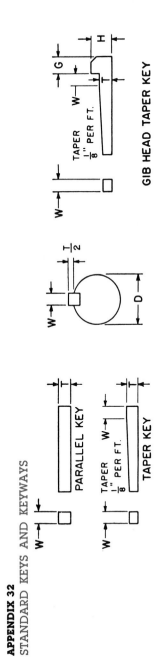

PARALLEL KEY

TAPER $\frac{1}{8}$" PER FT.
TAPER KEY

$\frac{T}{2}$
D
W

TAPER $\frac{1}{8}$" PER FT.
GIB HEAD TAPER KEY

SPROCKET BORE (= SHAFT DIAM.) INCHES D	KEYWAY DIMENSIONS — INCHES				KEY DIMENSIONS — INCHES				TOLERANCE ON W AND T (−)	GIB HEAD DIMENSIONS — INCHES				KEY TOLERANCES TAPER AND GIB HEAD	
	FOR SQUARE KEY		FOR FLAT KEY		SQUARE		FLAT			SQUARE KEY		FLAT KEY		W (−)	T (+)
	WIDTH W	DEPTH T/2	WIDTH W	DEPTH T/2	WIDTH W × HEIGHT T		WIDTH W × HEIGHT T			H	G	H	G		
$\frac{1}{2}$ — $\frac{9}{16}$	$\frac{1}{8}$	$\frac{1}{16}$	$\frac{1}{8}$	$\frac{3}{64}$	$\frac{1}{8}$ × $\frac{1}{8}$		$\frac{1}{8}$ × $\frac{3}{32}$		0.002	$\frac{1}{4}$	$\frac{7}{32}$	$\frac{3}{16}$	$\frac{1}{8}$	0.002	0.002
$\frac{5}{8}$ — $\frac{7}{8}$	$\frac{3}{16}$	$\frac{3}{32}$	$\frac{3}{16}$	$\frac{1}{16}$	$\frac{3}{16}$ × $\frac{3}{16}$		$\frac{3}{16}$ × $\frac{1}{8}$		0.002	$\frac{5}{16}$	$\frac{9}{32}$	$\frac{1}{4}$	$\frac{3}{16}$	0.002	0.002
$\frac{13}{16}$ — $1\frac{1}{4}$	$\frac{1}{4}$	$\frac{1}{8}$	$\frac{1}{4}$	$\frac{3}{32}$	$\frac{1}{4}$ × $\frac{1}{4}$		$\frac{1}{4}$ × $\frac{3}{16}$		0.002	$\frac{7}{16}$	$\frac{11}{32}$	$\frac{5}{16}$	$\frac{1}{4}$	0.002	0.002
$1\frac{3}{16}$ — $1\frac{3}{8}$	$\frac{5}{16}$	$\frac{5}{32}$	$\frac{5}{16}$	$\frac{1}{8}$	$\frac{5}{16}$ × $\frac{5}{16}$		$\frac{5}{16}$ × $\frac{1}{4}$		0.002	$\frac{9}{16}$	$\frac{13}{32}$	$\frac{3}{8}$	$\frac{5}{16}$	0.002	0.002
$1\frac{7}{16}$ — $1\frac{3}{4}$	$\frac{3}{8}$	$\frac{3}{16}$	$\frac{3}{8}$	$\frac{1}{8}$	$\frac{3}{8}$ × $\frac{3}{8}$		$\frac{3}{8}$ × $\frac{1}{4}$		0.002	$\frac{11}{16}$	$\frac{15}{32}$	$\frac{7}{16}$	$\frac{3}{8}$	0.002	0.002
$1\frac{13}{16}$ — $2\frac{1}{4}$	$\frac{1}{2}$	$\frac{1}{4}$	$\frac{1}{2}$	$\frac{3}{16}$	$\frac{1}{2}$ × $\frac{1}{2}$		$\frac{1}{2}$ × $\frac{3}{8}$		0.0025	$\frac{7}{8}$	$\frac{19}{32}$	$\frac{5}{8}$	$\frac{1}{2}$	0.0025	0.0025
$2\frac{5}{16}$ — $2\frac{3}{4}$	$\frac{5}{8}$	$\frac{5}{16}$	$\frac{5}{8}$	$\frac{7}{32}$	$\frac{5}{8}$ × $\frac{5}{8}$		$\frac{5}{8}$ × $\frac{7}{16}$		0.0025	$1\frac{1}{16}$	$\frac{23}{32}$	$\frac{3}{4}$	$\frac{5}{8}$	0.0025	0.0025
$2\frac{7}{8}$ — $3\frac{1}{4}$	$\frac{3}{4}$	$\frac{3}{8}$	$\frac{3}{4}$	$\frac{1}{4}$	$\frac{3}{4}$ × $\frac{3}{4}$		$\frac{3}{4}$ × $\frac{1}{2}$		0.0025	$1\frac{1}{4}$	$\frac{7}{8}$	$\frac{7}{8}$	$\frac{3}{4}$	0.0025	0.0025
$3\frac{3}{8}$ — $3\frac{3}{4}$	$\frac{7}{8}$	$\frac{7}{16}$	$\frac{7}{8}$	$\frac{5}{16}$	$\frac{7}{8}$ × $\frac{7}{8}$		$\frac{7}{8}$ × $\frac{5}{8}$		0.003	$1\frac{1}{2}$	1	$1\frac{1}{16}$	$\frac{7}{8}$	0.003	0.003
$3\frac{7}{8}$ — $4\frac{1}{2}$	1	$\frac{1}{2}$	1	$\frac{3}{8}$	1 × 1		1 × $\frac{3}{4}$		0.003	$1\frac{3}{4}$	$1\frac{3}{16}$	$1\frac{1}{4}$	1	0.003	0.003
$4\frac{3}{4}$ — $5\frac{1}{2}$	$1\frac{1}{4}$	$\frac{5}{8}$	$1\frac{1}{4}$	$\frac{7}{16}$	$1\frac{1}{4}$ × $1\frac{1}{4}$		$1\frac{1}{4}$ × $\frac{7}{8}$		0.003	2	$1\frac{7}{16}$	$1\frac{1}{2}$	$1\frac{1}{4}$	0.003	0.003
$5\frac{3}{4}$ — $7\frac{3}{8}$	$1\frac{1}{2}$	$\frac{3}{4}$	$1\frac{1}{2}$	$\frac{1}{2}$	$1\frac{1}{2}$ × $1\frac{1}{2}$		$1\frac{1}{2}$ × 1		0.003	$2\frac{1}{2}$	$1\frac{3}{4}$	$1\frac{3}{4}$	$1\frac{1}{2}$	0.003	0.003
$7\frac{1}{2}$ — $9\frac{7}{8}$	$1\frac{3}{4}$	$\frac{7}{8}$	. .	. .	$1\frac{3}{4}$ × $1\frac{3}{4}$		. . × . .		0.004	3	2	. .	. .	0.004	0.004
10 — $12\frac{1}{2}$	2	1	. .	. .	2 × 2		. . × . .		0.004	$3\frac{1}{2}$	$2\frac{3}{8}$	. .	. .	0.004	0.004

Standard Keyway Tolerances: Straight Keyway — Width (W) $+ .005$ Depth (T/2) $+ .010$
                                                          $- .000$ $- .000$

Taper Keyway — Width (W) $+ .005$ Depth (T/2) $+ .000$
                                                       $- .000$ $- .010$

APPENDIX 33
WOODRUFF KEYS

USA STANDARD

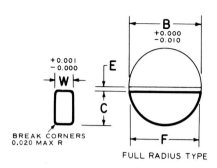

FULL RADIUS TYPE

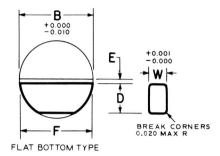

FLAT BOTTOM TYPE

Woodruff Keys

Key No.	Nominal Key Size W × B	Actual Length F +0.000-0.010	Height of Key				Distance Below Center E
			C		D		
			Max	Min	Max	Min	
202	1/16 × 1/4	0.248	0.109	0.104	0.109	0.104	1/64
202.5	1/16 × 5/16	0.311	0.140	0.135	0.140	0.135	1/64
302.5	3/32 × 5/16	0.311	0.140	0.135	0.140	0.135	1/64
203	1/16 × 3/8	0.374	0.172	0.167	0.172	0.167	1/64
303	3/32 × 3/8	0.374	0.172	0.167	0.172	0.167	1/64
403	1/8 × 3/8	0.374	0.172	0.167	0.172	0.167	1/64
204	1/16 × 1/2	0.491	0.203	0.198	0.194	0.188	3/64
304	3/32 × 1/2	0.491	0.203	0.198	0.194	0.188	3/64
404	1/8 × 1/2	0.491	0.203	0.198	0.194	0.188	3/64
305	3/32 × 5/8	0.612	0.250	0.245	0.240	0.234	1/16
405	1/8 × 5/8	0.612	0.250	0.245	0.240	0.234	1/16
505	5/32 × 5/8	0.612	0.250	0.245	0.240	0.234	1/16
605	3/16 × 5/8	0.612	0.250	0.245	0.240	0.234	1/16
406	1/8 × 3/4	0.740	0.313	0.308	0.303	0.297	1/16
506	5/32 × 3/4	0.740	0.313	0.308	0.303	0.297	1/16
606	3/16 × 3/4	0.740	0.313	0.308	0.303	0.297	1/16
806	1/4 × 3/4	0.740	0.313	0.308	0.303	0.297	1/16
507	5/32 × 7/8	0.866	0.375	0.370	0.365	0.359	1/16
607	3/16 × 7/8	0.866	0.375	0.370	0.365	0.359	1/16
707	7/32 × 7/8	0.866	0.375	0.370	0.365	0.359	1/16
807	1/4 × 7/8	0.866	0.375	0.370	0.365	0.359	1/16
608	3/16 × 1	0.992	0.438	0.433	0.428	0.422	1/16
708	7/32 × 1	0.992	0.438	0.433	0.428	0.422	1/16
808	1/4 × 1	0.992	0.438	0.433	0.428	0.422	1/16
1008	5/16 × 1	0.992	0.438	0.433	0.428	0.422	1/16
1208	3/8 × 1	0.992	0.438	0.433	0.428	0.422	1/16
609	3/16 × 1 1/8	1.114	0.484	0.479	0.475	0.469	5/64
709	7/32 × 1 1/8	1.114	0.484	0.479	0.475	0.469	5/64
809	1/4 × 1 1/8	1.114	0.484	0.479	0.475	0.469	5/64
1009	5/16 × 1 1/8	1.114	0.484	0.479	0.475	0.469	5/64

(Courtesy of ANSI; B17.2-1967.)

APPENDIX 34
WOODRUFF KEYSEATS

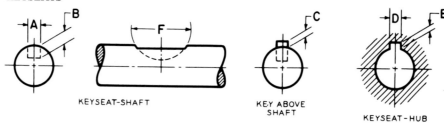

KEYSEAT-SHAFT KEY ABOVE SHAFT KEYSEAT-HUB

Keyseat Dimensions

Key Number	Nominal Size Key	Keyseat — Shaft					Key Above Shaft	Keyseat — Hub	
		Width A*		Depth B	Diameter F		Height C	Width D	Depth E
		Min	Max	+0.005 -0.000	Min	Max	+0.005 -0.005	+0.002 -0.000	+0.005 -0.000
202	1/16 × 1/4	0.0615	0.0630	0.0728	0.250	0.268	0.0312	0.0635	0.0372
202.5	1/16 × 5/16	0.0615	0.0630	0.1038	0.312	0.330	0.0312	0.0635	0.0372
302.5	3/32 × 5/16	0.0928	0.0943	0.0882	0.312	0.330	0.0469	0.0948	0.0529
203	1/16 × 3/8	0.0615	0.0630	0.1358	0.375	0.393	0.0312	0.0635	0.0372
303	3/32 × 3/8	0.0928	0.0943	0.1202	0.375	0.393	0.0469	0.0948	0.0529
403	1/8 × 3/8	0.1240	0.1255	0.1045	0.375	0.393	0.0625	0.1260	0.0685
204	1/16 × 1/2	0.0615	0.0630	0.1668	0.500	0.518	0.0312	0.0635	0.0372
304	3/32 × 1/2	0.0928	0.0943	0.1511	0.500	0.518	0.0469	0.0948	0.0529
404	1/8 × 1/2	0.1240	0.1255	0.1355	0.500	0.518	0.0625	0.1260	0.0685
305	3/32 × 5/8	0.0928	0.0943	0.1981	0.625	0.643	0.0469	0.0948	0.0529
405	1/8 × 5/8	0.1240	0.1255	0.1825	0.625	0.643	0.0625	0.1260	0.0685
505	5/32 × 5/8	0.1553	0.1568	0.1669	0.625	0.643	0.0781	0.1573	0.0841
605	3/16 × 5/8	0.1863	0.1880	0.1513	0.625	0.643	0.0937	0.1885	0.0997
406	1/8 × 3/4	0.1240	0.1255	0.2455	0.750	0.768	0.0625	0.1260	0.0685
506	5/32 × 3/4	0.1553	0.1568	0.2299	0.750	0.768	0.0781	0.1573	0.0841
606	3/16 × 3/4	0.1863	0.1880	0.2143	0.750	0.768	0.0937	0.1885	0.0997
806	1/4 × 3/4	0.2487	0.2505	0.1830	0.750	0.768	0.1250	0.2510	0.1310
507	5/32 × 7/8	0.1553	0.1568	0.2919	0.875	0.895	0.0781	0.1573	0.0841
607	3/16 × 7/8	0.1863	0.1880	0.2763	0.875	0.895	0.0937	0.1885	0.0997
707	7/32 × 7/8	0.2175	0.2193	0.2607	0.875	0.895	0.1093	0.2198	0.1153
807	1/4 × 7/8	0.2487	0.2505	0.2450	0.875	0.895	0.1250	0.2510	0.1310
608	3/16 × 1	0.1863	0.1880	0.3393	1.000	1.020	0.0937	0.1885	0.0997
708	7/32 × 1	0.2175	0.2193	0.3237	1.000	1.020	0.1093	0.2198	0.1153
808	1/4 × 1	0.2487	0.2505	0.3080	1.000	1.020	0.1250	0.2510	0.1310
1008	5/16 × 1	0.3111	0.3130	0.2768	1.000	1.020	0.1562	0.3135	0.1622
1208	3/8 × 1	0.3735	0.3755	0.2455	1.000	1.020	0.1875	0.3760	0.1935
609	3/16 × 1 1/8	0.1863	0.1880	0.3853	1.125	1.145	0.0937	0.1885	0.0997
709	7/32 × 1 1/8	0.2175	0.2193	0.3697	1.125	1.145	0.1093	0.2198	0.1153
809	1/4 × 1 1/8	0.2487	0.2505	0.3540	1.125	1.145	0.1250	0.2510	0.1310
1009	5/16 × 1 1/8	0.3111	0.3130	0.3228	1.125	1.145	0.1562	0.3135	0.1622

(Courtesy of ANSI; B17.2–1967.)

APPENDIX 35
TAPER PINS

Number	7/0	6/0	5/0	4/0	3/0	2/0	0	1	2	3	4	5	6	7	8	9	10
Size (large end)	0.0625	0.0780	0.0940	0.1090	0.1250	0.1410	0.1560	0.1720	0.1930	0.2190	0.2500	0.2890	0.3410	0.4090	0.4920	0.5910	0.7060
Length, L																	
0.375	X	X															
0.500	X	X	X	X	X												
0.625	X	X	X	X	X	X											
0.750		X	X	X	X	X	X	X									
0.875					X	X	X	X	X								
1.000			X	X			X	X	X	X							
1.250						X	X	X	X	X	X	X					
1.500						X	X	X	X	X	X	X	X				
1.750							X	X	X	X	X	X	X				
2.000								X	X	X	X	X	X	X			
2.250									X	X	X	X	X	X	X		
2.500									X	X	X	X	X	X	X		
2.750										X	X	X	X	X	X		
3.000										X	X	X	X	X	X	X	
3.250												X	X	X	X	X	
3.500													X	X	X	X	X
3.750													X	X	X	X	X
4.000														X	X	X	X
4.250															X	X	X
4.500															X	X	X
4.750															X	X	X
5.000															X	X	X
5.250																X	X
5.500																X	X
5.750																X	X
6.000																X	X

All dimensions are given in inches.

Standard reamers are available for pins given above the line.

Pins Nos. 11 (size 0.8600), 12 (size 1.032), 13 (size 1.241), and 14 (1.523) are special sizes—hence their lengths are special.

To find small diameter of pin, multiply the length by 0.2083 and subtract the result from the large diameter.

(Courtesy of ANSI; B5.20–1958.)

APPENDIX 39

AMERICAN STANDARD RUNNING AND SLIDING FITS

Limits are in thousandths of an inch.
Limits for hole and shaft are applied algebraically to the basic size to obtain the limits of size for the parts.
Data in bold face are in accordance with ABC agreements.
Symbols H5, g5, etc., are Hole and Shaft designations used in ABC System.

Nominal Size Range Inches Over	To	Class RC 1 Limits of Clearance	Standard Limits Hole H5	Shaft g4	Class RC 2 Limits of Clearance	Standard Limits Hole H6	Shaft g5	Class RC 3 Limits of Clearance	Standard Limits Hole H7	Shaft f6	Class RC 4 Limits of Clearance	Standard Limits Hole H8	Shaft f7
0	− 0.12	0.1 / 0.45	+ 0.2 / 0	− 0.1 / − 0.25	0.1 / 0.55	+ 0.25 / 0	− 0.1 / − 0.3	0.3 / 0.95	+ 0.4 / 0	− 0.3 / − 0.55	0.3 / 1.3	+ 0.6 / 0	− 0.3 / − 0.7
0.12	− 0.24	0.15 / 0.5	+ 0.2 / 0	− 0.15 / − 0.3	0.15 / 0.65	+ 0.3 / 0	− 0.15 / − 0.35	0.4 / 1.12	+ 0.5 / 0	− 0.4 / − 0.7	0.4 / 1.6	+ 0.7 / 0	− 0.4 / − 0.9
0.24	− 0.40	0.2 / 0.6	0.25 / 0	− 0.2 / − 0.35	0.2 / 0.85	+ 0.4 / 0	− 0.2 / − 0.45	0.5 / 1.5	+ 0.6 / 0	− 0.5 / − 0.9	0.5 / 2.0	+ 0.9 / 0	− 0.5 / − 1.1
0.40	− 0.71	0.25 / 0.75	+ 0.3 / 0	− 0.25 / − 0.45	0.25 / 0.95	+ 0.4 / 0	− 0.25 / − 0.55	0.6 / 1.7	+ 0.7 / 0	− 0.6 / − 1.0	0.6 / 2.3	+ 1.0 / 0	− 0.6 / − 1.3
0.71	− 1.19	0.3 / 0.95	+ 0.4 / 0	− 0.3 / − 0.55	0.3 / 1.2	+ 0.5 / 0	− 0.3 / − 0.7	0.8 / 2.1	+ 0.8 / 0	− 0.8 / − 1.3	0.8 / 2.8	+ 1.2 / 0	− 0.8 / − 1.6
1.19	− 1.97	0.4 / 1.1	+ 0.4 / 0	− 0.4 / − 0.7	0.4 / 1.4	+ 0.6 / 0	− 0.4 / − 0.8	1.0 / 2.6	+ 1.0 / 0	− 1.0 / − 1.6	1.0 / 3.6	+ 1.6 / 0	− 1.0 / − 2.0
1.97	− 3.15	0.4 / 1.2	+ 0.5 / 0	− 0.4 / − 0.7	0.4 / 1.6	+ 0.7 / 0	− 0.4 / − 0.9	1.2 / 3.1	+ 1.2 / 0	− 1.2 / − 1.9	1.2 / 4.2	+ 1.8 / 0	− 1.2 / − 2.4
3.15	− 4.73	0.5 / 1.5	+ 0.6 / 0	− 0.5 / − 0.9	0.5 / 2.0	+ 0.9 / 0	− 0.5 / − 1.1	1.4 / 3.7	+ 1.4 / 0	− 1.4 / − 2.3	1.4 / 5.0	+ 2.2 / 0	− 1.4 / − 2.8
4.73	− 7.09	0.6 / 1.8	+ 0.7 / 0	− 0.6 / − 1.1	0.6 / 2.3	+ 1.0 / 0	− 0.6 / − 1.3	1.6 / 4.2	+ 1.6 / 0	− 1.6 / − 2.6	1.6 / 5.7	+ 2.5 / 0	− 1.6 / − 3.2
7.09	− 9.85	0.6 / 2.0	+ 0.8 / 0	− 0.6 / − 1.2	0.6 / 2.6	+ 1.2 / 0	− 0.6 / − 1.4	2.0 / 5.0	+ 1.8 / 0	− 2.0 / − 3.2	2.0 / 6.6	+ 2.8 / 0	− 2.0 / − 3.8
9.85	−12.41	0.8 / 2.3	+ 0.9 / 0	− 0.8 / − 1.4	0.8 / 2.9	+ 1.2 / 0	− 0.8 / − 1.7	2.5 / 5.7	+ 2.0 / 0	− 2.5 / − 3.7	2.5 / 7.5	+ 3.0 / 0	− 2.5 / − 4.5
12.41	−15.75	1.0 / 2.7	+ 1.0 / 0	− 1.0 / − 1.7	1.0 / 3.4	+ 1.4 / 0	− 1.0 / − 2.0	3.0 / 6.6	+ / 0	− 3.0 / − 4.4	3.0 / 8.7	+ 3.5 / 0	− 3.0 / − 5.2
15.75	−19.69	1.2 / 3.0	+ 1.0 / 0	− 1.2 / − 2.0	1.2 / 3.8	+ 1.6 / 0	− 1.2 / − 2.2	4.0 / 8.1	+ 1.6 / 0	− 4.0 / − 5.6	4.0 / 10.5	+ 4.0 / 0	− 4.0 / − 6.5
19.69	−30.09	1.6 / 3.7	+ 1.2 / 0	− 1.6 / − 2.5	1.6 / 4.8	+ 2.0 / 0	− 1.6 / − 2.8	5.0 / 10.0	+ 3.0 / 0	− 5.0 / − 7.0	5.0 / 13.0	+ 5.0 / 0	− 5.0 / − 8.0
30.09	−41.49	2.0 / 4.6	+ 1.6 / 0	− 2.0 / − 3.0	2.0 / 6.1	+ 2.5 / 0	− 2.0 / − 3.6	6.0 / 12.5	+ 4.0 / 0	− 6.0 / − 8.5	6.0 / 16.0	+ 6.0 / 0	− 6.0 / −10.0
41.49	−56.19	2.5 / 5.7	+ 2.0 / 0	− 2.5 / − 3.7	2.5 / 7.5	+ 3.0 / 0	− 2.5 / − 4.5	8.0 / 16.0	+ 5.0 / 0	− 8.0 / −11.0	8.0 / 21.0	+ 8.0 / 0	− 8.0 / −13.0
56.19	−76.39	3.0 / 7.1	+ 2.5 / 0	− 3.0 / − 4.6	3.0 / 9.5	+ 4.0 / 0	− 3.0 / − 5.5	10.0 / 20.0	+ 6.0 / 0	−10.0 / −14.0	10.0 / 26.0	+10.0 / 0	−10.0 / −16.0
76.39	−100.9	4.0 / 9.0	+ 3.0 / 0	− 4.0 / − 6.0	4.0 / 12.0	+ 5.0 / 0	− 4.0 / − 7.0	12.0 / 25.0	+ 8.0 / 0	−12.0 / −17.0	12.0 / 32.0	+12.0 / 0	−12.0 / −20.0
100.9	−131.9	5.0 / 11.5	+ 4.0 / 0	− 5.0 / − 7.5	5.0 / 15.0	+ 6.0 / 0	− 5.0 / − 9.0	16.0 / 32.0	+10.0 / 0	−16.0 / −22.0	16.0 / 36.0	+16.0 / 0	−16.0 / −26.0
131.9	−171.9	6.0 / 14.0	+ 5.0 / 0	− 6.0 / − 9.0	6.0 / 19.0	+ 8.0 / 0	− 6.0 / −11.0	18.0 / 38.0	+ 8.0 / 0	−18.0 / −26.0	18.0 / 50.0	+20.0 / 0	−18.0 / −30.0
171.9	−200	8.0 / 18.0	+ 6.0 / 0	− 8.0 / −12.0	8.0 / 22.0	+10.0 / 0	− 8.0 / −12.0	22.0 / 48.0	+16.0 / 0	−22.0 / −32.0	22.0 / 63.0	+25.0 / 0	−22.0 / −38.0

(Courtesy of USASI; B4.1–1955.)

Cont.

APPENDIX 39
AMERICAN STANDARD RUNNING AND SLIDING FITS (Cont.)

Values in each cell are given as upper value / lower value (in thousandths of an inch).

Class RC 5 Lim. Clear.	RC5 Hole H8	RC5 Shaft e7	RC 6 Lim. Clear.	RC6 Hole H9	RC6 Shaft e8	RC 7 Lim. Clear.	RC7 Hole H9	RC7 Shaft d8	RC 8 Lim. Clear.	RC8 Hole H10	RC8 Shaft c9	RC 9 Lim. Clear.	RC9 Hole H11	RC9 Shaft	Nominal Size Range Over	To
0.6 / 1.6	+0.6 / −0	−0.6 / −1.0	0.6 / 2.2	+1.0 / −0	−0.6 / −1.2	1.0 / 2.6	+1.0 / 0	−1.0 / −1.6	2.5 / 5.1	+1.6 / 0	−2.5 / −3.5	4.0 / 8.1	+2.5 / 0	−4.0 / −5.6	0	0.12
0.8 / 2.0	+0.7 / −0	−0.8 / −1.3	0.8 / 2.7	+1.2 / −0	−0.8 / −1.5	1.2 / 3.1	+1.2 / 0	−1.2 / −1.9	2.8 / 5.8	+1.8 / 0	−2.8 / −4.0	4.5 / 9.0	+3.0 / 0	−4.5 / −6.0	0.12	0.24
1.0 / 2.5	+0.9 / −0	−1.0 / −1.6	1.0 / 3.3	+1.4 / −0	−1.0 / −1.9	1.6 / 3.9	+1.4 / 0	−1.6 / −2.5	3.0 / 6.6	+2.2 / 0	−3.0 / −4.4	5.0 / 10.7	+3.5 / 0	−5.0 / −7.2	0.24	0.40
1.2 / 2.9	+1.0 / −0	−1.2 / −1.9	1.2 / 3.8	+1.6 / −0	−1.2 / −2.2	2.0 / 4.6	+1.6 / 0	−2.0 / −3.0	3.5 / 7.9	+2.8 / 0	−3.5 / −5.1	6.0 / 12.8	+4.0 / −0	−6.0 / −8.8	0.40	0.71
1.6 / 3.6	+1.2 / −0	−1.6 / −2.4	1.6 / 4.8	+2.0 / −0	−1.6 / −2.8	2.5 / 5.7	+2.0 / 0	−2.5 / −3.7	4.5 / 10.0	+3.5 / 0	−4.5 / −6.5	7.0 / 15.5	+5.0 / 0	−7.0 / −10.5	0.71	1.19
2.0 / 4.6	+1.6 / −0	−2.0 / −3.0	2.0 / 6.1	+2.5 / −0	−2.0 / −3.6	3.0 / 7.1	+2.5 / 0	−3.0 / −4.6	5.0 / 11.5	+4.0 / 0	−5.0 / −7.5	8.0 / 18.0	+6.0 / 0	−8.0 / −12.0	1.19	1.97
2.5 / 5.5	+1.8 / −0	−2.5 / −3.7	2.5 / 7.3	+3.0 / −0	−2.5 / −4.3	4.0 / 8.8	+3.0 / 0	−4.0 / −5.8	6.0 / 13.5	+4.5 / 0	−6.0 / −9.0	9.0 / 20.5	+7.0 / 0	−9.0 / −13.5	1.97	3.15
3.0 / 6.6	+2.2 / −0	−3.0 / −4.4	3.0 / 8.7	+3.5 / −0	−3.0 / −5.2	5.0 / 10.7	+3.5 / 0	−5.0 / −7.2	7.0 / 15.5	+5.0 / 0	−7.0 / −10.5	10.0 / 24.0	+9.0 / 0	−10.0 / −15.0	3.15	4.73
3.5 / 7.6	+2.5 / −0	−3.5 / −5.1	3.5 / 10.0	+4.0 / −0	−3.5 / −6.0	6.0 / 12.5	+4.0 / 0	−6.0 / −8.5	8.0 / 18.0	+6.0 / 0	−8.0 / −12.0	12.0 / 28.0	+10.0 / 0	−12.0 / −18.0	4.73	7.09
4.0 / 8.6	+2.8 / −0	−4.0 / −5.8	4.0 / 11.3	+4.5 / 0	−4.0 / −6.8	7.0 / 14.3	+4.5 / 0	−7.0 / −9.8	10.0 / 21.5	+7.0 / 0	−10.0 / −14.5	15.0 / 34.0	+12.0 / 0	−15.0 / −22.0	7.09	9.85
5.0 / 10.0	+3.0 / 0	−5.0 / −7.0	5.0 / 13.0	+5.0 / 0	−5.0 / −8.0	8.0 / 16.0	+5.0 / 0	−8.0 / −11.0	12.0 / 25.0	+8.0 / 0	−12.0 / −17.0	18.0 / 38.0	+12.0 / 0	−18.0 / −26.0	9.85	12.41
6.0 / 11.7	+3.5 / 0	−6.0 / −8.2	6.0 / 15.5	+6.0 / 0	−6.0 / −9.5	10.0 / 19.5	+6.0 / 0	−10.0 / 13.5	14.0 / 29.0	+9.0 / 0	−14.0 / −20.0	22.0 / 45.0	+14.0 / 0	−22.0 / −31.0	12.41	15.75
8.0 / 14.5	+4.0 / 0	−8.0 / −10.5	8.0 / 18.0	+6.0 / 0	−8.0 / −12.0	12.0 / 22.0	+6.0 / 0	−12.0 / −16.0	16.0 / 32.0	+10.0 / 0	−16.0 / −22.0	25.0 / 51.0	+16.0 / 0	−25.0 / −35.0	15.75	19.69
10.0 / 18.0	+5.0 / 0	−10.0 / −13.0	10.0 / 23.0	+8.0 / 0	−10.0 / −15.0	16.0 / 29.0	+8.0 / 0	−16.0 / −21.0	20.0 / 40.0	+12.0 / 0	−20.0 / −28.0	30.0 / 62.0	+20.0 / 0	−30.0 / −42.0	19.69	30.09
12.0 / 22.0	+6.0 / 0	−12.0 / −16.0	12.0 / 28.0	+10.0 / 0	−12.0 / −18.0	20.0 / 36.0	+10.0 / 0	−20.0 / −26.0	25.0 / 51.0	+16.0 / 0	−25.0 / −35.0	40.0 / 81.0	+25.0 / 0	−40.0 / −56.0	30.09	41.49
16.0 / 29.0	+8.0 / 0	−16.0 / −21.0	16.0 / 36.0	+12.0 / 0	−16.0 / −24.0	25.0 / 45.0	+12.0 / 0	−25.0 / −33.0	30.0 / 62.0	+20.0 / 0	−30.0 / −42.0	50.0 / 100	+30.0 / 0	−50.0 / −70.0	41.49	56.19
20.0 / 36.0	+10.0 / 0	−20.0 / −26.0	20.0 / 46.0	+16.0 / 0	−20.0 / −30.0	30.0 / 56.0	+16.0 / 0	−30.0 / −40.0	40.0 / 81.0	+25.0 / 0	−40.0 / −56.0	60.0 / 125	+40.0 / 0	−60.0 / −85.0	56.19	76.39
25.0 / 45.0	+12.0 / 0	−25.0 / −33.0	25.0 / 57.0	+20.0 / 0	−25.0 / −37.0	40.0 / 72.0	+20.0 / 0	−40.0 / −52.0	50.0 / 100	+30.0 / 0	−50.0 / −70.0	80.0 / 160	+50.0 / 0	−80.0 / −110	76.39	100.9
30.0 / 56.0	+16.0 / 0	−30.0 / −40.0	30.0 / 71.0	+25.0 / 0	−30.0 / −46.0	50.0 / 91.0	+25.0 / 0	−50.0 / −66.0	60.0 / 125	+40.0 / 0	−60.0 / −85.0	100 / 200	+60.0 / 0	−100 / −140	100.9	131.9
35.0 / 67.0	+20.0 / 0	−35.0 / −47.0	35.0 / 85.0	+30.0 / 0	−35.0 / −55.0	60.0 / 110.0	+30.0 / 0	−60.0 / −80.0	80.0 / 160	+50.0 / 0	−80.0 / −110	130 / 260	+80.0 / 0	−130 / −180	131.9	171.9
45.0 / 86.0	+25.0 / 0	−45.0 / −61.0	45.0 / 110.0	+40.0 / 0	−45.0 / −70.0	80.0 / 145.0	+40.0 / 0	−80.0 / −105.0	100 / 200	+60.0 / 0	−100 / −140	150 / 310	+100 / 0	−150 / −210	171.9	200

(Courtesy of ANSI; B4.1–1955.)

APPENDIX 40

AMERICAN STANDARD CLEARANCE LOCATIONAL FITS

Limits are in thousandths of an inch.

Limits for hole and shaft are applied algebraically to the basic size to obtain the limits of size for the parts.

Data in bold face are in accordance with ABC agreements.

Symbols H9,f8, etc., are Hole and Shaft designations used in ABC System.

Nominal Size Range Inches Over	To	LC 1 Limits of Clearance	LC 1 Hole H6	LC 1 Shaft h5	LC 2 Limits of Clearance	LC 2 Hole H7	LC 2 Shaft h6	LC 3 Limits of Clearance	LC 3 Hole H8	LC 3 Shaft h7	LC 4 Limits of Clearance	LC 4 Hole H10	LC 4 Shaft h9	LC 5 Limits of Clearance	LC 5 Hole H7	LC 5 Shaft g6
0 —	0.12	0 / 0.45	+0.25 / −0	+0 / −0.2	0 / 0.65	+0.4 / −0	+0 / −0.25	0 / 1	+0.6 / −0	+0 / −0.4	0 / 2.6	+1.6 / −0	+0 / −1.0	0.1 / 0.75	+0.4 / −0	−0.1 / −0.35
0.12—	0.24	0 / 0.5	+0.3 / −0	+0 / −0.2	0 / 0.8	+0.5 / −0	+0 / −0.3	0 / 1.2	+0.7 / −0	+0 / −0.5	0 / 3.0	+1.8 / −0	+0 / −1.2	0.15 / 0.95	+0.5 / −0	−0.15 / −0.45
0.24—	0.40	0 / 0.65	+0.4 / −0	+0 / −0.25	0 / 1.0	+0.6 / −0	+0 / −0.4	0 / 1.5	+0.9 / −0	+0 / −0.6	0 / 3.6	+2.2 / −0	+0 / −1.4	0.2 / 1.2	+0.6 / −0	−0.2 / −0.6
0.40—	0.71	0 / 0.7	+0.4 / −0	+0 / −0.3	0 / 1.1	+0.7 / −0	+0 / −0.4	0 / 1.7	+1.0 / −0	+0 / −0.7	0 / 4.4	+2.8 / −0	+0 / −1.6	0.25 / 1.35	+0.7 / −0	−0.25 / −0.65
0.71—	1.19	0 / 0.9	+0.5 / −0	+0 / −0.4	0 / 1.3	+0.8 / −0	+0 / −0.5	0 / 2	+1.2 / −0	+0 / −0.8	0 / 5.5	+3.5 / −0	+0 / −2.0	0.3 / 1.6	+0.8 / −0	−0.3 / −0.8
1.19—	1.97	0 / 1.0	+0.6 / −0	+0 / −0.4	0 / 1.6	+1.0 / −0	+0 / −0.6	0 / 2.6	+1.6 / −0	+0 / −1	0 / 6.5	+4.0 / −0	+0 / −2.5	0.4 / 2.0	+1.0 / −0	−0.4 / −1.0
1.97—	3.15	0 / 1.2	+0.7 / −0	+0 / −0.5	0 / 1.9	+1.2 / −0	+0 / −0.7	0 / 3	+1.8 / −0	+0 / −1.2	0 / 7.5	+4.5 / −0	+0 / −3	0.4 / 2.3	+1.2 / −0	−0.4 / −1.1
3.15—	4.73	0 / 1.5	+0.9 / −0	+0 / −0.6	0 / 2.3	+1.4 / −0	+0 / −0.9	0 / 3.6	+2.2 / −0	+0 / −1.4	0 / 8.5	+5.0 / −0	+0 / −3.5	0.5 / 2.8	+1.4 / −0	−0.5 / −1.4
4.73—	7.09	0 / 1.7	+1.0 / −0	+0 / −0.7	0 / 2.6	+1.6 / −0	+0 / −1.0	0 / 4.1	+2.5 / −0	+0 / −1.6	0 / 10	+6.0 / −0	+0 / −4	0.6 / 3.2	+1.6 / −0	−0.6 / −1.6
7.09—	9.85	0 / 2.0	+1.2 / −0	+0 / −0.8	0 / 3.0	+1.8 / −0	+0 / −1.2	0 / 4.6	+2.8 / −0	+0 / −1.8	0 / 11.5	+7.0 / −0	+0 / −4.5	0.6 / 3.6	+1.8 / −0	−0.6 / −1.8
9.85—	12.41	0 / 2.1	+1.2 / −0	+0 / −0.9	0 / 3.2	+2.0 / −0	+0 / −1.2	0 / 5	+3.0 / −0	+0 / −2.0	0 / 13	+8.0 / −0	+0 / −5	0.7 / 3.9	+2.0 / −0	−0.7 / −1.9
12.41—	15.75	0 / 2.4	+1.4 / −0	+0 / −1.0	0 / 3.6	+2.2 / −0	+0 / −1.4	0 / 5.7	+3.5 / −0	+0 / −2.2	0 / 15	+9.0 / −0	+0 / −6	0.7 / 4.3	+2.2 / −0	−0.7 / −2.1
15.75—	19.69	0 / 2.6	+1.6 / −0	+0 / −1.0	0 / 4.1	+2.5 / −0	+0 / −1.6	0 / 6.5	+4 / −0	+0 / −2.5	0 / 16	+10.0 / −0	+0 / −6	0.8 / 4.9	+2.5 / −0	−0.8 / −2.4
19.69—	30.09	0 / 3.2	+2.0 / −0	+0 / −1.2	0 / 5.0	+3 / −0	+0 / −2	0 / 8	+5 / −0	+0 / −3	0 / 20	+12.0 / −0	+0 / −8	0.9 / 5.9	+3.0 / −0	−0.9 / −2.9
30.09—	41.49	0 / 4.1	+2.5 / −0	+0 / −1.6	0 / 6.5	+4 / −0	+0 / −2.5	0 / 10	+6 / −0	+0 / −4	0 / 26	+16.0 / −0	+0 / −10	1.0 / 7.5	+4.0 / −0	−1.0 / −3.5
41.49—	56.19	0 / 5.0	+3.0 / −0	+0 / −2.0	0 / 8.0	+5 / −0	+0 / −3	0 / 13	+8 / −0	+0 / −5	0 / 32	+20.0 / −0	+0 / −12	1.2 / 9.2	+5.0 / −0	−1.2 / −4.2
56.19—	76.39	0 / 6.5	+4.0 / −0	+0 / −2.5	0 / 10	+6 / −0	+0 / −4	0 / 16	+10 / −0	+0 / −6	0 / 41	+25.0 / −0	+0 / −16	1.2 / 11.2	+6.0 / −0	−1.2 / −5.2
76.39—	100.9	0 / 8.0	+5.0 / −0	+0 / −3.0	0 / 13	+8 / −0	+0 / −5	0 / 20	+12 / −0	+0 / −8	0 / 50	+30.0 / −0	+0 / −20	1.4 / 14.4	+8.0 / −0	−1.4 / −6.4
100.9 —	131.9	0 / 10.0	+6.0 / −0	+0 / −4.0	0 / 16	+10 / −0	+0 / −6	0 / 26	+16 / −0	+0 / −10	0 / 65	+40.0 / −0	+0 / −25	1.6 / 17.6	+10.0 / −0	−1.6 / −7.6
131.9 —	171.9	0 / 13.0	+8.0 / −0	+0 / −5.0	0 / 20	+12 / −0	+0 / −8	0 / 32	+20 / −0	+0 / −12	0 / 8	+50.0 / −0	+0 / −30	1.8 / 21.8	+12.0 / −0	−1.8 / −9.8
171.9 —	200	0 / 16.0	+10.0 / −0	+0 / −6.0	0 / 26	+16 / −0	+0 / −10	0 / 41	+25 / −0	+0 / −16	0 / 100	+60.0 / −0	+0 / −40	1.8 / 27.8	+16.0 / −0	−1.8 / −11.8

(Courtesy of USASI; B4.1–1955.)

Cont.

APPENDIX 40
AMERICAN STANDARD CLEARANCE LOCATIONAL FITS (Cont.)

Class LC 6			Class LC 7			Class LC 8			Class LC 9			Class LC 10			Class LC 11			Nominal Size Range Inches	
Limits of Clearance	Hole H9	Shaft f8	Limits of Clearance	Hole H10	Shaft e9	Limits of Clearance	Hole H10	Shaft d9	Limits of Clearance	Hole H11	Shaft c10	Limits of Clearance	Hole H12	Shaft	Limits of Clearance	Hole H13	Shaft	Over	To
0.3 / 1.9	+1.0 / 0	−0.3 / −0.9	0.6 / 3.2	+1.6 / 0	−0.6 / −1.6	1.0 / 3.6	+1.6 / −0	−1.0 / −2.0	2.5 / 6.6	+2.5 / −0	−2.5 / −4.1	4 / 12	+4 / −0	−4 / −8	5 / 17	+6 / −0	−5 / −11	0	0.12
0.4 / 2.3	+1.2 / 0	−0.4 / −1.1	0.8 / 3.8	+1.8 / 0	−0.8 / −2.0	1.2 / 4.2	+1.8 / −0	−1.2 / −2.4	2.8 / 7.6	+3.0 / −0	−2.8 / −4.6	4.5 / 14.5	+5 / −0	−4.5 / −9.5	6 / 20	+7 / −0	−6 / −13	0.12	0.24
0.5 / 2.8	+1.4 / 0	−0.5 / −1.4	1.0 / 4.6	+2.2 / 0	−1.0 / −2.4	1.6 / 5.2	+2.2 / −0	−1.6 / −3.0	3.0 / 8.7	+3.5 / −0	−3.0 / −5.2	5 / 17	+6 / −0	−5 / −11	7 / 25	+9 / −0	−7 / −16	0.24	0.40
0.6 / 3.2	+1.6 / 0	−0.6 / −1.6	1.2 / 5.6	+2.8 / 0	−1.2 / −2.8	2.0 / 6.4	+2.8 / −0	−2.0 / −3.6	3.5 / 10.3	+4.0 / −0	−3.5 / −6.3	6 / 20	+7 / −0	−6 / −13	8 / 28	+10 / −0	−8 / −18	0.40	0.71
0.8 / 4.0	+2.0 / 0	−0.8 / −2.0	1.6 / 7.1	+3.5 / 0	−1.6 / −3.6	2.5 / 8.0	+3.5 / −0	−2.5 / −4.5	4.5 / 13.0	+5.0 / −0	−4.5 / −8.0	7 / 23	+8 / −0	−7 / −15	10 / 34	+12 / −0	−10 / −22	0.71	1.19
1.0 / 5.1	+2.5 / 0	−1.0 / −2.6	2.0 / 8.5	+4.0 / 0	−2.0 / −4.5	3.0 / 9.5	+4.0 / −0	−3.0 / −5.5	5 / 15	+6 / −0	−5 / −9	8 / 28	+10 / −0	−8 / −18	12 / 44	+16 / −0	−12 / −28	1.19	1.97
1.2 / 6.0	+3.0 / 0	−1.2 / −3.0	2.5 / 10.0	+4.5 / 0	−2.5 / −5.5	4.0 / 11.5	+4.5 / −0	−4.0 / −7.0	6 / 17.5	+7 / −0	−6 / −10.5	10 / 34	+12 / −0	−10 / −22	14 / 50	+18 / −0	−14 / −32	1.97	3.15
1.4 / 7.1	+3.5 / 0	−1.4 / −3.6	3.0 / 11.5	+5.0 / 0	−3.0 / −6.5	5.0 / 13.5	+5.0 / −0	−5.0 / −8.5	7 / 21	+9 / −0	−7 / −12	11 / 39	+14 / −0	−11 / −25	16 / 60	+22 / −0	−16 / −38	3.15	4.73
1.6 / 8.1	+4.0 / 0	−1.6 / −4.1	3.5 / 13.5	+6.0 / 0	−3.5 / −7.5	6 / 16	+6 / −0	−6 / −10	8 / 24	+10 / −0	−8 / −14	12 / 44	+16 / −0	−12 / −28	18 / 68	+25 / −0	−18 / −43	4.73	7.09
2.0 / 9.3	+4.5 / 0	−2.0 / −4.8	4.0 / 15.5	+7.0 / 0	−4.0 / −8.5	7 / 18.5	+7 / −0	−7 / −11.5	10 / 29	+12 / −0	−10 / −17	16 / 52	+18 / −0	−16 / −34	22 / 78	+28 / −0	−22 / −50	7.09	9.85
2.2 / 10.2	+5.0 / 0	−2.2 / −5.2	4.5 / 17.5	+8.0 / 0	−4.5 / −9.5	7 / 20	+8 / −0	−7 / −12	12 / 32	+12 / −0	−12 / −20	20 / 60	+20 / −0	−20 / −40	28 / 88	+30 / −0	−28 / −58	9.85	12.41
2.5 / 12.0	+6.0 / 0	−2.5 / −6.0	5.0 / 20.0	+9.0 / 0	−5 / −11	8 / 23	+9 / −0	−8 / −14	14 / 37	+14 / −0	−14 / −23	22 / 66	+22 / −0	−22 / −44	30 / 100	+35 / −0	−30 / −65	12.41	15.75
2.8 / 12.8	+6.0 / 0	−2.8 / −6.8	5.0 / 21.0	+10.0 / 0	−5 / −11	9 / 25	+10 / −0	−9 / −15	16 / 42	+16 / −0	−16 / −26	25 / 75	+25 / −0	−25 / −50	35 / 115	+40 / −0	−35 / −75	15.75	19.69
3.0 / 16.0	+8.0 / 0	−3.0 / −8.0	6.0 / 26.0	+12.0 / 0	−6 / −14	10 / 30	+12 / −0	−10 / −18	18 / 50	+20 / −0	−18 / −30	28 / 88	+30 / −0	−28 / −58	40 / 140	+50 / −0	−40 / −90	19.69	30.09
3.5 / 19.5	+10.0 / 0	−3.5 / −9.5	7.0 / 33.0	+16.0 / −0	−7 / −17	12 / 38	+16 / −0	−12 / −22	20 / 61	+25 / −0	−20 / −36	30 / 110	+40 / −0	−30 / −70	45 / 165	+60 / −0	−45 / −105	30.09	41.49
4.0 / 24.0	+12.0 / 0	−4.0 / −12.0	8.0 / 40.0	+20.0 / 0	−8 / −20	14 / 46	+20 / −0	−14 / −26	25 / 75	+30 / −0	−25 / −45	40 / 140	+50 / −0	−40 / −90	60 / 220	+80 / −0	−60 / −140	41.49	56.19
4.5 / 30.5	+16.0 / 0	−4.5 / −14.5	9.0 / 50.0	+25.0 / −0	−9 / −25	16 / 57	+25 / −0	−16 / −32	30 / 95	+40 / −0	−30 / −55	50 / 170	+60 / −0	−50 / −110	70 / 270	+100 / −0	−70 / −170	56.19	76.39
5.0 / 37.0	+20.0 / 0	−5 / −17	10.0 / 60.0	+30.0 / −0	−10 / −30	18 / 68	+30 / −0	−18 / −38	35 / 115	+50 / −0	−35 / −65	50 / 210	+80 / −0	−50 / −130	80 / 330	+125 / −0	−80 / −205	76.39	100.9
6.0 / 47.0	+25.0 / 0	−6 / −22	12.0 / 67.0	+40.0 / −0	−12 / −27	20 / 85	+40 / −0	−20 / −45	40 / 140	+60 / −0	−40 / −80	60 / 260	+100 / −0	−60 / −160	90 / 410	+160 / −0	−90 / −250	100.9	131.9
7.0 / 57.0	+30.0 / 0	−7 / −27	14.0 / 94.0	+50.0 / −0	−14 / −44	25 / 105	+50 / −0	−25 / −55	50 / 180	+80 / −0	−50 / −100	80 / 330	+125 / −0	−80 / −205	100 / 500	+200 / −0	−100 / −300	131.9	171.9
7.0 / 72.0	+40.0 / 0	−7 / −32	14.0 / 114.0	+60.0 / −0	−14 / −54	25 / 125	+60 / −0	−25 / −65	50 / 210	+100 / −0	−50 / −110	90 / 410	+160 / −0	−90 / −250	125 / 625	+250 / −0	−125 / −375	171.9	200

(Courtesy of ANSI; B4.1–1955.)

APPENDIX 41
AMERICAN STANDARD TRANSITION LOCATIONAL FITS

Limits are in thousandths of an inch.

Limits for hole and shaft are applied algebraically to the basic size to obtain the limits of size for the mating parts.

Data in bold face are in accordance with ABC agreements.

"Fit" represents the maximum interference (minus values) and the maximum clearance (plus values).

Symbols H7, js6, etc., are Hole and Shaft designations used in ABC System.

Nominal Size Range Inches (Over – To)	Class LT 1 Fit	LT 1 Hole H7	LT 1 Shaft js6	Class LT 2 Fit	LT 2 Hole H8	LT 2 Shaft js7	Class LT 3 Fit	LT 3 Hole H7	LT 3 Shaft k6	Class LT 4 Fit	LT 4 Hole H8	LT 4 Shaft k7	Class LT 5 Fit	LT 5 Hole H7	LT 5 Shaft n6	Class LT 6 Fit	LT 6 Hole H7	LT 6 Shaft n7
0 – 0.12	−0.10 / +0.50	+0.4 / −0	+0.10 / −0.10	−0.2 / +0.8	+0.6 / −0	+0.2 / −0.2							−0.5 / +0.15	+0.4 / −0	+0.5 / +0.25	−0.65 / +0.15	+0.4 / −0	+0.65 / +0.25
0.12 – 0.24	−0.15 / +0.65	+0.5 / −0	+0.15 / −0.15	−0.25 / +0.95	+0.7 / −0	+0.25 / −0.25							−0.6 / +0.2	+0.5 / −0	+0.6 / +0.3	−0.8 / +0.2	+0.5 / −0	+0.8 / +0.3
0.24 – 0.40	−0.2 / +0.8	+0.6 / −0	+0.2 / −0.2	−0.3 / +1.2	+0.9 / −0	+0.3 / −0.3	−0.5 / +0.5	+0.6 / −0	+0.5 / +0.1	−0.7 / +0.8	+0.9 / −0	+0.7 / +0.1	−0.8 / +0.2	+0.6 / −0	+0.8 / +0.4	−1.0 / +0.2	+0.6 / −0	+1.0 / +0.4
0.40 – 0.71	−0.2 / +0.9	+0.7 / −0	+0.2 / −0.2	−0.35 / +1.35	+1.0 / −0	+0.35 / −0.35	−0.5 / +0.6	+0.7 / −0	+0.5 / +0.1	−0.8 / +0.9	+1.0 / −0	+0.8 / +0.1	−0.9 / +0.2	+0.7 / −0	+0.9 / +0.5	−1.2 / +0.2	+0.7 / −0	+1.2 / +0.5
0.71 – 1.19	−0.25 / +1.05	+0.8 / −0	+0.25 / −0.25	−0.4 / +1.6	+1.2 / −0	+0.4 / −0.4	−0.6 / +0.7	+0.8 / −0	+0.6 / +0.1	−0.9 / +1.1	+1.2 / −0	+0.9 / +0.1	−1.1 / +0.2	+0.8 / −0	+1.1 / +0.6	−1.4 / +0.2	+0.8 / −0	+1.4 / +0.6
1.19 – 1.97	−0.3 / +1.3	+1.0 / −0	+0.3 / −0.3	−0.5 / +2.1	+1.6 / −0	+0.5 / −0.5	−0.7 / +0.9	+1.0 / −0	+0.7 / +0.1	−1.1 / +1.5	+1.6 / −0	+1.1 / +0.1	−1.3 / +0.3	+1.0 / −0	+1.3 / +0.7	−1.7 / +0.3	+1.0 / −0	+1.7 / +0.7
1.97 – 3.15	−0.3 / +1.5	+1.2 / −0	+0.3 / −0.3	−0.6 / +2.4	+1.8 / −0	+0.6 / −0.6	−0.8 / +1.1	+1.2 / −0	+0.8 / +0.1	−1.3 / +1.7	+1.8 / −0	+1.3 / +0.1	−1.5 / +0.4	+1.2 / −0	+1.5 / +0.8	−2.0 / +0.4	+1.2 / −0	+2.0 / +0.8
3.15 – 4.73	−0.4 / +1.8	+1.4 / −0	+0.4 / −0.4	−0.7 / +2.9	+2.2 / −0	+0.7 / −0.7	−1.0 / +1.3	+1.4 / −0	+1.0 / +0.1	−1.5 / +2.1	+2.2 / −0	+1.5 / +0.1	−1.9 / +0.4	+1.4 / −0	+1.9 / +1.0	−2.4 / +0.4	+1.4 / −0	+2.4 / +1.0
4.73 – 7.09	−0.5 / +2.1	+1.6 / −0	+0.5 / −0.5	−0.8 / +3.3	+2.5 / −0	+0.8 / −0.8	−1.1 / +1.5	+1.6 / −0	+1.1 / +0.1	−1.7 / +2.4	+2.5 / −0	+1.7 / +0.1	−2.2 / +0.4	+1.6 / −0	+2.2 / +1.2	−2.8 / +0.4	+1.6 / −0	+2.8 / +1.2
7.09 – 9.85	−0.6 / +2.4	+1.8 / −0	+0.6 / −0.6	−0.9 / +3.7	+2.8 / −0	+0.9 / −0.9	−1.4 / +1.6	+1.8 / −0	+1.4 / +0.2	−2.0 / +2.6	+2.8 / −0	+2.0 / +0.2	−2.6 / +0.4	+1.8 / −0	+2.6 / +1.4	−3.2 / +0.4	+1.8 / −0	+3.2 / +1.4
9.85 – 12.41	−0.6 / +2.6	+2.0 / −0	+0.6 / −0.6	−1.0 / +4.0	+3.0 / −0	+1.0 / −1.0	−1.4 / +1.8	+2.0 / −0	+1.4 / +0.2	−2.2 / +2.8	+3.0 / −0	+2.2 / +0.2	−2.6 / +0.6	+2.0 / −0	+2.6 / +1.4	−3.4 / +0.6	+2.0 / −0	+3.4 / +1.4
12.41 – 15.75	−0.7 / +2.9	+2.2 / −0	+0.7 / −0.7	−1.0 / +4.5	+3.5 / −0	+1.0 / −1.0	−1.6 / +2.0	+2.2 / −0	+1.6 / +0.2	−2.4 / +3.3	+3.5 / −0	+2.4 / +0.2	−3.0 / +0.6	+2.2 / −0	+3.0 / +1.6	−3.8 / +0.6	+2.2 / −0	+3.8 / +1.6
15.75 – 19.69	−0.8 / +3.3	+2.5 / −0	+0.8 / −0.8	−1.2 / +5.2	+4.0 / −0	+1.2 / −1.2	−1.8 / +2.3	+2.5 / −0	+1.8 / +0.2	−2.7 / +3.8	+4.0 / −0	+2.7 / +0.2	−3.4 / +0.7	+2.5 / −0	+3.4 / +1.8	−4.3 / +0.7	+2.5 / −0	+4.3 / +1.8

(Courtesy of ANSI: B4.1–1955.)

APPENDIX 42

AMERICAN STANDARD INTERFERENCE LOCATIONAL FITS

Limits are in thousandths of an inch.
Limits for hole and shaft are applied algebraically to the
basic size to obtain the limits of size for the parts.
Data in bold face are in accordance with ABC agreements,
Symbols H7, p6, etc., are Hole and Shaft designations
used in ABC System.

Nominal Size Range Inches Over	To	Class LN 1 Limits of Interference	Standard Limits Hole H6	Shaft n5	Class LN 2 Limits of Interference	Standard Limits Hole H7	Shaft p6	Class LN 3 Limits of Interference	Standard Limits Hole H7	Shaft r6
0	0.12	0 / 0.45	+0.25 / −0	+0.45 / +0.25	0 / 0.65	+0.4 / −0	+0.65 / +0.4	0.1 / 0.75	+0.4 / −0	+0.75 / +0.5
0.12	0.24	0 / 0.5	+0.3 / −0	+0.5 / +0.3	0 / 0.8	+0.5 / −0	+0.8 / +0.5	0.1 / 0.9	+0.5 / 0	+0.9 / +0.6
0.24	0.40	0 / 0.65	+0.4 / −0	+0.65 / +0.4	0 / 1.0	+0.6 / −0	+1.0 / +0.6	0.2 / 1.2	+0.6 / −0	+1.2 / +0.8
0.40	0.71	0 / 0.8	+0.4 / −0	+0.8 / +0.4	0 / 1.1	+0.7 / −0	+1.1 / +0.7	0.3 / 1.4	+0.7 / −0	+1.4 / +1.0
0.71	1.19	0 / 1.0	+0.5 / −0	+1.0 / +0.5	0 / 1.3	+0.8 / −0	+1.3 / +0.8	0.4 / 1.7	+0.8 / −0	+1.7 / +1.2
1.19	1.97	0 / 1.1	+0.6 / −0	+1.1 / +0.6	0 / 1.6	+1.0 / −0	+1.6 / +1.0	0.4 / 2.0	+1.0 / −0	+2.0 / +1.4
1.97	3.15	0.1 / 1.3	+0.7 / −0	+1.3 / +0.7	0.2 / 2.1	+1.2 / −0	+2.1 / +1.4	0.4 / 2.3	+1.2 / −0	+2.3 / +1.6
3.15	4.73	0.1 / 1.6	+0.9 / −0	+1.6 / +1.0	0.2 / 2.5	+1.4 / −0	+2.5 / +1.6	0.6 / 2.9	+1.4 / −0	+2.9 / +2.0
4.73	7.09	0.2 / 1.9	+1.0 / −0	+1.9 / +1.2	0.2 / 2.8	+1.6 / −0	+2.8 / +1.8	0.9 / 3.5	+1.6 / −0	+3.5 / +2.5
7.09	9.85	0.2 / 2.2	+1.2 / −0	+2.2 / +1.4	0.2 / 3.2	+1.8 / −0	+3.2 / +2.0	1.2 / 4.2	+1.8 / −0	+4.2 / +3.0
9.85	12.41	0.2 / 2.3	+1.2 / −0	+2.3 / +1.4	0.2 / 3.4	+2.0 / −0	+3.4 / +2.2	1.5 / 4.7	+2.0 / −0	+4.7 / +3.5
12.41	15.75	0.2 / 2.6	+1.4 / −0	+2.6 / +1.6	0.3 / 3.9	+2.2 / −0	+3.9 / +2.5	2.3 / 5.9	+2.2 / −0	+5.9 / +4.5
15.75	19.69	0.2 / 2.8	+1.6 / −0	+2.8 / +1.8	0.3 / 4.4	+2.5 / −0	+4.4 / +2.8	2.5 / 6.6	+2.5 / −0	+6.6 / +5.0
19.69	30.09		+2.0 / −0		0.5 / 5.5	+3 / −0	+5.5 / +3.5	4 / 9	+3 / −0	+9 / +7
30.09	41.49		+2.5 / −0		0.5 / 7.0	+4 / −0	+7.0 / +4.5	5 / 11.5	+4 / −0	+11.5 / +9
41.49	56.19		+3.0 / −0		1 / 9	+5 / −0	+9 / +6	7 / 15	+5 / −0	+15 / +12
56.19	76.39		+4.0 / −0		1 / 11	+6 / −0	+11 / +7	10 / 20	+6 / −0	+20 / +16
76.39	100.9		+5.0 / −0		1 / 14	+8 / −0	+14 / +9	12 / 25	+8 / −0	+25 / +20
100.9	131.9		+6.0 / −0		2 / 18	+10 / −0	+18 / +12	15 / 31	+10 / −0	+31 / +25
131.9	171.9		+8.0 / −0		4 / 24	+12 / −0	+24 / +16	18 / 38	+12 / −0	+38 / +30
171.9	200		+10.0 / −0		4 / 30	+16 / −0	+30 / +20	24 / 50	+16 / −0	+50 / +40

(Courtesy of ANSI; B4.1–1955.)

APPENDIX 43
AMERICAN STANDARD FORCE AND SHRINK FITS

Limits are in thousandths of an inch.
Limits for hole and shaft are applied algebraically to the basic size to obtain the limits of size for the parts.
Data in bold face are in accordance with ABC agreements.
Symbols H7, s6, etc., are Hole and Shaft designations used in ABC System.

Nominal Size Range Inches Over — To	Class FN 1 Limits of Interference	Class FN 1 Hole H6	Class FN 1 Shaft	Class FN 2 Limits of Interference	Class FN 2 Hole H7	Class FN 2 Shaft s6	Class FN 3 Limits of Interference	Class FN 3 Hole H7	Class FN 3 Shaft t6	Class FN 4 Limits of Interference	Class FN 4 Hole H7	Class FN 4 Shaft u6	Class FN 5 Limits of Interference	Class FN 5 Hole H8	Class FN 5 Shaft x7
0 — 0.12	0.05 / 0.5	+0.25 / −0	+0.5 / +0.3	0.2 / 0.85	+0.4 / −0	+0.85 / +0.6				0.3 / 0.95	+0.4 / −0	+0.95 / +0.7	0.3 / 1.3	+0.6 / −0	+1.3 / +0.9
0.12 — 0.24	0.1 / 0.6	+0.3 / −0	+0.6 / +0.4	0.2 / 1.0	+0.5 / −0	+1.0 / +0.7				0.4 / 1.2	+0.5 / −0	+1.2 / +0.9	0.5 / 1.7	+0.7 / −0	+1.7 / +1.2
0.24 — 0.40	0.1 / 0.75	+0.4 / −0	+0.75 / +0.5	0.4 / 1.4	+0.6 / −0	+1.4 / +1.0				0.6 / 1.6	+0.6 / −0	+1.6 / +1.2	0.5 / 2.0	+0.9 / −0	+2.0 / +1.4
0.40 — 0.56	0.1 / 0.8	−0.4 / −0	+0.8 / +0.5	0.5 / 1.6	+0.7 / −0	+1.6 / +1.2				0.7 / 1.8	+0.7 / −0	+1.8 / +1.4	0.6 / 2.3	+1.0 / −0	+2.3 / +1.6
0.56 — 0.71	0:2 / 0.9	+0.4 / −0	+0.9 / +0.6	0.5 / 1.6	+0.7 / −0	+1.6 / +1.2				0.7 / 1.8	+0.7 / −0	+1.8 / +1.4	0.8 / 2.5	+1.0 / −0	+2.5 / +1.8
0.71 — 0.95	0.2 / 1.1	+0.5 / −0	+1.1 / +0.7	0.6 / 1.9	+0.8 / −0	+1.9 / +1.4				0.8 / 2.1	+0.8 / −0	+2.1 / +1.6	1.0 / 3.0	+1.2 / −0	+3.0 / +2.2
0.95 — 1.19	0.3 / 1.2	+0.5 / −0	+1.2 / +0.8	0.6 / 1.9	+0.8 / −0	+1.9 / +1.4	0.8 / 2.1	+0.8 / −0	+2.1 / +1.6	1.0 / 2.3	+0.8 / −0	+2.3 / +1.8	1.3 / 3.3	+1.2 / −0	+3.3 / +2.5
1.19 — 1.58	0.3 / 1.3	+0.6 / −0	+1.3 / +0.9	0.8 / 2.4	+1.0 / −0	+2.4 / +1.8	1.0 / 2.6	+1.0 / −0	+2.6 / +2.0	1.5 / 3.1	+1.0 / −0	+3.1 / +2.5	1.4 / 4.0	+1.6 / −0	+4.0 / +3.0
1.58 — 1.97	0.4 / 1.4	+0.6 / −0	+1.4 / +1.0	0.8 / 2.4	+1.0 / −0	+2.4 / +1.8	1.2 / 2.8	+1.0 / −0	+2.8 / +2.2	1.8 / 3.4	+1.0 / −0	+3.4 / +2.8	2.4 / 5.0	+1.6 / −0	+5.0 / +4.0
1.97 — 2.56	0.6 / 1.8	+0.7 / −0	+1.8 / +1.3	0.8 / 2.7	+1.2 / −0	+2.7 / +2.0	1.3 / 3.2	+1.2 / −0	+3.2 / +2.5	2.3 / 4.2	+1.2 / −0	+4.2 / +3.5	3.2 / 6.2	+1.8 / −0	+6.2 / +5.0
2.56 — 3.15	0.7 / 1.9	+0.7 / −0	+1.9 / +1.4	1.0 / 2.9	+1.2 / −0	+2.9 / +2.2	1.8 / 3.7	+1.2 / −0	+3.7 / +3.0	2.8 / 4.7	+1.2 / −0	+4.7 / +4.0	4.2 / 7.2	+1.8 / −0	+7.2 / +6.0
3.15 — 3.94	0.9 / 2.4	+0.9 / −0	+2.4 / +1.8	1.4 / 3.7	+1.4 / −0	+3.7 / +2.8	2.1 / 4.4	+1.4 / −0	+4.4 / +3.5	3.6 / 5.9	+1.4 / −0	+5.9 / +5.0	4.8 / 8.4	+2.2 / −0	+8.4 / +7.0
3.94 — 4.73	1.1 / 2.6	+0.9 / −0	+2.6 / +2.0	1.6 / 3.9	+1.4 / −0	+3.9 / +3.0	2.6 / 4.9	+1.4 / −0	+4.9 / +4.0	4.6 / 6.9	+1.4 / −0	+6.9 / +6.0	5.8 / 9.4	+2.2 / −0	+9.4 / +8.0
4.73 — 5.52	1.2 / 2.9	+1.0 / −0	+2.9 / +2.2	1.9 / 4.5	+1.6 / −0	+4.5 / +3.5	3.4 / 6.0	+1.6 / −0	+6.0 / +5.0	5.4 / 8.0	+1.6 / −0	+8.0 / +7.0	7.5 / 11.6	+2.5 / −0	+11.6 / +10.0
5.52 — 6.30	1.5 / 3.2	+1.0 / −0	+3.2 / +2.5	2.4 / 5.0	+1.6 / −0	+5.0 / +4.0	3.4 / 6.0	+1.6 / −0	+6.0 / +5.0	5.4 / 8.0	+1.6 / −0	+8.0 / +7.0	9.5 / 13.6	+2.5 / −0	+13.6 / +12.0
6.30 — 7.09	1.8 / 3.5	+1.0 / −0	+3.5 / +2.8	2.9 / 5.5	+1.6 / −0	+5.5 / +4.5	4.4 / 7.0	+1.6 / −0	+7.0 / +6.0	6.4 / 9.0	+1.6 / −0	+9.0 / +8.0	9.5 / 13.6	+2.5 / −0	+13.6 / +12.0
7.09 — 7.88	1.8 / 3.8	+1.2 / −0	+3.8 / +3.0	3.2 / 6.2	+1.8 / −0	+6.2 / +5.0	5.2 / 8.2	+1.8 / −0	+8.2 / +7.0	7.2 / 10.2	+1.8 / −0	+10.2 / +9.0	11.2 / 15.8	+2.8 / −0	+15.8 / +14.0
7.88 — 8.86	2.3 / 4.3	+1.2 / −0	+4.3 / +3.5	3.2 / 6.2	+1.8 / −0	+6.2 / +5.0	5.2 / 8.2	+1.8 / −0	+8.2 / +7.0	8.2 / 11.2	+1.8 / −0	+11.2 / +10.0	13.2 / 17.8	+2.8 / −0	+17.8 / +16.0
8.86 — 9.85	2.3 / 4.3	+1.2 / −0	+4.3 / +3.5	4.2 / 7.2	+1.8 / −0	+7.2 / +6.0	6.2 / 9.2	+1.8 / −0	+9.2 / +8.0	10.2 / 13.2	+1.8 / −0	+13.2 / +12.0	13.2 / 17.8	+2.8 / −0	+17.8 / +16.0
9.85 — 11.03	2.8 / 4.9	+1.2 / −0	+4.9 / +4.0	4.0 / 7.2	+2.0 / −0	+7.2 / +6.0	7.0 / 10.2	+2.0 / −0	+10.2 / +9.0	10.0 / 13.2	+2.0 / −0	+13.2 / +12.0	15.0 / 20.0	+3.0 / −0	+20.0 / +18.0
11.03 — 12.41	2.8 / 4.9	+1.2 / −0	+4.9 / +4.0	5.0 / 8.2	+2.0 / −0	+8.2 / +7.0	7.0 / 10.2	+2.0 / −0	+10.2 / +9.0	12.0 / 15.2	+2.0 / −0	+15.2 / +14.0	17.0 / 22.0	+3.0 / −0	+22.0 / +20.0
12.41 — 13.98	3.1 / 5.5	+1.4 / −0	+5.5 / +4.5	5.8 / 9.4	+2.2 / −0	+9.4 / +8.0	7.8 / 11.4	+2.2 / −0	+11.4 / +10.0	13.8 / 17.4	+2.2 / −0	+17.4 / +16.0	18.5 / 24.2	+3.5 / +0	+24.2 / +22.0
13.98 — 15.75	3.6 / 6.1	+1.4 / −0	+6.1 / +5.0	5.8 / 9.4	+2.2 / −0	+9.4 / +8.0	9.8 / 13.4	+2.2 / −0	+13.4 / +12.0	15.8 / 19.4	+2.2 / −0	+19.4 / +18.0	21.5 / 27.2	+3.5 / −0	+27.2 / +25.0
15.75 — 17.72	4.4 / 7.0	+1.6 / −0	+7.0 / +6.0	6.5 / 10.6	+2.5 / −0	+10.6 / +9.0	9.5 / 13.6	+2.5 / −0	+13.6 / +12.0	17.5 / 21.6	+2.5 / −0	+21.6 / +20.0	24.0 / 30.5	+4.0 / −0	+30.5 / +28.0
17.72 — 19.69	4.4 / 7.0	+1.6 / −0	+7.0 / +6.0	7.5 / 11.6	+2.5 / −0	+11.6 / +10.0	11.5 / 15.6	+2.5 / −0	+15.6 / +14.0	19.5 / 23.6	+2.5 / −0	+23.6 / +22.0	26.0 / 32.5	+4.0 / −0	+32.5 / +30.0

(Courtesy of ANSI; B4.1–1955.)

APPENDIX 44
THE INTERNATIONAL TOLERANCE GRADES (ANSI B4.2)

Dimensions are in mm.

Basic sizes		Tolerance grades[3]																	
Over	Up to and including	IT01	IT0	IT1	IT2	IT3	IT4	IT5	IT6	IT7	IT8	IT9	IT10	IT11	IT12	IT13	IT14	IT15	IT16
0	3	0.0003	0.0005	0.0008	0.0012	0.002	0.003	0.004	0.006	0.010	0.014	0.025	0.040	0.060	0.100	0.140	0.250	0.400	0.600
3	6	0.0004	0.0006	0.001	0.0015	0.0025	0.004	0.005	0.008	0.012	0.018	0.030	0.048	0.075	0.120	0.180	0.300	0.480	0.750
6	10	0.0004	0.0006	0.001	0.0015	0.0025	0.004	0.006	0.009	0.015	0.022	0.036	0.058	0.090	0.150	0.220	0.360	0.580	0.900
10	18	0.0005	0.0008	0.0012	0.002	0.003	0.005	0.008	0.011	0.018	0.027	0.043	0.070	0.110	0.180	0.270	0.430	0.700	1.100
18	30	0.0006	0.001	0.0015	0.0025	0.004	0.006	0.009	0.013	0.021	0.033	0.052	0.084	0.130	0.210	0.330	0.520	0.840	1.300
30	50	0.0006	0.001	0.0015	0.0025	0.004	0.007	0.011	0.016	0.025	0.039	0.062	0.100	0.160	0.250	0.390	0.620	1.000	1.600
50	80	0.0008	0.0012	0.002	0.003	0.005	0.008	0.013	0.019	0.030	0.046	0.074	0.120	0.190	0.300	0.460	0.740	1.200	1.900
80	120	0.001	0.0015	0.0025	0.004	0.006	0.010	0.015	0.022	0.035	0.054	0.087	0.140	0.220	0.350	0.540	0.870	1.400	2.200
120	180	0.0012	0.002	0.0035	0.005	0.008	0.012	0.018	0.025	0.040	0.063	0.100	0.160	0.250	0.400	0.630	1.000	1.600	2.500
180	250	0.002	0.003	0.0045	0.007	0.010	0.014	0.020	0.029	0.046	0.072	0.115	0.185	0.290	0.460	0.720	1.150	1.850	2.900
250	315	0.0025	0.004	0.006	0.008	0.012	0.016	0.023	0.032	0.052	0.081	0.130	0.210	0.320	0.520	0.810	1.300	2.100	3.200
315	400	0.003	0.005	0.007	0.009	0.013	0.018	0.025	0.036	0.057	0.089	0.140	0.230	0.360	0.570	0.890	1.400	2.300	3.600
400	500	0.004	0.006	0.008	0.010	0.015	0.020	0.027	0.040	0.063	0.097	0.155	0.250	0.400	0.630	0.970	1.550	2.500	4.000
500	630	0.0045	0.006	0.009	0.011	0.016	0.022	0.030	0.044	0.070	0.110	0.175	0.280	0.440	0.700	1.100	1.750	2.800	4.400
630	800	0.005	0.007	0.010	0.013	0.018	0.025	0.035	0.050	0.080	0.125	0.200	0.320	0.500	0.800	1.250	2.000	3.200	5.000
800	1000	0.0055	0.008	0.011	0.015	0.021	0.029	0.040	0.056	0.090	0.140	0.230	0.360	0.560	0.900	1.400	2.300	3.600	5.600
1000	1250	0.0065	0.009	0.013	0.018	0.024	0.034	0.046	0.066	0.105	0.165	0.260	0.420	0.660	1.050	1.650	2.600	4.200	6.600
1250	1600	0.008	0.011	0.015	0.021	0.029	0.040	0.054	0.078	0.125	0.195	0.310	0.500	0.780	1.250	1.950	3.100	5.000	7.800
1600	2000	0.009	0.013	0.018	0.025	0.035	0.048	0.065	0.092	0.150	0.230	0.370	0.600	0.920	1.500	2.300	3.700	6.000	9.200
2000	2500	0.011	0.015	0.022	0.030	0.041	0.057	0.077	0.110	0.175	0.280	0.440	0.700	1.100	1.750	2.800	4.400	7.000	11.000
2500	3150	0.013	0.018	0.026	0.036	0.050	0.069	0.093	0.135	0.210	0.330	0.540	0.860	1.350	2.100	3.300	5.400	8.600	13.500

[3] IT Values for tolerance grades larger than IT16 can be calculated by using the following formulas:
IT17 = IT12 × 10; IT18 = IT13 × 10; etc.

APPENDIX 45
PREFERRED HOLE BASIS CLEARANCE FITS—CYLINDRICAL FITS (ANSI B4.2)

AMERICAN NATIONAL STANDARD
PREFERRED METRIC LIMITS AND FITS

ANSI B4.2-1978

Dimensions in mm.

BASIC SIZE		LOOSE RUNNING Hole H11	LOOSE RUNNING Shaft c11	LOOSE RUNNING Fit	FREE RUNNING Hole H9	FREE RUNNING Shaft d9	FREE RUNNING Fit	CLOSE RUNNING Hole H8	CLOSE RUNNING Shaft f7	CLOSE RUNNING Fit	SLIDING Hole H7	SLIDING Shaft g6	SLIDING Fit	LOCATIONAL CLEARANCE Hole H7	LOCATIONAL CLEARANCE Shaft h6	LOCATIONAL CLEARANCE Fit
1	MAX	1.060	0.940	0.180	1.025	0.980	0.070	1.014	0.994	0.030	1.010	0.998	0.018	1.010	1.000	0.016
	MIN	1.000	0.880	0.060	1.000	0.955	0.020	1.000	0.984	0.006	1.000	0.992	0.002	1.000	0.994	0.000
1.2	MAX	1.260	1.140	0.180	1.225	1.180	0.070	1.214	1.194	0.030	1.210	1.198	0.018	1.210	1.200	0.016
	MIN	1.200	1.080	0.060	1.200	1.155	0.020	1.200	1.184	0.006	1.200	1.192	0.002	1.200	1.194	0.000
1.6	MAX	1.660	1.540	0.180	1.625	1.580	0.070	1.614	1.594	0.030	1.610	1.598	0.018	1.610	1.600	0.016
	MIN	1.600	1.480	0.060	1.600	1.555	0.020	1.600	1.584	0.006	1.600	1.592	0.002	1.600	1.594	0.000
2	MAX	2.060	1.940	0.180	2.025	1.980	0.070	2.014	1.994	0.030	2.010	1.998	0.018	2.010	2.000	0.016
	MIN	2.000	1.880	0.060	2.000	1.955	0.020	2.000	1.984	0.006	2.000	1.992	0.002	2.000	1.994	0.000
2.5	MAX	2.560	2.440	0.180	2.525	2.480	0.070	2.514	2.494	0.030	2.510	2.498	0.018	2.510	2.500	0.016
	MIN	2.500	2.380	0.060	2.500	2.455	0.020	2.500	2.484	0.006	2.500	2.492	0.002	2.500	2.494	0.000
3	MAX	3.060	2.940	0.180	3.025	2.980	0.070	3.014	2.994	0.030	3.010	2.998	0.018	3.010	3.000	0.016
	MIN	3.000	2.880	0.060	3.000	2.955	0.020	3.000	2.984	0.006	3.000	2.992	0.002	3.000	2.994	0.000
4	MAX	4.075	3.930	0.220	4.030	3.970	0.090	4.018	3.990	0.040	4.012	3.996	0.024	4.012	4.000	0.020
	MIN	4.000	3.855	0.070	4.000	3.940	0.030	4.000	3.978	0.010	4.000	3.988	0.004	4.000	3.992	0.000
5	MAX	5.075	4.930	0.220	5.030	4.970	0.090	5.018	4.990	0.040	5.012	4.996	0.024	5.012	5.000	0.020
	MIN	5.000	4.855	0.070	5.000	4.940	0.030	5.000	4.978	0.010	5.000	4.988	0.004	5.000	4.992	0.000
6	MAX	6.075	5.930	0.220	6.030	5.970	0.090	6.018	5.990	0.040	6.012	5.996	0.024	6.012	6.000	0.020
	MIN	6.000	5.855	0.070	6.000	5.940	0.030	6.000	5.978	0.010	6.000	5.988	0.004	6.000	5.992	0.000
8	MAX	8.090	7.920	0.260	8.036	7.960	0.112	8.022	7.987	0.050	8.015	7.995	0.029	8.015	8.000	0.024
	MIN	8.000	7.830	0.080	8.000	7.924	0.040	8.000	7.972	0.013	8.000	7.986	0.005	8.000	7.991	0.000
10	MAX	10.090	9.920	0.260	10.036	9.960	0.112	10.022	9.987	0.050	10.015	9.995	0.029	10.015	10.000	0.024
	MIN	10.000	9.830	0.080	10.000	9.924	0.040	10.000	9.972	0.013	10.000	9.986	0.005	10.000	9.991	0.000
12	MAX	12.110	11.905	0.315	12.043	11.950	0.136	12.027	11.984	0.061	12.018	11.994	0.035	12.018	12.000	0.029
	MIN	12.000	11.795	0.095	12.000	11.907	0.050	12.000	11.966	0.016	12.000	11.983	0.006	12.000	11.989	0.000
16	MAX	16.110	15.905	0.315	16.043	15.950	0.136	16.027	15.984	0.061	16.018	15.994	0.035	16.018	16.000	0.029
	MIN	16.000	15.795	0.095	16.000	15.907	0.050	16.000	15.966	0.016	16.000	15.983	0.006	16.000	15.989	0.000
20	MAX	20.130	19.890	0.370	20.052	19.935	0.169	20.033	19.980	0.074	20.021	19.993	0.041	20.021	20.000	0.034
	MIN	20.000	19.760	0.110	20.000	19.883	0.065	20.000	19.959	0.020	20.000	19.980	0.007	20.000	19.987	0.000
25	MAX	25.130	24.890	0.370	25.052	24.935	0.169	25.033	24.980	0.074	25.021	24.993	0.041	25.021	25.000	0.034
	MIN	25.000	24.760	0.110	25.000	24.883	0.065	25.000	24.959	0.020	25.000	24.980	0.007	25.000	24.987	0.000
30	MAX	30.130	29.890	0.370	30.052	29.935	0.169	30.033	29.980	0.074	30.021	29.993	0.041	30.021	30.000	0.034
	MIN	30.000	29.760	0.110	30.000	29.883	0.065	30.000	29.959	0.020	30.000	29.980	0.007	30.000	29.987	0.000

Cont.

APPENDIX 45

PREFERRED HOLE BASIS CLEARANCE FITS—CYLINDRICAL FITS (Cont.)

AMERICAN NATIONAL STANDARD
PREFERRED METRIC LIMITS AND FITS

ANSI B4.2-1978

Dimensions in mm.

BASIC SIZE		LOOSE RUNNING Hole H11	Shaft c11	Fit	FREE RUNNING Hole H9	Shaft d9	Fit	CLOSE RUNNING Hole H8	Shaft f7	Fit	SLIDING Hole H7	Shaft g6	Fit	LOCATIONAL CLEARANCE Hole H7	Shaft h6	Fit
40	MAX	40.160	39.880	0.440	40.062	39.920	0.204	40.039	39.975	0.089	40.025	39.991	0.050	40.025	40.000	0.041
	MIN	40.000	39.720	0.120	40.000	39.858	0.080	40.000	39.950	0.025	40.000	39.975	0.009	40.000	39.984	0.000
50	MAX	50.160	49.870	0.450	50.062	49.920	0.204	50.039	49.975	0.089	50.025	49.991	0.050	50.025	50.000	0.041
	MIN	50.000	49.710	0.130	50.000	49.858	0.080	50.000	49.950	0.025	50.000	49.975	0.009	50.000	49.984	0.000
60	MAX	60.190	59.860	0.520	60.074	59.900	0.248	60.046	59.970	0.106	60.030	59.990	0.059	60.030	60.000	0.049
	MIN	60.000	59.670	0.140	60.000	59.826	0.100	60.000	59.940	0.030	60.000	59.971	0.010	60.000	59.981	0.000
80	MAX	80.190	79.850	0.530	80.074	79.900	0.248	80.046	79.970	0.106	80.030	79.990	0.059	80.030	80.000	0.049
	MIN	80.000	79.660	0.150	80.000	79.826	0.100	80.000	79.940	0.030	80.000	79.971	0.010	80.000	79.981	0.000
100	MAX	100.220	99.830	0.610	100.087	99.880	0.294	100.054	99.964	0.125	100.035	99.988	0.069	100.035	100.000	0.057
	MIN	100.000	99.610	0.170	100.000	99.793	0.120	100.000	99.929	0.036	100.000	99.966	0.012	100.000	99.978	0.000
120	MAX	120.220	119.820	0.620	120.087	119.880	0.294	120.054	119.964	0.125	120.035	119.988	0.069	120.035	120.000	0.057
	MIN	120.000	119.600	0.180	120.000	119.793	0.120	120.000	119.929	0.036	120.000	119.966	0.012	120.000	119.978	0.000
160	MAX	160.250	159.790	0.710	160.100	159.855	0.345	160.063	159.957	0.146	160.040	159.986	0.079	160.040	160.000	0.065
	MIN	160.000	159.540	0.210	160.000	159.755	0.145	160.000	159.917	0.043	160.000	159.961	0.014	160.000	159.975	0.000
200	MAX	200.290	199.760	0.820	200.115	199.830	0.400	200.072	199.950	0.168	200.046	199.985	0.090	200.046	200.000	0.075
	MIN	200.000	199.470	0.240	200.000	199.715	0.170	200.000	199.904	0.050	200.000	199.956	0.015	200.000	199.971	0.000
250	MAX	250.290	249.720	0.860	250.115	249.830	0.400	250.072	249.950	0.168	250.046	249.985	0.090	250.046	250.000	0.075
	MIN	250.000	249.430	0.280	250.000	249.715	0.170	250.000	249.904	0.050	250.000	249.956	0.015	250.000	249.971	0.000
300	MAX	300.320	299.670	0.970	300.130	299.810	0.450	300.081	299.944	0.189	300.052	299.983	0.101	300.052	300.000	0.084
	MIN	300.000	299.350	0.330	300.000	299.680	0.190	300.000	299.892	0.056	300.000	299.951	0.017	300.000	299.968	0.000
400	MAX	400.360	399.600	1.120	400.140	399.790	0.490	400.089	399.938	0.208	400.057	399.982	0.111	400.057	400.000	0.093
	MIN	400.000	399.240	0.400	400.000	399.650	0.210	400.000	399.881	0.062	400.000	399.946	0.018	400.000	399.964	0.000
500	MAX	500.400	499.520	1.280	500.155	499.770	0.540	500.097	499.932	0.228	500.063	499.980	0.123	500.063	500.000	0.103
	MIN	500.000	499.120	0.480	500.000	499.615	0.230	500.000	499.869	0.068	500.000	499.940	0.020	500.000	499.960	0.000

APPENDIX 46

PREFERRED HOLE BASIS TRANSITION AND INTERFERENCE FITS—CYLINDRICAL FITS (ANSI B4.2)

AMERICAN NATIONAL STANDARD
PREFERRED METRIC LIMITS AND FITS

ANSI B4.2-1978

Dimensions in mm.

BASIC SIZE		LOCATIONAL TRANSN. Hole H7	Shaft k6	Fit	LOCATIONAL TRANSN. Hole H7	Shaft n6	Fit	LOCATIONAL INTERF. Hole H7	Shaft p6	Fit	MEDIUM DRIVE Hole H7	Shaft s6	Fit	FORCE Hole H7	Shaft u6	Fit
1	MAX	1.010	1.006	0.010	1.010	1.010	0.006	1.010	1.012	0.004	1.010	1.020	-0.004	1.010	1.024	-0.008
	MIN	1.000	1.000	-0.006	1.000	1.004	-0.010	1.000	1.006	-0.012	1.000	1.014	-0.020	1.000	1.018	-0.024
1.2	MAX	1.210	1.206	0.010	1.210	1.210	0.006	1.210	1.212	0.004	1.210	1.220	-0.004	1.210	1.224	-0.008
	MIN	1.200	1.200	-0.006	1.200	1.204	-0.010	1.200	1.206	-0.012	1.200	1.214	-0.020	1.200	1.218	-0.024
1.6	MAX	1.610	1.606	0.010	1.610	1.610	0.006	1.610	1.612	0.004	1.610	1.620	-0.004	1.610	1.624	-0.008
	MIN	1.600	1.600	-0.006	1.600	1.604	-0.010	1.600	1.606	-0.012	1.600	1.614	-0.020	1.600	1.618	-0.024
2	MAX	2.010	2.006	0.010	2.010	2.010	0.006	2.010	2.012	0.004	2.010	2.020	-0.004	2.010	2.024	-0.008
	MIN	2.000	2.000	-0.006	2.000	2.004	-0.010	2.000	2.006	-0.012	2.000	2.014	-0.020	2.000	2.018	-0.024
2.5	MAX	2.510	2.506	0.010	2.510	2.510	0.006	2.510	2.512	0.004	2.510	2.520	-0.004	2.510	2.524	-0.008
	MIN	2.500	2.500	-0.006	2.500	2.504	-0.010	2.500	2.506	-0.012	2.500	2.514	-0.020	2.500	2.518	-0.024
3	MAX	3.010	3.006	0.010	3.010	3.010	0.006	3.010	3.012	0.004	3.010	3.020	-0.004	3.010	3.024	-0.008
	MIN	3.000	3.000	-0.006	3.000	3.004	-0.010	3.000	3.006	-0.012	3.000	3.014	-0.020	3.000	3.018	-0.024
4	MAX	4.012	4.009	0.011	4.012	4.016	0.004	4.012	4.020	0.000	4.012	4.027	-0.007	4.012	4.031	-0.011
	MIN	4.000	4.001	-0.009	4.000	4.008	-0.016	4.000	4.012	-0.020	4.000	4.019	-0.027	4.000	4.023	-0.031
5	MAX	5.012	5.009	0.011	5.012	5.016	0.004	5.012	5.020	0.000	5.012	5.027	-0.007	5.012	5.031	-0.011
	MIN	5.000	5.001	-0.009	5.000	5.008	-0.016	5.000	5.012	-0.020	5.000	5.019	-0.027	5.000	5.023	-0.031
6	MAX	6.012	6.009	0.011	6.012	6.016	0.004	6.012	6.020	0.000	6.012	6.027	-0.007	6.012	6.031	-0.011
	MIN	6.000	6.001	-0.009	6.000	6.008	-0.016	6.000	6.012	-0.020	6.000	6.019	-0.027	6.000	6.023	-0.031
8	MAX	8.015	8.010	0.014	8.015	8.019	0.005	8.015	8.024	0.000	8.015	8.032	-0.008	8.015	8.037	-0.013
	MIN	8.000	8.001	-0.010	8.000	8.010	-0.019	8.000	8.015	-0.024	8.000	8.023	-0.032	8.000	8.028	-0.037
10	MAX	10.015	10.010	0.014	10.015	10.019	0.005	10.015	10.024	0.000	10.015	10.032	-0.008	10.015	10.037	-0.013
	MIN	10.000	10.001	-0.010	10.000	10.010	-0.019	10.000	10.015	-0.024	10.000	10.023	-0.032	10.000	10.028	-0.037
12	MAX	12.018	12.012	0.017	12.018	12.023	0.006	12.018	12.029	0.000	12.018	12.039	-0.010	12.018	12.044	-0.015
	MIN	12.000	12.001	-0.012	12.000	12.012	-0.023	12.000	12.018	-0.029	12.000	12.028	-0.039	12.000	12.033	-0.044
16	MAX	16.018	16.012	0.017	16.018	16.023	0.006	16.018	16.029	0.000	16.018	16.039	-0.010	16.018	16.044	-0.015
	MIN	16.000	16.001	-0.012	16.000	16.012	-0.023	16.000	16.018	-0.029	16.000	16.028	-0.039	16.000	16.033	-0.044
20	MAX	20.021	20.015	0.019	20.021	20.028	0.006	20.021	20.035	-0.001	20.021	20.048	-0.014	20.021	20.054	-0.020
	MIN	20.000	20.002	-0.015	20.000	20.015	-0.028	20.000	20.022	-0.035	20.000	20.035	-0.048	20.000	20.041	-0.054
25	MAX	25.021	25.015	0.019	25.021	25.028	0.006	25.021	25.035	-0.001	25.021	25.048	-0.014	25.021	25.061	-0.027
	MIN	25.000	25.002	-0.015	25.000	25.015	-0.028	25.000	25.022	-0.035	25.000	25.035	-0.048	25.000	25.048	-0.061
30	MAX	30.021	30.015	0.019	30.021	30.028	0.006	30.021	30.035	-0.001	30.021	30.048	-0.014	30.021	30.061	-0.027
	MIN	30.000	30.002	-0.015	30.000	30.015	-0.028	30.000	30.022	-0.035	30.000	30.035	-0.048	30.000	30.048	-0.061

Cont.

APPENDIX 46
PREFERRED HOLE BASIS TRANSITION AND INTERFERENCE FITS—CYLINDRICAL FITS (ANSI B4.2) (Cont.)

AMERICAN NATIONAL STANDARD
PREFERRED METRIC LIMITS AND FITS

ANSI B4.2–1978

Dimensions in mm.

BASIC SIZE		LOCATIONAL TRANSN. Hole H7	Shaft k6	Fit	LOCATIONAL TRANSN. Hole H7	Shaft n6	Fit	LOCATIONAL INTERF. Hole H7	Shaft p6	Fit	MEDIUM DRIVE Hole H7	Shaft s6	Fit	FORCE Hole H7	Shaft u6	Fit
40	MAX	40.025	40.018	0.023	40.025	40.033	0.008	40.025	40.042	-0.001	40.025	40.059	-0.018	40.025	40.076	-0.035
	MIN	40.000	40.002	-0.018	40.000	40.017	-0.033	40.000	40.026	-0.042	40.000	40.043	-0.059	40.000	40.060	-0.076
50	MAX	50.025	50.018	0.023	50.025	50.033	0.008	50.025	50.042	-0.001	50.025	50.059	-0.018	50.025	50.086	-0.045
	MIN	50.000	50.002	-0.018	50.000	50.017	-0.033	50.000	50.026	-0.042	50.000	50.043	-0.059	50.000	50.070	-0.086
60	MAX	60.030	60.021	0.028	60.030	60.039	0.010	60.030	60.051	-0.002	60.030	60.072	-0.023	60.030	60.106	-0.057
	MIN	60.000	60.002	-0.021	60.000	60.020	-0.039	60.000	60.032	-0.051	60.000	60.053	-0.072	60.000	60.087	-0.106
80	MAX	80.030	80.021	0.028	80.030	80.039	0.010	80.030	80.051	-0.002	80.030	80.078	-0.029	80.030	80.121	-0.072
	MIN	80.000	80.002	-0.021	80.000	80.020	-0.039	80.000	80.032	-0.051	80.000	80.059	-0.078	80.000	80.102	-0.121
100	MAX	100.035	100.025	0.032	100.035	100.045	0.012	100.035	100.059	-0.002	100.035	100.093	-0.036	100.035	100.146	-0.089
	MIN	100.000	100.003	-0.025	100.000	100.023	-0.045	100.000	100.037	-0.059	100.000	100.071	-0.093	100.000	100.124	-0.146
120	MAX	120.035	120.025	0.032	120.035	120.045	0.012	120.035	120.059	-0.002	120.035	120.101	-0.044	120.035	120.166	-0.109
	MIN	120.000	120.003	-0.025	120.000	120.023	-0.045	120.000	120.037	-0.059	120.000	120.079	-0.101	120.000	120.144	-0.166
160	MAX	160.040	160.028	0.037	160.040	160.052	0.013	160.040	160.068	-0.003	160.040	160.125	-0.060	160.040	160.215	-0.150
	MIN	160.000	160.003	-0.028	160.000	160.027	-0.052	160.000	160.043	-0.068	160.000	160.100	-0.125	160.000	160.190	-0.215
200	MAX	200.046	200.033	0.042	200.046	200.060	0.015	200.046	200.079	-0.004	200.046	200.151	-0.076	200.046	200.265	-0.190
	MIN	200.000	200.004	-0.033	200.000	200.031	-0.060	200.000	200.050	-0.079	200.000	200.122	-0.151	200.000	200.236	-0.265
250	MAX	250.046	250.033	0.042	250.046	250.060	0.015	250.046	250.079	-0.004	250.046	250.169	-0.094	250.046	250.313	-0.238
	MIN	250.000	250.004	-0.033	250.000	250.031	-0.060	250.000	250.050	-0.079	250.000	250.140	-0.169	250.000	250.284	-0.313
300	MAX	300.052	300.036	0.048	300.052	300.066	0.018	300.052	300.088	-0.004	300.052	300.202	-0.118	300.052	300.382	-0.298
	MIN	300.000	300.004	-0.036	300.000	300.034	-0.066	300.000	300.056	-0.088	300.000	300.170	-0.202	300.000	300.350	-0.382
400	MAX	400.057	400.040	0.053	400.057	400.073	0.020	400.057	400.098	-0.005	400.057	400.244	-0.151	400.057	400.471	-0.378
	MIN	400.000	400.004	-0.040	400.000	400.037	-0.073	400.000	400.062	-0.098	400.000	400.208	-0.244	400.000	400.435	-0.471
500	MAX	500.063	500.045	0.058	500.063	500.080	0.023	500.063	500.108	-0.005	500.063	500.292	-0.189	500.063	500.580	-0.477
	MIN	500.000	500.005	-0.045	500.000	500.040	-0.080	500.000	500.068	-0.108	500.000	500.252	-0.292	500.000	500.540	-0.580

APPENDIX 47

PREFERRED SHAFT BASIS CLEARANCE FITS—CYLINDRICAL FITS (ANSI B4.2)

AMERICAN NATIONAL STANDARD
PREFERRED METRIC LIMITS AND FITS

ANSI B4.2-1978

Dimensions in mm.

Basic Size		Loose Running — Hole C11	Shaft h11	Fit	Free Running — Hole D9	Shaft h9	Fit	Close Running — Hole F8	Shaft h7	Fit	Sliding — Hole G7	Shaft h6	Fit	Locational Clearance — Hole H7	Shaft h6	Fit
1	MAX	1.120	1.000	0.180	1.045	1.000	0.070	1.020	1.000	0.030	1.012	1.000	0.018	1.010	1.000	0.016
	MIN	1.060	0.940	0.060	1.020	0.975	0.020	1.006	0.990	0.006	1.002	0.994	0.002	1.000	0.994	0.000
1.2	MAX	1.320	1.200	0.180	1.245	1.200	0.070	1.220	1.200	0.030	1.212	1.200	0.018	1.210	1.200	0.016
	MIN	1.260	1.140	0.060	1.220	1.175	0.020	1.206	1.190	0.006	1.202	1.194	0.002	1.200	1.194	0.000
1.6	MAX	1.720	1.600	0.180	1.645	1.600	0.070	1.620	1.600	0.030	1.612	1.600	0.018	1.610	1.600	0.016
	MIN	1.660	1.540	0.060	1.620	1.575	0.020	1.606	1.590	0.006	1.602	1.594	0.002	1.600	1.594	0.000
2	MAX	2.120	2.000	0.180	2.045	2.000	0.070	2.020	2.000	0.030	2.012	2.000	0.018	2.010	2.000	0.016
	MIN	2.060	1.940	0.060	2.020	1.975	0.020	2.006	1.990	0.006	2.002	1.994	0.002	2.000	1.994	0.000
2.5	MAX	2.620	2.500	0.180	2.545	2.500	0.070	2.520	2.500	0.030	2.512	2.500	0.018	2.510	2.500	0.016
	MIN	2.560	2.440	0.060	2.520	2.475	0.020	2.506	2.490	0.006	2.502	2.494	0.002	2.500	2.494	0.000
3	MAX	3.120	3.000	0.180	3.045	3.000	0.070	3.020	3.000	0.030	3.012	3.000	0.018	3.010	3.000	0.016
	MIN	3.060	2.940	0.060	3.020	2.975	0.020	3.006	2.990	0.006	3.002	2.994	0.002	3.000	2.994	0.000
4	MAX	4.145	4.000	0.220	4.060	4.000	0.090	4.028	4.000	0.040	4.016	4.000	0.024	4.012	4.000	0.020
	MIN	4.070	3.925	0.070	4.030	3.970	0.030	4.010	3.988	0.010	4.004	3.992	0.004	4.000	3.992	0.000
5	MAX	5.145	5.000	0.220	5.060	5.000	0.090	5.028	5.000	0.040	5.016	5.000	0.024	5.012	5.000	0.020
	MIN	5.070	4.925	0.070	5.030	4.970	0.030	5.010	4.988	0.010	5.004	4.992	0.004	5.000	4.992	0.000
6	MAX	6.145	6.000	0.220	6.060	6.000	0.090	6.028	6.000	0.040	6.016	6.000	0.024	6.012	6.000	0.020
	MIN	6.070	5.925	0.070	6.030	5.970	0.030	6.010	5.988	0.010	6.004	5.992	0.004	6.000	5.992	0.000
8	MAX	8.170	8.000	0.260	8.076	8.000	0.112	8.035	8.000	0.050	8.020	8.000	0.029	8.015	8.000	0.024
	MIN	8.080	7.910	0.080	8.040	7.964	0.040	8.013	7.985	0.013	8.005	7.991	0.005	8.000	7.991	0.000
10	MAX	10.170	10.000	0.260	10.076	10.000	0.112	10.035	10.000	0.050	10.020	10.000	0.029	10.015	10.000	0.024
	MIN	10.080	9.910	0.080	10.040	9.964	0.040	10.013	9.985	0.013	10.005	9.991	0.005	10.000	9.991	0.000
12	MAX	12.205	12.000	0.315	12.093	12.000	0.136	12.043	12.000	0.061	12.024	12.000	0.035	12.018	12.000	0.029
	MIN	12.095	11.890	0.095	12.050	11.957	0.050	12.016	11.982	0.016	12.006	11.989	0.006	12.000	11.989	0.000
16	MAX	16.205	16.000	0.315	16.093	16.000	0.136	16.043	16.000	0.061	16.024	16.000	0.035	16.018	16.000	0.029
	MIN	16.095	15.890	0.095	16.050	15.957	0.050	16.016	15.982	0.016	16.006	15.989	0.006	16.000	15.989	0.000
20	MAX	20.240	20.000	0.370	20.117	20.000	0.169	20.053	20.000	0.074	20.028	20.000	0.041	20.021	20.000	0.034
	MIN	20.110	19.870	0.110	20.065	19.948	0.065	20.020	19.979	0.020	20.007	19.987	0.007	20.000	19.987	0.000
25	MAX	25.240	25.000	0.370	25.117	25.000	0.169	25.053	25.000	0.074	25.028	25.000	0.041	25.021	25.000	0.034
	MIN	25.110	24.870	0.110	25.065	24.948	0.065	25.020	24.979	0.020	25.007	24.987	0.007	25.000	24.987	0.000
30	MAX	30.240	30.000	0.370	30.117	30.000	0.169	30.053	30.000	0.074	30.028	30.000	0.041	30.021	30.000	0.034
	MIN	30.110	29.870	0.110	30.065	29.948	0.065	30.020	29.979	0.020	30.007	29.987	0.007	30.000	29.987	0.000

Cont.

APPENDIX 47

PREFERRED SHAFT BASIS CLEARANCE FITS—CYLINDRICAL FITS (Cont.)

AMERICAN NATIONAL STANDARD
PREFERRED METRIC LIMITS AND FITS

ANSI B4.2-1978

Dimensions in mm.

BASIC SIZE		LOOSE RUNNING Hole C11	Shaft h11	Fit	FREE RUNNING Hole D9	Shaft h9	Fit	CLOSE RUNNING Hole F8	Shaft h7	Fit	SLIDING Hole G7	Shaft h6	Fit	LOCATIONAL CLEARANCE Hole H7	Shaft h6	Fit
40	MAX	40.280	40.000	0.440	40.142	40.000	0.204	40.064	40.000	0.089	40.034	40.000	0.050	40.025	40.000	0.041
	MIN	40.120	39.840	0.120	40.080	39.938	0.080	40.025	39.975	0.025	40.009	39.984	0.009	40.000	39.984	0.000
50	MAX	50.290	50.000	0.450	50.142	50.000	0.204	50.064	50.000	0.089	50.034	50.000	0.050	50.025	50.000	0.041
	MIN	50.130	49.840	0.130	50.080	49.938	0.080	50.025	49.975	0.025	50.009	49.984	0.009	50.000	49.984	0.000
60	MAX	60.330	60.000	0.520	60.174	60.000	0.248	60.076	60.000	0.106	60.040	60.000	0.059	60.030	60.000	0.049
	MIN	60.140	59.810	0.140	60.100	59.926	0.100	60.030	59.970	0.030	60.010	59.981	0.010	60.000	59.981	0.000
80	MAX	80.340	80.000	0.530	80.174	80.000	0.248	80.076	80.000	0.106	80.040	80.000	0.059	80.030	80.000	0.049
	MIN	80.150	79.810	0.150	80.100	79.926	0.100	80.030	79.970	0.030	80.010	79.981	0.010	80.000	79.981	0.000
100	MAX	100.390	100.000	0.610	100.207	100.000	0.294	100.090	100.000	0.125	100.047	100.000	0.069	100.035	100.000	0.057
	MIN	100.170	99.780	0.170	100.120	99.913	0.120	100.036	99.965	0.036	100.012	99.978	0.012	100.000	99.978	0.000
120	MAX	120.400	120.000	0.620	120.207	120.000	0.294	120.090	120.000	0.125	120.047	120.000	0.069	120.035	120.000	0.057
	MIN	120.180	119.780	0.180	120.120	119.913	0.120	120.036	119.965	0.036	120.012	119.978	0.012	120.000	119.978	0.000
160	MAX	160.460	160.000	0.710	160.245	160.000	0.345	160.106	160.000	0.146	160.054	160.000	0.079	160.040	160.000	0.065
	MIN	160.210	159.750	0.210	160.145	159.900	0.145	160.043	159.960	0.043	160.014	159.975	0.014	160.000	159.975	0.000
200	MAX	200.530	200.000	0.820	200.285	200.000	0.400	200.122	200.000	0.168	200.061	200.000	0.090	200.046	200.000	0.075
	MIN	200.240	199.710	0.240	200.170	199.885	0.170	200.050	199.954	0.050	200.015	199.971	0.015	200.000	199.971	0.000
250	MAX	250.570	250.000	0.860	250.285	250.000	0.400	250.122	250.000	0.168	250.061	250.000	0.090	250.046	250.000	0.075
	MIN	250.280	249.710	0.280	250.170	249.885	0.170	250.050	249.954	0.050	250.015	249.971	0.015	250.000	249.971	0.000
300	MAX	300.650	300.000	0.970	300.320	300.000	0.450	300.137	300.000	0.189	300.069	300.000	0.101	300.052	300.000	0.084
	MIN	300.330	299.680	0.330	300.190	299.870	0.190	300.056	299.948	0.056	300.017	299.968	0.017	300.000	299.968	0.000
400	MAX	400.760	400.000	1.120	400.350	400.000	0.490	400.151	400.000	0.208	400.075	400.000	0.111	400.057	400.000	0.093
	MIN	400.400	399.640	0.400	400.210	399.860	0.210	400.062	399.943	0.062	400.018	399.964	0.018	400.000	399.964	0.000
500	MAX	500.880	500.000	1.280	500.385	500.000	0.540	500.165	500.000	0.228	500.083	500.000	0.123	500.063	500.000	0.103
	MIN	500.480	499.600	0.480	500.230	499.845	0.230	500.068	499.937	0.068	500.020	499.960	0.020	500.000	499.960	0.000

APPENDIX 48

PREFERRED SHAFT BASIS TRANSITION AND INTERFERENCE FITS—CYLINDRICAL FITS

AMERICAN NATIONAL STANDARD
PREFERRED METRIC LIMITS AND FITS

ANSI B4.2–1978

Dimensions in mm.

BASIC SIZE		LOCATIONAL TRANSN. Hole K7	Shaft h6	Fit	LOCATIONAL TRANSN. Hole N7	Shaft h6	Fit	LOCATIONAL INTERF. Hole P7	Shaft h6	Fit	MEDIUM DRIVE Hole S7	Shaft h6	Fit	FORCE Hole U7	Shaft h6	Fit
1	MAX	1.000	1.000	0.006	0.996	1.000	0.002	0.994	1.000	0.000	0.986	1.000	-0.008	0.982	1.000	-0.012
	MIN	0.990	0.994	-0.010	0.986	0.994	-0.014	0.984	0.994	-0.016	0.976	0.994	-0.024	0.972	0.994	-0.028
1.2	MAX	1.200	1.200	0.006	1.196	1.200	0.002	1.194	1.200	0.000	1.186	1.200	-0.008	1.182	1.200	-0.012
	MIN	1.190	1.194	-0.010	1.186	1.194	-0.014	1.184	1.194	-0.016	1.176	1.194	-0.024	1.172	1.194	-0.028
1.6	MAX	1.600	1.600	0.006	1.596	1.600	0.002	1.594	1.600	0.000	1.586	1.600	-0.008	1.582	1.600	-0.012
	MIN	1.590	1.594	-0.010	1.586	1.594	-0.014	1.584	1.594	-0.016	1.576	1.594	-0.024	1.572	1.594	-0.028
2	MAX	2.000	2.000	0.006	1.996	2.000	0.002	1.994	2.000	0.000	1.986	2.000	-0.008	1.982	2.000	-0.012
	MIN	1.990	1.994	-0.010	1.986	1.994	-0.014	1.984	1.994	-0.016	1.976	1.994	-0.024	1.972	1.994	-0.028
2.5	MAX	2.500	2.500	0.006	2.496	2.500	0.002	2.494	2.500	0.000	2.486	2.500	-0.008	2.482	2.500	-0.012
	MIN	2.490	2.494	-0.010	2.486	2.494	-0.014	2.484	2.494	-0.016	2.476	2.494	-0.024	2.472	2.494	-0.028
3	MAX	3.000	3.000	0.006	2.996	3.000	0.002	2.994	3.000	0.000	2.986	3.000	-0.008	2.982	3.000	-0.012
	MIN	2.990	2.994	-0.010	2.986	2.994	-0.014	2.984	2.994	-0.016	2.976	2.994	-0.024	2.972	2.994	-0.028
4	MAX	4.003	4.000	0.011	3.996	4.000	0.004	3.992	4.000	0.000	3.985	4.000	-0.007	3.981	4.000	-0.011
	MIN	3.991	3.992	-0.009	3.984	3.992	-0.016	3.980	3.992	-0.020	3.973	3.992	-0.027	3.969	3.992	-0.031
5	MAX	5.003	5.000	0.011	4.996	5.000	0.004	4.992	5.000	0.000	4.985	5.000	-0.007	4.981	5.000	-0.011
	MIN	4.991	4.992	-0.009	4.984	4.992	-0.016	4.980	4.992	-0.020	4.973	4.992	-0.027	4.969	4.992	-0.031
6	MAX	6.003	6.000	0.011	5.996	6.000	0.004	5.992	6.000	0.000	5.985	6.000	-0.007	5.981	6.000	-0.011
	MIN	5.991	5.992	-0.009	5.984	5.992	-0.016	5.980	5.992	-0.020	5.973	5.992	-0.027	5.969	5.992	-0.031
8	MAX	8.005	8.000	0.014	7.996	8.000	0.005	7.991	8.000	0.000	7.983	8.000	-0.008	7.978	8.000	-0.013
	MIN	7.990	7.991	-0.010	7.981	7.991	-0.019	7.976	7.991	-0.024	7.968	7.991	-0.032	7.963	7.991	-0.037
10	MAX	10.005	10.000	0.014	9.996	10.000	0.005	9.991	10.000	0.000	9.983	10.000	-0.008	9.978	10.000	-0.013
	MIN	9.990	9.991	-0.010	9.981	9.991	-0.019	9.976	9.991	-0.024	9.968	9.991	-0.032	9.963	9.991	-0.037
12	MAX	12.006	12.000	0.017	11.995	12.000	0.006	11.989	12.000	0.000	11.979	12.000	-0.010	11.974	12.000	-0.015
	MIN	11.988	11.989	-0.012	11.977	11.989	-0.023	11.971	11.989	-0.029	11.961	11.989	-0.039	11.956	11.989	-0.044
16	MAX	16.006	16.000	0.017	15.995	16.000	0.006	15.989	16.000	0.000	15.979	16.000	-0.010	15.974	16.000	-0.015
	MIN	15.988	15.989	-0.012	15.977	15.989	-0.023	15.971	15.989	-0.029	15.961	15.989	-0.039	15.956	15.989	-0.044
20	MAX	20.006	20.000	0.019	19.993	20.000	0.006	19.986	20.000	0.000	19.973	20.000	-0.014	19.967	20.000	-0.020
	MIN	19.985	19.987	-0.015	19.972	19.987	-0.028	19.965	19.987	-0.035	19.952	19.987	-0.048	19.946	19.987	-0.054
25	MAX	25.006	25.000	0.019	24.993	25.000	0.006	24.986	25.000	0.000	24.973	25.000	-0.014	24.960	25.000	-0.027
	MIN	24.985	24.987	-0.015	24.972	24.987	-0.028	24.965	24.987	-0.035	24.952	24.987	-0.048	24.939	24.987	-0.061
30	MAX	30.006	30.000	0.019	29.993	30.000	0.006	29.986	30.000	0.000	29.973	30.000	-0.014	29.960	30.000	-0.027
	MIN	29.985	29.987	-0.015	29.972	29.987	-0.028	29.965	29.987	-0.035	29.952	29.987	-0.048	29.939	29.987	-0.061

Cont.

APPENDIX 48
PREFERRED SHAFT BASIS TRANSITION AND INTERFERENCE FITS—CYLINDRICAL FITS (Cont.)

AMERICAN NATIONAL STANDARD
PREFERRED METRIC LIMITS AND FITS

ANSI B4.2-1978

Dimensions in mm.

BASIC SIZE		LOCATIONAL TRANSN. Hole K7	Shaft h6	Fit	LOCATIONAL TRANSN. Hole N7	Shaft h6	Fit	LOCATIONAL INTERF. Hole P7	Shaft h6	Fit	MEDIUM DRIVE Hole S7	Shaft h6	Fit	FORCE Hole U7	Shaft h6	Fit
40	MAX	40.007	40.000	0.023	39.992	40.000	0.008	39.983	40.000	-0.001	39.966	40.000	-0.018	39.949	40.000	-0.035
	MIN	39.982	39.984	-0.018	39.967	39.984	-0.033	39.958	39.984	-0.042	39.941	39.984	-0.059	39.924	39.984	-0.076
50	MAX	50.007	50.000	0.023	49.992	50.000	0.008	49.983	50.000	-0.001	49.966	50.000	-0.018	49.939	50.000	-0.045
	MIN	49.982	49.984	-0.018	49.967	49.984	-0.033	49.958	49.984	-0.042	49.941	49.984	-0.059	49.914	49.984	-0.086
60	MAX	60.009	60.000	0.028	59.991	60.000	0.010	59.979	60.000	-0.002	59.958	60.000	-0.023	59.924	60.000	-0.057
	MIN	59.979	59.981	-0.021	59.961	59.981	-0.039	59.949	59.981	-0.051	59.928	59.981	-0.072	59.894	59.981	-0.106
80	MAX	80.009	80.000	0.028	79.991	80.000	0.010	79.979	80.000	-0.002	79.952	80.000	-0.029	79.909	80.000	-0.072
	MIN	79.979	79.981	-0.021	79.961	79.981	-0.039	79.949	79.981	-0.051	79.922	79.981	-0.078	79.879	79.981	-0.121
100	MAX	100.010	100.000	0.032	99.990	100.000	0.012	99.976	100.000	-0.002	99.942	100.000	-0.036	99.889	100.000	-0.089
	MIN	99.975	99.978	-0.025	99.955	99.978	-0.045	99.941	99.978	-0.059	99.907	99.978	-0.093	99.854	99.978	-0.146
120	MAX	120.010	120.000	0.032	119.990	120.000	0.012	119.976	120.000	-0.002	119.934	120.000	-0.044	119.869	120.000	-0.109
	MIN	119.975	119.978	-0.025	119.955	119.978	-0.045	119.941	119.978	-0.059	119.899	119.978	-0.101	119.834	119.978	-0.166
160	MAX	160.012	160.000	0.037	159.988	160.000	0.013	159.972	160.000	-0.003	159.915	160.000	-0.060	159.825	160.000	-0.150
	MIN	159.972	159.975	-0.028	159.948	159.975	-0.052	159.932	159.975	-0.068	159.875	159.975	-0.125	159.785	159.975	-0.215
200	MAX	200.013	200.000	0.042	199.986	200.000	0.015	199.967	200.000	-0.004	199.895	200.000	-0.076	199.781	200.000	-0.190
	MIN	199.967	199.971	-0.033	199.940	199.971	-0.060	199.921	199.971	-0.079	199.849	199.971	-0.151	199.735	199.971	-0.265
250	MAX	250.013	250.000	0.042	249.986	250.000	0.015	249.967	250.000	-0.004	249.877	250.000	-0.094	249.733	250.000	-0.238
	MIN	249.967	249.971	-0.033	249.940	249.971	-0.060	249.921	249.971	-0.079	249.831	249.971	-0.169	249.687	249.971	-0.313
300	MAX	300.016	300.000	0.048	299.986	300.000	0.018	299.964	300.000	-0.004	299.850	300.000	-0.118	299.670	300.000	-0.298
	MIN	299.964	299.968	-0.036	299.934	299.968	-0.066	299.912	299.968	-0.088	299.798	299.968	-0.202	299.618	299.968	-0.382
400	MAX	400.017	400.000	0.053	399.984	400.000	0.020	399.959	400.000	-0.005	399.813	400.000	-0.151	399.586	400.000	-0.378
	MIN	399.960	399.964	-0.040	399.927	399.964	-0.073	399.902	399.964	-0.098	399.756	399.964	-0.244	399.529	399.964	-0.471
500	MAX	500.018	500.000	0.058	499.983	500.000	0.023	499.955	500.000	-0.005	499.771	500.000	-0.189	499.483	500.000	-0.477
	MIN	499.955	499.960	-0.045	499.920	499.960	-0.080	499.892	499.960	-0.108	499.708	499.960	-0.292	499.420	499.960	-0.580

APPENDIX 49

HOLE SIZES FOR NON-PREFERRED DIAMETERS

Basic Size	C11	D9	F8	G7	H7	H8	H9	H11	K7	N7	P7	S7	U7
OVER 0 TO 3	+0.120 +0.060	+0.045 +0.020	+0.020 +0.006	+0.012 +0.002	+0.010 0.000	+0.014 0.000	+0.025 0.000	+0.060 0.000	0.000 −0.010	−0.004 −0.014	−0.006 −0.016	−0.014 −0.024	−0.018 −0.028
OVER 3 TO 6	+0.145 +0.070	+0.060 +0.030	+0.028 +0.010	+0.016 +0.004	+0.012 0.000	+0.018 0.000	+0.030 0.000	+0.075 0.000	+0.003 −0.009	−0.004 −0.016	−0.008 −0.020	−0.015 −0.027	−0.019 −0.031
OVER 6 TO 10	+0.170 +0.080	+0.076 +0.040	+0.035 +0.013	+0.020 +0.005	+0.015 0.000	+0.022 0.000	+0.036 0.000	+0.090 0.000	+0.005 −0.010	−0.004 −0.019	−0.009 −0.024	−0.017 −0.032	−0.022 −0.037
OVER 10 TO 14	+0.205 +0.095	+0.093 +0.050	+0.043 +0.016	+0.024 +0.006	+0.018 0.000	+0.027 0.000	+0.043 0.000	+0.110 0.000	+0.006 −0.012	−0.005 −0.023	−0.011 −0.029	−0.021 −0.039	−0.026 −0.044
OVER 14 TO 18	+0.205 +0.095	+0.093 +0.050	+0.043 +0.016	+0.024 +0.006	+0.018 0.000	+0.027 0.000	+0.043 0.000	+0.110 0.000	+0.006 −0.012	−0.005 −0.023	−0.011 −0.029	−0.021 −0.039	−0.026 −0.044
OVER 18 TO 24	+0.240 +0.110	+0.117 +0.065	+0.053 +0.020	+0.028 +0.007	+0.021 0.000	+0.033 0.000	+0.052 0.000	+0.130 0.000	+0.006 −0.015	−0.007 −0.028	−0.014 −0.035	−0.027 −0.048	−0.033 −0.054
OVER 24 TO 30	+0.240 +0.110	+0.117 +0.065	+0.053 +0.020	+0.028 +0.007	+0.021 0.000	+0.033 0.000	+0.052 0.000	+0.130 0.000	+0.006 −0.015	−0.007 −0.028	−0.014 −0.035	−0.027 −0.048	−0.040 −0.061
OVER 30 TO 40	+0.280 +0.120	+0.142 +0.080	+0.064 +0.025	+0.034 +0.009	+0.025 0.000	+0.039 0.000	+0.062 0.000	+0.160 0.000	+0.007 −0.018	−0.008 −0.033	−0.017 −0.042	−0.034 −0.059	−0.051 −0.076
OVER 40 TO 50	+0.290 +0.130	+0.142 +0.080	+0.064 +0.025	+0.034 +0.009	+0.025 0.000	+0.039 0.000	+0.062 0.000	+0.160 0.000	+0.007 −0.018	−0.008 −0.033	−0.017 −0.042	−0.034 −0.059	−0.061 −0.086
OVER 50 TO 65	+0.330 +0.140	+0.174 +0.100	+0.076 +0.030	+0.040 +0.010	+0.030 0.000	+0.046 0.000	+0.074 0.000	+0.190 0.000	+0.009 −0.021	−0.009 −0.039	−0.021 −0.051	−0.042 −0.072	−0.076 −0.106
OVER 65 TO 80	+0.340 +0.150	+0.174 +0.100	+0.076 +0.030	+0.040 +0.010	+0.030 0.000	+0.046 0.000	+0.074 0.000	+0.190 0.000	+0.009 −0.021	−0.009 −0.039	−0.021 −0.051	−0.048 −0.078	−0.091 −0.121
OVER 80 TO 100	+0.390 +0.170	+0.207 +0.120	+0.090 +0.036	+0.047 +0.012	+0.035 0.000	+0.054 0.000	+0.087 0.000	+0.220 0.000	+0.010 −0.025	−0.010 −0.045	−0.024 −0.059	−0.058 −0.093	−0.111 −0.146

Cont.

APPENDIX 49
HOLE SIZES FOR NON-PREFERRED DIAMETERS (Cont.)

Basic Size	c11	d9	f8	g7	h7	h8	h9	h11	k7	n7	p7	s7	u7
OVER 100 TO 120	+0.400 +0.180	+0.207 +0.120	+0.090 +0.036	+0.047 +0.012	+0.035 0.000	+0.054 0.000	+0.087 0.000	+0.220 0.000	+0.010 −0.025	−0.010 −0.045	−0.024 −0.059	−0.066 −0.101	−0.131 −0.166
OVER 120 TO 140	+0.450 +0.200	+0.245 +0.145	+0.106 +0.043	+0.054 +0.014	+0.040 0.000	+0.063 0.000	+0.100 0.000	+0.250 0.000	+0.012 −0.028	−0.012 −0.052	−0.028 −0.068	−0.077 −0.117	−0.155 −0.195
OVER 140 TO 160	+0.460 +0.210	+0.245 +0.145	+0.106 +0.043	+0.054 +0.014	+0.040 0.000	+0.063 0.000	+0.100 0.000	+0.250 0.000	+0.012 −0.028	−0.012 −0.052	−0.028 −0.068	−0.085 −0.125	−0.175 −0.215
OVER 160 TO 180	+0.480 +0.230	+0.245 +0.145	+0.106 +0.043	+0.054 +0.014	+0.040 0.000	+0.063 0.000	+0.100 0.000	+0.250 0.000	+0.012 −0.028	−0.012 −0.052	−0.028 −0.068	−0.093 −0.133	−0.195 −0.235
OVER 180 TO 200	+0.530 +0.240	+0.285 +0.170	+0.122 +0.050	+0.061 +0.015	+0.046 0.000	+0.072 0.000	+0.115 0.000	+0.290 0.000	−0.013 −0.033	−0.014 −0.060	−0.033 −0.079	−0.105 −0.151	−0.219 −0.265
OVER 200 TO 225	+0.550 +0.260	+0.285 +0.170	+0.122 +0.050	+0.061 +0.015	+0.046 0.000	+0.072 0.000	+0.115 0.000	+0.290 0.000	+0.013 −0.033	−0.014 −0.060	−0.033 −0.079	−0.113 −0.159	−0.241 −0.287
OVER 225 TO 250	+0.570 +0.280	+0.285 +0.170	+0.122 +0.050	+0.061 +0.015	+0.046 0.000	+0.072 0.000	+0.115 0.000	+0.290 0.000	+0.013 −0.033	−0.014 −0.060	−0.033 −0.079	−0.123 −0.169	−0.267 −0.313
OVER 250 TO 280	+0.620 +0.300	+0.320 +0.190	+0.137 +0.056	+0.069 +0.017	+0.052 0.000	+0.081 0.000	+0.130 0.000	+0.320 0.000	+0.016 −0.036	−0.014 −0.066	−0.036 −0.088	−0.138 −0.190	−0.295 −0.347
OVER 280 TO 315	+0.650 +0.330	+0.320 +0.190	+0.137 +0.056	+0.069 0.017	+0.052 0.000	+0.081 0.000	+0.130 0.000	+0.320 0.000	+0.016 −0.036	−0.014 −0.066	−0.036 −0.088	−0.150 −0.202	−0.330 −0.382
OVER 315 TO 355	+0.720 +0.360	+0.350 +0.210	+0.151 +0.062	+0.075 +0.018	+0.057 0.000	+0.089 0.000	+0.140 0.000	+0.360 0.000	+0.017 −0.040	−0.016 −0.073	−0.041 −0.058	−0.169 −0.226	−0.369 −0.426
OVER 355 TO 400	+0.760 +0.400	+0.350 +0.210	+0.151 +0.062	+0.075 +0.018	+0.057 0.000	+0.089 0.000	+0.140 0.000	+0.360 0.000	+0.017 −0.040	−0.016 −0.073	−0.041 −0.058	−0.187 −0.244	−0.414 −0.471
OVER 400 TO 450	+0.840 +0.440	+0.385 +0.230	+0.165 +0.068	+0.083 +0.020	+0.063 0.000	+0.097 0.000	+0.155 0.000	+0.400 0.000	+0.018 −0.045	−0.017 −0.080	−0.045 −0.108	−0.209 −0.272	−0.467 −0.530
OVER 450 TO 500	+0.880 +0.480	+0.385 +0.230	+0.165 +0.068	+0.083 +0.020	+0.063 0.000	+0.097 0.000	+0.155 0.000	+0.400 0.000	+0.018 −0.045	−0.017 −0.080	−0.045 −0.108	−0.229 −0.292	−0.517 −0.580

APPENDIX 50
SHAFT SIZES FOR NON-PREFERRED DIAMETERS

Basic Size		c11	d9	f7	g6	h6	h7	h9	h11	k6	n6	p6	s6	u6
OVER	0	−0.060	−0.020	−0.006	−0.002	0.000	0.000	0.000	0.000	+0.006	+0.010	+0.012	+0.020	+0.024
TO	3	−0.120	−0.045	−0.016	−0.008	−0.006	−0.010	−0.025	−0.060	0.000	+0.004	+0.006	+0.014	+0.018
OVER	3	−0.070	−0.030	−0.010	−0.004	0.000	0.000	0.000	0.000	+0.009	+0.016	+0.020	+0.027	+0.031
TO	6	−0.145	−0.060	−0.022	−0.012	−0.008	−0.012	−0.030	−0.075	+0.001	+0.008	+0.012	+0.019	+0.023
OVER	6	−0.080	−0.040	−0.013	−0.005	0.000	0.000	0.000	0.000	+0.010	+0.019	+0.024	+0.032	+0.037
TO	10	−0.170	−0.076	−0.028	−0.014	−0.009	−0.015	−0.036	−0.090	+0.001	+0.010	+0.015	+0.023	+0.028
OVER	10	−0.095	−0.050	−0.016	−0.006	0.000	0.000	0.000	0.000	+0.012	+0.023	+0.029	+0.039	+0.044
TO	14	−0.205	−0.093	−0.034	−0.017	−0.011	−0.018	−0.043	−0.110	+0.001	+0.012	+0.018	+0.028	+0.033
OVER	14	−0.095	−0.050	−0.016	−0.006	0.000	0.000	0.000	0.000	+0.012	+0.023	+0.029	+0.039	+0.044
TO	18	−0.205	−0.093	−0.034	−0.017	−0.011	−0.018	−0.043	−0.110	+0.001	+0.012	+0.018	+0.028	+0.033
OVER	18	−0.110	−0.065	−0.020	−0.007	0.000	0.000	0.000	0.000	+0.015	+0.028	+0.035	+0.048	+0.054
TO	24	−0.240	−0.117	−0.041	−0.020	−0.013	−0.021	−0.052	−0.130	+0.002	+0.015	+0.022	+0.035	+0.041
OVER	24	−0.110	−0.065	−0.020	−0.007	0.000	0.000	0.000	0.000	+0.015	+0.028	+0.035	+0.048	+0.061
TO	30	−0.240	−0.117	−0.041	−0.020	−0.013	−0.021	−0.052	−0.130	+0.002	+0.015	+0.022	+0.035	+0.048
OVER	30	−0.120	−0.080	−0.025	−0.009	0.000	0.000	0.000	0.000	+0.018	+0.033	+0.042	+0.059	+0.076
TO	40	−0.280	−0.142	−0.050	−0.025	−0.016	−0.025	−0.062	−0.160	+0.002	+0.017	+0.026	+0.043	+0.060
OVER	40	−0.130	−0.080	−0.025	−0.009	0.000	0.000	0.000	0.000	+0.018	+0.033	+0.042	+0.059	+0.086
TO	50	−0.290	−0.142	−0.050	−0.025	−0.016	−0.025	−0.062	−0.160	+0.002	+0.017	+0.026	+0.043	+0.070
OVER	50	−0.140	−0.100	−0.030	−0.010	0.000	0.000	0.000	0.000	+0.021	+0.039	+0.051	+0.072	+0.106
TO	65	−0.330	−0.174	−0.060	−0.029	−0.019	−0.030	−0.074	−0.190	+0.002	+0.020	−0.032	+0.053	+0.087
OVER	65	−0.150	−0.100	−0.030	−0.010	0.000	0.000	0.000	0.000	+0.021	+0.039	+0.051	+0.078	+0.121
TO	80	−0.340	−0.174	−0.060	−0.029	−0.019	−0.030	−0.074	−0.190	+0.002	+0.020	+0.032	+0.059	+0.102
OVER	80	−0.170	−0.120	−0.036	−0.012	0.000	0.000	0.000	0.000	+0.025	+0.045	+0.059	+0.093	+0.146
TO	100	−0.390	−0.207	−0.071	−0.034	−0.022	−0.035	−0.087	−0.220	+0.003	+0.023	+0.037	+0.071	+0.124

Cont.

APPENDIX 50
SHAFT SIZES FOR NON-PREFERRED DIAMETERS (Cont.)

Basic Size	c11	d9	f7	g6	h6	h7	h9	h11	k6	n6	p6	s6	u6
OVER 100 TO 120	−0.180 / −0.400	−0.120 / −0.207	−0.036 / −0.071	−0.012 / −0.034	0.000 / −0.022	0.000 / −0.035	0.000 / −0.087	0.000 / −0.220	+0.025 / +0.003	+0.045 / +0.023	+0.059 / +0.037	+0.101 / +0.079	+0.166 / +0.144
OVER 120 TO 140	−0.200 / −0.450	−0.145 / −0.245	−0.043 / −0.083	−0.014 / −0.039	0.000 / −0.025	0.000 / −0.040	0.000 / −0.100	0.000 / −0.250	+0.028 / +0.003	+0.052 / +0.027	+0.068 / +0.043	+0.117 / +0.092	+0.195 / +0.170
OVER 140 TO 160	−0.210 / −0.460	−0.145 / −0.245	−0.043 / −0.083	−0.014 / −0.039	0.000 / −0.025	0.000 / −0.040	0.000 / −0.100	0.000 / −0.250	+0.028 / +0.003	+0.052 / +0.027	+0.068 / +0.043	+0.125 / +0.100	+0.215 / +0.190
OVER 160 TO 180	−0.230 / −0.480	−0.145 / −0.245	−0.043 / −0.083	−0.014 / −0.039	0.000 / −0.025	0.000 / −0.040	0.000 / −0.100	0.000 / −0.250	+0.028 / +0.003	+0.052 / +0.027	+0.068 / +0.043	+0.133 / +0.108	+0.235 / +0.210
OVER 180 TO 200	−0.240 / −0.530	−0.170 / −0.285	−0.050 / −0.096	−0.015 / −0.044	0.000 / −0.029	0.000 / −0.046	0.000 / −0.115	0.000 / −0.290	+0.033 / +0.004	+0.060 / +0.031	+0.079 / +0.050	+0.151 / +0.122	+0.265 / +0.236
OVER 200 TO 225	−0.260 / −0.550	−0.170 / −0.285	−0.050 / −0.096	−0.015 / −0.044	0.000 / −0.029	0.000 / −0.046	0.000 / −0.115	0.000 / −0.290	+0.033 / +0.004	+0.060 / +0.031	+0.079 / +0.050	+0.159 / +0.130	+0.287 / +0.258
OVER 225 TO 250	−0.280 / −0.570	−0.170 / −0.285	−0.050 / −0.096	−0.015 / −0.044	0.000 / −0.029	0.000 / −0.046	0.000 / −0.115	0.000 / −0.290	+0.033 / +0.004	+0.060 / +0.031	+0.079 / +0.050	+0.169 / +0.140	+0.313 / +0.284
OVER 250 TO 280	−0.300 / −0.620	−0.190 / −0.320	−0.056 / −0.108	−0.017 / −0.049	0.000 / −0.032	0.000 / −0.052	0.000 / −0.130	0.000 / −0.320	+0.036 / +0.004	+0.066 / +0.034	+0.088 / +0.056	+0.190 / +0.158	+0.347 / +0.315
OVER 280 TO 315	−0.330 / −0.650	−0.190 / −0.320	−0.056 / −0.108	−0.017 / −0.049	0.000 / −0.032	0.000 / −0.052	0.000 / −0.130	0.000 / −0.320	+0.036 / +0.004	+0.066 / +0.034	+0.088 / +0.056	+0.202 / +0.170	+0.382 / +0.350
OVER 315 TO 355	−0.360 / −0.720	−0.210 / −0.350	−0.062 / −0.119	−0.018 / −0.054	0.000 / −0.036	0.000 / −0.057	0.000 / −0.140	0.000 / −0.360	+0.040 / +0.004	+0.073 / +0.037	+0.098 / +0.062	+0.226 / +0.190	+0.426 / +0.390
OVER 355 TO 400	−0.400 / −0.760	−0.210 / −0.350	−0.062 / −0.119	−0.018 / −0.054	0.000 / −0.036	0.000 / −0.057	0.000 / −0.140	0.000 / −0.360	+0.040 / +0.004	+0.073 / +0.037	+0.098 / +0.062	+0.244 / +0.208	+0.471 / +0.435
OVER 400 TO 450	−0.440 / −0.840	−0.230 / −0.385	−0.068 / −0.131	−0.020 / −0.060	0.000 / 0.040	0.000 / 0.063	0.000 / −0.155	0.000 / 0.400	+0.045 / +0.005	+0.080 / +0.040	+0.108 / +0.068	+0.272 / +0.232	+0.530 / +0.490
OVER 450 TO 500	−0.480 / −0.880	−0.230 / −0.385	−0.068 / −0.131	−0.020 / −0.060	0.000 / −0.040	0.000 / −0.063	0.000 / −0.155	0.000 / −0.400	+0.045 / +0.005	+0.080 / +0.040	+0.108 / +0.068	+0.292 / +0.252	+0.580 / +0.540

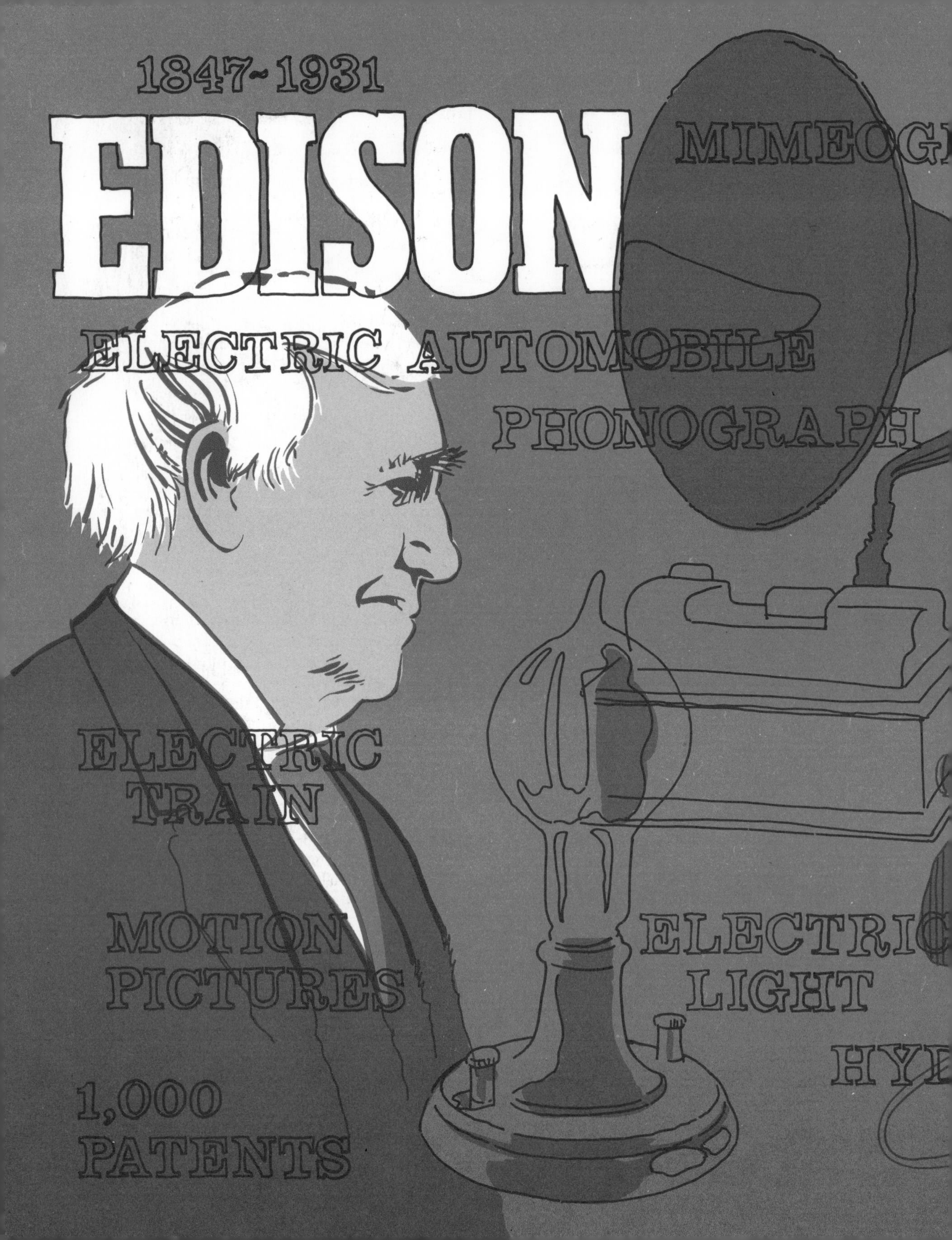

1847-1931
EDISON
MIMEOG
ELECTRIC AUTOMOBILE
PHONOGRAPH
ELECTRIC
TRAIN
MOTION
PICTURES
ELECTRI
LIGHT
HY
1,000
PATENTS